7168

D0152892

PLASTICS
MATERIALS

PLASTICS
MATERIALS

J. A. Brydson

Senior Lecturer in Polymer Technology,
National College of Rubber Technology, London

LONDON
ILIFFE BOOKS LTD

PRINCETON, NEW JERSEY
D. VAN NOSTRAND COMPANY INC.

First published in 1966 by
Iliffe Books Ltd., Dorset House
Stamford Street, London S.E.1

Published in the U.S.A. by
D. Van Nostrand Company, Inc. 120 Alexander Street
Princeton, New Jersey

Printed and bound in England by
The Chapel River Press Ltd
Andover, Hants

Contents

Preface

There are at the present time many thousands of grades of commercial plastics materials offered for sale throughout the world. Only rarely are the properties of any two of these grades identical, for although the number of chemically distinct species (e.g. polyethylenes, polystyrenes) is limited, there are many variations within each group. Such variations can arise through differences in molecular structure, differences in physical form, the presence of impurities and also in the nature and amount of additives which may have been incorporated into the base polymer. One of the aims of this book is to show how the many different materials arise, to discuss their properties and to show how these properties can to a large extent be explained by consideration of the composition of a plastics material and in particular the molecular structure of the base polymer employed.

After a brief historical review in Chapter 1 the following five chapters provide a short summary of the general methods of preparation of plastics materials and follow on by showing how properties are related to chemical structure. These particular chapters are largely qualitative in nature and are aimed not so much at the theoretical physical chemist but rather at the polymer technologist and the organic chemist who will require this knowledge in the practice of polymer and compound formulation.

Subsequent chapters deal with individual classes of plastics. In each case a review is given of the preparation, structure and properties of the material. In order to prevent the book from becoming too large I have omitted detailed discussion of processing techniques. Instead, with each major class of material an indication is given of the main processing characteristics. The applications of the various materials are considered in the light of the merits and the demerits of the material.

The title of the book requires that a definition of plastics materials be given. This is however very difficult. For the purpose of this book I eventually used as a working definition 'Those materials which are considered to be plastics materials by common acceptance'. Not a positive definition but one which is probably less capable of being criticised than any other definition I have seen. Perhaps a rather more useful definition but one which requires clarification is 'Plastics materials are processable compositions based on macromolecules'. In most cases (certainly with all synthetic materials) the macromolecules are polymers, large molecules made by the joining together of many smaller ones. Such a definition does however include rubbers, surface coatings, fibres and

7

glasses and these, largely for historical reasons, are not generally regarded as plastics. While we may arbitrarily exclude the above four classes of material the borderlines remain undefined. How should we classify the flexible poly-urethane foams—as rubbers or as plastics? What about nylon tennis racquet filament?—or polyethylene dip coatings? Without being tied by definition I have for convenience included such materials in this book but have only given brief mention to coatings, fibres and glasses generally. The rubbers I have treated as rather a special case considering them as plastics materials that show reversible high elasticity. For this reason I have briefly reviewed the range of elastomeric materials commercially available.

I hope that this book will prove to be of value to technical staff who are involved in the development and use of plastics materials and who wish to obtain a broader picture of these products than they could normally obtain in their everyday work. Problems that are encountered in technical work can generally be classified into three groups; problems which have already been solved elsewhere, problems whose solutions are suggested by a knowledge of the way in which similar problems have been tackled elsewhere and finally completely novel problems. In practice most industrial problems fall into the first two categories so that the technologist who has a good background knowledge to his subject and who knows where to look for details of original work has an enhanced value to industry. It is hoped that in a small way the text of this book will help to provide some of the background knowledge required and that the references, particularly to more detailed monographs, given at the end of each chapter will provide signposts along the pathways of the ever thickening jungle of technical literature.

1965   J.A.B.

Acknowledgments

Whilst it is not possible to acknowledge the help of everyone who has assisted me in the preparation of this book there are a number of people to whom credit must be given.　First I would like to thank those who have commented on the manuscript, in particular M. Kaufman (Chapters 1 and 11), G. L. Duncan (Chapters 2–5), N. H. Langton (Chapter 6), V. G. Kendall (Chapter 7), G. Campbell (Chapter 8), D. J. Dowrick (Chapter 9), G. W. Bowley (Chapter 10), J. M. J. Estevez (Chapter 12), E. Lloyd (Chapter 15), S. J. Barker (Chapter 16), J. A. Rhys (Chapter 17), C. H. Hall (Chapter 20), T. T. Healy (Chapter 24).　I should also like to thank Messrs T. C. Moorshead and J. A. Rhys for help with Chapter 9, Dr. C. L. Child for his generous assistance in a number of matters, Dr. C. A. Redfarn for his valuable advice when the book was in the embryo stage and Mr. E. R. Yescombe, Librarian of the National College of Rubber Technology for his help in providing me with books and other technical literature.

Finally, I should like to thank two ladies, Mrs. M. Hall for her patience and skill in deciphering and typing my notes and my wife for her patience and encouragement when the book was being written.

1

The Historical Development of Plastics Materials

1.1. NATURAL PLASTICS

Historians frequently classify the early ages of man according to the materials that he used for making his implements and other basic necessities. The most well known of these periods are the Stone Age, the Iron Age and the Bronze Age. Such a system of classification cannot be used to describe subsequent periods for with the passage of time man learnt to use other materials and by the time of the ancient civilisations of Egypt and Babylonia he was employing a range of metals, stones, woods, ceramics, glasses, skins, horns and fibres. Until the nineteenth century man's inanimate possessions, his home, his tools, his furniture, were made from varieties of these eight classes of material.

During the last century and a half, two new closely related classes of material have been introduced which have not only challenged the older materials for their well-established uses but have also made possible new products which have helped to extend the range of activities of mankind. Without these two groups of materials, rubbers and plastics, it is difficult to conceive how such everyday features of modern life such as the motor car, the telephone and the television set could ever have been developed.

Whereas the use of natural rubber was well-established by the turn of the present century, the major growth period of the plastics industry has been since 1930. This is not to say that some of the materials now classified as plastics were unknown before this time since the use of the natural plastics may be traced well into antiquity.

In the book of *Exodus* (Chapter 2) we read that the mother of Moses 'when she could no longer hide him, she took for him an ark of bullrushes and daubed it with slime and with pitch, and put the child therein and she laid it in the flags by the river's brink'. Biblical commentaries indicate that slime is the same as bitumen but whether or not this is so we have here the precursor of our modern fibre-reinforced plastics boat.

The use of bitumen is mentioned even earlier. In the book of *Genesis* (Chapter 11) we read that the builders in the plain of Shinar (i.e. Babylonia) 'had brick for stone and slime they had for mortar'. In *Genesis* (Chapter 14) we read that 'the vale of Siddim was full of slimepits; and the Kings of Sodom and Gomorrah fled, and fell there; and they that remained fled to the mountain'.

In Ancient Rome, Pliny the Elder (*c.* A.D. 23–79) dedicated thirty-seven volumes of *Natural History* to the emperor Titus. In the last of these books, dealing with gems and precious stones, he describes the properties of the fossil resin, amber. The ability of amber to attract dust was recognised and in fact the word electricity is derived from *elektron*, the Greek for amber.

Further east another natural resin, lac, had already been used for at least a thousand years before Pliny was born. Lac is mentioned in early Vedic writings and also in the *Kama Sutra* of Vatsyayona. In 1596 John Huyglen von Linschoeten undertook a scientific mission to India at the instance of the King of Portugal. In his report he describes the process of covering objects with shellac, now known as Indian turnery and still practised:

> 'Thence they dresse their besteds withall, that is to say, in turning of the woode, they take a peece of Lac of what colour they will, and as they turne it when it commeth to his fashion they spread the Lac upon the whole peece of woode which presently, with the heat of the turning (melteth the waxe) so that it entreth into the crestes and cleaveth unto it, about the thickness of a mans naile: then they burnish it (over) with a broad straw or dry rushes so (cunningly) that all the woode is covered withall, and it shineth like glasse, most pleasant to behold, and continueth as long as the woode being well looked unto: in this sort they cover all kind of household stuffe in India, as Bedsteddes, Chaires, stooles, etc. . . .'

Early records also indicate that cast mouldings were prepared from shellac by the ancient Indians. In Europe the use of sealing wax based on shellac can be traced back to the Middle Ages. The first patents for shellac mouldings were taken out in 1868.

The introduction to western civilisation of another natural resin from the east took place in the middle of the 17th century. To John Tradescant (1608–1662) the English traveller and gardener is given the credit of introducing gutta percha. The material became of substantial importance as a cable insulation material and for general moulding purposes during the 19th century and it is only since 1940 that this material has been replaced by synthetic materials in undersea cable insulation.

Prior to the eastern adventures of Linschoeten and Tradescant, the sailors of Columbus had discovered the natives of Central America playing with lumps of natural rubber. Obtained, like gutta percha, by coagulation from a latex, the first recorded reference to natural rubber was in Valdes *La historia natural y general de las Indias* published in Seville (1535–57). In 1731 la Condamine, leading an expedition on behalf of the French government to study the shape of the earth, sent back from the Amazon basin rubber-coated cloth prepared by native tribes and used in the manufacture of waterproof shoes and flexible bottles.

The coagulated rubber was a highly elastic material and could not be shaped by moulding or extrusion. In 1820 an Englishman, Thomas

Hancock, discovered that if the rubber was highly sheared or masticated, it became plastic and hence capable of flow. This is now known to be due to severe reduction in molecular weight on mastication. In 1839 an American, Charles Goodyear, found that rubber heated with sulphur retained its elasticity over a wider range of temperature than the raw material and that it had greater resistance to solvents. Thomas Hancock also subsequently found that the plastic masticated rubber could be regenerated into an elastic material by heating with molten sulphur. The rubber–sulphur reaction was termed vulcanisation by William Brockendon, a friend of Hancock. Although the work of Hancock was subsequent to, and to some extent a consequence of, that of Goodyear, the former patented the discovery in 1843 in England whilst Goodyear's first (American) patent was taken out in 1844.

In extensions of this work on vulcanisation, which normally involved only a few per cent of sulphur, both Goodyear and Hancock found that if rubber was heated with larger quantities of sulphur (about 50 parts per 100 parts of rubber) a hard product was obtained. This subsequently became known variously as ebonite, vulcanite and hard rubber. A patent for producing hard rubber was taken out by Nelson Goodyear in 1851.

The discovery of ebonite is usually considered as a milestone in the history of the rubber industry. Its importance in the history of plastics materials, of which it obviously is one, is generally neglected. Its significance lies in the fact that ebonite was the first thermosetting plastics material to be prepared and also the first plastics material which involved a distinct chemical modification of a natural material. The exploitation of the ebonite process did not however occur for some years after its discovery and for this reason its historical importance has become somewhat blurred.

1.2. PARKESINE AND CELLULOID

While Hancock and Goodyear were developing the basic processes of rubber technology, other important discoveries were taking place in Europe. Following earlier work by Pelouze, Schönbein was able to establish conditions for controlled nitration of cellulose. The product soon became of interest as an explosive and in the manufacture of collodion, a solution in an alcohol–ether mixture. In the 1850's the English inventor Alexander Parkes 'observed after much research, labour and investigation that the solid residue left on the evaporation of the solvent of photographic collodion produced a hard, horny elastic and waterproof substance'. In 1856 he patented the process of waterproofing woven fabrics by the use of such solutions.

In 1862 the Great International Exhibition was held in London and was visited by six million people. At this exhibition a bronze medal was awarded to Parkes for his exhibit Parkesine. This was obtained by first preparing a suitable cellulose nitrate and dissolving it in a minimum of solvent. The mixture was then put on a heated rolling machine from which some of the solvent was then removed. While still in the plastic state the material was then shaped by 'dies or pressure'. In 1866 the Parkesine Co., Ltd. was formed but it failed in 1868. This appears in part due to the fact that in trying to reduce production costs products inferior to those exhibited in 1862 were produced. Although the Parkesine Company suffered an economic failure, credit must go to Parkes as the first man to attempt the commercial exploitation of a chemically modified polymer as a plastics material.

One year after the failure of the Parkesine Company a collaborator of Parkes, Daniel Spill, formed the Xylonite Company to process materials similar to Parkesine. Once again economic failure resulted and the Company was wound up in December 1874. Undaunted, Spill moved to a new site, established the Daniel Spill Company and working in a modest way continued production of Xylonite and Ivoride.

In America developments were also taking place in the use of cellulose nitrate. In 1865 John Wesley Hyatt who, like Parkes and Spill, had had no formal scientific training, but possessed that all-important requirement of a plastics technologist—inventive ingenuity, became engrossed in devising a method for producing billiard balls from materials other than ivory. Originally using mixtures of cloth, ivory dust and shellac, in 1869 he patented the use of collodion for coating billiard balls. The inflammability of collodion was quickly recognised. In his history of plastics, Kaufman[1] tells how Hyatt received a letter from a Colorado billiard saloon proprietor commenting that occasionally the violent contact of the balls would produce a mild explosion like a percussion guncap. This in itself he did not mind but each time this happened 'instantly every man in the room pulled a gun'.

Products made up to this time both in England and the United States suffered from the high shrinkage due to the evaporation of the solvent. In 1870 J. W. Hyatt and his brother took out U.S. Patent 105338 for a process of producing a horn-like material using cellulose nitrate and camphor. Although Parkes and Spill had mentioned camphor in their work it was left to the Hyatt brothers to appreciate the unique value of camphor as a plasticiser for cellulose nitrate. In 1872 the term celluloid was first used to describe the product which quickly became a commercial success. The validity of Hyatts patents was challenged by Spill and a number of court actions took place between 1877 and 1884. In the final action it was found that Spill had no claim on the Hyatt brothers, the judge opining that the true inventor of the process was in fact Alexander Parkes since he had mentioned the use of both camphor and alcohol in his patents. There was thus no restriction on the use of these processes and any company, including the Hyatts Celluloid Manufacturing Company, were free to use them. As a result of this decision the Celluloid Manufacturing Company prospered, changed its name to the American Celluloid and Chemical Corporation and eventually became absorbed by the Celanese Corporation.

It is interesting to note that during this period L. P. Merriam and Spill collaborated in their work and this led to the formation in 1877 of the British Xylonite Company. Although absorbed by the Distillers organisation in 1961, this company remains an important force in the British plastics industry.

1.3. 1900–1930

By 1900 the only plastics materials available were shellac, gutta percha, ebonite and celluloid (and the bitumens and amber if they are considered as plastics). Early experiments leading to other materials had however been carried out. The first group of these to bear fruit were those which had been involved with the milk protein, casein. About 1897 there was a demand in German schools for what may only be described as a white blackboard. As a result of efforts to obtain such a product, Krische and Spitteler were able to take out patents describing the manufacture of casein plastics by reacting casein with formaldehyde. The material soon became established under the well known trade names of Galalith and later Erinoid and today casein plastics still remain of interest to the button industry.

The ability of formaldehyde to form resinous substances had been observed by chemists in the second half of the 19th century. In 1859 Butlerov described formaldehyde polymers while in 1872 Adolf Bayer reported that phenols and aldehydes react to give resinous substances. In 1899 Arthur Smith took out British Patent 16274, the first dealing with phenol–aldehyde resins, in this case for use as an ebonite substitute in electrical insulation. During the next decade the phenol–aldehyde reaction was investigated, mainly for purely academic reasons, but on occasion, in the hope of commercial exploitation. In due course Leo Hendrik Baekeland discovered techniques of so controlling and modifying the reaction that useful products could be made. The first of his 119 patents on phenol–aldehyde plastics was taken out in 1907, and in 1910 the General Bakelite Company was formed in the United States. Within a very few years the material had been established in many fields, in particular for electrical insulation. When Baekeland died in 1944 world production of phenolic resins was of the order of 175,000 tons per annum and today annual consumption of the resins still continues to increase.

Whereas celluloid was the first plastics material obtained by chemical modification of a polymer to be exploited, the phenolics were the first commercially successful fully synthetic resins. It is interesting to note that in 1963, by a merger of two subsidiary companies of the Union Carbide and the Distillers organisations, there was formed the Bakelite Xylonite Company, an intriguing marriage of two of the earliest names in the plastics industry.

The success of phenol–formaldehyde mouldings stimulated research with other resins. In 1918 Hans John prepared resins by reacting urea with formaldehyde. The reaction was studied more fully by Pollak and Ripper in an unsuccessful attempt to produce an organic glass during the period 1920–1924. At the end of this period the British Cyanides Company (later to become British Industrial Plastics) who were in financial difficulties, were looking around for profitable outlets for their products.

E. C. Rossiter suggested that they might investigate the condensation of thiourea, which they produced, with formaldehyde. Although at the time neither thiourea–formaldehyde nor urea–formaldehyde resins proved of value, resins using urea and thiourea with formaldehyde were made which were successfully used in the manufacture of moulding powders. Unlike the phenolics, these materials could be moulded into light-coloured articles and they rapidly achieved commercial success. In due course the use of thiourea was dropped as improvements were made in the simpler urea–formaldehyde materials. Today these resins are used extensively for moulding powders, adhesives and textile and paper finishing whilst the related melamine–formaldehyde materials are also used in decorative laminates.

During the time of the development of the urea-based resins, a thermoplastic, cellulose acetate, was making its debut. The material had earlier been extensively used an an aircraft dope and for artificial fibres. The discovery of suitable plasticisers in 1927 led to the introduction of this material as a non-inflammable counterpart of celluloid. During the next ten years the material became increasingly used for injection moulding and it retained its pre-eminent position in this field until the early 1950's.

1.4. THE EVOLUTION OF THE ETHENOID PLASTICS

The decade 1930–1940 saw the initial industrial development of today's major thermoplastics: polystyrene, poly(vinyl chloride) (p.v.c.), the polyolefins and poly(methyl methacrylate). Since all these materials can be considered formally as derivatives of ethylene they are said to be members of the ethenoid family of thermoplastics.

About 1930 I.G. Farben, in Germany, first produced polystyrene, whilst at the same time the Dow Chemical Company commenced their ultimately successful development of the material.

Commercial interest in p.v.c. also commenced at about this time. The Russian, I. Ostromislensky, had patented the polymerisation of vinyl chloride and related substances in 1912, but the high decomposition rate at processing temperatures proved an insurmountable problem for over 15 years. Today p.v.c. is one of the two largest tonnage plastics materials, the other being polyethylene.

The discovery and development of polyethylene provides an excellent lesson in the value of observing and following up an unexpected experimental result. In 1931 the research laboratories of the Alkali Division of Imperial Chemical Industries designed an apparatus to investigate the effect of pressures up to 3,000 atmospheres on binary and ternary organic systems. Many systems were investigated but the results of the experiments did not show immediate promise. However E. W. Fawcett and R. O. Gibson, the chemists who carried out the research programme, noticed that in one of the experiments in which ethylene was being used, a small amount of a white waxy solid had been formed. On analysis this was found to be a polymer of ethylene.

In due course attempts were made to reproduce this polymer. It was eventually discovered that a trace of oxygen was necessary to bring about the formation of polyethylene. In the original experiment this had been present accidentally, owing to a leak in the apparatus. Investigation of the product showed that it was an excellent electrical insulator and that it had very good chemical resistance. At the suggestion of B. J. Habgood its value as a submarine cable insulator was investigated with the assistance of J. N. Dean (now Sir John Dean) and H. F. Wilson of the Telegraph Construction and Maintenance Company (Telcon).

Polyethylene was soon seen to have many properties suitable for this purpose and manufacture on a commercial scale was authorised. The polyethylene plant came on stream on 1st September 1939, just before the outbreak of the Second World War.

During this period, the I.C.I. laboratories were also making their other great contribution to the range of plastics materials—the product which they marketed as Perspex, poly(methyl methacrylate). As a result of work by two of their chemists, R. Hill and J. W. C. Crawford, it was found that a rigid transparent thermoplastics material could be produced at a commercially feasible cost. The material became invaluable during the Second World War for aircraft glazing and to a lesser extent in the manufacture of dentures. Today poly(methyl methacrylate) is produced in many countries and used for a wide variety of applications particularly where transparency and/or good weathering resistance are important.

1.5. DEVELOPMENTS SINCE 1939

The advent of war brought plastics more into demand, largely as substitutes for materials, such as natural rubber and gutta percha, which were in short supply. In the United States the crash programme leading to the large-scale production of synthetic rubbers resulted in extensive research into the chemistry underlying the formation of polymers. A great deal of experience was also obtained on the large scale production of such materials.

New materials also emerged. Nylon, developed brilliantly by W. H. Carothers and his team of research workers for Du Pont as a fibre in the mid-1930's, was first used as a moulding material in 1941. Also in 1941 a patent taken out by Kinetic Chemical Inc. described how R. J. Plunkett had first discovered polytetrafluoroethylene. This happened, when on one occasion, it was found that on opening the valve of a supposedly full cylinder of the gas tetrafluoroethylene no gas issued out. On subsequently cutting up the cylinder it was found that a white solid, polytetrafluoroethylene (p.t.f.e.) had been deposited on the inner walls of the cylinder. The process was developed by Du Pont and, in 1943, pilot plant to produce their product Teflon came on stream.

Interesting developments were also taking place in the field of thermosetting resins. The melamine–formaldehyde materials appeared commercially in 1940 whilst soon afterwards in the United States the first 'contact resins' were used. With these materials, the forerunners of today's polyester laminating resins, it was found possible to produce laminates without the need for application of external pressure. The first experiments in epoxide resins were also taking place during this period.

The first decade after the War saw the establishment of the newer synthetics in many applications.   Materials such as polyethylene and polystyrene, originally rather expensive special purpose materials, were produced in large tonnages at low cost and these started to oust some of the older materials from established uses.   The new materials were

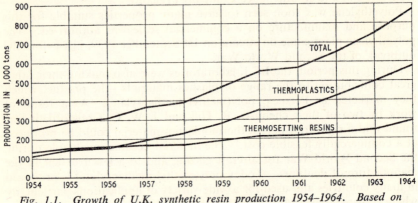

Fig. 1.1.   *Growth of U.K. synthetic resin production 1954–1964.   Based on Board of Trade figures*

however not only competitive with the older plastics but with the more traditional materials such as metals, woods, glasses and leathers.   In some instances the use of plastics materials was unwise but in others the use of plastics was of great value both technically and economically.   The occasional misuse of plastics was damaging to the industry and plastics became surrounded with an aura of disrepute for many years.   In due course it was appreciated that it was unfair to blame the plastics themselves.   Slowly there has developed an understanding of the advantages and limitations of the individual plastics in the way that we have for many years appreciated the good and bad features of our traditional materials. Wood warps and rots, iron rusts and much glass is brittle yet no one disputes the enormous value of these materials.

In the period 1945–1955 while there was a noticeable improvement in the quality of existing plastics materials and an increase in the range of grades of such materials, few new plastics were introduced commercially.   The only important newcomer was high-impact polystyrene and, at the time of its introduction this was a much inferior material to the variants available today.

In the mid 1950's a number of new thermoplastics with some very valuable properties became available.   High-density polyethylenes produced by the Phillips process and the Ziegler process were marketed and these were shortly followed by the discovery and rapid exploitation of polypropylene.   These polyolefins soon became large tonnage thermoplastics.   Somewhat more specialised materials were the acetal resins, first introduced by Du Pont, and the polycarbonates, developed simultaneously but independently in the United States and Germany.   Further developments in high-impact polystyrenes led to the development of ABS polymers.

Although the likelihood of discovering new important general purpose plastics is remote, special purpose materials continue to be introduced. Recent examples are the polypyromellitimides, fluorinated ethylene–propylene copolymers, the phenoxies and the ionomers. At the present time we have somewhat less than a dozen major tonnage plastics materials, each in a variety of grades and these are supplemented by a large number of additional plastics materials each of which have their own rather specialised fields of application.

1.6. PLASTICS CONSUMPTION AND RAW MATERIALS

While few important plastics materials have been introduced in the past five to ten years the growth rate of plastics materials in tonnage consumption during the period has been high in comparison with other basic materials. Some idea of the expansion in the use of plastics materials can be obtained by Tables 1.1 and 1.2 and Fig. 1.1.

Table 1.1 WORLD, U.S.A. AND BRITISH PRODUCTION OF PLASTICS MATERIALS
(long tons)

Year	World	U.S.A.	Britain
1939	300,000	90,000	45,000 (est.)
1951	2,000,000	810,000	160,000
1954	2,500,000	1,390,000	272,900
1957	4,600,000	1,920,000	337,600
1960	6,200,000	2,480,000	562,000
1963	8,500,000	3,730,000	746,000

U.S.A. figures: Estimates given in *Modern Plastics*. British figures: Board of Trade except 1939 Global figures: Based on various sources of data.

Table 1.2 BRITISH PRODUCTION OF MAJOR PLASTICS MATERIALS

	1951	1957	1963
Polyolefins	4,000	55,000 (est.)	197,900
P.V.C.	18,000	68,000 (est.)	152,000 (est.)
Polystyrene	8,000	31,000	76,400
Poly(methyl methacrylate)	7,000	20,000 (est.)	30,000 (est.)
Cellulosics	15,000	12,300	13,800
Phenolics	30,500	66,700	57,700
Aminoplastics	33,000	50,500	95,800
Polyesters (laminating)	Negligible	3,200	13,700

1957 and 1963 figures from Board of Trade except where stated. 1951 figures from various estimates.

As a result of this expansion the plastics industry has become a prominent user of chemical raw materials. Whereas before the Second World War the plastics industry was sometimes cynically referred to as a 'scavenger of raw materials' no such accusation is tenable today. Thirty years ago the main sources of intermediates for plastics were vegetable products (for cellulose plastics), animal and insect products (casein and shellac) and coal tar (a source of phenols). Today the plastics industry is becoming more and more based on petrochemicals.

Coal tar, one of four products of the destructive distillation of coal (the others are coke, coal gas and ammonia) may be fractionated to give a range of aromatic hydrocarbons such as benzene, toluene, xylenes, naphthalene and related products, such as phenols and cresols. The hydrocarbons can be converted into such products as phenol, styrene, adipic acid, hexamethylene diamine, caprolactam and phthalic anhydride and these reacted to form such polymers as the phenolics, polystyrene, the nylons and the polyesters.

A second product of coal distillation, coke, yields calcium carbide from which can be obtained calcium cyanamide and acetylene. The former is useful in the manufacture of melamine resins whilst for many years acetylene has been important as an intermediate in the production of vinyl chloride and vinyl acetate. During the Second World War a number of routes from acetylene to such materials as the polyurethanes, the polyamides and butadiene polymers were developed in Germany by Reppe and his co-workers. These routes have however generally proved uneconomic where petroleum was readily available as a raw material.

The development of the petrochemical industry is probably the greatest single contributing factor in the growth of the plastics industry, the two industries today having a remarkable degree of interdependence. In the first instance the growth potential of plastics stimulated research into the production of monomers and other intermediates from petroleum. As a result there became available cheap and abundant intermediates which in turn stimulated further growth of the plastics industry in a way which would not have been possible if the industry had been dependent on coal alone.

The petrochemical industry was founded shortly after the First World War to produce solvents from olefins which were then merely waste-products of 'cracking'—the process of breaking down higher molecular weight petroleum fractions into lower molecular weight products, such as gasoline. By the advent of the Second World War petrochemicals were also being used to produce ethylene dichloride, vinyl chloride, ethylene glycol, ethylene oxide and styrene. During the Second World War the large synthetic rubber industry created in the United States used butadiene and styrene, the former entirely, and the latter partly derived from petroleum. In its early days polyethylene was produced from molasses via ethyl alcohol and ethylene, today ethylene used for polymerisation, and for other purposes, is obtained almost entirely from petroleum.

With each succeeding year there has been a swing away from coal and vegetable sources of raw materials towards petroleum. Today such products as terephthalic acid, styrene, benzene, formaldehyde, vinyl acetate and acrylonitrile are being produced from petroleum sources. Large industrial concerns that had been built on acetylene chemistry are becoming based on petrochemicals whilst today coal tar is no longer an indispensable source of aromatics.

There are three general routes for producing intermediates from petroleum:

1. Separation of individual saturated hydrocarbons from the petroleum fractions and subsequent conversion to more useful products. Important examples are *n*-butane to butadiene and cyclohexane to nylon intermediates.
2. Separation of olefins produced by cracking operations and subsequent conversion. This is the major route to aliphatic petrochemicals.
3. Formation of aromatic structures such as benzene and its homologues by 'platforming' and other processes. This route is of ever increasing importance in the production of aromatic materials.

These three initial classes of product may then be converted to other chemicals by oxidation, halogenation, alkylation, hydration, carbonylation, telomerisation and many other reactions. There are nowadays few intermediates for plastics that cannot be produced more cheaply from petroleum than from other sources. The choice is not so much petroleum *versus* coal but which route from petroleum.

1.7. THE FUTURE OF PLASTICS

Plastics materials are now well established, not as cheap substitutes or as novelties but alongside other materials such as metals and woods. There is no doubt the usage of plastics materials will continue to increase rapidly because of the increased market for present applications and because of their development in new outlets. It is also very likely that the bulk of the growth will occur with existing materials and closely related polymers and compounds which may be developed from them. Whilst new materials will be developed which may have such desirable features as excellent high temperature resistance and/or high mechanical strength, the present indications are that the initial cost of these materials will restrict use to specialised fields such as aviation and astronautical equipment.

In some outlets, such as in the electrical industry, plastics are well established and future growth will depend largely on the growth of the particular industry. In other industries plastics are only just beginning to replace more traditional materials. Outstanding examples for which plastics materials have a large potential outlet include building, packaging, transportation, 'paper' and clothing.

In the building industry real advances in the use of plastics have been rather obscured by publicity given to a number of all-plastics structures, particularly those made by polymer manufacturers. These often incorporate features which at the present time seem to the ordinary man to be fanciful and futuristic although closer inspection may show this not necessarily to be the case. That plastics are not purely of novelty value can be seen from their increased use for piping, drainage, ducting and insulation. Whilst these specific uses could account for a large increase in plastics consumption, the establishment of plastics for partitioning, wall cladding (both external and internal) and for window frames would greatly increase the tonnage consumed.

In packaging, plastics find uses as bottles, bottle caps, bags, sacks, sack interliners, carton liners, blister packs and moulded containers. Tonnage consumption nevertheless is still quite small compared with traditional materials in most of these applications. For example, glass dominates the bottle and jar industry, paper bags and sacks are still more common than those from plastics whilst 'tins' are more common than plastics containers. Continued development is now enabling plastics to make inroads into each of these markets.

The car provides another challenge to the plastics technologist. Today the car manufacturer uses plastics in interior trim, seating, electrical equipment, instrument gears, rear lamp housings and more recently in the fascia panels, door handles and fuel piping. Plastics are now being considered for gear boxes, bonnet and boot lids and for metal-to-metal bonding. The use of reinforced plastics for car bodies would appear to be limited to sports cars and other cars which have too small a market to be mass produced. Whilst mass production of reinforced plastics structures is possible, the structures have been generally more costly to produce than comparable metal stampings. In commercial road transport, where the production of any one model is often limited, reinforced plastics have been widely used, particularly for translucent roof panelling. Reinforced plastics boats continue to grow in popularity.

The widespread use of plastics film for printing paper appears to be only a remote possibility in the near future. There is however the prospect for the establishment of polyethylene film for use in service manuals, maps and other printed articles which are required to be resistant to water, oils and other liquids. An alternative contender in this field is expandable polystyrene paper produced by film extrusion techniques.

Whilst natural and synthetic fibres will no doubt remain the major materials for clothing, plastics will be more widely employed. Footwear will provide the major outlet when plastics are not only likely to be used in soles and uppers but in the increasing use of the all-plastics moulded shoe, which although unacceptable for general wear in Britain is in high demand in under-developed regions of the world. In rainwear, plastics and rubbers will continue to be used for waterproof lining and in the manufacture of the all-plastics packable mackintosh. Polyurethane foam will find increased use as the insulation layer in cold-weather apparel and for giving 'body' to clothing whilst other plastics will be widely used for stiffening of light fabrics. Plastics also find large and growing outlets in many other spheres. These include aviation, road building, refrigeration, toys and games, lighting equipment, chemical plant, horticulture and hospital equipment.

This widespread use of plastics has not been achieved without large scale investment in research and development by those concerned. Polymer properties have been closely studied and slowly a relationship has been built up between structure and properties of polymers. In some instances the properties of a polymer were predicted before it was ever prepared. Studies of polymerisation methods have enabled a greater control to be made of the properties and structure of established polymers and have also led to the production of new polymers. It is sometimes said that 'twenty years ago we polymerised monomers in solution, today we polymerise the solvent', for indeed polymers of acetone and tetrahydrofuran have been prepared. Many polymers would have remained of academic interest had not chemists devised new economic syntheses from raw materials such as petroleum. The polymers produced have been investigated by the technologist and methods of processing and compounding requirements developed. Mathematicians have assisted in interpreting the rheological and heat transfer data important for processing, engineers have developed machines of ever-increasing sophistication, whilst suggested new applications have been vigorously pursued by sales organisations, often in conjunction with experts in aesthetics and design.

In this way chemist, physicist, mathematician, technologist, engineer, salesman and designer have all played a vital part. In many instances the tasks of these experts overlap but even where there is a clearer delineation it is important that the expert in one field should have a knowledge of the work of his counterparts in other fields. It is hoped that this volume will be of some assistance in achieving this end.

REFERENCES

1. KAUFMAN, M., *The First Century of Plastics—Celluloid and its Sequel*, The Plastics Institute, London (1963)

BIBLIOGRAPHY

Historical

DINGLEY, C. S., *The Story of B.I.P.*, British Industrial Plastics, Birmingham (1963)
FIELDING, T. J., *History of Bakelite Ltd.* Bakelite Ltd., London (c. 1948)
HANCOCK, T., *Personal Narrative of the Origin and Progress of the Caoutchouc or India Rubber Manufacturers in England*, Longmans, London (1857). Centenary Edn. (1920)
HANCOCK, T., *Fourteen Patents*, Barclay, London (1853)
HAYNES, W., *Cellulose—the Chemical that Grows*, Doubleday, New York (1953)
KAUFMAN, M., *The First Century of Plastics—Celluloid and its Sequel*, The Plastics Institute, London (1963)
Booklets published by British Xylonite Ltd. on the 50th and 75th anniversaries of the Company
Landmarks of the Plastics Industry (1862–1962), I.C.I. Plastics Division, Welwyn Garden City (1962)
The Telcon Story, 1850–1950, Telegraph Construction and Maintenance Co. Ltd., London (1950)

Petrochemicals

STANLEY, H. M., 'The impact of petrochemical developments on the plastics industry', *Trans. Plastics Inst.*, **28**, 110 (1960)
WADDAMS, A. L., *Chemicals from Petroleum*. John Murray, London (1962)

2

The Chemical Nature of Plastics

2.1. INTRODUCTION

Although it is very difficult and probably of little value to produce an adequate definition of the word 'plastics', it is profitable to consider the chemical structure of known plastics materials and try to see if they have any features in common.

When this is done it is seen that in all cases plastics materials, before compounding with additives, consist of a mass of very large molecules. In the case of a few naturally-occurring materials, such as bitumen, shellac and amber, the compositions are heterogeneous and complex but in all other cases the plastics materials belong to a chemical family referred to as high polymers.

For most practical purposes a *polymer* may be defined as a large molecule built up by repetition of small, simple chemical units. In the case of most of the existing thermoplastics there is in fact only one species of unit involved. For example the polyethylene molecule consists essentially of a long chain of repeating $-(CH_2)-$ (methylene) groups, viz.

$$-CH_2-CH_2-CH_2-CH_2-CH_2-CH_2-$$

The lengths of these chains may be varied but in commercial polymers chains with from 1,000 to 10,000 of these methylene groups are generally encountered. These materials are of high molecular weight and hence are spoken of as *high polymers* or *macromolecules*.

As a further illustration of the concept of polymers Table 2.1 gives the repeating units of a number of other well-known plastics.

In addition to plastics materials, many fibres, surface coatings and rubbers are also basically high polymers, whilst in nature itself there is an abundance of polymeric material. Proteins, cellulose, starch, lignin and natural rubber are high polymers. The detailed structures of these materials are complex and highly sophisticated; in comparison the

14

Table 2.1 REPEATING UNITS OF SOME WELL-KNOWN POLYMERS

Polymer	Repeating unit
Poly(vinyl chloride)	$-CH_2-CHCl-$
Polystyrene	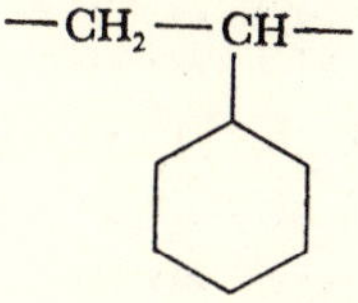 $-CH_2-CH-$
Polypropylene	$-CH_2-CH-$
	CH_3
Nylon 66	$\sim\sim(CH_2)_4CONH(CH_2)_6NHOC\sim\sim$
Acetal resin	$-CH_2-O-$

synthetic polymers produced by man are crude in the quality of their molecular architecture.

There are basically three ways by which polymers may be produced synthetically from simple starting materials. These techniques are referred to as addition polymerisation, condensation polymerisation and rearrangement polymerisation.

In *addition polymerisation* a simple, low molecular weight molecule, referred to in this context as a *monomer*, which possesses a double bond, is induced to break the double bond and the resulting free valences are able to join up to other similar molecules. For example poly(vinyl chloride) is produced by the double bonds of vinyl chloride molecules opening up and linking together (Fig. 2.1).

$$n\ CH_2{=}CH \longrightarrow \sim\sim(CH_2{-}CH_2)_n\sim\sim$$
$$\underset{Cl}{|} \qquad\qquad\qquad \underset{Cl}{|}$$

Monomer   Polymerisation   Polymer
Fig. 2.1

In these cases the monomer is converted to polymer, and no side products are formed. This approach is used with the major thermoplastics materials (Fig. 2.2) such as polyethylene (a polymer of ethylene), polystyrene (a polymer of styrene) and poly(methyl methacrylate) (a polymer of methyl methyacrylate).

An alternative technique is that of *condensation polymerisation*. A simple example of this is seen in the manufacture of linear polyesters. Here a dibasic acid is reacted with a dihydroxy compound, e.g. a glycol (Fig. 2.3).

In this case each acid group reacts with a hydroxyl group with the elimination of water to form an ester linkage. As each molecule has two ends that can react in this way long chain molecules are progressively built up. Condensation polymerisation differs from addition polymerisation in that some small molecule is split out during the reaction.

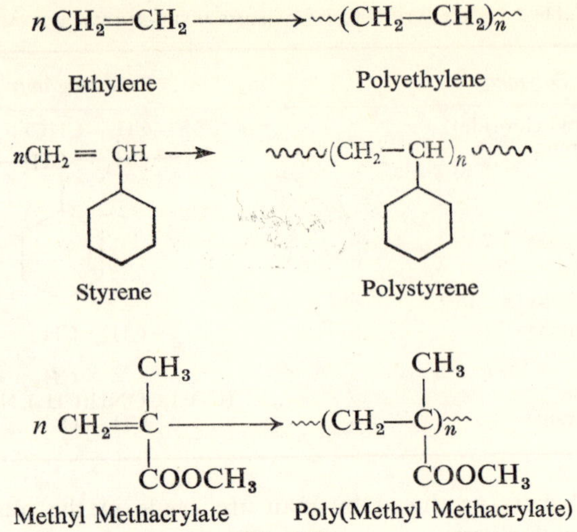

$$n\,CH_2{=}CH_2 \longrightarrow \text{\Large\textcurlywave}(CH_2{-}CH_2)_{\overset{\frown}{n}}$$

Ethylene Polyethylene

$$n\,CH_2{=}\,CH \longrightarrow \text{\Large\textcurlywave}(CH_2{-}CH)_n$$

Styrene Polystyrene

$$n\,CH_2{=}\underset{\underset{COOCH_3}{|}}{\overset{\overset{CH_3}{|}}{C}} \longrightarrow \text{\Large\textcurlywave}(CH_2{-}\underset{\underset{COOCH_3}{|}}{\overset{\overset{CH_3}{|}}{C}})_{\overset{\frown}{n}}$$

Methyl Methacrylate Poly(Methyl Methacrylate)

Fig. 2.2

$$\text{---}[HO]\,OCRCO\,[OH] + [H]O\;R_1\;O[H]\;[HO]\,OCRCO\,[OH]\text{---}$$

$$\longrightarrow \;\text{\textcurlywave}\,OCRCOOR_1OOCRCO\,\text{\textcurlywave} \qquad + H_2O$$

Fig. 2.3

Neither is it essential that the monomer should contain a double bond. Two further examples that may be given of condensation polymerisation are in the manufacture of polyamides and of polysulphides (Fig. 2.4).

$$HOOC\;R\;COOH + H_2N\;R_1\;NH_2 + HOOC\;R\;COOH$$

$$\longrightarrow \text{\textcurlywave}OC\;R\;CONH\;R_1\;NHOC\;R\;CO\text{\textcurlywave} + H_2O$$

Polyamide

$$n\,Cl{\cdot}R{\cdot}Cl + n\,Na_2S_x \longrightarrow \text{\textcurlywave}(R{-}S_x)_n\text{\textcurlywave} + 2n\,NaCl$$

Fig. 2.4

In the first case a dibasic acid is reacted with diamine to give a polyamide. A specific example is the formation of nylon 66 by the reaction of adipic acid and hexamethylene diamine.

Although the small molecule most commonly split out is water this is not necessarily the case. In the formation of polysulphides from dihalides and sodium polysulphide, sodium chloride is produced.

The third approach to synthetic polymers is of less commercial importance. There is in fact no universally accepted description for the route but the terms *rearrangement polymerisation* and *polyaddition* are commonly used. In many respects this process is intermediate between addition and condensation polymerisations. As with the former technique there is no molecule split out but the kinetics are akin to the latter. A typical example is the preparation of polyurethanes by interaction of diols (di-alcohols, glycols) with di-isocyanates (Fig. 2.5).

HOROH + OCNR₁NCO + HOROH

$$\longrightarrow \;—O \cdot ROOCNHR_1NHCOORO—$$

Fig. 2.5

These particular reactions are of importance in the manufacture of urethane foams.

It may also be mentioned that a number of commercial polymers are produced by chemical modification of other polymers, either natural or synthetic. Examples are cellulose acetate from the naturally-occurring polymer cellulose, poly(vinyl alcohol) from poly(vinyl acetate) and chlorosulphonated polyethylene (Hypalon) from polyethylene.

2.2. THERMOPLASTIC AND THERMOSETTING BEHAVIOUR

In all of the examples given so far in this chapter the product of polymerisation has been a long chain molecule, a linear polymer. With such materials it should be possible for the molecules to slide past each other under pressure above a certain temperature such that the molecules have enough energy to overcome the intermolecular attractions. In other words above a certain temperature the material is capable of flow, i.e. it is essentially plastic, whereas below this temperature it is to all intents and purposes a solid. Such materials are referred to as *thermoplastics* and

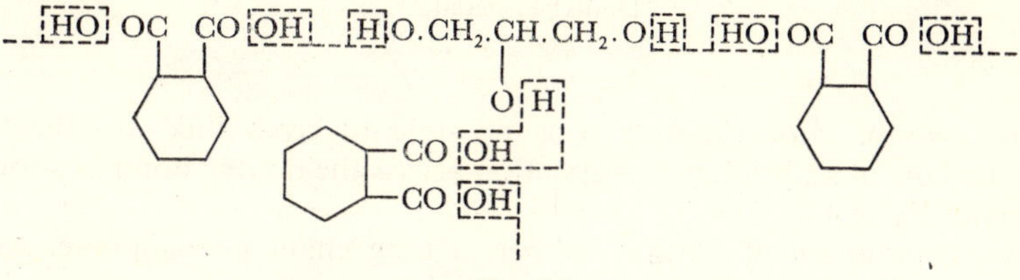

Fig. 2.6

today these may be considered to be the most important class of plastics material commercially available.

It is however possible to produce other structures. For example if phthalic acid is condensed with glycerol, the glycerol will react at each point (Fig. 2.6).

This will lead initially to branched chain structures such as indicated schematically in Fig. 2.7, G indicating a glycerol residue and P a phthalic acid residue. In due course these branched molecules will join up leading to a *cross-linked* three-dimensional product.

It is quite easy to take such a reaction to the cross-linked stage in one step but in practice it is often more convenient to first produce relatively low molecular weight structures sometimes referred to as A-*stage resins*.

These small branched molecules, which are comparatively stable at room temperature, are first deformed to shape and then either under the influence of heat or catalysts the molecules join together and some cross linking occurs to yield C-*stage resins*.   These materials are usually referred to as *thermosetting plastics* and important commercial examples include the

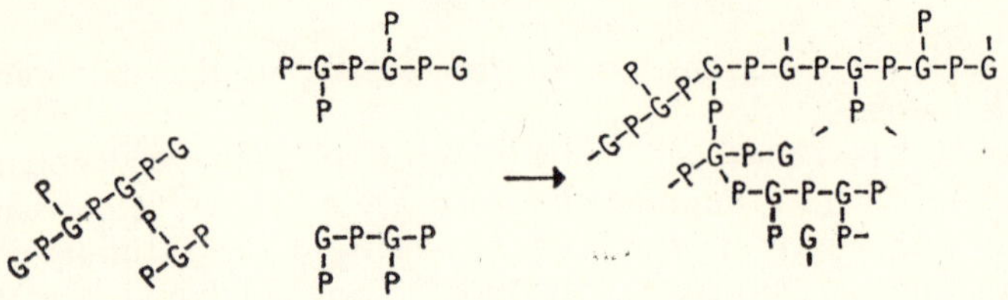

Fig. 2.7

phenolics, the aminoplastics, epoxy resins and many polyesters.   Although in fact most of these commercial polymers are made by condensation polymerisation this need not necessarily be the case.   For example, it is possible to polymerise diallyl phthalate (Fig. 2.8) through both of its double bonds and produce a thermoset polymer.

An alternative route to cross linking is to start with a linear polymer and then cross link the molecules by 'tying' the molecule through some

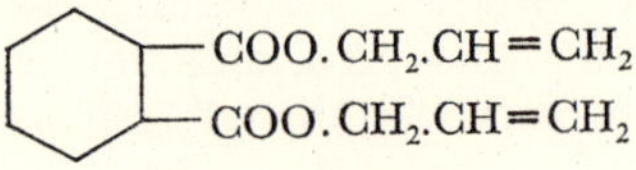

Diallyl Phthalate
Fig. 2.8

reactive group.   For example it is possible to cross link unsaturated polyesters by an addition polymerisation across the double bond as shown schematically in Fig. 2.9.

The *vulcanisation* of natural rubber, a long chain polyisoprene, with sulphur involves a similar type of cross linking.

2.3. FURTHER CONSIDERATION OF ADDITION POLYMERISATION

Addition polymerisation is effected by the activation of the double bond of a vinyl monomer thus enabling it to link up to other molecules.   It has been shown that this reaction occurs in the form of a chain addition process with initiation, propagation and termination steps.

The *initiation* stage may be activated by free radical, or ionic systems. In the following example a free radical system will be discussed.   In this case a material which can be made to decompose into free radicals on warming, or in the presence of a promoter or by irradiation with ultra violet light is added to the monomer and radicals are formed.   Two

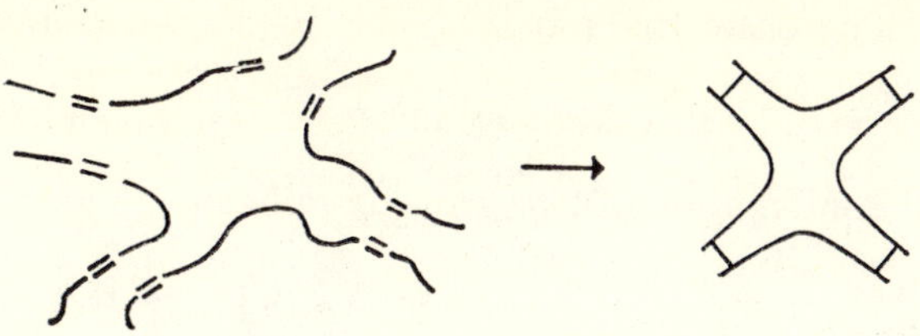

Fig. 2.9

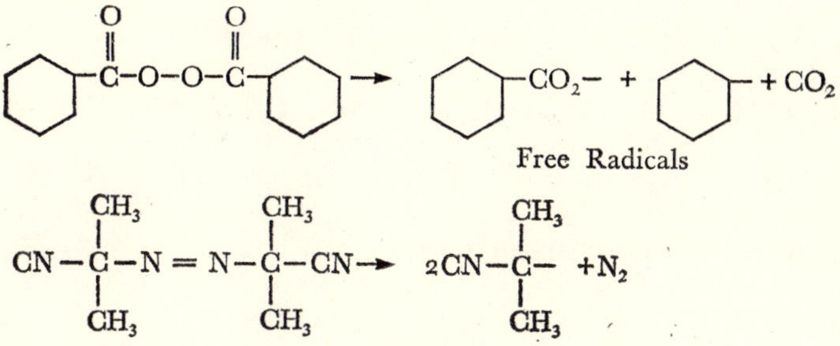

Free Radicals

Fig. 2.10

examples of such materials are benzoyl peroxide and azo-di-isobuty-ronitrile which decompose as indicated in Fig. 2.10.

Such free radical formation may be generally indicated as

$$I\!-\!I \longrightarrow 2I\!-$$

The rate of formation of radicals will depend on a number of features including the concentration of initiator, temperature and the presence of other agents. Since subsequent stages of polymer growth occur almost instantaneously it is the relative slowness of this stage which causes the overall conversion times in most polymerisations to be at least 30 minutes and sometimes much longer.

The radicals formed may then react with a monomer molecule by addition, producing another radical

$$I\!- + CH_2\!\!=\!\!CH \longrightarrow I\!-\!CH_2\!-\!CH\!-$$
$$\qquad\qquad X \qquad\qquad\qquad X$$

This radical then reacts with a further molecule of monomer, generating yet another free radical of the same order of reactivity.

$$I\!-\!CH_2\!-\!CH\!- + CH_2\!\!=\!\!CH \longrightarrow I\!-\!CH_2\!-\!CH\!-\!CH_2\!-\!CH\!-$$
$$\quad X \qquad\qquad X \qquad\qquad\qquad X \qquad\quad X$$

This reaction may then repeat itself many times so that several thousand monomer units are joined together in a time of the order of 1 second leading to a long chain free radical.   This is the *propagation* or *growth* stage.

Termination may be effected in a number of ways including:

1. Mutual combination of two growing radicals

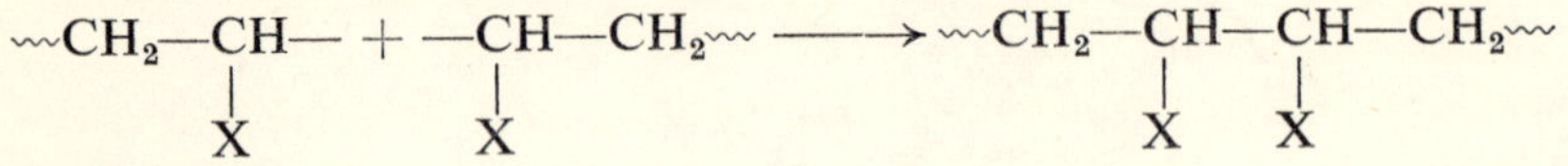

$$\text{\small$\sim$}CH_2{-}CH{-} + {-}CH{-}CH_2\text{\small$\sim$} \longrightarrow \text{\small$\sim$}CH_2{-}CH{-}CH{-}CH_2\text{\small$\sim$}$$

$$\underset{X}{|}\qquad\quad\underset{X}{|}\qquad\qquad\qquad\underset{X}{|}\quad\underset{X}{|}$$

2. Disproportionation between growing radicals

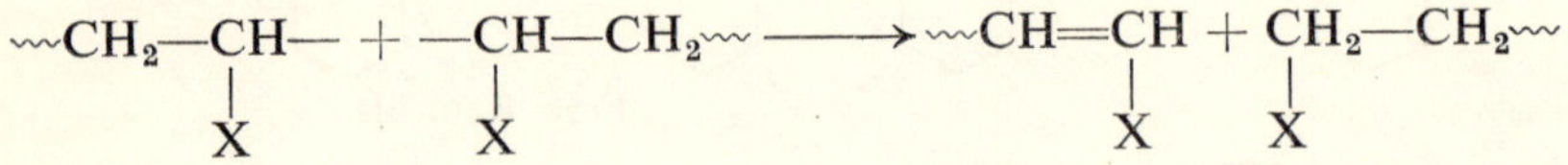

$$\text{\small$\sim$}CH_2{-}CH{-} + {-}CH{-}CH_2\text{\small$\sim$} \longrightarrow \text{\small$\sim$}CH{=}CH + CH_2{-}CH_2\text{\small$\sim$}$$

$$\underset{X}{|}\qquad\quad\underset{X}{|}\qquad\qquad\qquad\underset{X}{|}\quad\underset{X}{|}$$

3. Reaction with an initiator radical

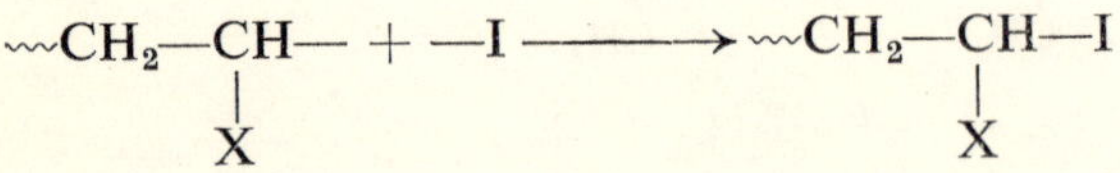

$$\text{\small$\sim$}CH_2{-}CH{-} + {-}I \longrightarrow \text{\small$\sim$}CH_2{-}CH{-}I$$

$$\underset{X}{|}\qquad\qquad\qquad\qquad\underset{X}{|}$$

4. Chain transfer with a modifier

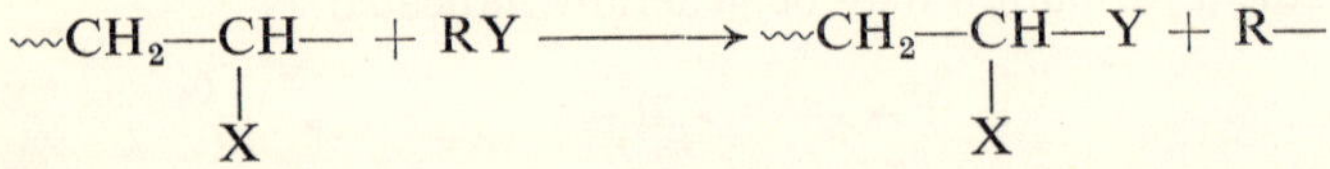

$$\text{\small$\sim$}CH_2{-}CH{-} + RY \longrightarrow \text{\small$\sim$}CH_2{-}CH{-}Y + R{-}$$

$$\underset{X}{|}\qquad\qquad\qquad\qquad\underset{X}{|}$$

(This reaction terminates growth of a chain but there is no net loss in the radical concentration and does not therefore affect the velocity of the reaction.)

5. Chain transfer with monomer (Fig. 2.11).

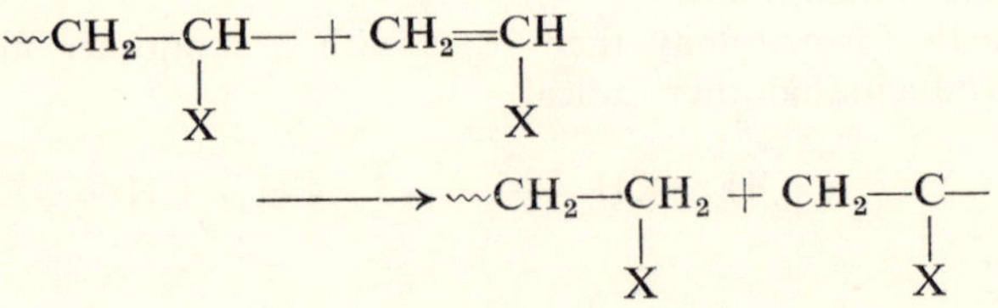

$$\text{\small$\sim$}CH_2{-}CH{-} + CH_2{=}CH$$

$$\underset{X}{|}\qquad\qquad\underset{X}{|}$$

$$\longrightarrow \text{\small$\sim$}CH_2{-}CH_2 + CH_2{-}C{-}$$

$$\underset{X}{|}\qquad\qquad\underset{X}{|}$$

Fig. 2.11

6. Reaction with a molecule to form a stable free radical, e.g. hydroquinone (Fig. 2.12).

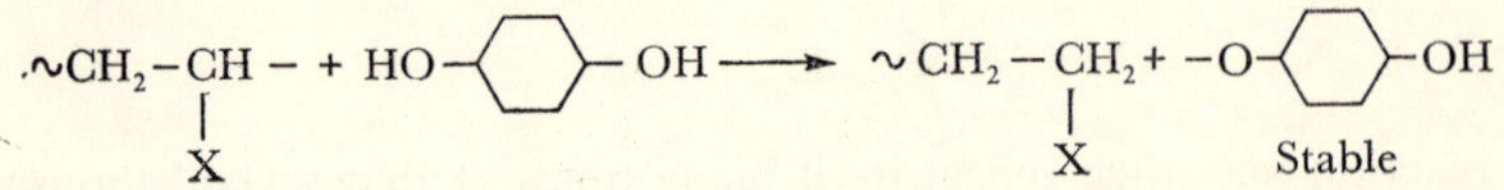

$$\cdot\text{\small$\sim$}CH_2{-}CH - + HO{-}\!\!\bigcirc\!\!{-}OH \longrightarrow \sim CH_2{-}CH_2 + {-}O{-}\!\!\bigcirc\!\!{-}OH$$

$$\underset{X}{|}\qquad\qquad\qquad\qquad\qquad\qquad\underset{X}{|}\qquad\text{Stable}$$

Fig. 2.12

Termination by mechanisms (1) and (2) are most common, whilst mechanisms (4) and (6) above are of particular technological importance. Although it is generally possible to reduce molecular weight to some extent by increasing the polymerisation temperature there is a limit to the amount that this can be done. In addition by raising the polymerisation temperature undesirable side reactions often occur. On the other hand by incorporating small quantities of a *modifier* a method of regulating the amount of chain growth is employed which does not interfere with the rate of the reaction. Such materials are also spoken of as *chain transfer agents*, and *regulators* and include chlorinated materials such as carbon tetra-chloride and trichlorethylene and mercaptans such as dodecyl mercaptan.

In the case of mechanism 6 there are materials available which completely prevent chain growth by reacting preferentially with free radicals formed to produce a stable product. These materials are known as *inhibitors* and include quinone, hydroquinone and tertiary butyl catechol. These materials are of particular value in preventing the premature polymerisation of monomer whilst in storage, or even during manufacture.

It may be noted here that it is frequently possible to polymerise two monomers together so that residues from both monomers occur together in the same polymer chain. In addition polymerisation this normally occurs in a somewhat random fashion and the product is known as a *copolymer*. It is possible to copolymerise more than two monomers together and in the case of three monomers the product is referred to as a *terpolymer*. The term *homopolymer* is sometimes used to refer to a polymer made from a single monomer.

Other copolymer forms, generally more of academic than commercial value, are *alternating copolymers, block copolymers* and *graft polymers*.

a   — AABAAABBABABBAAAB—   b   — ABABABABAB —

c   — AAAAAAAAABBBBBBBAAA—   d   — AAAAAAAAAAAA —
                                                      BBBBBB

Fig. 2.13. (a) Random copolymer, (b) alternating copolymer, (c) block copolymer, (d) graft copolymer

Figure 2.13 illustrates the possible forms in which two monomers A and B can be combined together in one chain.

Polymerisation may be carried out in bulk, in solution in a suitable solvent, in suspension or emulsion. Detailed considerations with individual polymers are made in later chapters but a number of general points may be made here. *Bulk* polymerisation is, in theory, comparatively straightforward and will give products of as good a clarity and electrical insulation characteristics as can be expected of a given material. However because polymerisation reactions are exothermic and because of the very low thermal conductivity of polymers there are very real dangers of the reactants overheating and the reaction getting out of control.

Reactions in bulk are used commercially but careful control of temperature is required. Polymerisation in a suitable solvent will dilute the concentration of reacting material and this together with the capability for convective movement or stirring of the reactant reduces exotherm problems. There is now however the necessity to remove solvent and this leads to problems of solvent recovery. Fire and toxicity hazards may also be increased.

An alternative approach to solving the exotherm problem is to polymerise *in suspension*. In this case the monomer is vigorously stirred in water to form tiny droplets. To prevent these droplets from cohering at

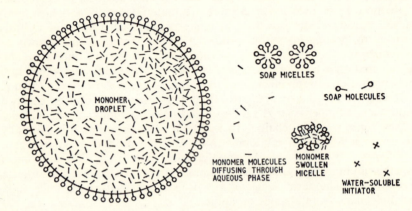

Fig. 2.14.   *Structures present during emulsion polymerisation*

the stage when the droplet is a sticky mixture of polymer and monomer, suspension or dispersion agents such as talc, poly(vinyl alcohol) or gelatine are added to provide a protective coating for each droplet. Polymerisation occurs within each droplet, providing a monomer-soluble initiator is employed, and the polymer is produced as small beads reasonably free from contaminants.

The reaction is considerably modified if the so-called *emulsion polymerisation* technique is used. In this process the reaction mixture contains about 5% soap and a water-soluble initiator system. The monomer, water, initiator, soap and other ingredients are stirred in the reaction vessel. The monomer forms into droplets which are emulsified by some of the soap molecules. Excess soap aggregates into micelles, of about 100 molecules, in which the polar ends of the soap molecules are turned outwards towards the water whilst the polar hydrocarbon ends are turned inwards (Fig. 2.14).

Monomer molecules, which have a low but finite solubility in water, diffuse through the water and drift into the soap micelles and swell them. The initiator decomposes into free radicals which also find their way into the micelles and activate polymerisation of a chain within the micelle. Chain growth proceeds until a second radical enters the micelle and starts the growth of a second chain. From kinetic considerations it can be shown that two growing radicals can only survive in the same micelle for a few thousandths of a second before mutual termination occurs. The micelles then remain inactive until a third radical enters the micelle, initiating growth of another chain which continues until a fourth radical comes into the miscelle. It is thus seen that statistically the micelle is active for half the time, and as a corollary at any one time half the micelles contain growing chains.

As reaction proceeds the micelles become swollen with monomer and polymer and they eject polymer particles. These particles which are stabilised with soap molecules taken from the micelles become the loci of further polymerisation, absorbing and being swollen by monomer molecules.

The final polymerised product is formed in particles much smaller (500–5000 Å) than produced with suspension polymerisation. Emulsion polymerisation can lead to rapid production of high molecular weight polymers but the unavoidable occlusion of large quantities of soap adversely affects the electrical insulation properties and the clarity of the polymer.

2.3.1 Elementary kinetics of free radical addition polymerisation

Polymerisation kinetics will only be dealt with here to an extent to be able to illustrate some points of technological significance. This will involve certain simplifications and the reader wishing to know more about this aspect of polymer chemistry should refer to more comprehensive studies.[1–4]

In a simple free radical indicated addition polymerisation the principal reactions involved are (assuming termination by combination for simplicity)

$$II \xrightarrow{k_d} 2I—$$
$$I— + M \xrightarrow{k_a} IM— \qquad \Big\} \text{ Initiation}$$

$$IM— + M \xrightarrow[R_p]{k_p} IMM— \quad \text{etc. Propagation}$$

$$\text{\scriptsize\textasciitilde}M— + —M\text{\scriptsize\textasciitilde} \xrightarrow[V_t]{k_t} \text{\scriptsize\textasciitilde}MM\text{\scriptsize\textasciitilde} \quad \text{Termination}$$

where M, I, M— and I— indicate monomers, initiators and their radicals respectively, each initiator yielding two radicals.

The rate of initiation, V_i, i.e. the rate of formation of growing polymer radicals can be shown to be given by

$$V_i = 2fk_d \,[I] \qquad\qquad (2.1)$$

where f is the fraction of radicals which initiate chains, i.e. the initiator efficiency, and $[I]$ is the initiator concentration.

The propagation rate is governed by the concentrations of growing chains $[M—]$ and of monomers $[M]$. Since this is in effect the rate of monomer consumption it also becomes the overall rate of polymerisation

$$R_p = k_p [M] [M—] \qquad (2.2)$$

In mutual termination the rate of reaction is determined by the concentration of growing radicals and since two radicals are involved in each termination the reaction is second order.

$$V_t = k_t [M—]^2 \qquad (2.3)$$

In practice it is found that the concentration of radicals rapidly reaches a constant value and the reaction takes place in the steady state. Thus the rate of radical formation V_i becomes equal to the rate of radical disappearance V_t. It is thus possible to combine Equations 2.1 and 2.3 to obtain on expression for $[M—]$ in terms of the rate constants

$$[M—] = \left(2f\frac{k_d}{k_t}[I]\right)^{1/2} \qquad (2.4)$$

This may then be substituted into Equation 2.2 to give

$$R_p = \left(2f\frac{k_d}{k_t}\right)^{1/2} k_p[M] [I]^{1/2} \qquad (2.5)$$

This equation indicates that the reaction rate is proportional to the square root of the initiator concentration and to the monomer concentration. It is found that the relationship with initiator concentration is commonly borne out in practice (see Fig. 2.15) but that deviations may occur with respect to monomer concentration. This may in some cases be attributed to the dependency of f on monomer concentration particularly at low efficiencies and to the effects of certain solvents in solution polymerisations.

The *average kinetic chain length r* is defined as the number of monomer units consumed per active centre formed and is given by R_p/V_i (or R_p/V_t).

Therefore combining Equations (2.1) and (2.5)

$$r = \frac{k_p}{(2fk_dk_t)^{1/2}} \cdot \frac{[M]}{[I]^{1/2}} \qquad (2.6)$$

The *number average degree of polymerisation* $\bar{x}_n$ is defined as the average number of monomer units per polymer chain. Therefore if termination is by disproportionation $r = \bar{x}$, but if by combination $r = \frac{1}{2}\bar{x}$.

It is seen from Equations (2.5) and (2.6) that while an increase in concentration of initiator *increases* the polymerisation rate it *decreases* the molecular weight.

In many technical polymerisations transfer reactions to modifier, solvent, monomer and even initiator may occur. In these cases whereas

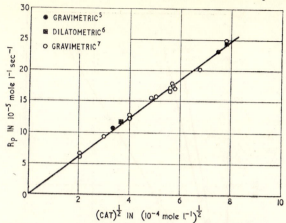

Fig. 2.15. *Rate of polymerisation R_p of methyl methacrylate with azobisisobutyronitrile at 60°C as measured by various workers.[7] (Copyright 1955 by the American Chemical Society and reprinted by permission of the copyright owner)*

the overall propagation rate is unaffected the additional ways of terminating a growing chain will cause a reduction in the degree of polymerisation.

The degree of polymerisation may also be expressed as

$$\bar{x}_n = \frac{\text{rate of propagation}}{\text{combined rate of all termination reactions}}$$

For modes of transfer with a single transfer reaction of the type

$$\sim\!\!\!\sim\!M\!-\! + SH \longrightarrow \sim\!\!\!\sim\!MH + S\!-\!$$

the rate equation, where [S] is the concentration of transfer agent SH, is

$$V_s = k_s[M\!-\!]\,[S] \tag{2.6}$$

Thus

$$\bar{x} = \frac{R_p}{V_t + V_s} = \frac{R_p}{V_i + V_s}$$

$$\frac{1}{\bar{x}} = r + \frac{V_s}{R_p} = r + \frac{k_s[S]}{k_p[M]} \tag{2.7}$$

Thus the greater the transfer rate constant and the concentration of the transfer agent the lower will be the molecular weight (Fig. 2.16).

An increase in temperature will increase the values of k_d, k_p and k_t. In practice it is observed that in free radical initiated polymerisations the overall rate of conversion is doubled per 10 degC rise in temperature (see Fig. 2.17). Since the molecular weight is inversely related to k_d and k_t it is observed in practice that this decreases with increases in temperature.

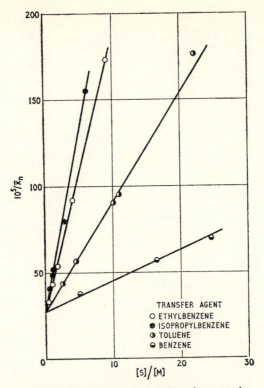

Fig. 2.16.   Effect of chain transfer solvents on the degree of polymerisation of polystyrene.   (After Gregg and Mayo[8])

The most important technological conclusions from these kinetic studies may be summarised as follows:

1. The formation of a polymer molecule takes place virtually instantaneously once an active centre is formed.   At any one time the reacting system will contain monomer and complete polymer with only a small amount of growing radicals.   Increase of reaction time will only increase the degree of conversion (of monomer to polymer) and to first approximation will not affect the degree of polymerisation.   (In fact at high conversions the high viscosity of the reacting medium may interfere with the ease of termination so that polymers formed towards the end of a reaction may have a somewhat higher molecular weight.)

2. An increase in initiator concentration or in temperature will increase the rate of conversion but decrease molecular weight.

3. Transfer reactions will reduce the degree of polymerisation without affecting the rate of conversion.

4. The statistical nature of the reaction leads to a distribution of polymer molecular weights.   Figures quoted for molecular weights

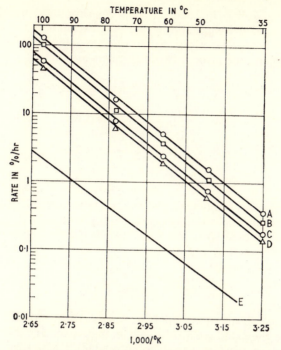

Fig. 2.17. *Rates of catalysed and uncatalysed polymerisation of styrene at different temperatures. Catalysts used (all at 0·0133 mole 1⁻¹). A, bis (2,4-dichloro-
benzoyl) peroxide: B, lauroyl peroxide: C, benzoyl
peroxide: D, bis (p-chlorobenzoyl) peroxide: E, none.
(After Boundy and Boyer[9])*

are thus averages of which different types exist.   The number average
molecular weight takes into account the numbers of molecules of
each size when assessing the average whereas the weight average
molecular weight takes into account the fraction of each size by
weight.   Thus the presence of 1% by weight of monomer would have
little effect on the weight average but since it had a great influence on
the number of molecules present per unit weight it would greatly
influence the number average.   The ratio of the two averages will
provide a measure of the molecular weight distribution.

In the case of emulsion polymerisation, half the micelles will be reacting
at any one time.   The conversion rate is thus virtually independent of
radical concentration (within limits) but dependent on the number of
micelles (or swollen polymer particles).

An increase in the rate of radical production in emulsion polymerisation
will reduce the molecular weight since it will increase the frequency of
termination.   An increase in the number of particles will however reduce
the rate of entry of radicals into a specific micelle and increase molecular weight.   Thus at constant initiator concentration and temperature

an increase in micelles (in effect in soap concentration) will lead to an increase in molecular weight and in rate of conversion.

The kinetics of copolymerisation are rather complex since four propagation reactions can take place if two monomers are present

$$\sim\!A\!-\; +\; A \xrightarrow{\quad k_{aa} \quad} \sim\!AA\!-$$

$$\sim\!A\!-\; +\; B \xrightarrow{\quad k_{ab} \quad} \sim\!AB\!-$$

$$\sim\!B\!-\; +\; B \xrightarrow{\quad k_{bb} \quad} \sim\!BB\!-$$

$$\sim\!B\!-\; +\; A \xrightarrow{\quad k_{ba} \quad} \sim\!BA\!-$$

Since these reactions rarely take place at the same rate one monomer will usually be consumed at a different rate from the other.

If k_{aa}/k_{ab} is denoted by r_a and k_{bb}/k_{ba} by r_b then it may be shown that the relative rates of consumption of the two monomers are given by

$$\frac{d\,[A]}{d\,[B]} = \frac{[A]}{[B]}\frac{r_a[A] + [B]}{r_b[B] + [A]} \tag{2.8}$$

When it is necessary that copolymers formed towards the end of the reaction have the same composition it is thus necessary that one of the monomers in the reaction vessel be continually replenished in order to maintain the relative rates of consumption. This is less necessary where r_1 and r_2 both approximate to unity and 50/50 compositions are desired.

An alternative approach is to copolymerise only up to a limited degree of conversion, say 40%. In such cases although there will be some variation in composition it will be far less than would occur if the reaction is taken to completion.

2.3.2 Ionic polymerisation

A number of important addition polymers are produced by ionic mechanisms. Although the process involves initiation, propagation and termination stages the growing unit is an ion rather than a radical.

The electron distribution around the carbon atom (marked with an asterisk in Fig. 2.18) of a growing chain may take a number of forms. In Fig. 2.18 (a) there is an unshared electron and it acts as a free radical. Fig. 2.18 (b) is a positively charged carbonium ion, unstable as it lacks a shared pair of electrons and Fig. 2.18 (c) is a negatively charged carbanion, unstable as there exists an unshared electron pair.

Both carbonium ions and carbanions may be used as the active centres for chain growth in polymerisation reactions (cationic polymerisation and anionic polymerisation respectively). The mechanisms of these reactions are less clearly understood than free radical polymerisations because here polymerisation often occurs at such a high rate that kinetic studies are difficult and because traces of certain ingredients (known in this context as

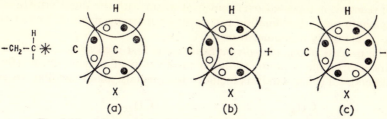

Fig. 2.18. (a) Free radical. (b) Carbonium ion. (c) Carbanion

cocatalysts) can have large effects on the reaction. Monomers which have electron donating groups attached to one of the double bond carbon atoms have a tendency to form carbonium ions in the presence of proton donors and may be polymerised by cationic methods whilst those with electron attracting substituents may be polymerised anionically. Free radical polymerisation is somewhat intermediate and is possible when substituents have moderate electron withdrawing characteristics. Many monomers may be polymerised by more than one mechanism.

Cationic polymerisation, used commercially with polyisobutylene and butyl rubber, is catalysed by Friedel–Crafts agents such as $AlCl_3$ and BF_3 (these being strong electron acceptors) in the presence of a cocatalyst. High molecular weight products may be obtained within a few seconds at $-100°C$. Although the reactions are not fully understood it is believed that the first stage involves the reaction of the catalyst with a cocatalyst (such as water) to produce a complex acid

$$TiCl_4 + RH \longrightarrow TiCl_4R^\ominus H^\oplus$$

This donates a proton to the monomer to produce a carbonium ion (Fig. 2.19).

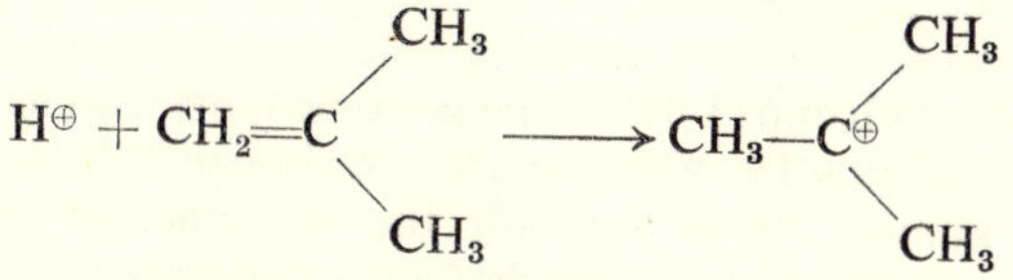

$$H^\oplus + CH_2{=}C\underset{CH_3}{\overset{CH_3}{<}} \longrightarrow CH_3{-}C^\oplus\underset{CH_3}{\overset{CH_3}{<}}$$

Fig. 2.19

In turn this ion reacts with a further monomer molecule to form another reactive carbonium ion (Fig. 2.20).

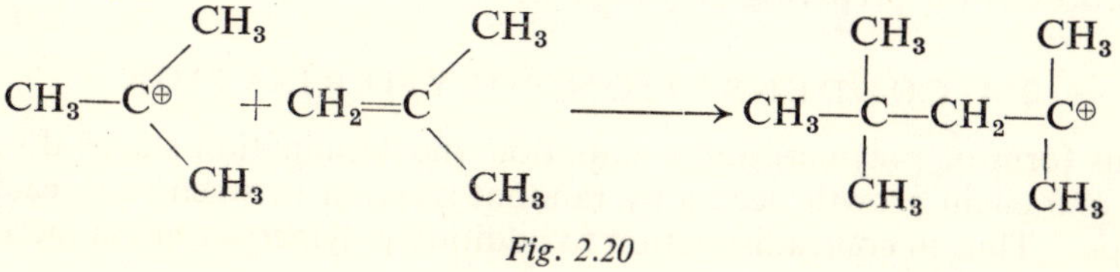

$$CH_3{-}C^\oplus\underset{CH_3}{\overset{CH_3}{<}} + CH_2{=}C\underset{CH_3}{\overset{CH_3}{<}} \longrightarrow CH_3{-}\underset{CH_3}{\overset{CH_3}{\underset{|}{\overset{|}{C}}}}{-}CH_2{-}\underset{CH_3}{\overset{CH_3}{\underset{|}{\overset{|}{C}}}}{}^\oplus$$

Fig. 2.20

The reaction is repeated over and over again with the rapid growth of a long chain ion. Termination can occur by rearrangement of the ion pair (Fig. 2.21) or by monomer transfer.

The process of *anionic polymerisation* was first used some fifty or more years ago in the sodium catalysed production of polybutadiene (Buna Rubbers). Typical catalysts include alkali metals, alkali metal alkyls and

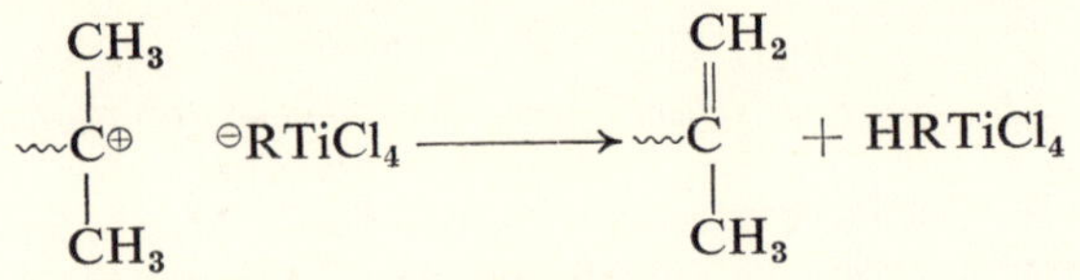

$$\underset{\overset{|}{CH_3}}{\overset{\overset{|}{CH_3}}{\sim\!\!\!\sim\!C^{\oplus}}} \quad ^{\ominus}RTiCl_4 \longrightarrow \underset{\overset{|}{CH_3}}{\overset{\overset{\|}{CH_2}}{\sim\!\!\!\sim\!C}} \; + \; HRTiCl_4$$

Fig. 2.21

sodium naphthalene, and these may be used for opening either a double bond or a ring structure to bring about polymerisation. Anionic polymerisation methods are of current interest in the preparation of certain diene rubbers.

As a result of the work of Ziegler in Germany, Natta in Italy and Pease and Roedel in the United States, the process of *co-ordination polymerisation*, a process related to ionic polymerisation, has become of significance. This process is today used in the commercial manufacture of polypropylene and polyethylene and has also been used in the laboratory for the manufacture of many novel polymers. In principle the catalyst system used governs the way in which a monomer and a growing chain approach each other and because of this it is possible to produce stereoregular polymers.

One way in which such stereospecificity occurs is by the growing polymer molecule forming a complex with a catalyst which is also complexed with a monomer molecule. In this way growing polymers and monomers are brought together in a highly specific fashion. The product of reaction of the growing polymer molecule and the monomer molecule is a further growing molecule which will then again complex itself with the catalyst and the cycle may be repeated.

The catalysts used are themselves complexes produced by interaction of alkyls of metals in Groups I–III of the periodic table with halides and other derivatives of Groups IV–VIII metals. Although soluble co-ordination catalysts are known, those used for the manufacture of stereoregular polymers are usually solid or absorbed on solid particles.

A number of olefins may be polymerised using certain metal oxides supported on the surface of an inert solid particle. The mechanism of these polymerisation reactions is little understood but is believed to be ionic in nature. Metal oxides are used in the Phillips and the Standard Oil processes for preparing polyethylene.

2.4 CONDENSATION POLYMERISATION

In this form of polymerisation, initiation and termination stages do not exist and chain growth occurs by random reaction between two reactive groups. Thus in contradistinction to addition polymerisation an increase in reaction time will produce a significant increase in average molecular weight. An increase in temperature and the use of appropriate catalysts, by increasing the reactivity, will also increase the degree of polymerisation achieved in a given time.

In the case of linear polymers it is often difficult to obtain high molecular weight polymers. The degree of polymerisation $\bar{x}$ will be given by the equation

$$\bar{x} = \frac{\text{No. of groups available for reaction}}{\text{No. of groups not reacted}} \qquad (2.9)$$

If p, the extent of reaction, is the fraction of groups that have reacted, then

$$\bar{x} = \frac{1}{1 - p} \qquad (2.10)$$

Thus when 95% of the groups have reacted ($p = 0.95$) the degree of polymerisation will only be 20.

Even lower molecular weights will be obtained where there is an excess of one reactive group, since these will eventually monopolise all the chain ends and prevent further reaction. The presence of monofunctional ingredients will have similar effects and are sometimes added deliberately to control molecular weight.

It is to be noted that only one condensation reaction is necessary to convert two molecules with values of $\bar{x} = 100$ to one molecule with $\bar{x} = 200$. A similar reaction between two dimers will only produce tetramers ($\bar{x} = 4$). Thus although the concentration of reactive groups

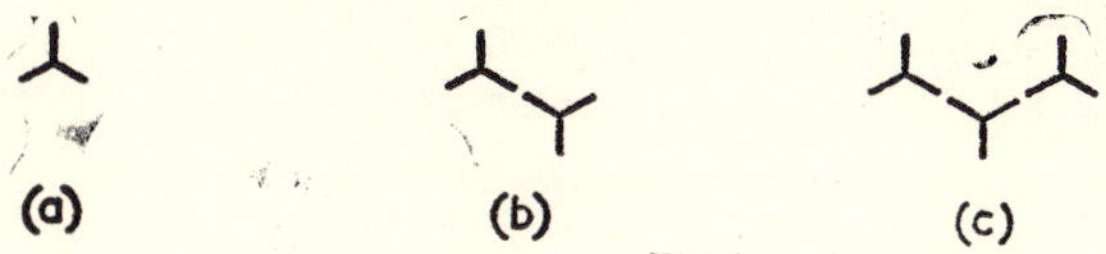

(a)　　　　　(b)　　　　　(c)

Fig. 2.22

may decrease during reaction, individual reactions at later stages of the reaction will have greater effect.

As with addition polymers, molecules with a range of molecular weights are produced. It may be shown that in the condensation of bifunctional monomers

$$\frac{\bar{x}_w}{\bar{x}_n} = (1 + p) \qquad (2.11)$$

where $\bar{x}_w$ and $\bar{x}_n$ are the weight average and number average degrees of polymerisation respectively. Thus as the reaction goes towards completion the ratio of the degrees of polymerisation and hence the molecular weights approaches 2.

In the case of trifunctional monomers the situation is more complex.

From the schematic diagrams (Fig. 2.22) it will be seen that the polymers have more functional groups than the monomers.

It is seen that the *functionality* (No. of reactive groups $= f$) is equal to $n + 2$ where n is the degree of polymerisation. Thus the chance of a specific 100-mer (102 reactive groups) reacting is over thirty times greater than a specific monomer (3 reactive groups) reacting. Large molecules therefore grow more rapidly than small ones and form even more reactive molecules. Thus 'infinitely' large, cross-linked molecules may suddenly be produced while many monomers have not even reacted. This corresponds to the 'gel point' observed with many processes using thermosetting resins. It may in fact be shown that at the gel point with a wholly trifunctional system that $\bar{x} = \infty$ whilst $\bar{x}_n$ is only 4.

REFERENCES

1. BILLMEYER, F. W., *Textbook of Polymer Science*, Interscience, New York (1962)
2. BAWN, C. E. H., *The Chemistry of High Polymers*, Interscience, New York (1948)
3. FLORY, P. J., *Principles of Polymer Chemistry*, Cornell University Press, Ithaca, New York (1953)
4. FRITH, E. M., and TUCKETT, R. F., *Linear Polymers*, Longmans Green, London (1951)
5. BAYSAL, B., and TOBOLSKY, A. V., *J. Polymer Sci.*, 8, 529 (1952)
6. BONSALL, E. P., VALENTINE L., and MELVILLE H. W., *Trans. Faraday Soc.* 48, 763 (1952)
7. O'BRIEN, J. L., and GORNICK, F., *J. Am. Chem. Soc.*, 77, 4757 (1955)
8. GREGG, R. A., and MAYO, F. R., *Disc. Faraday Soc.*, 2, 328 (1947)
9. BOUNDY, R. H., and BOYER, R. F., *Styrene, its Polymers, Copolymers and Derivatives*, Reinhold, New York (1952)

BIBLIOGRAPHY

ALFREY, F., *et al.*, *Copolymerisation*, John Wiley, New York (1952)
ALLEN, P. W., *Techniques of Polymer Characterisation*, Butterworths, London (1959)
BAWN, C. E. H., *The Chemistry of High Polymers*, Interscience, New York (1948)
BAWN, C. E. H., and LEDWITH, A., 'Stereoregular addition polymerisation', *Quart. Rev.* 16, 361 (1962)
BAMFORD, C. H., *et al. Kinetics of Vinyl Polymerisation by Radical Mechanisms*, Butterworths, London (1958)
BILLMEYER, F. W., *Textbook of Polymer Science*, Interscience, New York (1962)
BURLANT, W. J., and HOFFMAN, A. S., *Block and Graft Polymers*, Reinhold, New York (1960)
BURNETT, G. M., *Mechanism of Polymer Reactions*, John Wiley, New York (1954)
CERESA, R. J., *Block and Graft Copolymers*, Butterworths, London (1962)
FLORY, P. J., *Principles of Polymer Chemistry*, Cornell University Press, Ithaea, New York (1953)
FRITH, E. M., and TUCKETT, R. F., *Linear Polymers*, Longmans Green, London (1951)
GAYLORD, N. G., and MARK, H. F., *Linear and Stereoregular Addition Polymers*, Interscience, New York (1959)
GRASSIE, N., *Chemistry of High Polymer Degradation Processes*, Butterworths, London (1956)
MARK, H., and TOBOLSKY, A. V., *Physical Chemistry of High Polymeric Systems*, John Wiley, New York (2nd Edn.) (1960)
MELVILLE, SIR H., *Big Molecules* Bell, London (1958)
MEYER, K. H., *Natural and Synthetic High Polymers*, John Wiley, New York (2nd Edn.) (1950)
MOORE, W. R., *An Introduction to Polymer Chemistry*, University of London Press, London (1963)
PLESCH, P. H., *Cationic Polymerisation*, Pergamon Press, Oxford (1963)
SCHILDKNECHT, C. E., *Vinyl and Related Polymers*, John Wiley, New York (1952)
SCHILDKNECHT, C. E., *Polymer Processes*, John Wiley, New York (1956)
SCHMIDT, A. X., and MARLIES, C. A., *Principles of High Polymer Theory and Practice*, McGraw-Hill, New York (1948)
STILLE, J. K., *Introduction to Polymer Chemistry*, John Wiley, New York (1962)
Collected Papers of Wallace Hume Carothers on High Polymeric Substances (Eds. MARK, H., and WHITBY, G. S.), John Wiley, New York (1950)

3

States of Aggregation in Polymers

3.1 INTRODUCTION

In the previous chapter the various methods of synthesising polymers were briefly discussed. In this chapter the physical states of aggregation of these polymers will be considered, whilst in the three subsequent chapters the effect of molecular structure on the properties of polymers will be investigated.

Simple molecules like those of water, ethyl alcohol and sodium chloride can exist in any one of three physical states, i.e. the solid state, the liquid state and the gaseous, according to the ambient conditions. With some of these materials it may be difficult to achieve the gaseous state or even the liquid state because of thermal decomposition but in general these three phases with sharply defined boundaries, are discernable. Thus at a fixed ambient pressure, the melting point and the boiling point of a material such as pure water occur at a definite temperature. In polymers, changes of state are less well defined and may well occur over a finite temperature range. The behaviour of linear amorphous polymers, crystalline polymers and thermosetting structures will be considered in turn.

3.2. LINEAR AMORPHOUS POLYMERS

A specific linear amorphous polymer, such as poly(methyl methacrylate) or polystyrene, can exist in a number of states according to the temperature and the average molecular weight of the polymer. This is shown diagrammatically in Fig. 3.1. At low molecular weights (e.g. M_1) the polymer will be solid below some given temperature whilst above that temperature it will be liquid. The melting point for such polymers will be quite sharp and is the temperature above which the molecules have sufficient energy to move independently of each other, i.e. they are capable of viscous flow. Conversely below this temperature the molecules have insufficient energy for flow and the mass behaves as a rigid solid. At some temperature well above the melting point, the material will start to boil providing this is

33

below the decomposition temperature. In high polymers this is rarely, if ever, the case.

At high molecular weights (e.g. M_2) such a clearly defined melting point no longer occurs and a rubbery intermediate zone is often observed. In this case two transition temperatures may be observed, a rigid solid–rubber transition (usually known as the *glass transition temperature*) and secondly a generally very indefinite rubber–liquid transition, sometimes referred to as the flow temperature. (The term melting point will be reserved for crystalline polymers.)

It is instructive to consider briefly the three states and then to consider the processes which define the transition temperatures. In the solid state

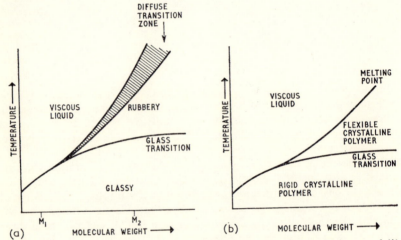

Fig. 3.1. *Temperature–molecular weight diagrams for* (*a*) *amorphous and* (*b*) *moderately crystalline polymers* (*with highly crystalline polymers the glass transition is less apparent*)

the polymer is hard and rigid. Amorphous polymers, under discussion in this section, are also transparent and thus the materials are glass-like and the state is sometimes referred to as the *glassy state*. Molecular movement other than bond rotation and vibrations is very limited. Above the glass transition temperature the molecule has more energy and movement of molecular segments becomes possible. It has been established that, above a given molecular weight, movement of the complete molecule as a unit does not take place. Flow will occur only when there is a co-operative movement of the molecular segments. In the rubbery range such co-operative motion leading to flow is very limited because of such features as entanglements and secondary (or even primary) cross linking. (In crystalline polymers, discussed in the next section, crystalline zones may also restrict flow.) In the rubbery state the molecules naturally take up a random, coiled conformation as a result of free rotation about single covalent bonds (usually C—C bonds) in the chain backbone. On application of a stress the molecules tend to uncoil and in the absence of crystallisation or premature rupture the polymer mass may be stretched until the molecules adopt the fully stretched conformation. In tension, elongations as high as 1,200% are possible with some rubbery polymers. On release of the stress the free rotations about the single bonds cause the molecule to coil-up once again. In commercial rubbery materials chain coiling and uncoiling processes are substantially complete within a small fraction of a second. They are, nevertheless, not instantaneous and the

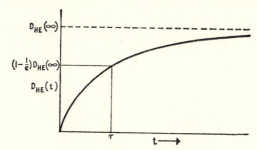

Fig. 3.2. *Application of stress to a highly elastic body.*
Rate of chain uncoiling with time

deformation changes lag behind the application and removal of stress. Thus the deformation characteristics are somewhat dependent on the rate of stressing.

Chain uncoiling, and the converse process of coiling, is conveniently considered as a unimolecular chemical reaction. It is assumed that the rate of uncoiling at any time after application of a stress is proportional to the molecules still coiled. The deformation $D_{HE}(t)$ at time t after application of stress can be shown to be related to the equilibrium deformation $D_{HE}(\infty)$ by the equation

$$D_{HE}(t) = D_{HE}(\infty)\,(1 - e^{-t/\tau_m}) \tag{3.1}$$

when τ_m, a reaction rate constant, is the time taken for the deformation to reach $(1 - 1/e)$ of its final value (Fig. 3.2). Since different molecules will vary in their orientation time depending on their initial disposition this value is an average time for all the molecules.

Whether or not a polymer is rubbery or glass-like depends on the relative values of t and τ_m. If t is much less than τ_m, the *orientation time*, then in the time available little deformation occurs and the rubber behaves like a solid. This is the case in tests normally carried out with a material such as polystyrene at room temperature where the orientation time has a large value, much greater than the usual time scale of an experiment. On the other hand if t is much greater than τ_m there will be time for deformation and the material will be rubbery, as is normally the case with tests carried out on natural rubber at room temperature. It is however vital to note the dependence on the time scale of the experiment. Thus a material which shows rubbery behaviour in normal tensile tests could appear to be quite stiff if it were subjected to very high frequency vibrational stresses.

The rate constant τ_m is a measure of the ease at which the molecule can uncoil through rotation about the C—C or other backbone bonds. This is found to vary with temperature by the exponential rate constant law so that

$$\tau_m = Ae^{E/RT} \tag{3.2}$$

If this is substituted into Equation 2.1 Equation 2.3 is obtained.

$$D_{HE}(t) = D_{HE}(\infty)\left[1 - \exp\left(\frac{-t}{Ae^{E/RT}}\right)\right] \tag{3.3}$$

In effect this equation indicates that the deformation can be critically dependent on temperature, and that the material will change from a rubbery to a glass-like state over a small drop in temperature. Frith and Tuckett[1] have illustrated (Fig. 3.3) how a polymer of $\tau_m = 100$ sec at 27°C, an activation energy E of 60 kcal will change from being rubbery to glass-like as the temperature is reduced from about 30°C to about 15°C. The time of stressing in this example was 100 sec.

It is now possible to understand the behaviour of real polymers and to interpret various measurements of the glass transition temperature. This last named property may be thus considered as the temperature at which molecular segment rotations do not occur within the time-scale of the experiment. There are many properties which are affected by the

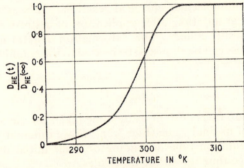

Fig. 3.3. *The ratio $D_{HE}(t)/D_{HE}(\infty)$ and its variation with temperature. (After Frith and Tuckett,[1] reproduced by permission of Longmans, Green and Co. Ltd.)*

transition from the rubbery to the glass-like state as a result of changes in the molecular mobility. Properties which show significant changes include specific volume, specific heat, thermal conductivity, power factor (see Chapter 6), nuclear magnetic resonance, dynamic modulus and simple stress–strain characteristics. The fact that measurements of the effect of temperature on these properties indicate different glass transition temperatures is simply due to the fact that the glass temperature is dependent on the time-scale of the experiment. This is illustrated by results obtained for a polyoxacyclobutane (poly-3:3,-bischloromethyloxacyclobutane), showing how transition temperatures depend on the frequency (or speed) of the test (Table 3.1).[2]

Electrical and dynamic mechanical tests often reveal more than one transition temperature in a given polymer. These additional phenomena are believed to be associated with additional restrictions in molecular movement, such as in side-chain mobility, which may occur at temperatures below the main glass transition temperature. In descending order

Table 3.1 INFLUENCE OF EXPERIMENTAL TIME SCALE ON THE GLASS TRANSITION POINT OF A POLYOXACYCLOBUTANE[2]

	Frequency (c/s)	*Glass temperature* (°C)
Electrical tests	1,000	32
Mechanical vibration	89	25
Slow tensile	3	15
Dilatometry	10^{-2}	7

of temperature these transitions are labelled as α, β, γ, δ, etc. In most cases the α-transition is dominant and corresponds to the 'glass transition temperature'. An exception occurs with polypropylene where the dominating transition at 0°C, below which brittleness is a prevalent phenomenon, has been found by dynamic tests to be the β-transition.

If an amorphous polymer is heated through the rubbery range a point will be reached where the molecular energy is sufficient to overcome restrictions through secondary bonding or entanglements and flow can occur (providing that this is below the decomposition temperature). The point of onset of flow is not clearly defined and in practice there is a transition from elastic to viscous properties through a visco-elastic range. In the case of most amorphous polymers processing is carried out under conditions in which elastic properties are far from being absent.

Orientation in linear amorphous polymers

If a sample of an amorphous polymer is heated to a temperature above its glass transition point and then subjected to a tensile stress the molecules will tend to align themselves in the general direction of the stress. If the mass is then cooled below its transition temperature while the molecule is still under stress the molecules will become frozen whilst in an oriented state. Such an orientation can have significant effects on the properties of the polymer mass. Thus if a filament of polystyrene is heated, stretched and frozen in this way a thinner filament will be produced with aligned molecules. The resultant filament has a tensile strength which may be five times that of the unoriented material because on application of stress much of the strain is taken up by covalent bonds forming the chain backbone. On the other hand the tensile strength will be lower in the directions perpendicular to the orientation. The polymer is thus anisotropic.

Anistropic behaviour is also exhibited in optical properties and orientation effects can be observed and to some extent measured by birefringence methods.   In such oriented materials the molecules are in effect frozen in an unstable state and they will normally endeavour to take-up a more coiled conformation due to rotation about the single bonds.   If an oriented sample is heated up the molecules will start to coil as soon as they possess sufficient energy and the mass will often distort.   Because of this oriented materials usually have a lower heat distortion temperature than oriented polymers.

In addition to monoaxial orientation, biaxial stretching of amorphous polymers is possible.   For example if poly(methyl methacrylate) sheet is heated above its glass temperature and stretched in two directions simultaneously there will be a planar disposition of the molecules.   It has been

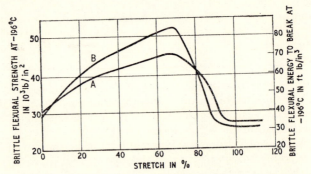

Fig. 3.4.   *Biaxial orientation of polymethyl methacrylate. Variation of (a) brittle flexural strength and (b) brittle flexural energy with percentage stretch.   (After Ladbury[3])*

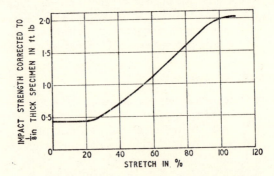

Fig. 3.5.   *Biaxial orientation of polymethyl methacrylate.   Variation of impact strength with percentage stretch.   (After Ladbury[3])*

found that with poly(methyl methacrylate) sheet such properties as tensile strength and brittle flexural strength increase with increased orientation up to a percentage stretch of about 70% (Fig. 3.4).[3]   Above this value there is a decrease in the numerical value of these properties presumably due to the increase in flaws between the layers of molecules.   Properties such as impact strength (Fig. 3.5)[3] and solvent crazing resistance which are less dependent on these flaws than other properties continue to improve with increased orientation.

In addition to the deliberate monoaxial or biaxial orientation carried out to produce oriented filament or sheet, orientation will often occur during polymer processing whether desired or not. Thus in injection moulding, extrusion or calendering the shearing of the melt during flow will cause molecular orientation. If the plastic mass 'sets' before the individual molecules have coiled then the product will contain frozen-in orientation with built-in, often undesirable, stresses. It is in order to reduce these frozen-in stresses that warm moulds and fast injection rates are preferred in injection moulding. In the manipulation of poly(methyl methacrylate) sheet to form baths, light fittings and other objects biaxial stretching will frequently occur. Such acrylic products produced by double curvature forming will revert completely to the original flat sheet from which they were prepared if they are heated above their glass transition temperature.

3.3 CRYSTALLINE POLYMERS

If a polymer molecule has a sufficiently regular structure it may be capable of some degree of crystallisation. The factors affecting regularity will be discussed in the next chapter but it may be said that crystallisation is limited to certain linear or slightly branched polymers with a high structural regularity. Well-known examples of crystalline polymers are polyethylene, acetal resins and polytetrafluoroethylene.

From a brief consideration of the properties of the above three polymers it will be realised that there are substantial differences between the crystallisation of simple molecules such as water and copper sulphate and of polymers such as polyethylene. The lack of rigidity, for example, of polyethylene indicates a much lower degree of crystallinity than in the simple molecules. In spite of this the presence of crystalline regions in a polymer has large effects on such properties as density, stiffness and clarity.

There have been, in the past few years profound changes in the theories of crystallisation in polymers. For many years it was believed that the crystallinity present was based on small crystallites of the order of a few hundred ångström units in length. This is very much less than the length of a high polymer molecule and it was believed that a single polymer molecule actually passed through several crystallites. The crystallites thus consisted of a bundle of segments from separate molecules which had packed together in a highly regular order. The method of packing was highly specific and could be ascertained from X-ray diffraction data. It was believed that in between the crystallites the polymer passed through amorphous regions in which molecular disposition was random. Thus there is the general picture of crystallites embedded in an amorphous matrix (Fig. 3.6). This theory known as the fringed micelle theory or fringed crystallite theory helped to explain many properties of crystalline polymers but it was difficult to explain the formation of certain larger structures such as spherulites which could possess a diameter as large as 0·1 mm.

As a result of work based initially on single polymer crystals prepared from solution, there is a growing school of thought that the fringed micelle theory is incorrect. Instead it is believed that polymer molecules fold upon themselves at intervals of about 100 Å to form lamellae which appear to be the fundamental structures in a crystalline polymer. Crystallisation

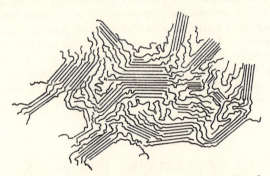

Fig. 3.6. *Two dimensional representation of molecules in a crystalline polymer according to the fringed micelle theory showing ordered regions (*crystallites*) embedded in an amorphous matrix. (After Bryant[4])*

spreads by the growth of individual lamellae as polymer molecules align themselves into position and start to fold. For a variety of reasons, such as a point of branching or some other irregularity in the structure of the molecule, growth would then tend to proceed in many directions. In effect this would mean an outward growth from the nucleus and the development of spherulites. In this concept it is seen that a spherulite is simply caused by growth of the initial crystal structure, whereas in the fringed micelle theory it was generally postulated that formation of a spherulite required considerable reorganisation of the disposition of the crystallites. Both theories are consistent with many observed effects in crystalline polymers. The closer packing of the molecules causes an increased density. The decreased intermolecular distances will increase the secondary forces holding the chain together and increase the value of properties such as tensile strength, stiffness and softening point. If it were not for crystallisation, polyethylene would be rubbery at room temperature and many grades would be quite fluid at 100°C.

The properties of a given polymer will very much depend on the way in which crystallisation has taken place. A polymer mass with relatively few large spherulitic structures will be very different in its properties to a polymer with far more, but smaller, spherulites. It is thus useful to consider the factors affecting the formation of the initial nuclei for crystallisation (nucleation) and on those which affect growth.

Homogeneous nucleation occurs when, as a result of statistically random segmental motion, a few segments have adopted the same conformation as they would have in a crystallite. The likelihood of the formation of such nuclei is greatest just above the glass transition temperature when the rate of segmental motion is low and when, for considerations of rotational energy (see Chapter 4), the molecules tend to take up an extended conformation. Above the melting point, nuclei do not form at all, and any crystalline structures present will also disappear. As opposed to nucleation rates, the growth rate increases with temperature, i.e. as the segmental motion increases, and thus the overall degree of crystallinity is a resultant of two effects of opposite temperature dependence. The net result is that crystallisation occurs between the glass transition temperature and the melting point with the maximum rate occurring at a temperature about half-way between these two points (Fig. 3.7).

There are certain differences between the properties of a polymer crystallised at a temperature just above the glass transition temperature

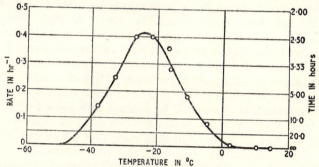

Fig. 3.7.   Rate of crystallisation of rubber as a function of temperature.   (After Wood[5])

compared with one crystallised just below the melting point. In the latter case the polymer develops large crystal structures which may be sufficiently large to interfere with light waves and cause opacity. It may also be somewhat brittle. In the former case the polymer mass, with smaller structures, is generally more transparent. The properties of the polymer will also depend on the time available for cooling. In the case of a polymer such as the bis-phenol A polycarbonate (Chapter 17) the glass temperature is about 140°C. There is in this case little time for crystallisation to develop in cooling from the melt after injection moulding and extrusion and transparent polymers are usually obtained.

On the other hand crystalline polymers with a glass temperature below that of the ambient temperature in which the polymer is to be used will continue to crystallise until equilibrium is reached. For this reason nylon 66 which has a glass temperature slightly below that of usual room temperature will exhibit after-shrinkage for up to 2 years after manufacture unless the sample has been specially annealed. In the case of the polyacetals (polyformaldehydes) the shrinkage is to all intents and purposes complete within 48 hours. The reason for this is that the glass transition point for the polyacetals is as low as −73°C. Therefore at the common ambient temperatures of about 20°C crystallisation rates are much faster for polyacetals than for nylon 66.   The problems of slow after-shrinkage of nylon 66 may be avoided by heating the polymer for a short period at a temperature at which crystallisation proceeds rapidly, i.e. at about 120°C.

Because polymers have a very low thermal conductivity, compared with metals, cooling from the melt proceeds unevenly, the surface cooling more rapidly than the centre.   This will be particularly marked with thick injection moulded sections and with piping and other extrusions which have been extruded into cold water.   In these cases the morphology (fine structure) of crystalline polymers will vary across the cooled polymer mass and hence the physical properties will also vary.   One consequence of this is that a surface produced by a machining operation may have quite different hardness, abrasion resistance and coefficient of friction than a moulded surface.

In many polymer products it is desirable to have a high degree of crystallinity but with small spherulite size.   A high homogeneous nucleation rate however requires the polymers to be held at a temperature only just above the transition temperature of the polymer.   Processing operations however demand a quick 'cooling' operation and it would be desirable to speed up the freezing operation.   High nucleation rates can be achieved together with high growth if *heterogeneous nucleation* is employed. In this case nucleation is initiated by seeding with some foreign particle. This can be of many types but is frequently a polymer of similar cohesive energy density (see Chapter 5) to that being crystallised but of a higher melting point.   Nucleating agents have been introduced recently in commercial products.   They have the overall effect of promoting rapid freezing, giving a high degree of crystallisation, good clarity in polymer films and reduce skin effects and formation of voids which can occur in conjunction with large morphological structures.

Mention may be made of the effect of the glass transition on the properties of a crystalline polymer.   In a highly crystalline polymer there is little scope for segmental motion since most of the segments are involved in a lattice formation in which they have low mobility.   Such polymers are comparatively rigid in the mass and there is little difference in properties immediately above and below the glass transition.   In fact with some highly crystalline polymers it is difficult to find the glass temperature. With less crystalline materials some distinction may be possible because of the higher quantity of segments in a less organised state.   Thus above the glass transition point the polymer may be flexible and below it quite stiff.

3.3.1   Orientation and crystallisation

If a rubbery polymer of regular structure (e.g. natural rubber) is stretched, the chain segments will be aligned and crystallisation is induced by orientation.   This crystallisation causes a pronounced stiffening in natural rubber on extension.   The crystalline structures are metastable and on retraction of the sample they disappear.

On the other hand if a polymer such as nylon 66 is stretched at some temperature well below its melting point but above its transition temperature, e.g. at room temperature, additional crystallisation will be induced and the crystalline structure will generally be aligned in the direction of extension. As a result, oriented crystalline filaments or fibres are much stronger than the unoriented product. This is the basis of the 'cold-drawing' process of the synthetic fibre industry. Poly(ethylene terephthalate) (Terylene) with a transition temperature of 67°C has to be cold-drawn at some higher temperature. The tensile strengths of nylon 66 and poly(ethylene terephthalate) fibres approach 10^5 lb in^{-2}, many times greater than those of the unoriented polymers.

Biaxial orientation effects are of importance in the manufacture of films and sheet. Biaxially stretched poly(ethylene terephthalate) (e.g. 'Melinex'), poly(vinylidene chloride) (Saran) and polypropylene films are strong films of high clarity, the orientation-induced crystallisation producing structures which do not interfere with the light waves.

3.4 CROSS-LINKED STRUCTURES

A cross-linked polymer can generally be placed into one of two groups

1. Lightly cross-linked materials.

2. Highly cross-linked materials.

In the lightly cross-linked polymers (e.g. the vulcanised rubbers) the main purpose of cross linking is to prevent the material deforming indefinitely under load. The chains can no longer slide past each other, and flow, in the usual sense of the word, is not possible without rupture of covalent bonds. Between the cross links, however, the molecular segments remain flexible. Thus under appropriate conditions of temperature the polymer mass may be rubbery or it may be rigid. It may also be capable of crystallisation in both the unstressed and the stressed state.

However if the degree of cross linking is increased the distance between cross links decreases and a tighter, less flexible network will be formed. Segmental motion will become more restricted as the degree of cross linking increases so that the transition temperature will eventually reach the decomposition temperature. In polymers of such a degree of cross linking only the amorphous rigid (glass-like) state will exist. This is the state commonly encountered with, for example, the technically important phenolic, aminoplastic and epoxide resins.

A novel form of cross linking is to be found in the 'ionomers' of Du Pont. In this case the chains are ionically linked and involve the use of sodium, potassium and similar ions. The ionic cross links are heat fugitive, i.e. they tend to disappear on heating, so that the ionomers behave like thermoplastics at processing temperatures but have some of the characteristics of cross-linked products, such as increased toughness and stiffness, at room temperature.

3.5 SUMMARY

Polymers can exist in a number of states.   They can be amorphous resins, rubbers or fluids; or they can be crystalline structures.   The molecules or the crystal structures can be monoaxially or biaxially oriented.   The form which a polymer will take at any given temperature will then depend on

1. The glass transition temperature.
2. The ability of the polymer to crystallise.
3. The crystalline melting point.
4. Any orientation of molecules or crystal structures that may have been induced.
5. The type and extent of cross linking (if any).

In the next chapter the structural features that control the first three of the above factors will be considered.

REFERENCES

1. FRITH, E. M., and TUCKETT, R. F., *Linear Polymers*, Longmans Green, London (1951)
2. SANDIFORD, D. J. H., *J. Appl. Chem.*, **8**, 186 (1958)
3. LADBURY, J. W., *Trans. Plastics Inst.*, **28**, 184 (1960)
4. BRYANT, W. M. D., *J. Polymer Sci.*, **2**, 547 (1947)
5. WOOD, L. A., *Recent Advances in Colloid Science*, Vol. 2, Interscience, New York, p. 58 (1946)

BIBLIOGRAPHY

BENTON, J., *Brit. Plastics*, **35**, 184, 251 (1962)
BILLMEYER, F. W., *Textbook of Polymer Science*, Interscience, New York (1962)
BRYANT, W. M. D., *J. Polymer Sci.*, **2**, 547 (1947)
FRITH, E. M., and TUCKETT, R. F., *Linear Polymers*, Longmans Green, London (1951)
GEIL, P. H., *Polymer Single Crystals*, Interscience, New York (1963)
GORDON, M., *High Polymers*, Iliffe Books Ltd., London (1963)
LADBURY, J. W., *Trans. Plastics Inst.*, **28**, 184 (1960)
MEARES, P., *Polymers: Structure and Bulk Properties*, Van Nostrand London (1965)
NIELSEN, L. E., *Mechanical Properties of Polymers*, Reinhold, New York (1963)
SANDIFORD, D. J. H., *J. Appl. Chem.*, **8**, 186 (1958)

4

Relation of Structure to Thermal and Mechanical Properties

4.1 INTRODUCTION

It is sometimes said that three factors determine whether a polymer is glassy, rubbery or fibre-forming under a given set of conditions. These are the chain flexibility, the interchain attraction and the regularity of the polymer. The relationship has been expressed diagrammatically by Swallow[1] (Fig. 4.1). The importance of these parameters arises from their influence on the glass transition temperature, the ability of a material to crystallise and, where relevant, the crystalline melting point. In this chapter specific influences which have a bearing on these last three properties will be discussed. At the end of the chapter there is a short discussion of structural features which determine certain selected properties.

4.2 FACTORS AFFECTING THE GLASS TRANSITION TEMPERATURE

There are a number of structural features which have a bearing on the value of the glass transition temperature. Since this temperature is that at which molecular rotation about single bonds becomes restricted, it is obvious that these features are ones which influence the ease of rotation. These can be divided into two groups:

1. Factors which affect the inherent or intrinsic mobility of a single chain considered on its own.
2. Those factors whose influence is felt because of the proximity and interaction of many polymer chains.

Before considering the special case of rotation about bonds in polymers it is useful to consider such rotations in simple molecules. Although reference is often made to the 'free rotation' about a single bond, in fact rotational energies of the order of 2 kcal/mole are required to overcome certain energy barriers in such simple hydrocarbons as ethane.   During rotation of one part of a molecule about another part the proximity of specific groups or atoms on one part to groups or atoms on the other part will vary.   When the specific groups or atoms come close to those on the adjacent carbon atom there is usually a repulsion effect due to steric hindrance, the presence of dipole forces or because of the electronic structure.   In some cases intramolecular hydrogen bonding may cause attraction rather than repulsion so that rotation away from this position may involve a large energy requirement.   Examples of some of these effects are given in Fig. 4.2.

Fig. 4.3 depicts the change in height of rotational energy barriers with bond rotation of methyl succinic acid.   This shows quite clearly the

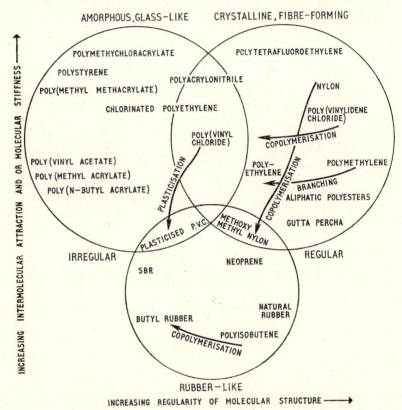

Fig. 4.1.   *Effect of structure regularity, molecular stiffness and inter-molecular attraction on polymer properties.   (After Swallow[1])*

repulsion effect which occurs when atoms or groups come into close proximity.   This effect is particularly great where the polar carboxyl groups come close together.   Such a position is less stable than others and is also the greater barrier to rotation.   The comparative stability of the staggered position is to be noted.   A consequence of these rotational

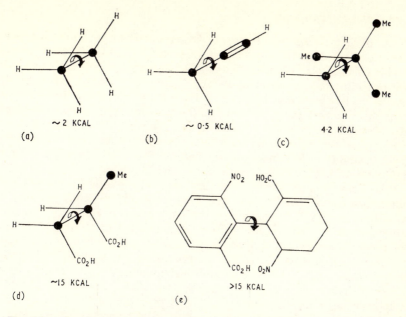

Fig. 4.2. Rotational-energy barriers as a function of substitution. The small barrier (∼ 2 kcal) in ethane (a) is lowered even further (∼ 0·5 kcal) if three bonds are 'tied back' by replacing three hydrogen atoms of a methyl group by a triply-bonded carbon, as in methylacetylene (b). The barrier is raised (4·2 kcal) when methyl groups replace the smaller hydrogen atoms, as in neopentane (c). Dipole forces raise the barrier further (∼ 15 kcal) in methylsuccinic acid (d) (cf. Fig. 4.3). Steric hindrance is responsible for the high barrier (> 15 kcal) in the diphenyl derivative (e). (After Gordon[2])

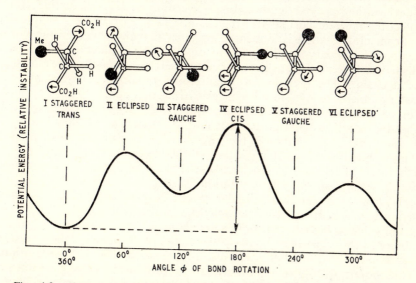

Fig. 4.3. Energy versus bond rotation in methylsuccinic acid (schematic). The diagram shows the greater stability of staggered as compared with eclipsed forms, and the effect of size and dipole moment of substituents on the barriers. The slope of the curve at any point represents the force opposing rotation there. (E = energy of activation of rotation). (After Gordon[2])

effects is that as the temperature is lowered certain conformations (the *trans* position in methyl succinic acid) become more likely.

This effect is also observed with high polymers. The *trans* form of a hydrocarbon chain requires an energy about 0·8 kcal/mole less than the *gauche*. The *trans* form leads to an extended molecule and in hydrocarbons this becomes more favoured as the temperature is lowered. Many hydrocarbons take up this conformation in the crystalline state.

From the preceding considerations it is appreciated that the intrinsic chain flexibility is determined by the nature of the chain backbone and by the nature of groups directly attached to the backbone.

It is generally considered that chains based on aliphatic C—C and C—O bonds are quite flexible. On the other hand the introduction of ring structures such as the *p*-phenylene group into the main chain has a marked stiffening effect. The glass transitions of poly(ethylene terephthalate) and the polycarbonate from bis-phenol A are much higher than those of their aliphatic counterparts because of the presence of such phenylene groups. Rotation about a single C—C bond is also impeded by the substitution of attached hydrogen atoms by methyl or other hydrocarbon groups. Polypropylene and the polymers of highly branched olefins have a higher glass transition than polyethylene for this reason. In the case of polymers of unbranched or lightly-branched olefins other factors come into play.

The size of the group attached to the main-chain carbon atom can influence the glass transition point. For example in polytetrafluoroethylene, which differs from polyethylene in having fluorine instead of hydrogen atoms attached to the backbone, the size of the fluorine atoms requires the molecule to take up a twisted zigzag configuration with the fluorine atoms packed tightly around the chain. In this case steric factors affect the inherent flexibility of the chain.

Segmental rotation is influenced by other polymer chains in the same region. Secondary bonding due to dipole forces, induction forces and dispersion forces or hydrogen bonding between chains can affect the mobility of a chain. The presence of polar groups or atoms such as chlorine will be a factor tending to raise the glass transition point. Thus the value for p.v.c. is much higher than for polyethylene. It is interesting to note that two identical polar groups attached to the same chain may lead to a lower glass transition than in the case of polymers with one polar group. This is the case with poly(vinylidene chloride) as compared with poly(vinyl chloride). This has been attributed to the reduction in dipole moment as a result of the symmetry of substitution.

Hydrogen bonding has a similar effect to that of polar groups. Thus nylon 6, which differs from polyethylene by the presence of —CONH— groups, has a higher transition point than the polyolefin because of its ability to form hydrogen bonds. The forces of attraction, which cause a reduction of chain mobility, will also be affected by the chain separation. Thus the introduction of *n*-alkyl groups in separating the chains will decrease the interchain attraction, more than offsetting the effect of increasing the inherent chain stiffness. However the presence of long side chains may cause entanglement and even permit some crystallisation. In some polymers, such as those of butadiene and the substituted butadienes, these effects cancel each other out. In the case of the methacrylates chain separation is more important than entanglement up to poly(*n*-dodecyl methacrylate) but with the higher substituted materials however, entanglement is the predominating influence causing an increase in the glass transition.

The molecular weight of a polymer will have some effect on the glass transition temperature. A low molecular weight polymer will have a greater number of 'chain ends' in a given volume than a high molecular weight polymer. The chain ends are less restrained and can become more active than segments in the centre of a molecule and cause the polymer mass to expand. This gives the molecules greater mobility and they can be taken to a lower temperature before the thermal energy of the molecules is too low for the segments to rotate. It has been found that, within limits, the glass transition temperature of polystyrene is proportional to the reciprocal of the molecular weight. At very high molecular weights the effect of the glass transition becomes negligible.

The glass transition of a random copolymer usually falls between those of the corresponding homopolymers. It can be shown that the glass transition of a copolymer may be given by the equation

$$C_2 = (T_g - T_g') [K(T_g'' - T_g) + T_g - T_g']$$

where C_2 is the weight fraction of the second component T_g, T_g', and T_g'' are glass transitions of the copolymer, the homopolymer of the first component, and the homopolymer of the second component, respectively. K is the ratio between the differences of the two expansion coefficients in the rubbery and the glassy state of the two homopolymers.

The restricting influence of cross linking on segmental mobility was pointed out in the previous chapter. The greater the degree of cross linking the higher the transition temperature. Table 4.1 shows how the transition temperature of natural rubber is related to the percentage of combined sulphur (a rough measure of the degree of cross linking)[3].

On the other hand the addition of liquids to the polymer will cause separation of the chains and increase their general mobility. This is the effect of plasticisation which will cause a marked reduction in the transition temperature. The addition of about 40% of diethyl hexyl phthalate to p.v.c. will reduce its glass transition temperature by about 100 degC.

Table 4.1   DEPENDENCE OF THE GLASS
TRANSITION OF RUBBER ON
THE SULPHUR CONTENT

Sulphur (%)	T_g(°C)
0	−65
0·25	−64
10	−40
20	−24

The factors which affect the glass transition are as follows:

1. Groups attached to the backbone which increase the energy required for rotation.
2. Rigid structures, e.g. phenylene groups, incorporated in the backbone of the molecule.
3. The packing of substituents around the main chain (c.f. p.t.f.e. with polyethylene).
4. Secondary bonding between chains, e.g. hydrogen bonding.
5. Primary bonding between chains, e.g. cross linking.
6. Length of side chains.
7. Molecular weight.
8. Copolymerisation.
9. Plasticisation.

The glass transition temperatures of a number of polymers are given in Table 4.2.

4.3 FACTORS AFFECTING THE ABILITY TO CRYSTALLISE

In the case of an amorphous polymer the glass transition temperature will define whether or not a material is glass-like or rubbery at a given temperature. If however the polymer will crystallise, rubbery behaviour may be limited since the orderly arrangement of molecules in the crystalline structure by necessity limits the chain mobility. In these circumstances the transition temperature is of less consequence in assessing the physical properties of the polymer.

The ability of a material to crystallise is determined by the regularity of its molecular structure. A regular structure is potentially capable of crystallinity whilst an irregular structure will tend to give amorphous polymers. Structural irregularities can occur in the following ways:

1. By copolymerisation.
2. By introduction of groups in an irregular manner.
3. By chain branching.
4. By lack of stereoregularity.
5. By differences in geometrical isomerism.

Copolymerisation provides a very effective way of reducing the regularity and hence the ability to crystallise. Polyethylene is a crystalline material but with random ethylene–propylene copolymers crystallisation becomes a difficult process and a rubbery material results. The introduction of side groups in a random manner has a similar effect. If polyethylene is partially chlorinated, the regularity of the structure is reduced and a rubbery polymer will result. A chlorinated polyethylene which also contains a few sulphonyl chloride groups is a commercially available rubber (Hypalon). The treatment of poly(vinyl alcohol) with formaldehyde to give poly(vinyl formal) will also prevent crystallisation because of the structural irregularities produced (Fig. 4.4).

Table 4.2 GLASS TRANSITION TEMPERATURE T_g OF SELECTED POLYMERS

Polymer	Structure	$T_g(°C)$	Feature described in text
Poly(dimethyl siloxane)	$-O \cdot Si(CH_3)_2-$	-123	siloxane bonds very flexible
Polybutadiene	$-CH_2 \cdot CH=CH \cdot CH_2-$	-85	C—C bonds flexible
*Cis*polyisoprene	$-CH_2 \cdot C(CH_3)=CH \cdot CH_2-$	-70	stiffening effect of methyl group
Poly(*n*-heptyl butadiene)	$-CH_2 \cdot C \cdot (C_7H_{15})=CH \cdot CH_2-$	-83	chain separation balanced by entanglement
Poly(*n*-decyl butadiene)	$-CH_2 \cdot C \cdot (C_{10}H_{21})=CH \cdot CH_2-$	-53	side chain entanglement and crystallisation
Butadiene–styrene copolymer	75% Bu–25% St.	-55	copolymerisation
Polypropylene	$-CH_2 \cdot CH(CH_3)-$	-27	stiffening effect of methyl group
Polyisobutylene	$-CH_2 \cdot C(CH_3)_2-$	-65	increased flexibility through reduction of dipole moment
Poly(3-methyl butene-1)	$-CH_2 \cdot CH \cdot [CH(CH_3)_2]-$	$+50$	stiffening effect of isopropyl group
Polyoxymethylene	$-CH_2 \cdot O-$	-73	flexibility of C—O bonds
Poly(ethylene adipate)	$-(CH_2)_2OOC \cdot (CH_2)_4COO-$	-70	flexibility of C—O and C—C bonds
Poly(ethylene terephthalate)	$-(CH_2)_2OOC \cdot C_6H_4COO-$	$+67$	stiffening effect of phenylene group in backbone
Polycarbonate of (bisphenol A)	$-C_6H_4 \cdot C(CH_3)_2 \cdot C_6H_4 \cdot O \cdot CO \cdot CO \cdot O$	$+149$	stiffening effect of phenylene groups
Poly(vinyl chloride)	$-CH_2 \cdot CH(Cl)-$	$+80$	dipole attraction of chlorine atoms
Poly(vinylidene chloride)	$-CH_2 \cdot C \cdot (Cl)_2-$	-17	reduction of dipole moment
Polystyrene	$-CH_2 \cdot C(C_6H_5)-$	$+100$	stiffening effect of attached benzene ring

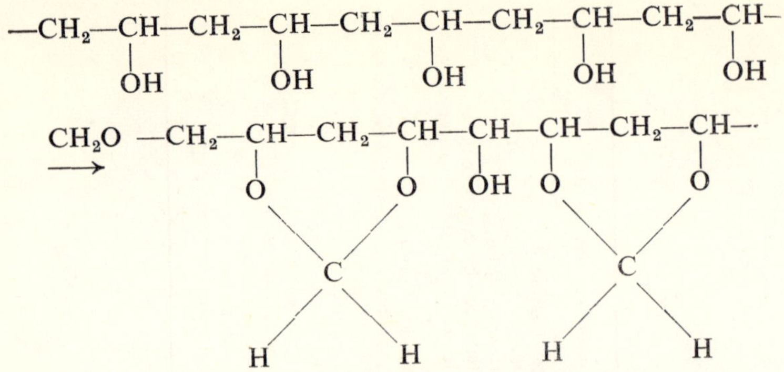

$$-CH_2-CH-CH_2-CH-CH_2-CH-CH_2-CH-CH_2-CH-$$

with OH groups below each CH, then:

$$CH_2O \longrightarrow -CH_2-CH-CH_2-CH-CH-CH-CH_2-CH---$$

Fig. 4.4

Branching can to some extent reduce the ability to crystallise. The frequent, but irregular, presence of side groups will interfere with the ability to pack. Branched polyethylenes, such as are made by high pressure processes (e.g. Alkathene), are less crystalline and of lower density than less branched structures prepared using metal oxide catalysts (e.g. Rigidex). In extreme cases crystallisation could be almost completely inhibited. (Crystallisation in high pressure polyethylenes is restricted more by the frequent short branches rather than by the occasional long branch.)

In recent years the significance of stereoregularity has become more appreciated. In vinyl compounds, for instance, different structures may arise for similar reasons that optical isomers are produced in simple organic chemicals. Covalent bonds linking one atom with others are not all in the same plane but form certain angles with each other. A carbon atom attached to four hydrogen atoms or to other carbon atoms has bonds subtended at angles of about 109°. As conventionally represented on paper it would appear that linear polypropylene can only take one form (Fig. 4.5).

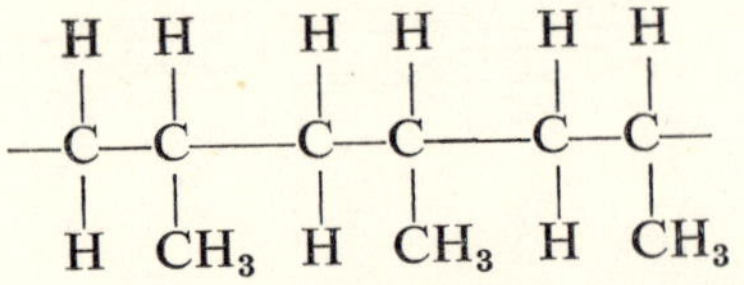

$$-\overset{\overset{\displaystyle H}{|}}{\underset{\underset{\displaystyle H}{|}}{C}}-\overset{\overset{\displaystyle H}{|}}{\underset{\underset{\displaystyle CH_3}{|}}{C}}-\overset{\overset{\displaystyle H}{|}}{\underset{\underset{\displaystyle H}{|}}{C}}-\overset{\overset{\displaystyle H}{|}}{\underset{\underset{\displaystyle CH_3}{|}}{C}}-\overset{\overset{\displaystyle H}{|}}{\underset{\underset{\displaystyle H}{|}}{C}}-\overset{\overset{\displaystyle H}{|}}{\underset{\underset{\displaystyle CH_3}{|}}{C}}-$$

Fig. 4.5

If however it is remembered that the actual molecule, even when fully extended, is not planar, then it will be seen that different forms can arise.

In Fig. 4.6 three idealised cases are given. Fig. 4.6 (a) depicts an isotactic structure in which all of the methyl groups are on the same side of the main chain. An alternating system is shown in Fig. 4.6 (b) which is known as a syndiotactic structure whilst Fig. 4.6 (c) depicts a random system—known as an atactic structure. Thus the isotactic and syndiotactic structures are regular whilst the atactic is irregular. It is not possible to convert from one form to another simply by rotating the molecule about a chain C—C bond. This is easier to see using a molecular model rather than by looking at a formula written on a piece of paper. Since the structures are dissimilar, it is to be expected that the bulk properties of the polymer will differ. (In practice perfect isotactic and syndiotactic structures are not usually obtained and a polymer molecule may be part atactic and part isotactic or syndiotactic. Furthermore in the mass,

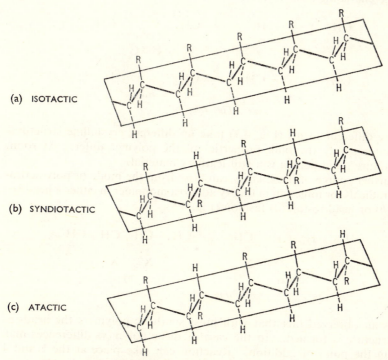

(a) ISOTACTIC

(b) SYNDIOTACTIC

(c) ATACTIC

Fig. 4.6. *Relationship between isotactic, syndiotactic and atactic forms in head-to-tail vinyl polymers. (For simplicity of comparison the main chain in each case is shown stretched in the planar all-trans zigzag form)*

molecules of differing tacticity may exist. In the case of isotactic polypropylene in the crystalline state the molecules take up a helical arrangement which can either have a clockwise or anticlockwise twist.)

The regular syndiotactic and isotactic structures are capable of crystallisation whereas the atactic polymer cannot normally do so. In the case of polypropylene the isotactic material is a crystalline fibre-forming material. It is also a useful thermoplastic which can withstand boiling water for prolonged periods. Atactic polypropylene is a 'dead' amorphous material. Commercial polystyrene is atactic and glass-like, but the isotactic polymer, which has been prepared in the laboratory, is crystalline. The only occasions when an atactic polymer can crystallise are when the atoms or groups attached to the asymmetric carbon atom are of a similar size. For example poly(vinyl alcohol) is atactic but as the hydroxy group is small enough for the polymer to pack into the same lattice as polyethylene a crystalline material results. (Most commercial grades of poly(vinyl alcohol) contain acetate groups so that crystallinity in these materials is somewhat limited.)

Related to stereoregularity is the possiblity of *cis, trans* isomerism. The molecule of natural rubber is a *cis*-1, 4-polyisoprene whilst that of gutta percha is the *trans* isomer.

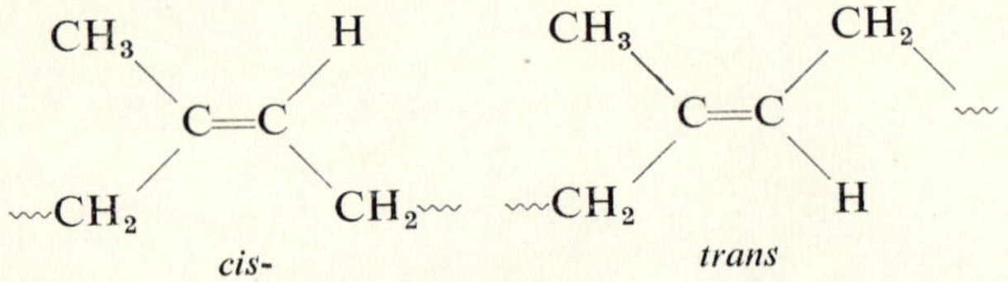

Fig. 4.7

These different forms (Fig. 4.7) take up different crystalline structures and consequently the bulk properties of the polymer differ. At room temperature gutta percha is a stiff leathery material.

A further source of irregularity can arise from the mode of polymerisation; radicals, or ions, can in theory add to a monomer in either a head-to-head (*b*) or head-to-tail (*a*) fashion (Fig. 4.8).

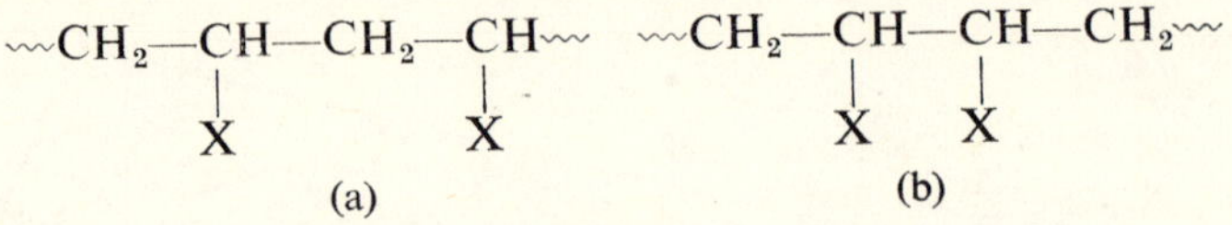

Fig. 4.8

It is an observed fact that with most synthetic polymers the head-to-tail structure is formed. In the case of diene polymers differences may arise in the point of addition. Reaction can take place at the 1 and 4 positions, the 1 and 2 positions or the 3 and 4 positions to give the structures indicated in Fig. 4.9.

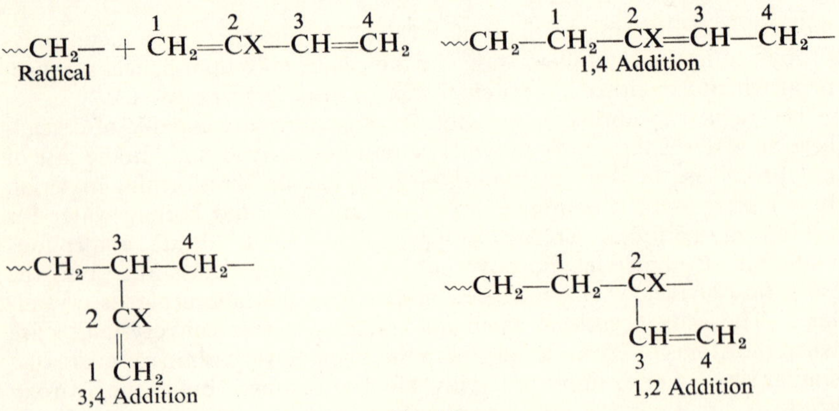

Fig. 4.9

The presence of pendant reactive vinyl groups through 1,2 and 3,4 addition provides a site for branching and cross linking since these may be involved in other chain reactions. Because of this a 1,4 polymer is generally to be desired.

4.4 FACTORS AFFECTING THE CRYSTALLINE MELTING POINT

To a large extent the factors which determine the position of the glass transition temperature of a polymer (chain stiffness and intermolecular forces) also determine the melting point of a crystalline polymer. In Fig. 4.10 a rough correlation is seen between the glass transition and melting points of a number of crystalline polymers. The glass transition temperature of many polymers is about $\frac{1}{2}$–$\frac{2}{3}$ that of the crystalline melting point when measured in degK. An important exception to this occurs with copolymers. Whereas the glass transition of a copolymer is usually intermediate between those of the corresponding homopolymers this is not commonly the case with the melting points. Fig. 4.11 shows the effect of copolymerising hexamethylene sebacamide with hexamethylene terephthalamide. Only when the monomer units are isomorphous, so that the molecules can take up the same structure, is there a linear relationship between melting point and composition (as with hexamethylene adipamide and hexamethylene terephthalamide).

Further information on the effect of polymer structure on melting points has been obtained in recent years by considering the heats and entropies of fusion. The relationship between free energy change ΔF with change in heat content ΔH and entropy change ΔS at constant temperature is given by the equation

$$\Delta F = \Delta H - T\Delta S$$

In thermodynamic language it is said that a reaction will occur if there is a decrease in the free energy, i.e. ΔF is negative. Since at the melting point melting and crystallising processes are balanced ΔF is zero and the expression may be written

$$T_m = \frac{\Delta H_m}{\Delta S_m}$$

where T_m is the melting point

ΔH_m the heat of fusion

ΔS_m the entropy of fusion

The entropy term is a measure of the degree of freedom of the molecules and thus a measure of its flexibility. Measurement of the heats and entropies of fusion has provided interesting information on the relative importance of various factors influencing the melting point of specific polymers. The linear aliphatic polyesters have low melting points and

P.M.—5

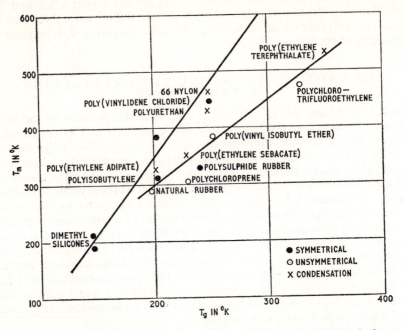

Fig. 4.10. *Relationship between crystalline melting point* (I_m) *and glass transition point* (I_g) *for various polymers.* (*After Boyer*[4])

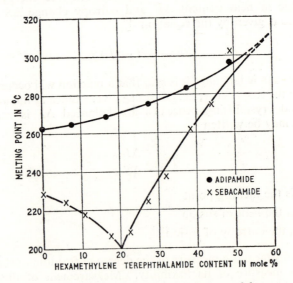

Fig. 4.11. *Melting points of copolymers of hexamethylene adipamide and terephthalamide, and of hexamethylene sebacamide and terephthalamide.* (*After Edgar and Hill*[5])

this has been attributed to the high flexibility of the C—O bond. This would suggest a high entropy of fusion but in fact it is observed that both the heat and entropy terms are lower than in the case of polyethylene These observations remain to be fully explained.

4.5 SOME INDIVIDUAL PROPERTIES

By a knowledge of the glass transition, the ability to crystallise and, where relevant, the crystalline melting point general statements may be made regarding the properties of a given polymer at a specified temperature. From this information it is possible to say that the material is either a rubber, a glass-like resin, a melt, a hard crystalline solid (below T_g or with a high degree of crystallinity) or a flexible crystalline solid (polymer above T_g and with a moderate low degree of crystallinity). Each class of materials has characteristically different properties. In the case of rubbers and glass-like amorphous resins T_g will indicate the range of minimum and maximum service temperatures respectively. A polymer with T_g close to room temperature might be expected to have only a limited value as its properties would be liable to large changes with changes in ambient temperature. Paradoxically many plasticised p.v.c. compounds are in such a position and yet they form one of the most important classes of plastics materials.

In the case of a crystalline polymer the maximum service temperature will be largely dependent on the crystalline melting point. When the polymer possesses a low degree of crystallinity the glass transition temperature will remain of paramount importance. This is the case with unplasticised p.v.c. and the polycarbonate of bis-phenol A.

4.5.1 Melt viscosity

The melt viscosity of a polymer at a given temperature is a measure of the rate at which chains can move relative to each other. This will be controlled by the ease of rotation about the backbone bonds, i.e. the chain flexibility, and on the degree of entanglement. Because of their low chain flexibility, polymers such as polytetrafluoroethylene, the aromatic polyimides, the aromatic polycarbonates and to a less extent poly(vinyl chloride) and poly(methyl methacrylate) are highly viscous in their melting range as compared with polyethylene and polystyrene.

For a specific polymer the melt viscosity is considerably dependent on the (weight average) molecular weight. The higher the molecular weight the greater the entanglements and the greater the melt viscosity. Natural rubber and certain poly(methyl methacrylate) products (e.g. Perspex) which have molecular weights of the order of 10^6, cannot be melt processed until the chains have been broken down into smaller units by mastication processes. Chain branching also has an effect. In the case of polyethylene and the silicones the greater the branching, at constant weight-average molecular weight, the lower the melt viscosity. However in poly(vinyl acetate) the melt viscosity increases with an increase in branching. It has been suggested that the branch length may be the controlling influence in this.

4.5.2   Yield strength and modulus

On comparison of the yield strengths and elastic moduli of amorphous polymers well below their glass transition temperature it is observed that the differences between polymers are quite small. Yield strengths are of the order of 8,000 lb in^{-2} and tension modulus values are of the order of 500,000 lb in^{-2}. In the molecular weight range in which these materials are used differences in molecular weight have little effect.

In the case of commercial crystalline polymers wider differences are to be noted. Many polyethylenes have a yield strength below 2,000 lb in^{-2} whilst the nylons may have a value of 12,000 lb in^{-2}. In these polymers the intermolecular attraction, the molecular weight and the type and amount of crystalline structure all influence the mechanical properties.

4.5.3   Specific gravity

This, the mass per unit volume, is a function of the weight of individual molecules and the way they pack. The hydrocarbons do not possess 'heavy' atoms and therefore the mass of the molecule per unit volume is rather low. Amorphous hydrocarbon polymers generally have specific gravities of 0·86–1·05. Where large atoms are present, e.g. chlorine atoms, the mass per unit volume is higher and so p.v.c., a substantially amorphous polymer, has a specific gravity of about 1·4.

If a polymer can crystallise then molecular packing is much more efficient and higher densities can be achieved. The high densities of p.t.f.e. (about 2·2) and poly(vinylidene chloride) (about 1·7) are partially attributable to this fact. Polyethylenes made by different processes often differ in the degree of branching and thus can crystallise or pack to varying extents. For this reason polyethylenes produced by a high pressure process (e.g. Alkathene) have a lower density than those produced using supported metal oxide catalysts (e.g. Rigidex). The amorphous ethylene–propylene rubbers have lower densities than either polyethylene or isotactic polypropylene, which are both capable of crystallising.

The conformation adopted by a molecule in the crystalline structure will also affect the density. Whereas polyethylene adopts a planar zigzag conformation, because of steric factors a polypropylene molecule adopts a helical conformation in the crystalline zone. This requires somewhat more space and isotactic polypropylene has a lower density than polyethylene.

4.5.4   Impact strength

Familiarity with a given plastics material under normal conditions of use leads to it being considered as either a brittle or a tough material. Thus polystyrene, poly(methyl methacrylate) and unmodified unplasticised p.v.c. are normally rated as brittle, breaking with a sharp fracture, whereas low density polythene and plasticised p.v.c. are considered to be tough. Whether a material exhibits brittle fracture or appears tough depends on the temperature and the rate of striking, that is it is a function of the rate of deformation. One object of current research into the physical properties of plastics material is to determine the locations of tough–brittle transitions for commercial polymers. As with other physical properties the position of the glass transition temperature and the facility with which crystallisation can take place are fundamental to the impact strength of a material. Well below the glass transition temperature amorphous polymers break with a brittle fracture but they become tougher as the glass transition temperature is approached. A rubbery state will develop above the glass transition and the term impact strength will cease to have significance. In the case of crystalline materials the toughness will depend on the degree of crystallinity, large degrees of crystallinity will lead to inflexible masses with only moderate impact strengths. The size of the crystalline structures formed will also be a significant factor, large spherulitic structures leading to masses with low impact strength. As indicated in the previous chapter spherulitic size may be controlled by varying the ratio of nucleation to growth rates.

One development of recent years is to incorporate rubbery materials into glass-like resins in order to enhance the toughness. This technique has been successfully used on a large scale with polystyrene and to a lesser extent with p.v.c.

REFERENCES

1. SWALLOW, J. C., *J. Roy. Soc. Arts*, **99,** 355 (1951)
2. GORDON, M., *High Polymers*, Iliffe, London (1963)
3. JENCKEL, E., and UEBERREITER, K., *Z. Phys. Chem.*, **A182,** 361 (1939)
4. BOYER, R. F., *J. Appl. Phys.*, **25,** 825 (1954)
5. EDGAR, O. B., and HILL, R., *J. Polymer Sci.*, **8,** 1 (1952)

BIBLIOGRAPHY

BILLMEYER, F. W., *Textbook of Polymer Science*, Interscience, New York (1962)
BUECHE, F., *Physical Properties of Polymers*, Interscience, New York (1962)
FLORY, P. J., *J. Chem. Phys.*, **17,** 223 (1949)
GORDON, M., *High Polymers,* Iliffe, London (1963)
MEARES, P., *Polymers: Structure and Bulk Properties*, Van Nostrand, London (1965)
RITCHIE, P. D. (Ed), *Physics of Plastics*, Iliffe, London (1965)
TOBOLSKY, A. V., *Properties and Structure of Polymers*, John Wiley, New York (1960)

5

Relation of Structure to Chemical Properties

5.1 INTRODUCTION

It is sometimes stated that a given material has 'a good chemical resistance', or alternatively the material may be stated to be poor or excellent in this respect. Such an all-embracing statement can be little more than a rough generalisation particularly since there are many facets to the behaviour of polymers in chemical environments.

There are a number of properties of a polymer about which information is required before detailed statements can be made about its chemical properties. The most important of these are

1. The solubility characteristics.
2. The effect of specific chemicals on molecular structure, particularly in so far as they lead to degradation and cross-linking reactions.
3. The effect of specific chemicals and environments on polymer properties at elevated temperatures.
4. The effect of high energy irradiation.
5. The aging and weathering of the material.
6. Permeability and diffusion characteristics.
7. Toxicity.

Before dealing with each of these aspects, it is useful to consider, very briefly, the types of bonds which hold atoms and molecules together.

5.2 CHEMICAL BONDS

The atoms of a molecule are held together by primary bonds. The attractive forces which act between molecules are usually referred to as secondary bonds, secondary valence forces, intermolecular forces or van der Waals forces.

Primary bond formation takes place by various interactions between electrons in the outermost shell of two atoms resulting in the production of a more stable state. The three main basic types of primary bond are ionic, covalent and co-ordinate.

An *ionic bond* is formed by the donation of an electron by one atom to another so that in each there is a stable number of electrons in the outermost shell (eight in the case of most atoms). An example is the reaction of sodium and chlorine (Fig. 5.1).

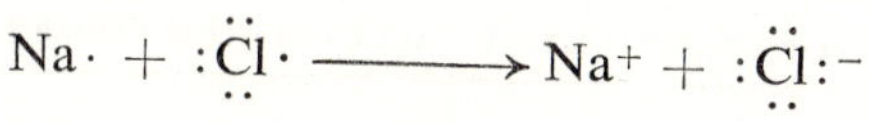

$$Na\cdot \ + \ :\overset{\cdot\cdot}{\underset{\cdot\cdot}{Cl}}\cdot \ \longrightarrow \ Na^+ \ + \ :\overset{\cdot\cdot}{\underset{\cdot\cdot}{Cl}}:^-$$

Fig. 5.1

The stable sodium ion has a positive charge because it is deficient of one electron and the chlorine atom is negatively charged for the converse reason. Ionic bonds are seldom found in polymers of current interest as plastics materials although the ionic bond is important in ion-exchange resins and in the 'ionomers' (see Chapter 8).

The most important interatomic bond in polymers, and indeed in organic chemistry, is the *covalent bond*. This is formed by the sharing of one or more pairs of electrons between two atoms. An example is the bonding of carbon and hydrogen to form methane (Fig. 5.2).

$$\overset{\cdot}{\underset{\cdot}{\cdot C \cdot}} \ + \ 4H\cdot \ \longrightarrow \ H:\overset{H}{\underset{H}{\overset{\cdot\cdot}{C}}}:H$$

Fig. 5.2

In the case of carbon the stable number of electrons for the outer shell is eight and for hydrogen, two. Thus all the atoms possess or share the number of electrons required for stability. Where a pair of electrons is shared between two atoms, it is stated that the atoms are bound by a single bond. If there are two pairs a double bond is formed and if there are three pairs a triple bond.

The third main type of bond is the *co-ordinate bond*, in which both of the shared electrons come from one atom. Examples of interest in polymer science are the addition compounds of boron trifluoride (Fig. 5.3).

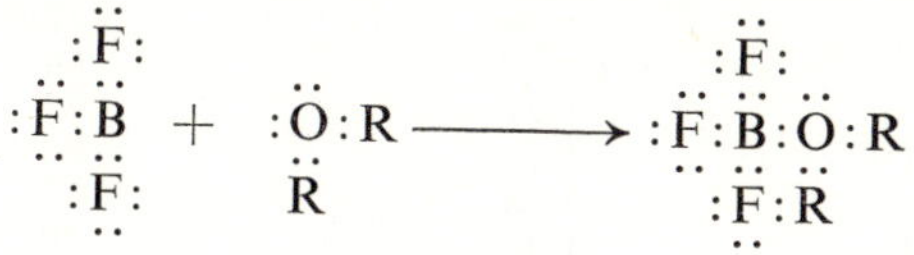

$$\underset{:\overset{\cdot\cdot}{\underset{\cdot\cdot}{F}}:}{\overset{:\overset{\cdot\cdot}{\underset{\cdot\cdot}{F}}:}{:F:B}} \ + \ \underset{R}{:\overset{\cdot\cdot}{O}:R} \ \longrightarrow \ \underset{:\overset{\cdot\cdot}{\underset{\cdot\cdot}{F}}:R}{\overset{:\overset{\cdot\cdot}{\underset{\cdot\cdot}{F}}:}{:F:B:\overset{\cdot\cdot}{O}:R}}$$

Fig. 5.3

In the covalent bond, the primary bond of greatest importance in high polymers, the electron pair is seldom equally shared between the two atoms. Thus one end of the bond has a small negative charge and the other end has a slight positive charge. Such a bond is said to be polar and the strength and direction of the polarity is determined by the atoms forming the bond.

An estimate of the polarity of a bond between two atoms may be obtained by reference to the electronegativity scale. The electronegativity values of some common elements are given in Table 5.1. The higher the value the greater the electronegativity.

Table 5.1   ELECTRONEGATIVITY VALUES OF SOME COMMON ELEMENTS

Element	Electronegativity	Element	Electronegativity
Caesium	0·7	Iodine	2·4
Potassium	0·8	Carbon	2·5
Sodium	0·9	Sulphur	2·5
Lithium	1·0	Bromine	2·8
Aluminium	1·5	Nitrogen	3·0
Silicon	1·8	Chlorine	3·0
Hydrogen	2·1	Oxygen	3·5
Phosphorus	2·1	Fluorine	4·0

The greater the difference in the electronegativity values of the atoms forming the bond the greater the polarity of the bond. Where the electro-negativity difference is greater than 2, electrovalent bonds are commonly formed, where it is less than 2 the bond is usually covalent but it may also be polar. Thus a carbon–fluorine bond will be more polar than a carbon–hydrogen bond. The electronegativity difference is approximately equal to the square root of the ionic resonance energy in electron volts (eV).

As a result of many observations on the energetics of the formation and dissociation of molecules it has been found possible to give typical bond energies and bond lengths to a number of bonds. Some of these are given in Table 5.2.

Although the primary bonds are important when considering the chemical reactivity and thermal stability of polymers it is the secondary bonds which

Table 5.2   TYPICAL BOND LENGTHS AND DISSOCIATION ENERGIES FOR SOME SELECTED PRIMARY BONDS

Bond	Bond length R	Dissociation energy kcal/mole
O—O	1·32	35
Si—Si	2·35	42·5
S—S	1·9–2·1	64
C—N	1·47	73
C—Cl	1·77	81
C—C	1·54	83
C—O	1·46	86
N—H	1·01	93
C—H	1·10	99
C—F	1·32–1·39	103–123
O—H	0·96	111
C=C	1·34	146
C=O	1·21	179
C≡N	1·15	213

are of dominant importance in determining the solubility of polymers. Although some of these secondary bonds act intramolecularly, it is the intermolecular forces which are of greatest importance. The inter-molecular forces can be of four types: dipole forces, induction forces, dispersion forces and the hydrogen bond.

Because of the polarity of many covalent bonds, different parts of a molecule may carry equal and opposite charges. At molecular distances a charged grouping of one polarity can attract a group of the opposite polarity on a neighbouring molecule. The dipole interaction leading to *dipole forces* between two polar molecules is shown in Fig. 5.4. The extent

Fig. 5.4

of mutual dipole alignment will be a predominant factor in determining the intermolecular attraction. Since this alignment is opposed by thermal motion, dipole forces are critically dependent on temperature.

A polar molecule can also induce a dipole on a neighbouring molecule that possesses no permanent dipole. The resultant intermolecular attraction between the permanent and the induced dipole is spoken of as the *induction force*. Its magnitude is small and independent of tempera-ture.

Although there are many molecules which appear to be non-polar, i.e. the centres of positive and negative charges appear coincident, all molecules, even the inert gases, have time-varying dipole moments which will depend on the position of the electrons at any given instant. These varying dipole moments average out to zero but they can lead to attractive forces between

Table 5.3   MAGNITUDE OF VARIOUS INTERMOLECULAR FORCES IN SOME SIMPLE MOLECULES

| Molecule | Intermolecular energy (kcal/mole) | | | |
	Dipole	Induction	Dispersion	Total
A	0	0	2·03	2·03
CO	0·0001	0·0002	2·09	2·09
HI	0·006	0·03	6·18	6·21
HCl	0·79	0·24	4·02	5·05
NH_3	3·18	0·37	3·52	7·07
H_2O	8·69	0·46	2·15	11·30

molecules. These are referred to as *dispersion forces* and they represent the bulk of the intermolecular forces which are present in the absence of strong permanent dipoles. The magnitudes of these three intermolecular forces in a few selected molecules are given in Table 5.3.

A special case is that of *hydrogen bonding* where the hydrogen atom attached to a proton donor group (e.g. carboxyl, hydroxyl, amine or amide group) is shared with a basic, proton accepting group (e.g. the oxygen in a carboxyl, ether or hydroxyl group or the nitrogen atom in amines or amides).

Intermolecular forces are generally less than 10 kcal/mole. In polymers, in the absence of hydrogen bonding, the intermolecular force is primarily due to dispersion effects.

5.3 POLYMER SOLUBILITY

A chemical will be a solvent for another material if the molecules of the two materials are compatible, i.e. they can co-exist on the molecular scale and there is no tendency to separate. This statement does not indicate the speed at which solution may take place since this will depend on additional

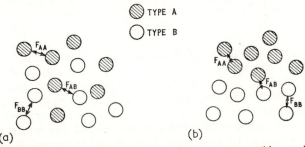

Fig. 5.5. *Schematic representation of compatible and incompatible systems.* (a) $F_{AB} \geqslant F_{AA}$; $F_{AB} \geqslant F_{BB}$. *Mixture compatible.* (b) F_{AA} or $F_{BB} > F_{AB}$. *Molecules separate*

considerations such as the molecular size of the potential solvent and the temperature. Molecules of two different species will be able to co-exist if the force of attraction between different molecules is not less than the forces of attraction between two like molecules of either species. If the force of attraction between dissimilar molecules A and B is F_{AB} and that between similar molecules of type B F_{BB} and between similar molecules of type A, F_{AA} then for compatibility $F_{AB} \geqslant F_{BB}$ and $F_{AB} \geqslant F_{AA}$. This is shown schematically in Fig. 5.5 (a).

If either F_{AA} or F_{BB} is greater than F_{AB} the molecules with the highest intermolecular attraction will tend to congregate or cohere and they will expel the dissimilar molecule with the result that two phases will be formed. These conditions are shown in Fig. 5.5 (b).

It is the aim of this part of the chapter to show how certain predictions may be made about the solubility of a given material such as a polymer in any given solvent. In order to be able to make predictions of solubility characteristics it will be necessary to consider three cases:

1. Amorphous polymers where there is no specific interaction between polymer and solvent.
2. Amorphous polymers where there is interaction.
3. Crystalline polymers.

It has been reasoned by a number of workers that where there is no specific interaction, e.g. hydrogen bonding, between solvent and solute, then the intermolecular attraction between the dissimilar molecules is intermediate between the intermolecular forces of like species, i.e.

$$F_{AA} < F_{AB} < F_{BB}$$

Therefore if F_{AA} and F_{BB} are similar then F_{AB} will be similar and the two materials should be soluble.

A measure of the intermolecular attraction of a material is the *cohesive energy density*, the energy of vaporisation of a component per molar volume. It may be shown that the cohesive energy density at a specified temperature $T°K$ is related to the latent heat of vaporisation at $T°K$, ΔH by the equation

$$\text{Cohesive energy density} = \frac{\Delta H - RT}{M/D}$$

where R is the gas constant (1·986)

M is the molecular weight
D is the density.

It is often more convenient in solubility studies to use the square root of the cohesive energy density. This is known as the *solubility parameter* and is denoted by the symbol δ, i.e.

$$\delta = \left(\frac{\Delta H - RT}{M/D}\right)^{1/2}$$

In practice it is found that unless there is specific interaction between dissimilar molecules, and in the absence of one species tending to crystallise, two materials will be compatible if their values are within unity of each other. Thus in general an amorphous polymer will dissolve in a solvent with a solubility parameter within unity of that of the polymer. Some typical values for the solubility parameters of a number of common polymers and solvents are given in Tables 5.4 and 5.5.

As examples from Tables 5.4 and 5.5 it is to be noted that natural rubber ($\delta = 8·3$) is dissolved by toluene ($\delta = 8·9$) and carbon tetrachloride ($\delta = 8·6$) but not by ethanol ($\delta = 12·7$). Cellulose diacetate ($\delta = 10·9$) is soluble in acetone ($\delta = 10·0$) but not methanol ($\delta = 14·5$) or toluene ($\delta = 8·9$).

In some cases polymers may be soluble in solvents of somewhat different solubility parameter. This is due to the fact that hydrogen bonding or some other forms of specific interaction can occur between the dissimilar molecules.

When a crystalline polymer is heated above its melting point and dissolved in a solvent with which no interaction takes place the polymer will tend to crystallise out on cooling. This is because crystallisation will be accompanied by a decrease in free energy. For the same reason no crystalline polymer will dissolve in a solvent unless there is some form of

Table 5.4 SOLUBILITY PARAMETERS OF POLYMERS

Polymer	$\delta(cal/sec)^{1/2}$
P.T.F.E.	6·2[c]
Poly(dimethyl siloxane)	7·3
Polyethylene	7·9° 8·1[c]
Polypropylene	7·9
Polyisobutylene	8·05° 7·7[c]
Styrene–butadiene rubber (25% styrene)	8·09–8·6° 8·54[c]
Polyisoprene (natural rubber)	7·9–8·35° 8·15[c]
Polybutadiene	8·4–8·6° 8·38[c]
Polysulphide rubbers	9·0–9·4
Polystyrene	8·5–9·7° 9·12[c]
Polychloroprene	9·18–9·25° 9·38[c]
Nitrile rubbers (25% acrylonitrile)	9·38–9·5° 9·25[c]
Poly(vinyl acetate)	9·4[c]
Poly(methyl methacrylate)	9·0–9·5° 9·2[c]
Poly(vinyl chloride)	9·38–9·7° 9·55[c]
Bisphenol A polycarbonate	9·5[c.m.]
Poly(vinylidene chloride)	9·8
Poly(ethylene terephthalate)	10·7
Cellulose (di)nitrate	10·56° 10·48[c]
Cellulose (di)acetate	11·35[c]
Epoxide resins	approx. 11·0
Acetal resin (homopolymer)	11·1°
Nylon 66	13·6
Polyacrylonitrile	15·4

o—observed, c—calculated from Small's data[1], c.m.—calculated but using obtained data with model organic carbonate

Table 5.5 SOLUBILITY PARAMETERS OF SOME COMMON SOLVENTS

	$\delta(cal/sec)^{1/2}$		$\delta(cal/sec)^{1/2}$
neo-pentane	6·3	Trichloroethylene	9·3
Isobutylene	6·7	Tetrachlorethane	9·4
n-hexane	7·3	Tetralin	9·5
Diethyl ether	7·4	Carbitol	9·6
n-octane	7·6	Methyl chloride	9·7
Methyl cyclohexane	7·8	Methylene chloride	9·7
Ethyl isobutyrate	7·9	Ethylene dichloride	9·8
Di-isopropyl ketone	8·0	Cyclohexanone	9·9
Methyl amyl acetate	8·0	Cellosolve	9·9
Turpentine	8·1	Dioxane	9·9
Cyclohexane	8·2	Carbon disulphide	10·0
2,2-dichloropropane	8·2	Acetone	10·0
Sec.amyl acetate	8·3	n-octanol	10·3
Methyl isobutyl ketone	8·4	Butyronitrile	10·5
Dipentene	8·5	n-hexanol	10·7
Amyl acetate	8·5	sec-butanol	10·8
Methyl isopropyl ketone	8·5	Pyridine	10·9
Carbon tetrachloride	8·6	Nitroethane	11·1
Pine oil	8·6	n-butanol	11·4
Piperidine	8·7	Cyclohexanol	11·4
Xylene	8·8	Isopropanol	11·5
Dimethyl ether	8·8	n-propanol	11·9
Toluene	8·9	Dimethyl formamide	12·1
Butyl cellosolve	8·9	Hydrogen cyanide	12·1
1,2-dichloropropane	9·0	Ethanol	12·7
Mesityl oxide	9·0	Cresol	13·3
Isophorone	9·1	Formic acid	13·5
Ethyl acetate	9·1	Methanol	14·5
Benzene	9·2	Phenol	14·5
Diacetone alcohol	9·2	Glycerol	16·5
Chloroform	9·3	Water	23·4

interaction. This is why at room temperature there are no solvents for polyethylene and polytetrafluorethylene although these polymers will dissolve at higher temperatures.

The polycarbonate from bis-phenol A ($\delta = 9 \cdot 5$) has a similar solubility parameter to poly(vinyl chloride) ($\delta = 9 \cdot 7$). Both polymers also crystallise to a limited extent but their solubility characteristics are quite different. For example chloroform ($\delta = 9 \cdot 3$) and methylene chloride ($\delta = 9 \cdot 7$) are good solvents for the bis-phenol A polycarbonate but not for poly(vinyl chloride). On the other hand tetrahydrofuran and cyclohexanone are good solvents for p.v.c. but only moderately poor solvents for the polycarbonate. This difference is believed to be due to the fact that while p.v.c. is a weak proton donor, the polycarbonate is a weak proton acceptor. Hence the p.v.c. will dissolve in proton acceptors such as cyclohexanone and the polycarbonate in proton donors, such as methylene chloride, because the specific interaction will overcome the free energy barrier that restricts solution of crystalline polymers. This effect is shown diagrammatically in Fig. 5.6.

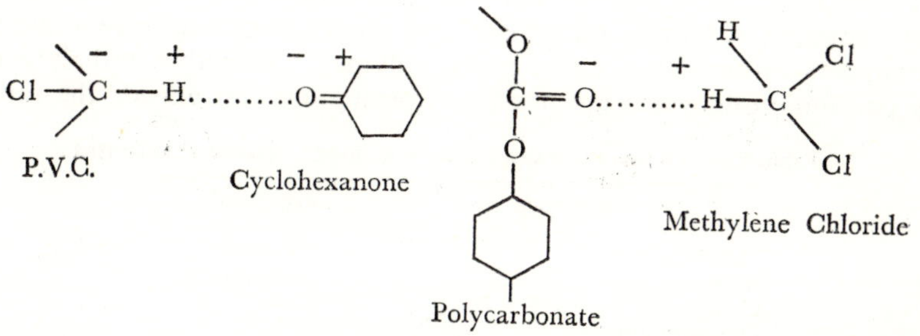

Fig. 5.6

Because p.v.c. and the polycarbonates have a very low level of crystallinity, solvents of similar solubility parameter which do not interact with the polymer may have some ability to dissolve it. However in highly polar, highly crystalline materials such as nylon 66 and polyacrylonitrile some form of interaction is necessary for solution to occur. Thus the polyamide will only dissolve in strong proton donors such as phenol and formic acid, and polyacrylonitrile in dimethyl formamide and tetramethylene sulphone.

5.3.1 Plasticisers

It has been found that the addition of certain liquids (and in rare instances solids) to a polymer will give a non-tacky product with a lower processing temperature and which is softer and more flexible than the polymer alone. As an example the addition of 70 parts of di-iso-octyl phthalate to p.v.c. will convert the polymer from a hard rigid solid at room temperature to a rubber-like material. Such liquids, which are referred to as plasticisers, are simply high boiling solvents for the polymer. Because it is important that such plasticisers should be non-volatile they have a molecular weight of at least 300. Hence because of their size they only dissolve into the polymer at a very slow rate at room temperature. For this reason they are blended (fluxed, gelled) with the polymer at elevated temperatures or in the presence of volatile solvents (the latter being removed at some subsequent stage of the operation).

For a material to act as a plasticiser it must conform to the following requirements:

1. It should have a molecular weight of at least 300.
2. It should have a similar solubility parameter to that of the polymer.
3. If the polymer has any tendency to crystallise, it should be capable of some specific interaction with the polymer.
4. It should not be a crystalline solid at the ambient temperature unless it is capable of specific interaction with the polymer.

The solubility parameters of a number of commercial plasticisers are given in Table 5.6.

From Table 5.6 it will be seen that plasticisers for p.v.c. such as the octyl phthalates, tritolyl phosphate and dioctyl sebacate have solubility parameters within unity of that of the polymer. Dimethyl phthalate and the paraffinic oils which are not p.v.c. plasticisers fall outside the range.

Table 5.6   SOLUBILITY PARAMETERS FOR SOME COMMON PLASTICISERS

Plasticiser	$\delta(\text{cal/sec})^{1/2}$
Paraffinic oils	7·5 approx.
Aromatic oils	8·0 approx.
Camphor	7·5
Diisoctyl adipate	8·7
Dioctyl sebacate	8·7
Diisodecyl phthalate	8·8
Dibutyl sebacate	8·9
Diethyl hexyl phthalate	8·9
Diisooctyl phthalate	8·9
Di-2-butoxyethyl phthalate	9·3
Dibutyl phthalate	9·4
Triphenyl phosphate	9·8
Tritolyl phosphate	9·8
Trixylyl phosphate	9·9
Dibenzyl ether	10·0
Triacetin	10·0
Dimethyl phthalate	10·5
Santicizer 8	11·0 approx.

Data obtained by Small's method[1] except for that of Santicizer 8 which was estimated from boiling point measurements.

It will be noted that tritolyl phosphate which gels the most rapidly with p.v.c. has the closest solubility parameter to the polymer. The sebacates which gel more slowly but give products which are flexible at lower temperatures than corresponding formulations from tritolyl phosphate have a lower solubility parameter. It is however likely that any difference in the effects of phthalate, phosphate and sebacate plasticisers in p.v.c. is due more to differences in hydrogen bonding or some other specific interaction.

It has been shown by Small[1] that the interaction of plasticiser and p.v.c. is greatest with the phosphate and lowest with the sebacate.

Comparisons of Tables 5.4 and 5.6 allows the prediction that aromatic oils will be plasticisers for natural rubber, that dibutyl phthalate will plasticise poly(methyl methacrylate), that tritolyl phosphate will plasticise nitrile rubbers, that dibenzyl ether will plasticise poly(vinylidene chloride) and that dimethyl phthalate will plasticise cellulose diacetate. These predictions are found to be correct. What is not predictable is that camphor should be an effective plasticiser for cellulose nitrate. It would seem that this crystalline material, which has to be dispersed into the polymer with the aid of liquids such as ethyl alcohol, is only compatible with the polymer because of some specific interaction between the carbonyl group present in the camphor with some group in the cellulose nitrate.

5.3.2 Extenders

In the formulation of p.v.c. compounds it is not uncommon to replace some of the plasticiser with an extender, a material that is not in itself a plasticiser but which can be tolerated up to a given concentration by a polymer-true plasticiser system. These materials, such as chlorinated waxes and refinery oils, are generally of lower solubility parameter than the true plasticisers and they do not appear to interact with the polymer. However, where the solubility parameter of a mixture of plasticiser and extender is within unity of that of the polymer the mixture of three components will be compatible. It may be shown that

$$\delta_{\text{mixture}} = X_1\delta_1 + X_2\delta_2$$

where δ_1 and δ_2 are the solubility parameters of two liquids

$\qquad X_1$ and X_2 are their mole fractions in the mixture.

Because the solubility parameter of tritolyl phosphate is higher than that of dioctyl sebacate, p.v.c.–tritolyl phosphate blends can tolerate more of a low solubility parameter extender than can a corresponding sebacate formulation.

5.3.3 Determination of solubility parameter

Since a knowledge of a solubility parameter of polymers and liquids is of value in assessing solubility and solvent power it is important that this may be easily assessed. A number of methods have been reviewed by Burrell[2] and of these two are of particular use.

From heat of vaporisation data

It has already been stated that

$$\delta = \left(\frac{\Delta E}{V}\right)^{1/2} = \left(\frac{\Delta H - RT}{M/D}\right)^{1/2}$$

where δ is the solubility parameter
  ΔE the energy of vaporisation
  ΔH the latent heat of vaporisation
  R the gas constant
  T the temperature in °K
  M the molecular weight
  D the density.

At 25°C, a common ambient temperature,

$$\Delta E_{25} = \Delta H_{25} - 592$$

Unfortunately values of ΔH at such low temperatures are not readily available and they have to be computed by means of the Clausius–Clapeyron equation or from the equation given by Hildebrand[3]

$$\Delta H_{25} = 23 \cdot 7 T_b + 0 \cdot 020 T_b^2 - 2,950$$

where T_b is the boiling point.

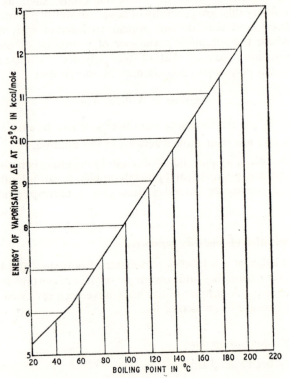

Fig. 5.7. *Relationship between ΔE and boiling point for use in calculating solubility parameters. (After Burrell[2])*

From this equation a useful curve relating ΔE and T_b has been compiled and from this the solubility parameter may easily be assessed (Fig. 5.7).

From structural formulae

The solubility parameter of high polymers cannot be obtained from latent heat of vaporisation data since such polymers cannot be vaporised without decomposition (there may be some exceptions to this generalisation for lower molecular weight materials and at very low pressures). It is therefore convenient to define the solubility parameter of a polymer 'as the same as that of a solvent in which the polymer will mix in all proportions without heat effect, volume change or without any reaction or specific association'. It is possible to estimate the value of δ for a given polymer by immersing samples in range of solvents of known δ and noting the δ value of the best solvents. In the case of cross-linked polymers the δ value can be obtained by finding the solvent which causes the greatest equilibrium swelling. Such a method is time consuming so that the additive method of Small[1] becomes of considerable value. By considering a number of simple molecules Small was able to compile a list of molar attraction constants G for the various parts of a molecule. By adding the molar attraction constants it was found possible to calculate δ by the relationship

$$\delta = \frac{D \Sigma G}{M}$$

where D is the density
  M is the molecular weight.

When applied to polymers it was found that good agreement was obtained with results obtained by immersion techniques except where hydrogen bonding was significant. The method is thus not suitable for alcohols, amines, carboxylic acids or other strongly hydrogen bonded compounds except where these form only a small part of the molecule. Where hydrogen bonding is insignificant accuracy to the first decimal place is claimed. The δ values given in Table 5.6 were computed by the author according to Small's method. The values in Tables 5.4 and 5.5 were obtained either by computation or from a diversity of sources.

Some molar attraction constants compiled by Small are given in Table 5.7.

As an example of the use of Small's table the solubility parameter of poly(methyl methacrylate) may be computed as follows:

The formula for the polymer is shown in Fig. 5.8.

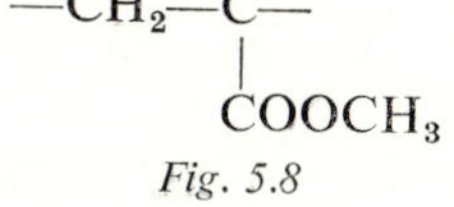

$$-CH_2-\underset{\underset{COOCH_3}{|}}{\overset{\overset{CH_3}{|}}{C}}-$$

Fig. 5.8

Table 5.7   MOLAR ATTRACTION CONSTANTS AT 25°C[1]

Group	G Molar attraction constant G
—CH$_3$	214
—CH$_2$— (single bonded)	133
—CH<	28
>C<	−93
CH$_2$=	190
—CH= (double bonded)	111
>C=	19
CH≡C—	285
—C≡C—	222
Phenyl	735
Phenylene (*o,m,p*)	658
Naphthyl	1146
Ring (5-membered)	105–115
Ring (6-membered)	95–105
Conjugation	20–30
H	80–100
O (ethers)	70
CO (ketones)	275
COO (esters)	310
CN	410
Cl single	270
Cl twinned as in >CCl$_2$	260
Cl triple as in —CCl$_3$	250
Br single	340
I single	425
CF$_2$ } in fluorocarbons only	150
CF$_3$	274
S sulphides	225
SH thiols	315
ONO$_2$ nitrates	~440
NO$_2$ (aliphatic)	~440
PO$_4$ (organic)	~500
†Si (in silicones)	~ 38

† Estimated by H. Burrell.

Small's formula is $\delta = D\Sigma G/M$ and the value of $\Sigma G/M$ will be the same for the repeating unit as for the polymer.

Now M (for repeating unit) = 100

D = 1·19

2 CH$_3$ at 214   428

1 CH$_2$ at 133   133

1 COO at 310   310

1 >C< at −93   −93
                        ————
ΣG = 778

$$\delta = \frac{D\Sigma G}{M} = \frac{1\cdot18 \times 778}{100} = 9\cdot2$$

In the case of crystalline polymers better results are obtained using an 'amorphous density' which can be extrapolated from data above the melting point, or from other sources. In the case of polyethylene the apparent amorphous density is in the range 0·84–0·86 at 25°C. This gives a calculated value of about 8·1 for the solubility parameter which is still slightly higher than observed values obtained by swelling experiments.

5.3.5 Thermodynamics and solubility

The first law of thermodynamics expresses the general principle of energy conservation. It may be stated as follows: 'In an energetically isolated system the total energy remains constant during any change which may occur in it'. Energy is the capacity to do work and units of energy are the product of an intensity factor and a capacity factor. Thus the unit of mechanical energy (erg) is the product of the unit of force (dyne) and the unit of distance (centimetre). Force is the intensity factor and distance the capacity factor. Similarly the unit of electrical energy (joule) is the product of an intensity factor (the potential measured in volts) and a capacity factor (the quantity of electricity measured in coulombs). Heat energy may, in the same way, be considered as the product of temperature (the intensity factor) and the quantity of heat, which is known as the entropy (the capacity factor).

It follows directly from the first law of thermodynamics that if a quantity of heat Q is absorbed by a body then part of that heat will do work W and part will be accounted for by a rise in the internal energy ΔE of that body, i.e.

$$Q = \Delta E + W$$
$$W = Q - \Delta E$$

This expression states that there will be energy free to do work when Q exceeds ΔE. Expressed in another way work can be done, that is an action can proceed, if $\Delta E - Q$ is negative. If the difference between ΔE and Q is given the symbol ΔF, then it can be said that a reaction will proceed if the value of ΔF is negative. Since the heat term is the product of temperature T and change of entropy ΔS, for reactions at constant temperature then

$$\Delta F = \Delta E - T\Delta S$$

This is the so-called free-energy equation, ΔF being the change in the free energy. It is in fact little more than a restatement of the fact that energy will only be available to do work when the heat absorbed exceeds the increase in internal energy. It is however a most useful equation and has in fact already been mentioned in the previous chapter in connection with melting points.   If applied to the mixing of molecules the equation indicates that mixing will occur if $T\Delta S$ is greater than ΔH.   Therefore

1. The higher the temperature the greater the likelihood of mixing (an observed fact).
2. The greater the increase in entropy the greater the likelihood of mixing.
3. The less the heat of mixing the greater the likelihood of mixing.

Now it may be shown that entropy is a measure of disorder or the degree of freedom of a molecule.   When mixing takes place it is to be expected that separation of polymer molecules by solvent will facilitate the movement of the polymer molecules and thus increase their degree of freedom and their degree of disorder.   This means that such a mixing process is bound to cause an increase in entropy.   A consequence of this is that as ΔS will always be positive during mixing, the term $T\Delta S$ will be positive and therefore solution will occur if ΔH, the heat of mixing is zero or at least less than $T\Delta S$.

It has been shown by Hildebrand[3] that, in the absence of specific interaction

$$\Delta H = V_m \left[\left(\frac{\Delta E_1}{V_1} \right)^{1/2} - \left(\frac{\Delta E_2}{V_2} \right)^{1/2} \right]^2 a_1 a_2$$

where

V_m is the total volume of the mixture in cm³.
ΔE is the energy of vaporisation in cal.
V the molar volume of each compound.
a the volume fraction of each compound.

Since we have defined the expression $(\Delta E/V)^{1/2}$ as the solubility parameter δ, the above equation may be written

$$\Delta H = V_m(\delta_1 - \delta_2)a_1 a_2$$

If δ_1 and δ_2 are identical then ΔH will be zero and so ΔF is bound to be negative and the compounds will mix.   Thus the intuitive arguments put forward in Section 5.3 concerning the solubility of amorphous polymers can be seen to be consistent with thermodynamical treatment.   The above discussion is, at best, an over-simplification of thermodynamics, particularly as applied to solubility.   Further information may be obtained from a number of authoritative sources[3-5].

5.4   CHEMICAL REACTIVITY

The chemical resistance of a plastics material is as good as its weakest point.   If it is intended that a plastics material is to be used in the presence of a certain chemical then each ingredient must be unaffected by the chemical.   In the case of a polymer molecule, its chemical reactivity will be determined by the nature of chemical groups present.   However by its very nature there are aspects of chemical reactivity which find no parallel in the chemistry of small molecules and these will be considered in due course.

In commercial plastics materials there are a comparatively limited number of chemical structures to be found and it is possible to make some general observations about chemical reactivity in the following tabulated list of examples:

1. Polyolefins such as polyethylene and polypropylene contain only C—C and C—H bonds and may be considered as high molecular weight paraffins. Like the simpler paraffins they are somewhat inert and their major chemical reaction is substitution, e.g. halogenation. In addition the branched polyethylenes and all the higher polyolefins contain tertiary carbon atoms which are reactive sites for oxidation. Because of this it is necessary to add antioxidants to stabilise the polymers against oxidation. The polyolefins may be cross linked by peroxides.

2. Polytetrafluoroethylene contains only C—C and C—F bonds. These are both very stable and the polymer is exceptionally inert. A number of other fluorine-containing polymers are available which may contain in addition C—H and C—Cl bonds. These are somewhat more reactive and those containing C—H bonds may be cross linked by peroxides and certain diamines and di-isocyanates.

3. Many polymers, such as the diene rubbers, contain double bonds. These will react with many agents such as oxygen, ozone, hydrogen halides and with halogens. Ozone, and in some instances oxygen, will lead to scission of the main chain at the site of the double bond and this will have a catastrophic effect on the molecular weight. The rupture of one such bond per chain will halve the number-average molecular weight.

4. Ester, amide and carbonate groups are susceptible to hydrolysis. When such groups are found in the main chain, their hydrolysis will also result in a reduction of molecular weight. Where hydrolysis occurs in a side chain the effect on molecular weight is usually insignificant. The presence of benzene rings adjacent to these groups may offer some protection against hydrolysis except where organophilic hydrolysing agents are employed.

5. Hydroxyl groups are extremely reactive. These occur attached to the backbone of the cellulose molecule and poly(vinyl alcohol). Chemically modified forms of these materials are dealt with in the appropriate chapters.

6. Benzene rings in both the skeleton structure and on the side groups can be subjected to substitution reactions. Such reactions do not normally cause great changes in the fundamental nature of the polymer, for example they seldom lead to chain scission or cross linking.

Polymer reactivity differs from the reactivity of simple molecules in two special respects. The first of these is due to the fact that a number of weak links exist in the chains of many polymer species.   These can form the site for chain scission or of some other chemical reaction.   The second reason for differences between polymers and small molecules is due to the fact that reactive groups occur repeatedly along a chain.   These adjacent groups can react with one another to form ring products such as poly(vinyl acetal) (Chapter 11) and cyclised rubbers (Chapter 26).   Further one-step reactions which take place in simple molecules can sometimes be replaced by chain reactions in polymers such as the 'zipper' reactions which cause the depolymerisation of polyacetals and poly(methyl methacrylate).

5.5   EFFECTS OF THERMAL, PHOTOCHEMICAL AND HIGH ENERGY RADIATION

Plastics materials are affected to varying extents by exposure to thermal, photochemical and high energy radiation.   These forms of energy may cause such effects as cross linking, chain scission, modifications to chain structure and modifications to the side group of the polymer, and they may also involve chemical changes in the other ingredients present.

In the absence of other active substances, e.g. oxygen, the heat stability is related to the bond energy of the chemical linkages present.   Table 5.2 gives typical values of bond dissociation energies and from them it is possible to make some assessment of the potential thermal stability of a polymer.   In practice there is some interaction between various linkages and so the assessment can only be considered as a guide.   Table 5.8[6] shows the values for T_h (the temperature at which a polymer loses half its weight *in vacuo* at 30 minutes preceded by 5 minutes preheating at that temperature) and K_{350} the rate constant (in %/min) for degradation at 350°C.

Table 5.8   THERMAL DEGRADATION OF SELECTED POLYMERS[6]

Polymer	$T_h(°C)$	K_{350} (%/min)
P.T.F.E.	509	0·0000052
Poly-*p*-xylene	432	0·002
Polymethylene	414	0·004
Polypropylene	387	0·069
Poly(methyl methacrylate)	327	5·2
Poly(vinyl chloride)	260	170

The high stability of p.t.f.e. is due to the fact that only C—C and C—F bonds are present, both of which are very stable.   It would also appear that the C—F bonds have a shielding effect on the C—C bonds.   Poly-*p*-xylene contains only the benzene ring structure (very stable thermally) and C—C and C—H bonds and these are also stable.   Polymethylene which contains only the repeating methylene groups, and hence only C—C and C—H bonds, is only slightly less stable.   Polypropylene has a somewhat lower value than polymethylene since the stability of the C—H at a tertiary carbon position is somewhat lower than that of a secondary carbon atom.   The lower stability of p.v.c. is partly explained by the lower dissociation energy of the C—Cl bond but also because of weak points which act as a site for chain reactions. The rather high thermal degradation rate of poly(methyl methacrylate) can be explained in the same way. Oxygen–oxygen and silicon–silicon bonds have a low dissociation energy and do not occur in polymers except possibly at weak points in some chains.

There is much evidence that weak links are present in the chains of most polymer species. These weak points may be at a terminal position and arise from the specific mechanism of chain termination or may be non-terminal and arise from a momentary aberration in the *modus operandi* of the polymerisation reaction. Because of these weak points it is found that polyethylene, polytetrafluoroethylene and poly(vinyl chloride), to take just three well-known examples, have a much lower resistance to thermal degradation than low molecular weight analogues. For similar reasons polyacrylonitrile and natural rubber may degrade whilst being dissolved in suitable solvents.

Weak links, particularly terminal weak links, can be the site of initiation of a chain 'unzipping' reaction.[7,8] A monomer or other simple molecule may be abstracted from the end of the chain in such a way that the new chain end is also unstable. The reaction repeats itself and the polymer depolymerises or otherwise degrades. This phenomenon occurs to a serious extent with polyacetals, poly(methyl methacrylate) and, it is believed, with p.v.c.

There are four ways in which these unzipping reactions may be moderated:

1. By preventing the initial formation of weak links. These will involve, amongst other things, the use of rigorously purified monomer.
2. By deactivating the active weak link. For example, commercial polyacetal (polyformaldehyde) resins have their chain ends capped by a stable grouping. (This will however be of little use where the initiation of chain degradation is not at the terminal group.)
3. By copolymerising with a small amount of second monomer which acts as an obstruction to the unzipping reaction, in the event of this being allowed to start. On the industrial scale methyl methacrylate is sometimes copolymerised with a small amount of ethyl acrylate, and formaldehyde copolymerised with ethylene oxide or 1,3-dioxolane for this very reason.
4. By the use of certain additives which divert or moderate the degradation reaction. A wide range of antioxidants and stabilisers function by this mechanism.

Most polymers are affected by exposure to light, particularly sunlight. This is the result of the absorption of radiant light energy by chemical structures. The lower the wavelength the higher the energy. Fortunately for most purposes, most of the light waves shorter than 3,000 Å are destroyed or absorbed before they reach the surface of the earth and for non-astronautical applications these short waves may be ignored and most damage appears to be done by rays of wavelength in the range

3,000–4,000 Å.   At 3,500 Å the light energy has been computed to be equal to 82 kcal/mole and it will be seen from Table 5.2 that this is greater than the dissociation energy of many bonds.   Whether or not damage is done to a polymer also depends on the absorption frequency of a bond. A C—C bond absorbs at 1,950 Å and at 2,300–2,500 Å and aldehyde and ketone carbonyl bonds at 1,870 Å and 2,800–3,200 Å.   Of these bonds it would only be expected that the carbonyl bond would cause much trouble

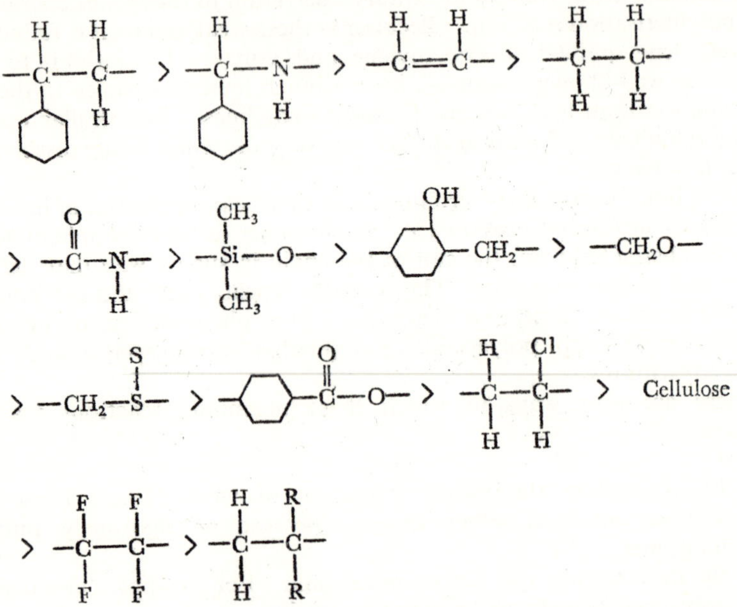

Fig. 5.9.   *Relative stabilities of various polymers to exposure by high energy sources.   (After Ballantine*[9]*)*

under normal terrestrial conditions.   P.T.F.E. and other fluorocarbon polymers would be expected to have good light stability because the linkages present normally have bond energies exceeding the light energy. Polyethylene and p.v.c. would also be expected to have good light stability because the linkages present do not absorb light at the damaging wavelength present on the earth's surface.   Unfortunately carbonyl and other groups which are present in processed polymer may prove to be a site for photochemical action and these two polymers only have limited light stability.   Antioxidants in polyethylene, used to improve heat stability, may in some instances prove to be a site at which a photochemical reaction can be initiated.   To some extent the light stability of a polymer may be improved by incorporating an additive that preferentially absorbs energy, at wavelengths that damage the polymer linkage.   It follows that an ultraviolet light absorber that is effective in one polymer may not be effective in another polymer. Common ultraviolet absorbers include certain salicylic esters such as phenyl salicylate, benzotriazole and benzophenones. Carbon black is found to be particularly effective in polyethylene and

Table 5.9 BEHAVIOUR OF POLYMERS SUBJECTED TO HIGH ENERGY RADIATION[10]

Polymers that cross link	Polymers that degrade
Polyethylene	Polyisobutylene
Polypropylene	Poly-α-methyl styrene
Poly(acrylic acid)	Poly(methyl methacrylate)
Poly(methyl acrylate)	Poly(methacrylic acid)
Polyacrylamide	Poly(vinylidene chloride)
Natural rubber	Polychlorotrifluoroethylene
Polychloroprene	Cellulose
Polydimethylsiloxanes	P.T.F.E.
Styrene–acrylonitrile copolymers	

acetal resins.   In the case of polyethylene it will reduce the efficiency of amine antioxidants.

In analogy with thermal and light radiations, high energy radiation may also lead to scission and cross linking.   The relative stabilities of various polymer structures are shown in Fig. 5.9.[9]   Whilst some materials cross link others degrade (i.e. are liable to chain scission).   Table 5.9 lists some polymers that cross link and some that degrade.

It is of interest to note that whereas most polymers of monosubstituted ethylene cross link, most polymers of disubstituted ethylenes degrade. P.V.C. is an important exception to this effect.

Also of interest is the different behaviour of both p.t.f.e. and poly(methyl methacrylate) when subjected to different types of radiation.   Although both polymers have a good stability to ultraviolet light they are both easily degraded by high energy radiation.

5.6 AGING AND WEATHERING

From the foregoing sections it will be realised that the ageing and weathering behaviour of a plastics material will be dependent on many factors. The following agencies may cause a change in the properties of a polymer:

1. Chemical environments, which may include atmospheric oxygen, acidic fumes and water.
2. Heat.
3. Ultraviolet light.
4. High energy radiation.

In a commercial plastics material there are also normally a number of other ingredients present and these may also be affected by the above agencies.   Furthermore they may interact with each other and with the polymer so that the effects of the above agencies may be more, or may be less, drastic.   Since different polymers and additives respond in different ways to the influence of chemicals and radiant energy weathering behaviour can be very specific.

A serious current problem for the plastics technologist is to be able to predict the aging and weathering behaviour of a polymer over a prolonged period of time, often 20 years or more.   For this reason it is desirable that some reliable accelerated weathering test should exist.   Unfortunately, accelerated tests have up until now only achieved very limited success. One reason is that when more than one deteriorating agency is present, the overall effect may be quite different from the sum of the individual effects of these agencies.   The effects of heat and light, or oxygen and light, in combination may be quite serious whereas individually their effect on a polymer may have been negligible.   It is also difficult to know how to accelerate a reaction.   Simply to carry out a test at higher temperature may be quite misleading since the temperature dependence of various reactions differ.   In an accelerated light aging test it is more desirable to subject the sample to the same light distribution as 'average daylight' but at greater intensity.   It is however difficult to obtain light sources which mimic the energy distribution.   Although some sources have been found that correspond well initially they often deteriorate quickly after some hours use and become unreliable.   Exposure to sources such as daylight, carbon arc lamps and xenon lamps can have quite different effects on plastics materials.

5.7 DIFFUSION AND PERMEABILITY

There are many instances where the diffusion of small molecules into, out of and through a plastics material are of importance in the processing and usage of the latter.   The solution of polymer in a solvent involves the diffusion of solvent into the polymer so that the polymer mass swells and eventually disintegrates.   The gelation of p.v.c. with a plasticiser such as tritolyl phosphate occurs through diffusion of plasticiser into the polymer mass.   Cellulose acetate film is produced by casting from solution and diffusion processes are involved in the removal of solvent.   The ease with which gases and vapours permeate through a polymer is of importance in packaging applications.   For example in the packaging of fruit the packaging film should permit diffusion of carbon dioxide through the film but restrain, as far as possible, the passage of oxygen.   Low air permeability is an essential requirement of an inner tube and a tubeless tyre and, in a somewhat less serious vein, a child's balloon.   Lubricants in many plastics compositions are chosen because of their incompatibility with the base polymers and they are required to diffuse out of the compound during processing and lubricate the interface of the compound and the metal surfaces of the processing equipment (e.g. mould surfaces and mill roll surfaces).   From the above examples it can be seen that a high diffusion and permeability is sometimes desirable but at other times undesirable.

Diffusion occurs as a result of natural processes that tend to equal out the concentration of a given species of particle (in the case under discussion, a molecule) in a given environment.   The diffusion coefficient of one material through another (D) is defined by the equation

$$F = - D\frac{\partial c}{\partial x}$$

where F is the weight of the diffusing material crossing unit area of the other material per unit time, and the differential is the concentration gradient in weight per ml. per cm at right angles to the unit area considered.

Diffusion through a polymer occurs by the small molecules passing through voids and other gaps between the polymer molecules. The diffusion rate will therefore depend to a large extent on the size of the small molecules and the size of the gaps. An example of the effect of molecular size is the difference in the effects of tetrahydrofuran and di-isooctyl phthalate on p.v.c. Both have similar solubility parameters but whereas tetrahydrofuran will diffuse sufficiently rapidly at room temperature to dissolve the polymer in a few hours the diffusion rate of the phthalate is so slow as to be almost insignificant at room temperature. (In p.v.c. pastes which are suspensions of polymer particles in plasticisers, the high interfacial areas allow sufficient diffusion for measurable absorption of plasticisers resulting in a rise of the paste viscosity.) The size of the gaps in the polymer will depend to a large extent on the physical state of the polymer, that is whether it is glassy, rubbery or crystalline. In the case of amorphous polymers above the glass transition temperature, i.e. in the rubbery state, molecular segments have considerable mobility and there is an appreciable 'free volume' in the mass of polymer. In addition, because of the segment mobility there is a high likelihood that a molecular segment will at some stage move out of the way of a diffusing small molecule and so diffusion rates are higher in rubbers than in other types of polymer.

Below the glass transition temperature the segments have little mobility and there is also a reduction in 'free volume'. This means that not only are there less voids but in addition a diffusing particle will have a much more tortuous path through the polymer to find its way through. About the glass transition temperature there are often complicating effects as diffusing particles may plasticise the polymers and thus reduce the effective glass transition temperature.

Crystalline structures have a much greater degree of molecular packing and the individual lamellae can be considered as almost impermeable so that diffusion can only occur in amorphous zones or through zones of imperfection. Hence crystalline polymers will tend to resist diffusion more than either rubbers or glassy-polymers.

Of particular interest in the usage of polymers is the permeability of a gas, vapour or liquid through a film. Permeation is a three-part process and involves solution of small molecules in polymer, migration or diffusion through the polymer according to the concentration gradient, and emergence of the small particle at the outer surface. Hence permeability is the product of solubility and diffusion and it is possible to write, where the solubility obeys Henry's law,

$$P = DS$$

where P is the permeability, D is the diffusion coefficient and S is the solubility coefficient.

Hence polyethylene will be more permeable to liquids of similar solubility parameter, e.g. hydrocarbons, than to liquids of different

Table 5.10 PERMEABILITY DATA FOR VARIOUS POLYMERS
VALUES FOR $P \times 10^{10}$ cm^3 sec^{-1} mm cm^{-2} cmHg^{-1}

Polymer	Permeability				Ratios to N_2 per $m = G$ value			Nature of polymer
	N_2 (30°C)	O_2 (30°C)	CO_2 (30°C)	90% R.H. H_2O (25°C)	P_{O_2}/P_{N_2}	P_{CO_2}/P_{N_2}	P_{H_2O}/P_{N_2}	
Poly(vinylidene chloride)	0·0094	0·053	0·29	14	5·6	31	1,400	crystalline
P.C.T.F.E.	0·03	0·10	0·72	2·9	3·3	24	97	crystalline
Poly(ethylene terephthalate)	0·05	0·22	1·53	1,300	4·4	31	26,000	crystalline
Rubber hydrochloride (Pliofilm ND)	0·08	0·30	1·7	240	3·8	21	3,000	crystalline
Nylon 6	0·10	0·38	1·6	7,000	3·8	16	70,000	crystalline
P.V.C. (unplasticised)	0·40	1·20	10	1,560	3·0	25	3,900	slight crystalline
Cellulose acetate	2·8	7·8	68	75,000	2·8	24	2,680	glassy
Polyethylene ($d = 0.954$–0.960)	2·7	10·6	35	130	3·9	13	48	crystalline
Polyethylene ($d = 0.922$)	19	55	352	800	2·9	19	42	some crystalline
Polystyrene	2·9	11	88	1,200	3·8	30	4,100	glassy
Polypropylene	—	23	92	680	—	—	—	crystalline
Butyl rubber	3·12	13·0	51·8	—	4·1	16·2	—	rubbery
Methyl rubber	4·8	21·1	75	—	4·4	15·6	—	rubbery
Polybutadiene	64·5	191	1,380	—	3·0	21·4	—	rubbery
Natural rubber	80·8	233	1,310	—	2·9	16·2	—	rubbery

solubility parameter but of similar size.   The permeabilities of a number of polymers to a number of gases are given in Table 5.10.[11,12]

Stannett and Szwarc[11] have argued that the permeability is a product of a factor F determined by the nature of the polymer, a factor G determined by the nature of the gas and an interaction factor H (considered to be of little significance and assumed to be unity).

Thus the permeability of polymer i to a gas k can be expressed as

$$P_{ik} = F_i G_k$$

Hence the ratio of the permeability of a polymer i to two gases k and l can be seen to be the same as the ratio between the two G factors

$$\frac{P_{ik}}{P_{il}} = \frac{G_k}{G_l}$$

Similarly between two polymers (i and j)

$$\frac{P_{ik}}{P_{kj}} = \frac{F_i}{F_j}$$

From a knowledge of various values of P it is possible to calculate F values for specific polymers and G values for specific gases if the G value for one of the gases, usually nitrogen, is taken as unity.   These values are generally found to be accurate within a factor of 2 for gases but unreliable with water vapour.   Some values are given in Table 5.11.   It will be realised that the F values correspond to the first column of Table 5.10 and the G values for oxygen and carbon dioxide are the averages of the P_{O_2}/P_{N_2} and P_{CO_2}/P_{N_2} ratios.

5.8 TOXICITY

No attempt will be made here to relate the toxicity of plastics materials to chemical structure.   Nevertheless this is a topic about which a few words must be said in a book of this nature.

A material may be considered toxic if it has an adverse effect on health. Although it is often not difficult to prove that a material is toxic it is almost impossible to prove that a material is not toxic.   Tobacco was smoked for many centuries before the dangerous effects of cigarette smoking were appreciated.   Whilst some materials may have an immediate effect others may take many years.   Some toxic materials are purged out of the body and providing they do not go above a certain concentration appear to cause little havoc; others accumulate and eventually a lethal dose may be present in the body.

Toxic chemicals can enter the body in various ways in particular by swallowing, inhalation and by skin absorption.   Skin absorption may lead to dermatitis and this can be a most annoying complaint.   Whereas some chemicals may have an almost universal effect on human beings, others may only attack a few persons.   A person who has worked with a chemical for

Table 5.11 F AND G CONSTANTS FOR POLYMERS AND GASES[11]

Polymer	F	Gas	G
Poly(vinylidene chloride) (Saran)	0·0094	N_2	1·0
P.C.T.F.E.	0·03		
Poly(ethylene terephthalate)	0·05	O_2	3·8
Rubber hydrochloride (Pliofilm)	0·08	H_2S	21·9
Nylon 6	0·1	CO_2	24·2
Nitrile rubber (Hycar OR—15)	2·35		
Butyl rubber	3·12		
Methyl rubber	4·8		
Cellulose acetate ($+15\%$ plasticiser)	5·0		
Polychloroprene	11·8		
Low density polyethylene	19·0		
Polybutadiene	64·5		
Natural rubber	80·8		
Plasticised ethyl cellulose	84		

some years may suddenly become sensitised to a given chemical and from then on be unable to withstand the slightest trace of that material in the atmosphere.   He may as a result also be sensitised not only to the specific chemical that caused the initial trouble but to a host of related products. Unfortunately a number of chemicals used in the plastics industry have a tendency to be dermatitic including certain halogenated aromatic materials, formaldehyde and aliphatic amines.

In addition many other chemicals used can attack the body, both externally and internally in many ways.   It is necessary that the effects of any material used should be known and appropriate precautions taken if trouble is to be avoided.   Amongst the materials used in the plastics industry for which special care should be taken are lead salts, phenol, aromatic hydrocarbons, isocyanates and aromatic amines.   In many plastics articles these toxic materials are often used only in trace doses. Providing they are surrounded by polymer or other inert material and they do not bleed or bloom and are not leached out under certain conditions of service it is sometimes possible to tolerate them.   This can however only be done with confidence after exhaustive testing.   The results of such testing of a chemical and the incidence of any adverse toxic effects should be readily available to all potential handlers of that chemical.   There is, unfortunately, in many countries a lack of an appropriate organisation which can collect and disseminate such information.   This is however a matter outside the scope of this book.

Most toxicity problems associated with the finished product arise from the nature of the additives and seldom from the polymer.   Mention should however be made of poly(vinyl carbazole) and the polychloro-acrylates which when monomer is present can cause unpleasant effects.

REFERENCES

1. SMALL, P. A. J., *J. Appl. Chem.*, **3**, 71 (1953)
2. BURRELL, H., *Interchem. Rev.*, **14**, 3 (1955)
3. HILDEBRAND, J., and SCOTT, R., *The Solubility of Non-Electrolytes*, Reinhold, New York, 3rd Edn (1949)

4. TOMPA, H., *Polymer Solutions*, Butterworths, London (1956)
5. BILLMEYER, F. W., *Textbook of Polymer Science*, John Wiley, New York (1962)
6. ACHHAMMER, B. G., TRYON, M., and KLINE, G. M., *Mod. Plastics*, **37**(4), 131 (1959)
7. GRASSIE, N., *Trans. Inst. Rubber Ind.*, **39**, 200 (1963)
8. GRASSIE, N., *Chemistry of High Polymer Degradation Processes*, Butterworths, London (1956)
9. BALLANTINE, D. S., *Mod. Plastics*, **32**(3), 131 (1954)
10. JONES, S. T., *Canad. Plastics*, April, 32 (1955)
11. STANNETT, V. T., and SZWARC, M., *J. Polymer Sci.*, **16**, 89 (1955)
12. PAINE, F. A., *J. Roy. Inst. Chem.*, **86**, 263 (1962)

BIBLIOGRAPHY

BILLMEYER, F. W., *Textbook of Polymer Science*, John Wiley, New York (1962)
GORDON, M., *High Polymers—Structure and Physical Properties*, Iliffe Books Ltd., London, 2nd Edn. (1963)
HILDEBRAND, J., and SCOTT, R., *The Solubility of Non-Electrolytes*, Reinhold, New York, 3rd Edn (1949)
PAULING, L., *The Nature of the Chemical Bond*, Cornell University Press, Ithaca, N.Y., 3rd Edn. (1960)
TOMPA, H., *Polymer Solutions*, Butterworths, London (1956)

6

Relation of Structure to Electrical and Optical Properties

6.1 INTRODUCTION

Most plastics materials may be considered as electrical insulators, i.e. they are able to withstand a potential difference between different points of a given piece of material with the passage of only a small electric current and a low dissipation energy. When assessing a potential insulating material, information on the following properties will be required:

1. Dielectric constant (specific inductive capacity, permittivity) over a wide range of temperature and frequency.
2. Power factor over a range of temperature and frequency.
3. Electrical strength (usually measured in V/0·001 in).
4. Volume resistivity (usually measured in Ω cm).
5. Surface resistivity (usually measured in Ω).
6. The influence of humidity variation on the above properties.

Typical properties for a selection of well known plastics materials are tabulated in Table 6.1.

6.2 DIELECTRIC CONSTANT, POWER FACTOR AND STRUCTURE

The materials in Table 6.1 may be roughly divided into two groups

1. Polymers with outstandingly high resistivity, low dielectric constant and negligible power factor, all substantially unaffected by temperature, frequency and humidity over the usual range of service conditions.
2. Moderate insulators with lower resistivity and higher dielectric constant and power factor affected further by the conditions of the test. These materials are often referred to as polar polymers.

It is not difficult to relate the differences between these two groups to molecular structure. In order to do this the structure and electrical

86

Table 6.1 TYPICAL ELECTRICAL PROPERTIES OF SOME SELECTED PLASTICS MATERIALS AT 20°C

Polymer	Volume resistivity (Ω cm)	Electric strength (V/0·001 in.) ($\frac{1}{8}$ in. sample)	Dielectric constant		Power factor	
			60 c/s	10^6 c/s	60 c/s	10^6 c/s
P.T.F.E.	$>10^{18}$	450	2·1	2·1	<0·0003	<0·0003
Polyethylene (L.D.)	10^{18}	450	2·3	2·3	<0·0003	<0·0003
Polystyrene	10^{18}	600	2·55	2·55	<0·0003	<0·0003
Polypropylene	$>10^{17}$	800	2·15	2·15	0·0008	0·0004
P.M.M.A.	10^{14}	370	3·7	3·0	0·06	0·02
P.V.C.	10^{15}	600	3·2	2·9	0·013	0·016
P.V.C. (plasticised)[a]	10^{15}	700	6·9	3·6	0·082	0·089
Nylon 66[b]	10^{13}	380	4·0	3·4	0·014	0·04
Polycarbonate[c]	10^{16}	400	3·17	2·96	0·0009	0·01
Phenolic[d]	10^{11}	250	5·0–9·0	5·0	0·08	0·04
Urea formaldehyde[d]	10^{12}	300	4·0	4·5	0·04	0·3

a P.V.C. 59% di-2-ethyl hexyl phthalate 30%, filler 5%, stabiliser 6%.
b 0·2% water content.
c Makrolon.
d General purpose moulding compositions.

properties of atoms, symmetrical molecules, simple polar molecules and polymeric polar molecules will be considered in turn.

An atom consists essentially of a positively charged nucleus surrounded by a cloud of light negatively charged electrons which are in motion around the nucleus. In the absence of an electric field, the centres of both negative and positive charges are coincident and there is no external effect of these two charges (Fig. 6.1 (a)). In a molecule we have a number of positive nuclei surrounded by overlapping electron clouds. In a truly covalent molecule the centres of negative and positive charges again coincide and there is no external effect.

If an atom or covalent molecule is placed in an electric field there will be a displacement of the light electron cloud in one direction and a considerably smaller displacement of the nucleus in the other direction (Fig. 6.1 (b)). The effect of the electron cloud displacement is known as *electron polarisation*. In these circumstances the centres of negative and positive charge are no longer coincident.

Let us now consider a condenser system in which a massive quantity of a species of atom or molecule is placed between the two plates, i.e. forming the dielectric. The condenser is a device for storing charge. If two parallel plates are separated by a vacuum and one of the plates is brought to a given potential it will become charged (a *conduction* charge). This conduction charge will induce an equal, but opposite charge on the second plate. For a condenser, the relationship between the charge Q and the potential difference V between the plates is given by

$$Q = CV \qquad (6.1)$$

where the constant of proportionality C is known as the *capacitance* and is a characteristic of a given condenser (see Fig. 6.1 (c)).

If the slab of dielectric composed of a mass of atoms or molecules is inserted between the plates, each of the atoms or molecules will be subject to electron polarisation.   In the centre of the dielectric there will be no apparent effect but the edges of the slab adjacent to the metal plates will have a resultant charge, known as a *polarisation charge*.   Since the charge near each metal plate is of opposite sign to the conduction charge, it tends to offset the conduction charge in electrostatic effects (Fig. 6.1 (d)).

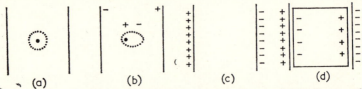

Fig. 6.1.   (a) *Atom not subject to external electric field.   Centre of electron cloud and nucleus coincident.   (b) Electron cloud displacement through application of external electric field.   (c) Charged condenser plates separated by vacuum.   (d) Condenser plates separated by dielectric*

This includes the potential difference V between the plates which is reduced when polarisation charges are present.   In any use of a condenser it is the conduction charge Q that is relevant and this is unaltered by the polarisation.   From Equation 6.1 it will be seen that since the insertion of a dielectric reduces the potential difference but maintains a constant conduction charge the capacitance of the system is increased.

The influence of a particular dielectric on the capacitance of a condenser is conveniently assessed by the dielectric constant, also known as the permittivity or specific inductive capacity.   This is defined as the ratio of the condenser capacity, using the given material as a dielectric, to the capacity of the same condenser, without dielectric, in a vacuum (or for all practical intents and purposes, air).

In the case of symmetrical molecules such as carbon tetrachloride, benzene, polyethylene and polyisobutylene the only polarisation effect is electronic and such materials have low dielectric constants.   Since electronic polarisation may be assumed to be instantaneous, the influence of frequency and temperature will be very small.   Furthermore, since the charge displacement is able to remain in phase with the alternating field there are negligible power losses.

Many intra-atomic bonds are not truly covalent and in a given linkage one atom may have a slight positive charge and the other a slight negative

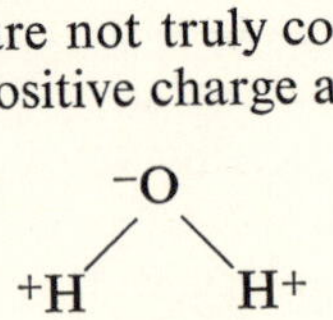

$$-O$$
$$+H \qquad H+$$

Fig. 6.2

charge.   Such a bond is said to be polar.   In a number of these molecules, such as carbon tetrachloride, the molecules are symmetrical and there is no external effect.   In the case of other molecules, the disposition of the polar linkage is unbalanced, as in the case of water (Fig. 6.2).

In the water molecule, the oxygen atom has a stronger attraction for the electrons than the hydrogen atoms and becomes negatively charged.

Since the angle between the O—H bonds is fixed (approx. 105°) the molecule is electrically unbalanced and the centres of positive and negative charge do not coincide. As a consequence the molecule will tend to turn in an electric field. The effect is known as *dipole polarisation*; it does not occur in balanced molecules where the centre of positive and negative charge are coincident.

In the dielectric of a condenser the dipole polarisation would increase the polarisation charge and such materials would have a higher dielectric constant than materials whose dielectric constant was only a function of electronic polarisation.

There is an important practical distinction between electronic and dipole polarisation: whereas the former only involves movement of electrons the latter entails movement of part of or even the whole of the molecule. Molecular movements take a finite time and complete orientation as induced by an alternating current may or may not be possible depending on the frequency of the change of direction of the electric field. Thus at zero frequency the dielectric constant will be at a maximum and this will remain approximately constant until the dipole orientation time is of the same order as the reciprocal of the frequency. Dipole movement will now be limited and the dipole polarisation effect and the dielectric constant will be reduced. As the frequency further increases, the dipole polarisation effect will tend to zero and the dielectric constant will tend to be dependent only on the electronic polarisation (Fig. 6.3). Where there are two dipole species differing in ease of orientation there will be two points of inflection in the dielectric constant–frequency curve.

The dielectric constant of unsymmetrical molecules containing dipoles (polar molecules) will be dependent on the internal viscosity of the

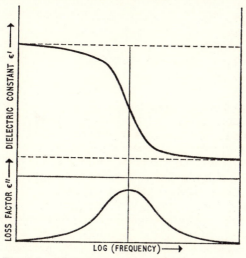

Fig. 6.3. The variation of dielectric constant ε′ and the loss factor ε″ with frequency. (After Frith and Tuckett,[1] reproduced by permission of Longmans, Green and Co. Ltd.)

dielectric. If very hard frozen ethyl alcohol is used as the dielectric the dielectric constant is approximately 3, at the melting point when the molecules are free to orient themselves the dielectric constant is about 55. Further heating reduces the ratio by increasing the energy of molecular motions which tend to disorient the molecules but at room temperature the dielectric constant is still as high as 35.

In addition to an enhanced dielectric constant dependent on temperature and frequency, polar molecules exhibit quite high dielectric power losses at certain frequencies, the maximum power loss corresponding to the point of inflection in the dielectric constant–frequency curve (Fig. 6.3). At very low frequencies, as already mentioned, the dipole movements are able to keep in phase with changes in the electric field and power losses are low. As the frequency is increased the point is reached when the dipole orientation cannot be completed in the time available and the dipole becomes out of phase. It is possible to have a mental picture of internal friction due to out-of-step motions of the dipoles leading to the generation of heat. Measures of the fraction of energy absorbed per cycle by the dielectric from the field are the *power factor* and *dissipation factor*. These terms arise by considering the delay between the changes in the field and the change in polarisation which in turn leads to a current in a condenser leading the voltage across it when a dielectric is present. The angle of lead is known as the phase angle and given the symbol θ. The value $90 - \theta$ is known as the loss angle and is given the symbol δ. The power factor is defined as $\cos \theta$ (or $\sin \delta$) and the dissipation factor as $\tan \delta$ (or $\cot \theta$). When δ is small the two are equivalent. Also quoted in the literature is the *loss factor* which is numerically the product of the dissipation factor and the dielectric constant.

At low frequencies when power losses are low these values are also low but they increase when such frequencies are reached that the dipoles cannot keep in phase. After passing through a peak at some characteristic frequency they fall in value as the frequency further increases. This is because at such high frequencies there is no time for substantial dipole movement and so the power losses are reduced. Because of the dependence of the dipole movement on the internal viscosity, the power factor like the dielectric constant, is strongly dependent on temperature.

In the case of polar polymers the situation is more complex, since there are a large number of dipoles attached to one chain. These dipoles may either be attached to the main chain (as with poly(vinyl chloride), polyesters and polycarbonates) or the polar groups may not be directly attached to the main chain and the dipoles may, to some extent, rotate independently of it, e.g. as with poly(methyl methacrylate).

In the first case, that is with dipoles integral with the main chain, in the absence of an electric field the dipoles will be randomly disposed but will be fixed by the disposition of the main chain atoms. On application of an electric field complete dipole orientation is not possible because of spatial requirements imposed by the chain structure. Furthermore in the polymeric system the different molecules are coiled in different ways and the time for orientation will be dependent on the particular disposition.

Thus whereas simple polar molecules have a sharply defined power loss maxima the power loss–frequency curve of polar polymers is broad, due to the dispersion of orientation times.

When dipoles are directly attached to the chain their movement will obviously depend on the ability of chain segments to move. Thus the dipole polarisation effect will be much less below the glass transition

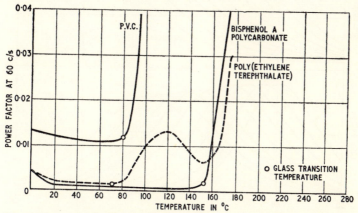

Fig. 6.4. *Power factor–temperature curves for three polar polymers whose polar groups are integral with or directly attached to the main chain. The rise in power factor above the glass transition point is clearly seen in these three examples*

temperature, than above it (Fig. 6.4). For this reason unplasticised p.v.c., poly(ethylene terephthalate) and the bis-phenol A polycarbonates are better high frequency insulators at room temperature, which is below the glass temperature of each of these polymers, than would be expected in polymers of similar polarity but with the polar groups in the side chains.

It was pointed out in Chapter 3 that the glass temperature is dependent on the time scale of the experiment and thus will be allocated slightly different values according to the method of measurement. One test carried out at a slower rate than in a second test will allow more time for segmental motion and thus lead to lower measured values of the glass temperature. In the case of electrical tests the lower the frequency of the alternating current the lower will be the temperature at the maxima of the power factor–temperature curve and of the temperature at the point of inflection in the dielectric constant–temperature curve (Fig. 6.5).

Since the incorporation of plasticisers into a polymer compound brings about a reduction in glass temperature they will also have an effect on the electrical properties. Plasticised p.v.c. with a glass temperature below that of the testing temperature will have a much higher dielectric constant than unplasticised p.v.c. at the same temperature (Fig. 6.6).

In the case of polymer molecules where the dipoles are not directly attached to the main chain, segmental movement of the chain is not essential for dipole polarisation and dipole movement is possible at temperatures below the glass transition temperature. Such materials

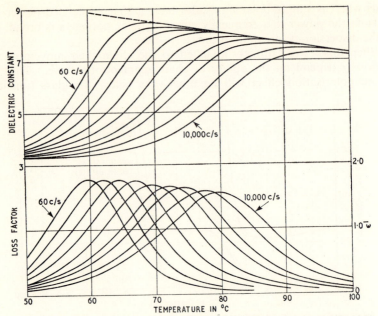

Fig. 6.5. *Electrical properties of polyvinyl acetate (Gelva 60) at 60, 120, 240, 500, 1,000, 2,000, 3,000, 6,000 and 10,000 cycles.*[2] *(Copyright 1941 by the American Chemical Society and reprinted by permission of the copyright holder)*

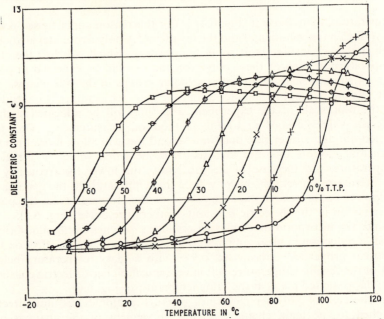

Fig. 6.6. *Effect of temperature on the 1,000 c/s dielectric constant of stabilised polyvinyl chloride–tritolyl phosphate systems.*[3] *(Copyright 1941 by the American Chemical Society and reprinted by permission of the copyright holder)*

are less effective as electrical insulators at temperatures in the glassy range. With many of these polymers, e.g. poly(methyl methacrylate), there are two or more maxima in the power factor–temperature curve for a given frequency. The presence of two such maxima is due to the different orientation times of the dipoles with and without associated segmental motion of the main chain.

The above discussion in so far as it applies to polymers may be summarised as follows:

1. For non-polar materials (i.e. materials free from dipoles or in which the dipoles are vectorially balanced) the dielectric constant is due to electronic polarisation only and will generally have a value of less than 3. Since polarisation is instantaneous the dielectric constant is independent of temperature and frequency. Power losses are also negligible irrespective of temperature and frequency.
2. With polar molecules the value of the dielectric constant is additionally dependent on dipole polarisation and commonly has values between 3·0 and 7·0. The extent of dipole polarisation will depend on frequency, an increase in frequency eventually leading to a reduction in dielectric constant. Power factor–frequency curves will go through a maximum.
3. The dielectric properties of polar materials will depend on whether or not the dipoles are attached to the main chain. When they are, dipole polarisation will depend on segmental mobility and is thus low at temperatures below the glass transition temperatures. Such polymers are therefore better insulators below the glass temperature than above it.

Finally mention may be made about the influence of humidity on the electrical insulating properties of plastics. Once again the polymers may be classified into two groups, those which do not absorb water and those which do. The non-absorbent materials are little affected by humidity whereas the insulation characteristics of the absorbent materials deteriorate seriously. These latter materials are generally certain polar materials which all appear capable of forming some sort of bond, probably a hydrogen bond, with water. Three reasons may be given for the deletrious effects of the water.

1. Its higher electrical conductivity, lowers the resistivity of the compound.
2. Its higher dielectric constant raises the overall volume of the polymer–water mixture.
3. Its plasticising effect on some polymers which increases segmental mobility and enhances the value of the dielectric constant of the polymer itself.

6.3 SOME QUALITATIVE RELATIONSHIPS OF DIELECTRICS

There are a number of properties of molecules that are additive to a reasonable approximation, i.e. the value of such a property of a given molecule is an approximate sum of the values of the properties of either the atoms or bonds present. It has been shown that the dielectric constant is related to some additive properties and it is thus possible to make some estimate of dielectric properties from consideration of molecular structure.

The total polarisation of a molecule in an electric field P is

$$P = P_E + P_A + P_O \tag{6.2}$$

where P_E is the electronic polarisation

P_A is the nuclear polarisation (considered to be negligible and hence ignored in further discussion)

P_O the dipole or orientation polarisation

P itself is defined by the Clausius–Mosotti Equation

$$P = \left(\frac{D-1}{D+2}\right)\frac{M}{\rho} \tag{6.3}$$

where D is the dielectric constant

M is the molecular weight

ρ is the density.

It may be shown that for electron polarisation

$$P_E = \frac{4}{3}\pi N\alpha_E \tag{6.4}$$

where N is the Avagadro number

α_E is the electron polarisability.

It has also been observed that

$$P_E = \left(\frac{n^2-1}{n^2+2}\right)\frac{M}{\rho} \tag{6.5}$$

where n is the refractive index.

It is thus seen that for polymers in which polarisations other than electronic ones are negligible (i.e. $P = P_E$) the dielectric constant is equal to the square of the refractive index (Table 6.2).

Where dipole polarisation occurs it may be shown that

$$P_O = \frac{4}{3}\pi N\left(\frac{\mu^2}{3KT}\right)$$

where μ is the permanent dipole moment

K is the Boltzmann constant

T the absolute temperature

Hence

$$P = \left(\frac{D-1}{D+2}\right)\frac{M}{\rho} = \frac{4}{3}\pi N\alpha_E + \frac{4}{3}\pi N\left(\frac{\mu^2}{3KT}\right)$$

Table 6.2

	D	n	n^2
Polyethylene	2·28	1·51	2·28
Polystyrene	2·55	1·60	2·56
P.T.F.E.	2·05	1·40	1·96
Polyisobutylene	2·30	1·51	2·28

This expression is known as the Debye equation. It is therefore obvious that if α_E and μ were to be additive properties then it would be possible to calculate the dielectric constant from a knowledge of molecular structure.

Now P_E is numerically equal to the molar refraction R which is an additive property. It has been shown that R is a property which can

Table 6.3 REFRACTIONS OF SOME ELECTRON GROUPS (MEASURED BY SODIUM D LINE)

Electron group	Refraction	Electron group	Refraction
C—H	1·705	C≡C	6·025
C—C	1·209	C—Cl	6·57
C=C	4·15	C—F	1·60

be calculated by adding the refractions of various electron groups. Six values for such partial molar refractions are given in Table 6.3.

If we assume that in polyethylene the polarisation is solely electronic, that the degree of polymerisation is r and that the repeating unit is as shown in Fig. 6.7.

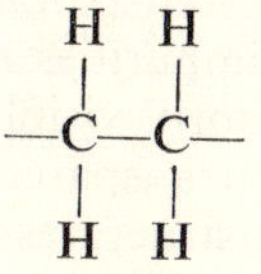

<pre>
    H   H
    | |
 —C —C—
    | |
    H   H
</pre>

Fig. 6.7

Then if M is the weight of the repeating unit

$$\left(\frac{D-1}{D+2}\right)\frac{Mr}{\rho} = r[4R_{C-H} + 2R_{C-C}]$$

then for polythene of density 0·92 the calculated dielectric constant is equal to 2·27 (c.f. observed value of 2·28 ± 0·01).

The calculated value for polypropylene (2·27) is also within the range of observed values (2·15–2·30) but the calculated value for p.t.f.e. (1·7) is less than the observed values of about 2·0.

The dipole moment of a molecule is another additive property since it arises from the difference in electronegativity of two atoms connected by a double bond. It should therefore be possible to associate a dipole moment with every linkage. Eucken and Meyer[4] have suggested the following moments for various linkages (in units of 10^{-18} e.s.cm)

$$\begin{array}{llll} \text{C—H} & 0\cdot4 & \text{C—Cl} & 1\cdot5 \\ \text{C—O} & 0\cdot7 & \text{C=O} & 2\cdot3 \\ & \text{H—O} & 1\cdot6 & \end{array}$$

In each case the left-hand atom of the pair as written is the least electronegative. Since dipole moments have direction as well as magnitude it is necessary to add the moments of each bond vectorially. For this reason the individual dipole moments cancel each other out in carbon tetrachloride but only partially in chloroform. In other molecules, such as that of water, it is necessary to know the bond angle to calculate the dipole moment. Alternatively since the dipole moment of the molecule is measurable the method may be used to compute the bond angle.

Computation of the dipole moment and hence the dielectric constant in polymers becomes complex but consideration of the bond dispositions allows useful qualitative prediction to be made.

6.4 OPTICAL PROPERTIES[5]

In addition to the refractive index (already seen to be closely linked with molecular structure) there are a number of other optical properties of importance with plastics materials. These include clarity, haze and birefringence, colour, transmittance and reflectance.

In order to achieve a product with a high *clarity* it is important that the refractive index is constant throughout the sample in the line of direction between the object in view and the eye. The presence of interfaces between regions of different refractive index will cause scatter of the light rays. This effect is easily demonstrated when fine fillers or even air bubbles are incorporated into an otherwise transparent polymer. Amorphous polymers free from fillers or other impurities are transparent unless chemical groups are present which absorb visible light radiation. Crystalline polymers may or may not be transparent, dependent on a number of factors. Where the crystalline structures such as spherulites are smaller than the wavelength of light then they do not interfere with the passage of light and the polymer is transparent. This occurs with rapidly quenched films of polyethylene. Where the structures formed are greater in diameter than the wavelength of light then the light waves will be scattered if the crystal structures have a different refractive index to that of the amorphous regions. Since this property is dependent on density it follows that where the crystalline and amorphous densities of polymers differ there will be a difference in refractive index. In the case of thick polyethylene objects fast quenching is not possible and as the spherulites formed have a significantly higher density (about 1·01) than the amorphous region (0·84–0·85) the polymer is opaque. In the case of polypropylene the difference is less marked (crystal density = 0·94, amorphous density = 0·85) and mouldings are more translucent. With poly(4-methyl pentene-1) amorphous and crystal densities are similar and the polymer is transparent even when large spherulites are present.

As the polarity across a molecule is different from the polarity along its length the refractive index of crystal structures depends on the direction in which it is measured (the crystal is said to be birefringent). Light scatter will then occur at the interface between structures aligned in different directions. By biaxial stretching, the crystal structures will be aligned into planes so that light travelling through films so oriented will pass through in a direction generally at right angles to the direction of the molecule. These light waves will thus not be affected by large changes in refractive index and the films will appear transparent. This phenomenon has been utilised in the manufacture of biaxially oriented polypropylene and poly(ethylene terephthalate) films of high clarity.

For transparent plastics materials *transparency* may be defined as the state permitting perception of objects through or beyond the specimen. It is often assessed as that fraction of the normally incident light transmitted with deviation from the primary beam direction of less than $0.1°$.

Some polymers although transparent may have a cloudy or milky appearance, generally known as *haze*. It is often measured quantitatively as the amount of light deviating by more than $2.5°$ from the transmitted beam direction. Haze is often the result of surface imperfections, particularly with thin films of low density polyethylene.

When light falls on a material some is transmitted, some is reflected and some absorbed. The *transmittance* is the ratio of the light passing through to the light incident on the specimens and the *reflectance* the ratio of the light reflected to the light incident. The *gloss* of a film is a function of the reflectance and the surface finish of a material. Where transmittance and reflectance do not add up to unity then some of the light waves are absorbed. This does not usually occur uniformly over the visible spectrum but is selective according to the chemical structures present. The uneven absorption of incident light results in the material being coloured. Comparatively few groupings present in commercial polymers are affected by visible radiation and so in the absence of impurities polymers are often water-white. Two important instances of the presence of colour-forming groups in polymers occur in the curing of novolak resins and in the dehydrochlorination of p.v.c. Although coloured polymers have been produced by the deliberate use of monomers containing chromophoric groups it is generally desirable to have a water-white polymer which may then be given any desired colour by the appropriate addition of dyes or pigments.

REFERENCES

1. FRITH, E. M., and TUCKETT, R. F., *Linear Polymers*, Longmans Green, London (1951)
2. MEAD, D. J., and FUOSS, R. M., *J. Am. Chem. Soc.*, **63**, 2832 (1941)
3. DAVIES, J. M., MILLER, R. F., and BUSSE, W. F., *J. Am. Chem. Soc.*, **63**, 361 (1941)
4. EUCKEN, A., and MEYER, L., *Phys. Z.*, **30**, 397 (1929)
5. PRITCHARD, R., *Soc. Plastics Engrs. Trans.*, **4**, 66 (1964)

BIBLIOGRAPHY

BAER, E., *Engineering Design for Plastics*, Reinhold, New York (1964)
FUOSS, R. M., *J. Am. Chem. Soc.*, **61**, 2334 (1938); **63**, 369, 378, 2401, 2410 (1941)
KIRKWOOD, J. G., and FUOSS, R. M., *J. Am. Chem. Soc.*, **63**, 385 (1941)
RITCHIE, P. D. (Ed.), *Physics of Plastics*, Iliffe, London (1965)
SMYTH, C. P., *Dielectric Behaviour and Structure*, McGraw-Hill, New York (1955)

7

Polyethylene

7.1 INTRODUCTION

It is sometimes difficult to realise that polyethylene, with one of the simplest molecular structures, one of the two largest tonnage plastics and a material about which more has probably been written than any other polymer, was first produced on a commercial scale as recently as 1939. The main attractive features of polyethylene, in addition to its low price, are excellent electrical insulation properties over a wide range of frequencies, very good chemical resistance, good processability, toughness, flexibility and, in thin films of certain grades, transparency.

Although the term polyethylene is chemically the more correct way of describing polymers of ethylene, these materials are also often known as polythene. Because this latter name is not chemically correct i.e. it does not fit accepted systems of nomenclature, this does not mean that this widely accepted alternative should not be used. Nobody considers it incorrect to refer to the *Betula pendula* as the silver birch. Similarly in the realm of plastics materials there should be no objection to the use of simple words or abbreviations providing the meaning is quite clear.

Although polyethylene is virtually defined by its very name as a polymer of ethylene produced by addition polymerisation, linear polymers with the formula $(CH_2)_n$ have also been prepared by condensation reactions. For example in 1898 von Pechmann[1] produced a white substance from an ethereal solution of diazomethane on standing. In 1900 Bamberger and Tschirner[2] analysed a similar product, found it to have the formula $(CH_2)_n$ and termed it 'polymethylene'. The reaction can be considered to be fundamentally

$$n\ CH_2\ \overset{\diagup\diagdown}{\underset{N=N}{}}\ \longrightarrow\ \sim\!\!\sim\!(CH_2)_n\!\sim\!\!\sim\ +\ n\ N_2$$

Since 1900 other methods have been devised for producing 'poly-methylene' including the use of boron trifluoride-diethyl ether catalysts at

0°C. Some of these methods give unbranched linear polymers, often of very high molecular weight, which are useful for comparing commercial polyethylenes which have molecules that are branched to varying extents.

Another condensation method was investigated by Carothers and co-workers[3] and reported in 1930. They reacted decamethylene dibromide with sodium in a Wurtz-type reaction but found it difficult to obtain polymers with molecular weights above 1,300.

$$n\,Br(CH_2)_{10}Br + 2n\,Na \longrightarrow \sim\!\!\sim(CH_2)_{10n}\!\!\sim\!\!\sim + 2n\,NaBr$$

Other routes have also been devised which are sometimes useful for research purposes and include

1. Modified Fischer–Tropsch reduction of carbon monoxide with hydrogen.[4]
2. Reduction of poly(vinyl chloride) with lithium aluminium hydride.
3. Hydrogenation of polybutadiene.

Commercially, polyethylene is produced from ethylene, the polymer being produced by this route in March 1933 and reported by Fawcett and Gibson in 1934.[5] The basic patent relating to the polymerisation of ethylene was applied for by I.C.I.[6] on February 4th 1936 and accepted on September 6th 1937.

Until the mid-1950's all commercial polyethylene was produced by high pressure processes developed from those described in the basic patent. These materials were somewhat branched materials and of moderate number average molecular weight, generally less than 50,000. However about 1954 two other routes were developed, one using metal oxide catalysts (e.g. the Phillips process) and the other aluminium alkyl or similar materials (the Ziegler process).[7] By these processes polymers could be prepared at lower temperature and pressures and with a modified structure. Because of these modifications these polymers had a higher density, were harder and had higher softening points.

Today the bulk of commercial polythene is still produced by high pressure processes and is used for a wide range of applications including packaging film, electrical insulation, chemical plant, domestic goods, toys, water piping and bottles. Polymers of the Phillips-type find important applications in mouldings, blown objects and electrical insulation. The Ziegler polyethylenes have been less developed, at least in Britain, to some extent because of the advent of polypropylene.

Some well-known trade names for the high pressure polyethylenes include Alathon (Du Pont), Alkathene (I.C.I.), Bakelite Polyethylene (Union Carbide and Bakelite-Xylonite), Monsanto Polyethylene (Monsanto), Petrothene (U.S.I. Chemicals) and Tenite Polyethylene (Eastman). Ziegler-type polymers include Hi-Fax (Hercules), Hostalen (Hoechst) Rotene (Montecatini) and Vestolen (Hüls). Examples of Phillips-type materials include Fortiflex, Fortilene (Celanese), Marlex (Phillips) and Rigidex (British Hydrocarbon Chemicals). The Shell Chemical Company use the trade name Carlona to cover their range of polyolefins (high pressure polyethylene, Ziegler-type polyethylene and polypropylene).

7.2 PREPARATION OF MONOMER

At one time ethylene for polymerisation was obtained largely from molasses, a by-product of the sugar industry. From molasses may be obtained ethyl alcohol and this may be dehydrated to yield ethylene. Today the bulk of ethylene is obtained from petroleum sources. When supplies of natural or petroleum gas are available the monomer is produced in high yield by high temperature cracking of ethane and propane. Good yields of ethylene may also be obtained if the gasoline ('petrol') fraction from primary distillation of oil is 'cracked'. The gaseous products of the reaction include a number of lower alkanes and olefins and the mixture may be separated by low temperature fractional distillation and by selective absorption. Olefins, in lower yield, are also obtained by cracking gas oil. At normal pressures (760 mmHg) ethylene is a gas boiling at $-103 \cdot 71°C$ and it has a very high heat of polymerisation (800–1,000 cal/g). In polymerisation reactions the heat of polymerisation must be carefully controlled particularly since decomposition reactions that take place at elevated temperatures are also exothermic and explosion can occur if the reaction gets out of control.

Since impurities can affect both the polymerisation reaction and the properties of the finished product (particularly electrical insulation properties and resistance to heat ageing) they must be rigorously removed. In particular carbon monoxide, acetylene, oxygen and moisture must be at a very low level. A number of patents require that the carbon monoxide content be less than $0 \cdot 02\%$.

7.3 POLYMERISATION

There are four quite distinct routes to the preparation of high polymers of ethylene:

1. High pressure processes.
2. Ziegler processes.
3. The Phillips process.
4. The Standard Oil (Indiana) process.

7.3.1 High pressure polymerisation

Although there are a number of publications dealing with the basic chemistry of ethylene polymerisation under high pressures little information has been made publicly available concerning details of current commercial processes. It may however be said that commercial high polymers are generally produced under conditions of high pressure (1,000–3,000 atm) and at temperatures of 80–300°C. A free-radical initiator such as benzoyl peroxide, azo-di-isobutyronitrile or oxygen is commonly used.

The process may be operated continuously by passing the reactants through narrow bore tubes or through stirred reactors or by a batch process in an autoclave. Because of the high heat of polymerisation care must be taken to prevent runaway reaction. This can be done by having a high cooling surface–volume ratio in the appropriate part of a continuous reactor and in addition by running water or a somewhat inert liquid such as benzene (which also helps to prevent tube blockage) through the tubes to dilute the exotherm. Local runaway reactions may be prevented by operating at a high flow velocity. In a typical process 10–30% of the monomer is converted to polymer. After a polymer–gas separation the polymer is extruded into a ribbon and then granulated. Film grades are often subjected to an homogenisation process in an internal mixer or a continuous compounder such as Werner-Pfleiderer ZSK machine to break up high molecular weight species present.

Although in principle the high pressure polymerisation of ethylene follows the free-radical-type mechanism discussed in Chapter 2 the reaction has two particular characteristics, the high exothermic reaction and a critical dependence on the monomer concentration.

The highly exothermic reaction has already been mentioned. It is particularly important to realise that at the elevated temperatures employed other reactions can occur leading to the formation of hydrogen, methane and graphite. These reactions are also exothermic and it is not at all difficult for the reaction to get out of hand. It is necessary to select conditions favourable to polymer formation and which allow a controlled reaction.

Most ethenoid monomers will polymerise by free-radical initiation over a wide range of monomer concentration. Methyl methacrylate can even be polymerised by photosensitised catalysts in the vapour phase at less than atmospheric pressure. In the case of ethylene only low molecular weight polymers are formed at low pressures but high molecular weights are possible at high pressures. It would appear that growing ethylene polymer radicals have a very limited life available for reaction with monomer. Unless they have reacted within a given interval they undergo changes which terminate their growth. Since the rate of reaction of radical with monomer is much greater with higher monomer concentration (higher pressure) it will be appreciated that the probability of obtaining high molecular weights is greater at high pressures than at low pressures.

At high reaction temperatures (e.g. 200°C) much higher pressures are required to obtain a given concentration or density of monomer than at temperatures of say 25°C and it might appear that better results would be obtained at lower reaction temperatures. This is in fact the case where a sufficiently active initiator is employed. This approach has an additional virtue in that side reactions leading to branching can be suppressed. For a given system the higher the temperature the faster the reaction and the lower the molecular weight.

By varying temperature, pressure, initiator type and composition, by incorporating chain transfer agents and by injecting the initiator into the reaction mixture at various points in the reactor it is possible to vary independently of each other polymer characteristics such as branching, molecular weight and molecular weight distribution over a wide range without necessarily the use of unduly long reaction times. In spite of the flexibility however, most high pressure polymers are of the lower density range for polyethylenes (0·915–0·94 g cm^{-3}) and usually also of the lower range of molecular weights.

7.3.2 Ziegler processes

As indicated by the title these processes are largely due to the work of Ziegler and co-workers. The type of polymerisation involved is sometimes referred to as co-ordination polymerisation since the mechanism involves a catalyst–monomer co-ordination complex or some other directing force that controls the way in which the monomer approaches the growing chain. The co-ordination catalysts are generally formed by the interaction of the alkyls of Group I–III metals with halides and other derivatives of transition metals in Groups IV–VIII of the Periodic Table. In a typical process the catalyst is prepared from titanium tetrachloride and aluminium triethyl or some related material.

In a typical process ethylene is fed under low pressure into the reactor which contains liquid hydrocarbon to act as diluent. The catalyst complex may be first prepared and fed into the vessel or may be prepared *in situ* by feeding the components directly into the main reactor. Reaction is carried out at some temperatures below 100°C (typically 70°C) in the absence of oxygen and water, both of which reduce the effectiveness of the catalyst. The catalyst remains suspended and the polymer as it is formed becomes precipitated from the solution and a slurry is formed which progressively thickens as the reaction proceeds. Before the slurry viscosity becomes high enough to interfere seriously with removing the heat of reaction, the reactants are discharged into a catalyst decomposition vessel. Here the catalyst is destroyed by the action of ethanol, water or caustic alkali. In order to reduce the amount of metallic catalyst fragments to the lowest possible value, the processes of catalyst decomposition, and subsequent purification are all important, particularly where the polymer is intended for use in high frequency electrical insulation. A number of variations in this stage of the process have been described in the literature.

The Ziegler polymers are intermediate in density (about 0·945 g cm^{-3}) between the high pressure polyethylenes and those produced by the Phillips and Standard Oil (Indiana) processes. A range of molecular weights may be obtained by varying the Al–Ti ratio in the catalyst, by introducing hydrogen as a chain transfer agent and by varying the reaction temperature.

7.3.3 The Phillips process

In this process ethylene, dissolved in a liquid hydrocarbon such as cyclohexane, is polymerised by a supported metal oxide catalyst at about

130–160°C and at about 200–500 lb in^{-2} pressure. The solvent serves to dissolve polymer as it is formed and as a heat transfer medium but is otherwise inert.

The preferred catalyst is one which contains 5% of chromium oxides, mainly CrO_3, on a finely divided silica–alumina catalyst (75–90% silica) and which has been activated by heating to about 250°C. After reaction the mixture is passed to a gas–liquid separator where the ethylene is flashed off, catalyst is then removed from the liquid product of the separator and the polymer separated from the solvent by either flashing off the solvent or by precipitating the polymer by cooling.

Polymers ranging in melt flow index (an inverse measure of molecular weight) from less than 0·1 to greater than 600 can be obtained by this process but commercial products have a melt flow index of only 0·2–5 and have the highest density of any commercial polyethylenes (~ 0.96 g cm^{-3}).

The polymerisation mechanism is largely unknown but no doubt occurs at or near the catalyst surface where monomer molecules are both concentrated and specifically oriented so that highly stereospecific polymers are obtained. It is found that the molecular weight of the product is critically dependent on temperature and in a typical process there is a forty-fold increase in melt flow index, and a corresponding decrease in molecular weight, in raising the polymerisation temperature from 140°C to just over 170°C. Above 400 lb in^{-2} the reaction pressure has little effect on either molecular weight or polymer yield but at lower pressures there is a marked decrease in yield and a measurable decrease in molecular weight. The catalyst activation temperature also has an effect on both yield and molecular weight. The higher the activation temperature the higher the yield and the lower the molecular weight. A number of materials including oxygen, acetylene, nitrogen and chlorine are catalyst poisons and very pure reactants must be employed.

7.3.4 Standard Oil Company (Indiana) process

This process has many similarities to the Phillips process and is based on the use of a supported transition metal oxide in combination with a promoter. Reaction temperatures are of the order of 230–270°C and pressures of 40–80 atm. Molybdenum oxide is a catalyst that figures in the literature and promoters include sodium and calcium as either metals or as hydrides. The reaction is carried out in a hydrocarbon solvent.

The products of the process have a density of about 0·96 g cm^{-3}, similar to the Phillips polymers. Another similarity between the processes is the marked effect of temperature on average molecular weight. The process is worked by the Furukawa Company of Japan and the product marketed as Staflen.

7.4 STRUCTURE AND PROPERTIES OF POLYETHYLENE

The relationship between structure and properties of polyethylene is largely in accord with the principles enunciated in the three previous

P.M.—8

chapters.   The polymer is essentially a long chain aliphatic hydrocarbon of the type

$$-CH_2-CH_2-CH_2-CH_2-$$

and would thus be thermoplastic.   The flexibility of the C—C bonds leads as might be expected to a low glass transition temperature of about $-120°C$ but this is masked by the crystallisation to be expected in such a regular polymer.   Some data on the crystalline structure of polyethylene are summarised in Table 7.1.   There are no strong intermolecular forces

Table 7.1   CRYSTALLINITY DATA FOR POLYETHYLENE

Molecular disposition	planar zigzag
Unit cell dimensions	$a = 7·36$ Å
	$b = 4·92$ Å
	$c = 2·54$ Å
Cell density (unbranched polymer) (25°C)	1·014
Amorphous density (20°C)	0·84

and most of the strength of the polymer is due to the fact that crystallisation allows close molecular packing.   The high crystallinity also leads to opaque structures except in the case of rapidly chilled film where the development of large crystalline structures is prevented.

Polyethylene, in essence a high molecular weight paraffin, would be expected to have a good resistance to chemical attack and this is found to be the case.

The polymer has a low cohesive energy density (the solubility parameter δ is about 7·9) and would be expected to be resistant to solvents of solubility parameter greater than 9·0.   Because it is a crystalline material and does not enter into specific interaction with any liquids there is no solvent at room temperature.   At elevated temperatures the thermodynamics are more favourable to solution and the polymer dissolves in a number of hydrocarbons of similar solubility parameter.

The polymer, in the absence of impurities, would also be expected to be an excellent high frequency insulator because of its non-polar nature. Once again fact is in accord with prediction.

Because of the low intermolecular forces the polymer would not be expected to have a high melting point.   Commercial polymers generally have melting points between 110°C and 132°C.

At the present time there are commercially available many hundreds of grades of polymer, most of which differ in their properties in one way or another.   These differences occur for four reasons.

1. Variation in the degree of branching in the polymer.
2. Variation in the average molecular weight.
3. Variation in molecular weight distribution.
4. The presence of impurities, some of which may be chemically combined into the polymer.

Further variation can also be obtained by compounding and cross linking the polymer but these aspects will not be considered at this stage.

Investigation of polyethylenes made by high pressure processes using infra-red spectroscopy indicates that in these polymers there are about 30 methyl groups per 1,000 carbon atoms in the chain which suggest that there must be some branching in the polymer.   Further work has indicated that the bulk of the branching is in the form of short ethyl and butyl groups attached to the chain but that there are also a few long chain branches present.   In Ziegler polyethylenes only ethyl side groups have been detected (5–7 groups per 1,000 carbon atoms) whilst in the metal oxide catalysed polymers no branching has been positively established although the number of methyl groups is rather greater than the main chain ends.

The presence of these branch points is bound to interfere with the ease of crystallisation and the difference is quite clearly shown in the differences between the polymers.   The branched high pressure polymers have the lowest density (since close-packing due to crystallisation is reduced) the least opacity (since the growth of large crystalline structures is impeded) and a lower melting point, yield point, surface hardness and Young's modulus in tension (these properties being dependent on the degree of crystallinity).   The greater the crystallinity the lower the permeability of the polymer to gases and vapour.   This is to be expected from the reasons given in Chapter 5.   For technological purposes the density is taken as a measure of the degree of short chain branching.

Differences in molecular weight will also give rise to differences in properties.   The higher the molecular weight the greater the number of points of attraction and entanglement between molecules.   Whereas differences in branching and hence degree of crystallinity largely affect properties characterised by small solid displacement, molecular weight differences will affect properties that involve large deformations such as ultimate tensile strength, elongation at break, melt viscosity and low temperature brittle point.   There is also an improvement in resistance to environmental stress cracking with increase in molecular weight.

Before the advent of Ziegler and Phillips polymers it was common practice to characterise the molecular weight for technological purposes by the melt flow index (M.F.I.), the weight in grams extruded under a standard load in a standard plastometer at 190°C in 10 minutes.   This test had also proved useful for quality control and as a very rough guide to processability.   From measurements of M.F.I. various workers have calculated the apparent viscosity of the polymer and correlated these figures with both number-average and weight-average molecular weight. (It should be noted that estimation of apparent viscosities from melt flow index data is rather hazardous since large corrections have to be made for end effects, pressure losses in the main cylinder and to friction of the plunger.   It would be better to use a high shear viscometer designed to minimise the sources of error and to compare results at equal shear rate.) Suffice it to say that the higher the melt flow index the lower the molecular weight.

With the availability of the higher density polymers the value of the melt flow index as a measure of molecular weight has diminished.   For example it has been found[8] that with two polymers of the same weight average molecular weight ($4 \cdot 2 \times 10^5$) the branched polymer (density = $0 \cdot 92$ g cm^{-3}) had only 1/50 the viscosity of the more or less unbranched polymer (density = $0 \cdot 96$ g cm^{-3}).

It has been found that the molecular weight distribution is increased by long chain branching. One measure of molecular weight spread is the ratio of weight-average molecular weight to number-average molecular weight ($\bar{M}_w/\bar{M}_n$). Whereas this ratio is about 2 in the case of poly-methylene, ratios of 20–50 are considered to be typical with low density polymers. With other structural factors constant a decrease in $\bar{M}_w/\bar{M}_n$ leads to an increase in impact strength, tensile strength, toughness, softening point and resistance to environmental stress cracking but with some loss in ease of processing. 'Sharkskin' effects can be prominent under these conditions.

The fourth variable in commercial polymers is the presence of impurities. These may be metallic fragments residual from Ziegler-type processes or they can be trace materials incorporated into the polymer chain. Certain impurities such as catalyst fragments and carbonyl groups incorporated into the chain can have a serious adverse influence on the power factor of the polymer whilst in other instances impurities can have an effect on aging behaviour.

Ethylene may also be copolymerised with a second monomer such as propylene or butene-1. Where small amounts only of a second olefin are used the effect is to produce a controlled degree of short chain branching and some retardation in the growth of large crystal structures. Copoly-mers produced by the Phillips process have better creep, environmental stress cracking and thermal stress cracking resistance than the correspond-ing homopolymer (see Chapter 8).

7.5 PROPERTIES OF POLYETHYLENE

Polyethylene is a wax-like thermoplastic softening at about 80–130°C with a density less than that of water. It is tough but has moderate tensile strength, is an excellent electrical insulator and has very good chemical resistance. In the mass it is translucent or opaque but thin films may be transparent.

7.5.1 Mechanical properties

The mechanical properties are very dependent on the molecular weight and on the degree of branching of the polymer. As with other polymers these properties are also dependent on the rate of testing, the temperature of test, the method of specimen preparation, the size and shape of the specimen and, to only a small degree with polyethylene, the conditioning of samples before testing. The data in Table 7.2, although not all obtained from the same source, has been obtained using only one test method for each property. The figures given show clearly the general effects of branching (density) and molecular weight on some polymer properties but it should be remembered that under different test conditions different

Table 7.2 EFFECT OF MOLECULAR WEIGHT AND DENSITY (BRANCHING) ON SOME MECHANICAL AND THERMAL PROPERTIES OF POLYETHYLENE

Property	Test	Density ≃0·92 g cm⁻³ (high pressure polymers)					Density ≃0·94 g cm⁻³ high pressure polymerisation	Density ≃0·95 g cm⁻³ Ziegler-type polymers			Density ≃0·96 g cm⁻³ Phillips-type polymer	Density ≃0·98 g cm⁻³ poly-methylene
Melt flow index	BS2782	0·3	2	7	20	70	0·7	0·02	0·2	2·0	1·5	—
Tensile strength (lb in⁻²)	BS903	2,200	1,800	1,500	1,300	—	3,000	3,200†	3,350†	3,350†	~4,000	~5,000
Elongation at break (%)	BS903	620	600	500	300	150	—	>800	380	20	500	~500
Izod impact strength (ft lb)	BS2782	~10	~10	~10	~10	~10	—	3·2	2·0	1·5	5·0	—
Vicat softening point (°C)	BS2782	98	90	85	81	77	116	124	122	121	—	—
Softening temperature (°C)	BS1493	—	—	—	—	—	—	110	110	106	122	—
Crystalline melting point (°C)	—	~108	~108	~108	~108	~108	125	~130	~130	~130	~133	136
Number-average molecular weight	—	48,000	32,000	28,000	24,000	20,000	—	—	—	—	—	—
CH₃ groups per 1,000 C atoms	—	20	23	28	31	33	—	5-7	5-7	5-7	<1·5	unbranched

† Yield strength

results may be obtained.   It should also be remembered that polymers of different density but with the same melt flow index do not have the same molecular weight.   The general effects of changing rate of testing, temperature and density on the tensile stress–strain curves are shown schematically in Fig. 7.1.   It is seen in particular that as the test temperature is lowered or the testing rate increased a pronounced 'hump' in the curve becomes apparent, the apex of the hump A being the yield point.

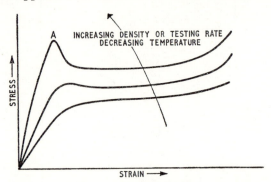

Fig. 7.1.   Effect of polymer density, testing rate and temperature on the shape of the stress–strain curve for polyethylene[9]

Up to the yield point deformations are recoverable and the polymer is almost Hookean in its behaviour.   Due to the working of the sample on stretching the polymer heats up and some crystal melting occurs accounting for the cold drawing phenomenon in which the polymer extends at constant stress.   Cold drawing however causes molecular orientation and induces crystallisation so that there is a stiffening of the sample and an upward sweep of the stress–strain curve.   The effect of temperature on a sample of low density polyethylene with an M.F.I. of 2 is shown in Fig. 7.2. The varying influence of rate of strain on tests results can be clearly shown from figures obtained with two commercial polyethylene samples (Table 7.3).   It is seen that in one case an increase in rate of strain is accompanied by an increase in tensile strength and in the other case, a reduction.

The elongation at break of polyethylene is strongly dependent on density (Fig. 7.3), the more highly crystalline high density materials being less ductile.   This lack of ductility results in high density polymers tending to be brittle, particularly with low molecular weight materials. The tough–brittle dependence on melt flow index and density is shown in Fig. 7.4.

Under load polyethylene will deform continuously with time ('creep'). A knowledge of creep behaviour is important when considering load-bearing applications, water piping being a case in point with polyethylene. In general there will be an increase in creep with increased load, increased temperature and decreased density.   A large amount of creep data has been made available in specialised monographs and in trade literature.

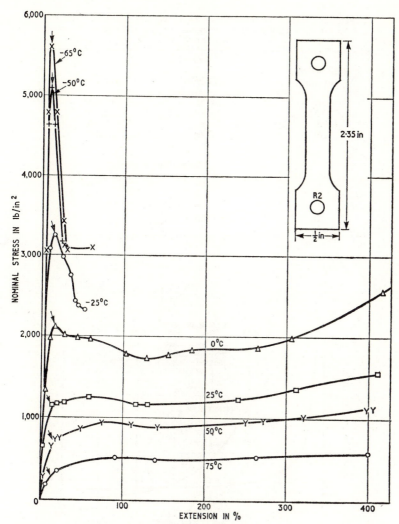

Fig. 7.2.   *Effect of temperature on the tensile stress–strain curve for polyethylene. (Low density polymer ~ 0.92 g cm³. M.F.I.2). Rate of extension 190% per minute*[10]

Table 7.3 EFFECT OF STRAINING RATE ON THE MEASURED TENSILE STRENGTH AND ELONGATION AT BREAK OF TWO SAMPLES OF POLY-ETHYLENE

Rate of strain (in/min)	Tensile strength (lb in^{-2})	
	Polymer A	Polymer B
6	2,680	1,600
12	2,750	1,580
18	2,900	1,500
30	3,200	1,400
	Elongation at break (%)	
6	380	450
12	300	490
18	200	490
30	180	500

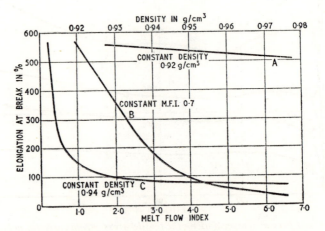

Fig. 7.3. Effect of density and melt flow index on elongation at break. (Separation rate 18in/min on specimen of 1 in gauge length). A, constant density (0·92 g/cm³). B, constant M.F.I. (0·7). C, constant density (0·94 g/cm³).[9] (Reproduced by permission of ICI Plastics Division)

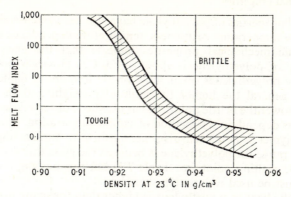

Fig. 7.4. *Effects of melt flow index and density on the room temperature tough–brittle transition of polyethylene.*[9] *(Reproduced by permission of ICI Plastics Division)*

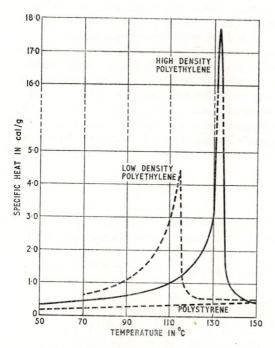

Fig. 7.5. *Specific heat–temperature relationships for low density polyethylene, high density polyethylene and polystyrene.*[11] *(The Distillers Company Ltd.)*

7.5.2 Thermal properties

The thermal properties of polyethylene have also been extensively reported. Data on low temperature brittleness, temperature dependence of modulus and dynamic mechanical properties indicate that the glass transition temperature of polyethylene is about $-120°C$. However both low and high density polyethylenes often show brittle fracture at higher temperatures.[3] In general the higher the molecular weight and the lower the density the lower the brittle point. It has also been observed that measured brittle points depend on the method of sample preparation, thus indicating that the polymer is notch sensitive, i.e. sensitive to surface imperfections. The service range for polyethylene is between the glass transition temperature and the melting point (see Table 7.2) but for many applications the polymer must be used well within these limits.

The specific heat of polyethylene is higher than for many other thermoplastics and is strongly dependent on temperature. Low density materials have a value of about 0·55 cal/g at room temperature and a value of 0·70 cal/g at 120–140°C. A somewhat schematic representation of the specific heat-temperature curve is given in Fig. 7.5. The peaks in these curves can be considered to be due to a form of latent heat of fusion of the crystalline zones. Melting point data for various polymers are given in Table 7.2 and are seen to vary with density.

The flow properties of polyethylene have been widely studied. Commercial polyethylenes, although differing widely amongst themselves, generally have viscosities at processing temperatures intermediate between the high viscosity melts, such as unplasticised p.v.c. and the polycarbonates, and low viscosity melts such as nylon 66. As is typical with thermoplastics, polyethylene melts are non-Newtonian in that their apparent viscosity decreases with increased rate of shear. In a Newtonian material there is a linear relationship between shear stress τ and shear rate $(d\gamma/dt)$

$$\tau = \mu \, d\gamma/dt$$

where the constant of proportionality is known as the coefficient of viscosity. Thermoplastics such as polyethylene do not obey this relationship, but follow, to a degree, a power relationship of the form.

$$\text{Shear stress at wall} = K' \left(\frac{\text{Apparent shear}}{\text{rate at wall}} \right)^{n'}$$

In particular for pipe flow

$$\frac{R\Delta P}{2L} = K' \left(\frac{4Q}{\pi R^3} \right)^{n'}$$

where R is the radius of the pipe, ΔP is the pressure drop along the pipe of length L and Q is the volume flow.

K' is sometimes referred to as the consistency index and n' the flow behaviour index. In a log–log plot of shear stress against shear rate the slope of the line is the flow behaviour index. A slope of unity would indicate a Newtonian material, one of less than unity would indicate a pseudoplastic (common with thermoplastics) and one of greater than unity a dilatant material.

The apparent viscosity (μ_a) at a given shear rate is defined as

$$\mu_a = \frac{R\Delta P/2L}{4Q/\pi R^3}$$

It should be noted that in practice the log–log plot is not quite linear showing that polymers only approximate in their behaviour to a power law fluid (Fig. 7.6).

In addition to the dependence on shear rate, polymers such as poly-ethylene deviate from Newtonian behaviour in that at high shear rates the

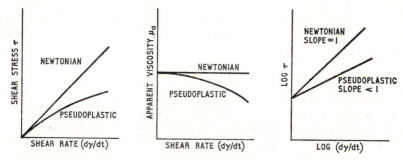

Fig. 7.6. Relationship between basic flow properties for Newtonian and pseudoplastic materials

extrudate takes on an irregular wavy or knobbly appearance. This is believed to be due to the fact that in the region of very high shear in the entry region to a die the melt tears or fractures allowing into the stream at that point polymer from less highly sheared regions and hence of higher apparent viscosity. The passage of such a melt of inhomogeneous viscosity through a die can well be expected to result in an irregular extrudate. This effect is known variously as 'bambooing', 'waviness', 'knobbliness', 'elastic turbulence', and 'melt fracture'. In polyethylene it normally occurs at shear stresses somewhat above 1.0×10^6 dyne cm^{-2}.

A further flow anomaly can sometimes be noted at much lower shear rates. Under certain conditions a mattness of the extrudate is noted and on closer examination ridges transverse to the extrudate direction are observed. This effect known as 'roughening' or 'sharkskin' is believed to be due to some form of slip-stick on the die and is very dependent on temperature and on the linear velocity of the melt through the die.

The influence of branching, molecular weight and molecular weight distribution on flow properties have been investigated.[12] It has been shown that branched polymers (low density polymers) have much lower melt viscosities than unbranched polymers of the same weight-average molecular weight. An unbranched (0·96 g cm^{-3}) polymer was found[8] to have a melt viscosity 50 times greater than that of a branched polymer of similar molecular weight. Experimental work indicates that polymer density in itself has little effect on the flow behaviour index although this value appears to be more temperature sensitive with low density polymers. It has also been indicated that polymers of higher densities can be processed at higher shear rates and extrusion pressures before melt irregularities occur.

For high density polyethylenes it has been found that the apparent viscosity (extrapolated to zero shear rate) is related to weight average by the equation

$$\log (\mu_a) = K + 3 \cdot 4 \log_{10} (M_w)$$

where M_w is the weight-average molecular weight.

This particular relationship has also been found to hold with many other thermoplastics such as polystyrene and polyisobutylene. In the case of branched (low density) polyethylenes better correlations have been obtained between viscosity and number-average molecular weight using the equation

$$\log (\mu_a) = A + BM_n^{1/2}$$

An increase in molecular weight increases susceptibility to 'melt fracture' and tends to decrease the flow behaviour index slightly. The effect of molecular weight distribution has been studied using both whole polymers and fractions. It has been found that with Phillips and Ziegler polymers molecular weight distribution has little effect on the flow properties at low shear rates but that at high shear rates whole polymers flow more readily than their fractions. This indicates that the wider the molecular weight distribution the lower the apparent viscosity. Fractionated polymers are found to have flow behaviour indices closer to unity than whole fractions. It was found that polymers of broad molecular weight distribution have higher critical shear rates before the onset of melt fracture and in addition

Table 7.4 EFFECT OF POLYMER STRUCTURE ON FLOW PROPERTIES

Effect of increase of:	*On viscosity*	*On flow behaviour index*	*On critical shear rate*	*On sharkskin*
Branching	decreases	little effect	decreases	?
Molecular weight	increases	slightly increases	decreases	?
Mol. Wt. distribution	decreases	decreases	increases	decreases

it has been noted that sharkskin effects are more prominent with polymers of narrow distribution and with a flow behaviour index approaching unity. Table 7.4 summarises the above observations.

The processing temperatures for polyethylene vary according to the polymer used and to the process employed and range from about 140°C to over 300°C in some paper-coating operations. In an inert atmosphere the polymer is stable at temperatures up to 300°C so that the high processing temperatures do not provide severe problems due to degradation providing contact of the melt with oxygen is reduced to a minimum.

7.5.3 Chemical properties

The chemical resistance of polyethylene is, to a large measure, that expected of a paraffin. It is not chemically attacked by non-oxidising acids, alkalis and many aqueous solutions. Nitric acid oxidises the polymer leading to a rise in power factor and to a deterioration in mechanical properties. As with the simple paraffins, halogens combine with the hydrocarbon by means of substitution mechanisms. When polyethylene is chlorinated in the presence of sulphur dioxide sulphonyl chloride as well as chlorine groups may be incorporated into the polymer (Fig. 7.7).

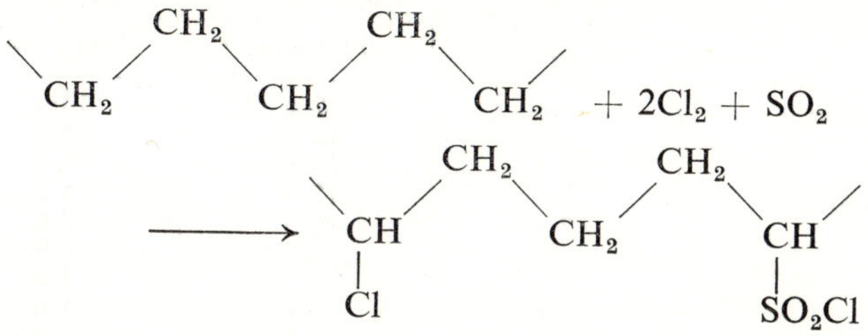

Fig. 7.7

This reaction is used to produce a useful elastomer (Hypalon, see Chapter 8).

Oxidation of polyethylene which leads to structural changes can occur to a measurable extent at temperatures as low as 50°C. Under the influence of ultraviolet light the reaction can occur at room temperature. The oxidation reactions can occur during processing and may initially cause a reduction in melt viscosity. Further oxidation can cause discoloration and streaking and in the case of polymers rolled for 1–2 hours on a two-roll mill at about 150°C the product becomes ropey and incapable of flow. It is rare that such drastic operating conditions occur but it is found that at a much earlier stage in the oxidation of the polymer there is a serious deterioration in power factor and for electrical insulation applications in particular it is necessary to incorporate antioxidants. It is to be expected that the less branched high density polythene, because of the small number of tertiary carbon atoms, would be more resistant to oxidation. That this is not always the case has been attributed to residual metallic impurities since purer samples of high density polymers are somewhat superior to the low density materials.

Since polyethylene is a crystalline hydrocarbon polymer incapable of specific interaction and with a melting point of about 100°C, there are no solvents at room temperature. Low density polymers will dissolve in benzene at about 60°C but the more crystalline high density polymers only dissolve at temperatures some 20–30 degC higher. Materials of similar solubility parameter and low molecular weight will however cause swelling, the more so in low density polymers (Table 7.5).

Low density polyethylene has a gas permeability in the range normally expected with rubbery materials (Table 5.10). This is because in the amorphous zones the free volume and segmental movements facilitate the passage of small molecules. Polymers of the Phillips type (density 0·96 g cm⁻³) have a permeability of about $\frac{1}{5}$ that of the low density materials.

Exposure of polyethylene to ultraviolet light causes eventual embrittlement of the polymer. This is believed to be due to the absorption of energy by carbonyl groups introduced into the chain during polymerisation

Table 7.5 ABSORPTION OF LIQUIDS BY POLYETHYLENES OF DENSITY 0·92 AND 0·96 g cm⁻³ AT 20°C AFTER 30 DAYS IMMERSION

Solvent	*Solubility parameter* δ *(cal cm⁻³)¹ᐟ²*	*% increase in weight in polymers*	
		0·92 g cm⁻³	*0·96 g cm⁻³*
Carbon tetrachloride	8·6	42·4	13·5
Benzene	9·2	14·6	5·0
Tetrahydrofuran	9·5	13·8	4·6
Petrol (B.P. 60–100°C)	—	12·8	5·8
Diethyl ether	7·4	8·5	2·6
Lubricating oil	—	4·9	0·95
Cyclohexanone	9·9	3·9	2·4
Ethyl acetate	9·1	2·9	1·6
Oleic acid	—	1·81	1·53
Acetone	10·0	1·24	0·79
Acetic acid	—	1·01	0·85
Ethanol	12·7	0·7	0·4
Water	23·4	<0·01	<0·01

and/or processing. The carbonyl groups absorb energy from wavelengths in range 2,200–3,200 Å. Fortunately very little energy from wavelengths below 3,000 Å strikes the earth's surface and so the atmosphere offers some protection. However in different climates and in different seasons there is some variation in the screening effect of the atmosphere and this can give rise to considerable variation in the outdoor weathering behaviour of the polymer.

When polyethylene is subjected to high energy irradiation gases such as hydrogen and some lower hydrocarbons are evolved, there is an increase in unsaturation and, most important, cross linking occurs by the formation of C—C bonds between molecules. The formation of cross-link points interferes with crystallisation and progressive radiation will eventually yield an amorphous but cross-linked polymer. Extensive exposure may lead to colour formation and in the presence of air surface oxidation will occur. Oxygen will cause polymer degradation during irradiation and this offsets the effects of cross linking. Long exposure to low radiation doses of thin film in the presence of oxygen may lead to serious degradation but with short exposure, high radiation doses and thicker specimens the degradation effects become less significant. Since cross linking is accompanied by a loss of crystallisation irradiation does not necessarily mean an increased tensile strength at room temperature. However at temperatures about 130°C irradiated polymer still has some strength (it is quite rubbery), whereas the untreated material will have negligible tenacity. It is found that incorporation of carbon black into polyethylene which is subsequently irradiated can give substantial reinforcement whereas corresponding quantities in the untreated product leads to brittleness.

If polyethylene is exposed to a mechanical stress in certain environments fracture of the sample occurs at stresses much lower than in the absence of the environment. As a corollary if a fixed stress, or alternatively fixed strain, is imposed on a sample the time for fracture is much less in the 'active environment' than in its absence. This phenomenon is referred to as *environmental stress cracking*. An example of this effect can be given by considering one of the tests used (the Bell Telephone Laboratory Test) to measure the resistance of a specific polymer to this effect. A small moulded rectangle is nicked to a fixed length and depth with a sharp blade and the nicked sample is then bent through 180° so that the nick is on the outside of the bend and at right angles to the line of the bend. The bent sample is held in a jig and immersed in a specific detergent, usually an alkyl aryl polyethylene glycol ether (e.g. Igepal CA) and placed in oven at 50°C. Low density polymers with an M.F.I. of 20 and above will often be observed to crack in an hour or two. Amongst materials which appear to be active environments are alcohols, liquid hydrocarbons, organic esters, metallic soap, sulphated and sulphonated alcohols, polyglycol ethers and silicone fluids. This is rather a formidible list and at one time it was thought that this would lead to some limitation in the use of polyethylene for bottles and other containers. However for a number of reasons this has not proved a problem except with high density homopolymers and the main reason for concern about the cracking phenomenon is in fact associated with cables when the polyethylene insulator is in contact with greases and oils.

The reason for the activity of the above named classes of liquids is not fully understood but it has been noted that the most active liquids are those which reduce the molecular cohesion to the greatest extent. It is also noticed that the effect is far more serious where biaxial stresses are involved (a condition which invariably causes a greater tendency to brittleness). Such stresses may be frozen-in as a result of molecular orientation during processing or may be due to distortion during use.

Different polyethylenes vary considerably in the environmental stress cracking resistance. It has been found that with low density polymers the Bell Test generally shows that the higher the molecular weight the greater the resistance, low density polymers with a melt flow index of 0·4 being immune to the common detergents. Narrow molecular weight distributions considerably improve resistance of a polymer of given density and average molecular weight. Large crystalline structures and molecular orientations appear to aggravate the problem. The effect of polymer density is somewhat complicated. The Bell Test is performed at constant strain and hence much higher stresses will be involved in the high density polymers. It is thus not surprising that these materials often appear to be inferior by this test but in constant stress tests different results may be expected. Paradoxically Phillips-type homopolymers have often been less satisfactory in service than indicated by the Bell Test.

It may seem surprising that low density, comparatively low molecular weight (M.F.I. 20) materials have been successfully used for detergent bottles in view of the stress cracking phenomenon. (Nevertheless higher molecular weight materials are usually used here, i.e. with an M.F.I. <0·7.) The reason for this lies in the fact that good processing conditions and good design result in low stresses being imparted to the products. Under these conditions stress cracking times are invariably longer than the required service life of the product.

7.5.4   Electrical properties

The insulating properties of polyethylene compare favourably with those of any other dielectric material. As it is a non-polar material properties such as power factor and dielectric constant are almost independent of temperature and frequency. Dielectric constant is linearly dependent on density and a reduction of density on heating leads to a small reduction in dielectric constant. Some typical data are given in Table 7.6.

Oxidation of polyethylene with the formation of carbonyl groups can lead to a serious increase in power factor. Antioxidants are incorporated into compounds for electrical applications in order to reduce the effect.

7.6   ADDITIVES

Although polyethylene can be, and indeed often is, used without additives a number may be blended into the polymer for various reasons. These additives can be classified as follows:

1. Fillers.
2. Pigments.
3. Flame retarders.
4. Slip agents.
5. Blowing agents.
6. Rubbers.
7. Cross-linking agents.
8. Antioxidants.
9. Carbon black.
10. Antistatic additives.

Fillers, important constituents of many plastics materials, are rarely used with polyethylene since they interfere with the crystallinity of the polymer and often give rather brittle products of low ductility. Carbon black has some reinforcing effect and is of use in cross-linked polymers. A number of *pigments* are available for use in polyethylene. The principal requirements of a pigment are that it should have a high covering power–cost ratio and that it should withstand processing and service conditions. In the case of polyethylene special care should be taken to ensure that the pigment does not catalyse oxidation, an effect observed with a number of pigments based on cobalt, cadmium and manganese. Other adverse effects have also been reported with hydrated chromic oxide, iron blues, ultramarine and anatase titanium dioxide. For electrical insulation

Table 7.6   ELECTRICAL PROPERTIES OF POLYETHYLENE

Volume resistivity	$>10^{20}$ Ω cm
Dielectric strength	700 kV/mm
Dielectric constant	
density $= 0.92$ g cm^{-3}	2·28
density $= 0.96$ g cm^{-3}	2·35
Power factor	~ 1–2×10^{-4}

applications pigments such as cobalt blues, which cause a rapid rise of power factor on ageing, should be avoided.

Polyethylene burns readily and a number of materials have been used as *flame retarders*. These include antimony trioxide and a number of halogenated materials.

Layers of low density polyethylene film often show high cohesion, or 'blocking', a feature which is often a nuisance on both processing and use. One way of overcoming this defect is to incorporate *slip agents* or *anti-blocking* agents such as oleamide. Polymers with densities of above 0·935 g cm^{-3} show good slip properties and slip agents are not normally required for these products.

Products with very low dielectric constant (about 1·45) can be obtained by the use of cellular polymers. *Blowing agents* such as pp'-oxy-bis-benzene sulphonyl hydrazide and azodicarbonamide are incorporated into the polymer. On extrusion the blowing agent decomposes with the evolution of gas and gives rise to a cellular extrudate. Cellular polyethylene is a useful dielectric in communication cables.

Although many *rubbery materials* show varying compatibility with polyethylene the only elastomeric materials used in commercial compounds are polyisobutylene (p.i.b.) and butyl rubber. Polyisobutylene was originally used as a 'plasticiser' for polyethylene but was later found also to improve the environmental stress cracking resistance. 10% p.i.b. in polyethylene gives a compound resistant to stress cracking as assessed by the severe B.T.L. test. Recent work[13] has indicated that within broad limits the higher the molecular weight of the p.i.b. the greater the beneficial effect. Very high molecular weight polyisobutylenes are however less effective possibly due to the difficulty in obtaining satisfactory blends. P.I.B. may or may not increase the 'ease of flow' of polyethylene this depending on the molecular weights of the two polymers. Because of its lower cost butyl rubber is preferred to polyisobutylene at the present time, its use in polyethylene being largely restricted to cable applications.

Vulcanised (cross-linked) polyethylene is being used for cable applications where service temperatures up to 90°C are encountered. Typical *cross-linking agents* for this purpose are peroxides such as dicumyl

Table 7.7 STRUCTURAL FORMULAE OF SOME ANTIOXIDANTS FOR POLYETHYLENE

2,6-tert.butyl-4-methyl phenol

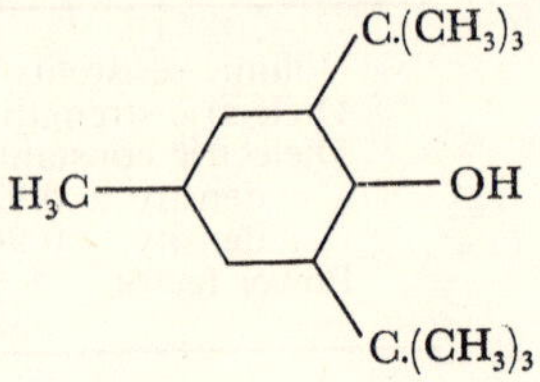

Di-o-cresylol propane

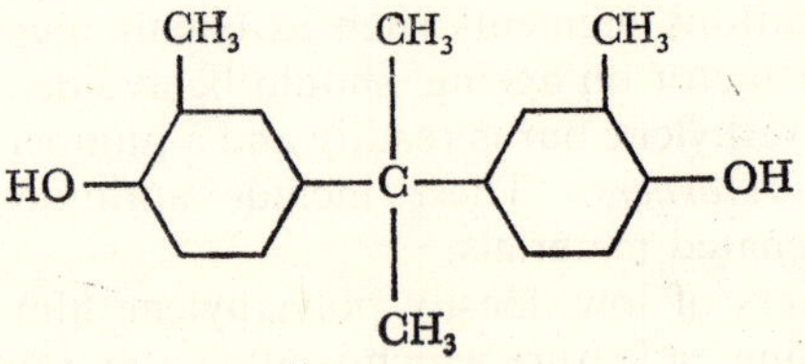

Di-β-naphthyl-p-phenylene diamine

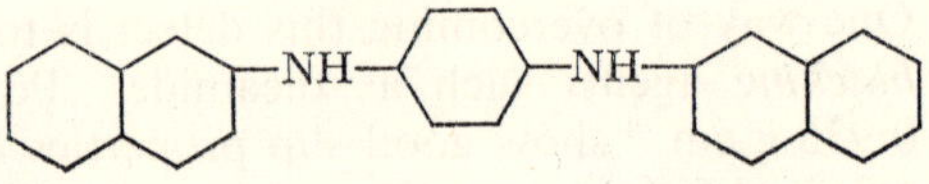

Diphenyl-p-phenylene diamine

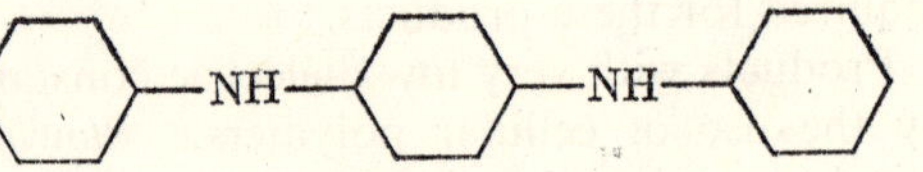

4:4-thio-bis(6-tert butyl-m-cresol)

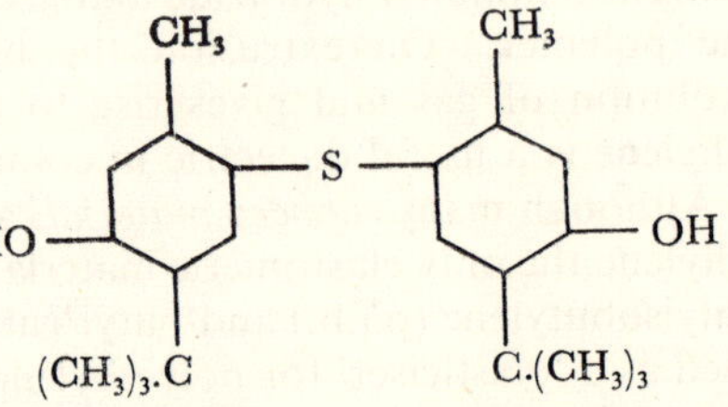

2.2-dihydroxy·3:3-di(α-methyl
cyclohexyl)-5:5-dimethyl diphenyl
methane
      (Nonox WSP—I.C.I.)

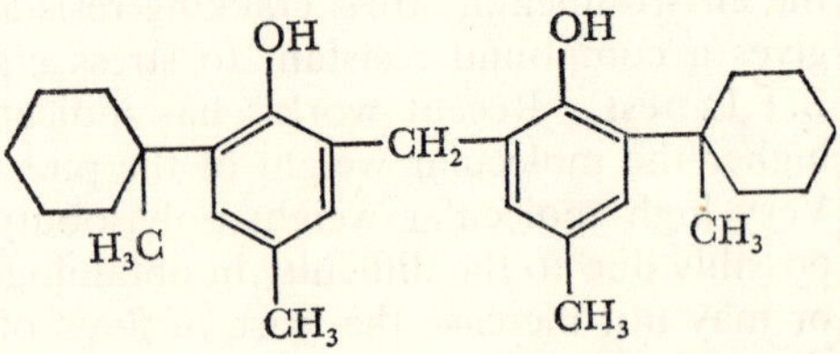

peroxide. The use of such agents is significantly cheaper than irradiation processes for the cross linking of the polymer.

When polyethylene is to be used in long term applications where a low power factor is to be maintained, and where it is desired to afford thermal protection during fabrication, *antioxidants* are incorporated into the polymer. These materials are of the type that were developed originally for diene rubbers and are basically amines or phenols. The four most commonly used at the present time are 2:2'-dihydroxy-3:3'-di-α-methyl cyclohexyl-5:5'-dimethyl-diphenyl methane (Nonox WSP), 2,6-tert.-butyl-4-methyl phenol, di-β-naphthyl-*p*-phenylene diamine and 4:4-thio-bis-6-tert.butyl-*m*-cresol. Other antioxidants important at one time but now obsolescent include di-*o*-cresylol propane and diphenyl-*p*-phenylene diamine (see Table 7.7).

Di-β-naphthyl-p-phenylene diamine has the lowest volatility of the above named antioxidants and is the most powerful antioxidant as measured by air oven ageing tests at 105°C. It does however colour the compound and is also liable to bleed out.

Nonox WSP is slightly more volatile and is less powerful as measured by air oven tests but more powerful when assessed by heating in aerated water at 93°C. It neither blooms, bleeds, nor discolours the compound. Whereas both of the above named antioxidants are adversely affected by the presence of carbon black the antioxidant efficiency of 4:4-thio-bis(6-tert.butyl *m*-cresol) is increased by the presence of the black. It is not a particularly good antioxidant as assessed by air aging at 105°C but in aerated water tests it is very similar to di-β-naphthyl-*p*-phenylene diamine.

A number of tests have been devised for measuring the efficiency of antioxidants. Samples may be aged by hot rolling at about 160°C, by air oven aging or by aging in aerated water. Changes in the polymer can be noted by measurements of such properties as carbonyl content (by infra-red measurements), gel content, melt flow index, oxygen uptake and power factor. Since the *raison d'etre* for incorporating antioxidants is largely to prevent an increase in power factor on ageing, power factor measurements are generally the most useful. In addition the property is very sensitive to oxidation. Fig. 7.8 shows the change in power factor of polyethylenes containing various antioxidants after air aging. Fig. 7.9 shows the effect of varying the antioxidant concentration. The sharp increase in power factor after an induction time during which little change occurs is to be noted particularly. In practice about 0·1% of antioxidant is employed in electrical grade compounds.

The weathering properties of polyethylene are improved by the incorporation of *carbon blacks*. Maximum protection is obtained using blacks with a particle size of 25 mμ and below. In practice finely divided channel or furnace blacks are used at 2–3% concentration and to be effective they must be very well dispersed into the polymer. The use of more than 3% black leads to little improvement in weathering resistance and may adversely affect other properties.

Antistatic additives are widely used to reduce dust attraction and also in films to improve handling behaviour on certain types of bag making and

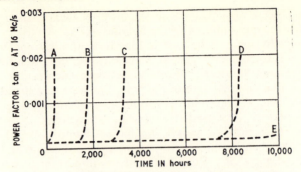

Fig. 7.8. *Oxidation of polyethylene in air at 105°C.*
Effect of adding 0·1% antioxidant on power factor.[1]
A, blank. B, diphenyl-p-phenylene diamine. C, 4:4-
thio-bis(6-tertbutyl-m-cresol). D, Nonox WSP. E,
di-β-naphthyl-p-phenylene diamine

packaging equipment. Whilst at one time it was more common to apply an antistatic agent to the surface by wiping, dipping or spraying, increasing use is now being made of antistatic agents which are incorporated into the polymer mass during normal compounding and which migrate to the surface with the passage of time. This approach has the advantage that the extra coating process is avoided and also that any layer of material removed by normal handling will be replaced by material which will migrate out of the mass. The selection of such antistats is critical and will depend particularly on the polymers used and on the thermal stability required. For a given polymer the agent should have a limited compatibility and a high diffusion rate in order to produce an antistatic layer as soon as possible after manufacture. Whereas quaternary ammonium compounds are widely used for polystyrene, polyethylene glycol alkyl esters would appear to be preferred to polyethylene compositions. The actual chemical composition of the rather small number of antistatic agents so far found suitable is rarely disclosed by the suppliers.

Polyethylene can be compounded on any of the standard types of mixing

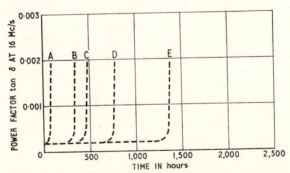

Fig. 7.9. *Oxidation of polyethylene in air at 105°C.*
Effect of antioxidant concentration (diphenyl-p-pheny-
lene diamine). A, blank. B, 50 p.p.m. C, 100 p.p.m.
D, 500 p.p.m. E, 1,000 p.p.m.

equipment used for visco-elastic materials. For laboratory purposes a two-roll mill is suitable; operating temperatures varying from about 90°C to about 140°C according to the type of polymer. Internal mixers such as the Banbury are widely used commercially but in recent years there has been increasing use made of continuous compounders such as the Werner and Pfleiderer ZSK.

7.7 PROCESSING

Although plastics materials may in principle be processed in a variety of physical states (in solution, in emulsion, as a paste or as a melt), melt processing is used almost exclusively with polyethylene. The main features to be borne in mind when processing the polymers are:

1. The low water absorption of the polymer avoids the necessity of pre-drying before processing except where hygroscopic additives are present.
2. The tendency of the material to oxidise in air particularly at melt temperatures means that the melt should come in contact with air as little as possible.
3. Although processing temperatures are low compared with many plastics the specific heat, which varies with temperature, is high.
4. The melt viscosity is highly non-Newtonian in that the apparent viscosity drops considerably with increasing shear rate. Melt viscosities are about the average encountered with plastics materials but there is a considerable variation between grades.
5. The high degree of crystallisation, which leads, among other things, to a high shrinkage on cooling.
6. The short polymer relaxation times (see page 127).
7. A rather sharp melting point.

Polyethylene is processed by large numbers of different techniques. There is insufficient space here to deal adequately with the principles of these processes or even with the behaviour of polyethylene in these processes. For this reason a list of books and other references dealing with both the processes and the fabrication of polyethylene products by these processes is given at the end of this chapter.

Compression moulding is only occasionally used with polyethylene. In this process the polymer is heated in a mould at about 150°C, compressed to shape and cooled. The process is slow since heating and cooling of the mould must be carried out in each cycle and it is only employed for the manufacture of large blocks and sheets; for relatively strain-free objects such as test-pieces, and where alternative processes cannot be used because of lack of equipment.

A very large number of products are produced by *injection moulding*. In this process the polymer is melted and injected into a mould which is at a temperature below the freezing point of the polymer so that the latter can harden. For mouldings with a minimum frozen-in strain operating conditions should be so selected that the mould cavity pressure drops to zero as the material sets.   Because of the tendency of the material to crystallise, high shrinkage values are observed ranging from 0·015–0·050 in/in with low density materials to 0·025–0·060 in/in with high density polymers.   High mould temperatures, desirable to reduce strains through freezing of oriented molecules, lead to increased shrinkage since there is more time available for crystallisation.   High cavity pressures reduce shrinkage since during much of the cooling point of the cycle there is only pressure reduction in the polymer and not physical contraction.   High cavity pressures also reduce packing near to the point of entry into the mould which can occur as the material in the cavity shrinks.   Since it is partially solidified material which is packed into the mould, and which often freezes whilst the molecules are oriented, a weakness of the part around the gate can occur and so 'packing' should be reduced as far as possible.   Melt temperatures are of the order of 160°C for low density polymers and up to 50 degC higher with high density materials.   In order to achieve these melt temperatures cylinder temperatures may be anything from 30 to 100 degC higher.

Many articles, bottles and containers in particular, are made by *blow moulding* techniques of which there are many variations.   In one typical process a hollow tube is extruded vertically downwards on to a spigot. Two mould halves close on to the extrudate (known in this context as the 'parison') and air is blown through the spigot to inflate the parison so that it takes up the shape of the mould.   As in injection moulding, polymers of low, intermediate and high density each find use according to the flexibility required of the finished product.

Another moulding process based on the extruder is '*extrusion moulding*'. Molten polymer is extruded into a mould where it sets.   Since satisfactory mouldings can be produced using low moulding pressures, cheap cast moulds can be used.   The process has been used to produce very large objects from polyethylene.   The techniques of screw-preplasticising with injection moulding can be considered as a development of this process.

Approximately three quarters of the polyethylene produced is formed into products by means of *extrusion* processes.   These processes will differ according to the product being made, i.e. according to whether the end-product is film, coated paper, sheet, tube, rod or wire covering.   In principle the extrusion process consists of metering polymer (usually in granular form) into a heated barrel in which a screw is rotating.   The rotation of the screw causes the granules to move up the barrel where they are compacted and plasticised.   The resultant melt is then forced under pressure through an orifice to give a product of constant cross section. Although the polymer may be processed on a variety of different machines screws usually have a length–diameter ratio in excess of 16:1 and a compression ratio of between 2·5:1 and 4:1.   (The compression ratio can be considered as the volume of one turn of the screw channel at the feed end to the volume of one turn at the delivery end of the barrel).   Since well over half the polyethylene extruded is converted into film, film extrusion processes will be considered in somewhat greater detail.   There are

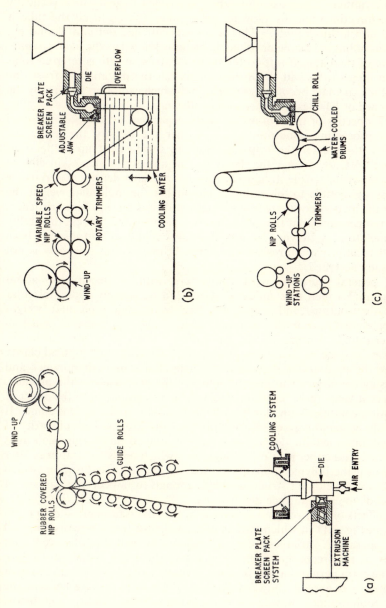

Fig. 7.10. Methods of producing polyethylene film: (a) tubular process using air cooling; (b) flat film process using water bath cooling; (c) flat film process using chill roll cooling[10]

basically two processes, the tubular process and the flat film process which are shown schematically in Fig. 7.10.

In the tubular process a thin tube is extruded (usually in a vertically upward direction) and by blowing air through the die head the tube is inflated into a thin bubble. This is cooled, flattened out and wound up. The ratio of bubble diameter to die diameter is known as the *blow-up ratio*, the ratio of the haul-off rate to the natural extrusion rate is referred to as the *draw-down ratio* and the distance between the die and the frost line (when the extrudate becomes solidified and which can often be seen by the appearance of haziness), the *freeze-line distance*.

The properties of the film are strongly dependent on the polymer used and on processing conditions. The higher the density the lower the flexibility, the greater the brittleness and to some extent up to densities of about 0·94, the greater the clarity (in the absence of sharkskin effects) and the lower the tensile strength at rupture.   Higher density materials are also less susceptible to 'blocking'. The higher the molecular weight the greater the melt viscosity, tensile strength and resistance to film brittleness at low temperatures but the lower the transparency and the less the ability of the melt to draw down.   Wide molecular weight distributions have been claimed to improve the resistance to film brittleness but this view is not universally accepted.

For general purpose work, polymers with a density of about 0·923 g cm^{-3} a melt flow index of about 2, a reasonably wide molecular weight distribution, and which are free from high molecular weight oxidised 'blobs', are most commonly used. The requirements of a film however differ according to the application. Whereas in some instances toughness may be of greater importance in other cases high optical clarity may be the paramount requirement. The choice of both polymer and processing conditions can greatly influence the properties of the product.

In recent years there has been extensive investigation[14] into the affect of processing conditions on clarity, haze and gloss of polyethylene film.   It can easily be demonstrated that the presence of haze and lack of clarity of low and intermediate density polymers is due to surface irregularities which can arise either as a melt roughness (which tends to disappear after extrusion if the polymer remains molten) or due to crystallisation (which though not developing structures large enough to impede the passage of light tends to cause surface distortions).

The effects of melt roughness and surface crystallisation are shown more clearly in Fig. 7.11 in which haze is plotted against freeze-line distance, all other operating variables being constant. It will be observed that initially as the freeze-line distance increases there is a reduction in haze since the extra period that the polymer is molten allows a reduction in melt roughness.   Eventually however there is an increase in haze because the longer cooling times allow larger crystal structures to build up and distort the surface.   An increase in extrusion temperature reduces haze because there is a reduction in melt imperfections and because the time available for crystallisation is reduced (although the total cooling time is the same the polymer will be above its melting point for a longer fraction of this time).   An increased output rate with constant freeze-line distance would increase melt imperfections but reduce surface crystallinity effects and thus shift the curve to the right.   The effects of most other operating variable can also be explained in terms of melt roughness and surface crystallinity.

The effect of extrusion conditions on the impact strength of tubular film has also been studied and found to be related to molecular orientation.

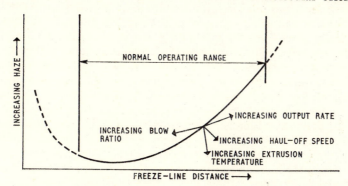

Fig. 7.11.   *Effect of freeze line distance and other operating variables on the haze of low density polyethylene film.*[9]   (*Reproduced by permission of ICI Plastics Division*)

Polyethylene molecules in the melt have a very short relaxation time (a measure of the time taken for molecules to coil after release of an orienting stress).   Thus in the tubular film process only molecules that have been oriented just before the melt freezes will remain in the oriented state. Because of this the order in which drawing down and transverse stretching of the film occur will affect the impact strength.   These factors can be adjusted by varying freeze-line distance, blow ratio and output rate, the shape of the bubble acting giving a guide to the sequence of events.

Although a large proportion of polyethylene film is made by the tubular process some film is produced by extruding flat film from a slit die either into a water bath or on to a chilled casting roll.   Although extrusion directly into water results in the most rapid quenching and tends to give products of highest clarity the presence of antistatic and slip additives tends to cause water to carry-over from the cooling bath and become trapped in the rolled film.   For this reason the chill-roll process is usually preferred.   Because of the higher cooling rates that are possible with chill-roll processes higher linear output rates can be achieved than with tubular processes.   In addition since high melt temperatures can be used (which reduce melt roughness) and quenching is carried out within an inch or two of the die lips (which reduces surface effects due to crystallinity) products of high clarity may be obtained.   In comparison the principal advantages of the tubular process are the ease with which bags and sacks may be made from the extruded tube and the ability to make wide sheet from fairly simple dies.

Calendering processes, of great importance in the production of sheet materials from p.v.c. compounds, are little used with polyethylene because of the difficulty in obtaining a smooth sheet.   Commercial products have however been made by calendering low density polymer containing a small amount of a peroxide such as benzoyl peroxide to give a stiff but crinkly sheet (Crinothene).   It is suitable for lampshades and other decorative applications.

Although the tonnage consumption is small compared with the amount of polyethylene consumed in injection moulding or extrusion, polyethylene is widely processed using a powder as the starting material.   This powder can be made either by precipitating the polymer by cooling a solution or by grinding processes.   The former method gives the finest powder whilst the latter is more economic.

The most well known of the *powder processes* is fluidised-bed coating. In this process a metal object which is to be coated with polyethylene is heated to about 160–250°C and then suspended in a fluidised bed of powdered polymer.   Fluidisation is brought about by blowing air through a porous base in the powder container so that in effect the individual particles are lubricated with a thin film of air.   Particles coming into contact with the hot metal fuse and adhere to the metal part.   This together with the adhering particles is then transferred to a second oven where the particles fuse together to give an even coating.   In a recent variation of this process the fluidised powder particles are electrically charged whilst the object to be coated is earthed.   This can then attract the charged particles without the need for preheating the part.

Larger objects may be coated by spraying techniques of which there are two basic variants, flame spraying and electrostatic spraying.   In the first process the powder is sprayed on to preheated metal with a flame spray gun and the coating is then heated to fuse the particles.   With electrostatic spraying the particles are charged as they leave the spray gun and are attracted to the object to be coated which is earthed.   It is claimed that thinner coatings are possible using electrostatic spraying techniques so that the method may in future years become a serious competitor of present day techniques of applying surface coatings.

Whereas dipping and spraying techniques require the use of fine powder, the coarser ground powders may be tolerated in powder moulding techniques.   Once again there are a number of variants of the techniques such as the Engel process but in principle they involve pouring powder into a heated low cost mould.   Some of the powder close to the mould fuses and sinters on to the mould while the excess is poured out.   The mould and the adhering powder are then heated until the powder forms a smooth continuous layer, the assembly is cooled and the resultant moulding removed from the mould.   This process is very suitable for large objects but because of the long process time is less suitable for small articles.   One variation of the process is rotational moulding.   In this case the required amount of powder is added to the mould which is completely closed and then rotated in an oven about two axes.   The powder melts and is distributed over the walls of the mould.   The mould is then cooled whilst the moulds are rotating.

Powdered polyethylene is also used to bond cloth interliners to garments.

Automobile carpets are now often made with a polyethylene rather than a latex backing because the polyethylene-backed carpet can be heat-formed so that the carpet will fit the shape of the car floor and cutting and sewing operations are consequently eliminated. It is common practice to print upon certain products such as films and bottles. Because polyethylene is chemically inert it is not possible to obtain a good ink adhesion directly on to the polymer surface. To overcome this problem the surface of the polymer is usually modified in order to provide an oxidised layer that will enable the ink to adhere. Modification may be carried out either by treatment with a naked flame or, more commonly, by subjecting the surface to a high voltage discharge. The presence of slip agents and antistatics may complicate the process and in the case of polyethylene film the modification is usually carried out immediately after extrusion before the wind-up stage and before an appreciable concentration of slip agent or antistatic additive has migrated to the surface. Once the surface has been modified it is possible to print on to the film without undue difficulty.

7.8 APPLICATIONS

Polyethylene was introduced initially as a special purpose dielectric material of particular value for high frequency insulation. With increasing availability the polymer subsequently began to be used for chemical plant and, to a small extent, for water piping. Since World War II there has been a considerable and continuing expansion in polyethylene production and this, together with increasing competition between manufacturers, has resulted in the material becoming available in a wide range of grades, most of which are sold in the lowest price bracket for plastics materials. The present position of polyethylene as a general purpose thermoplastic material is due in no small measure to the low cost and easy processability of the polymer.

The characteristics of polyethylene which lead to its widespread use may be summarised as follows:

1. Low cost.
2. Easy processability.
3. Excellent electrical properties.
4. Excellent chemical resistance.
5. Toughness and flexibility even at low temperatures.
6. Reasonable clarity of thin films.
7. Freedom from odour and toxicity.
8. A sufficiently low water vapour permeability for many packaging, building and agricultural applications.

To these could also be added the fact that a great quantity of information is available concerning the processing and properties of these materials and that it is a material whose properties are reasonably well known and understood by the public at large.

The limitations of the polymer are:

1. The low softening point.
2. The susceptibility of low molecular weight grades to environmental stress cracking.
3. The susceptibility to oxidation (however polyethylene is better in this respect than many other polymers).
4. The opacity of the material in bulk.
5. The wax-like appearance.
6. The poor scratch resistance.
7. The lack of rigidity (a limitation in some applications but a virtue in others).
8. The low tensile strength.
9. The high gas permeability.

For many purposes these limitations are not serious whilst in other cases the correct choice of polymer, additives, processing conditions and after-treatment can help considerably.

In 1964 it was estimated that about 223,000 tons of polyethylene were produced in Britain of which about 200,000 tons were of the low density type. The breakdown of end-use for that year in Britain has been estimated[15] and is given in Table 7.8.

From these figures it is seen that the largest end-use of the material is in the form of film and sheet. Polyethylene film has found ever-increasing use during the past few years for packaging, horticultural and building applications. More recently the all-polyethylene sack has become available and found to be particularly useful in that its superior weather resistance to paper sacks frequently enables filled polyethylene sacks to be stored out-of-doors. Another interesting development is the advent of polyethylene books and periodicals. At the development stage at the time of writing this outlet could be of use in the manufacture of maps, service manuals and other printed documents which may come into contact with grease, water or other chemicals. It is generally expected that film will be the fastest-growing polyethylene outlet during the next few years.

Polyethylene has also found wide use as an injection moulding material for such products as housewares, toys, chemical plant and electrical parts.

Table 7.8   END-USE OF POLYETHYLENE FOR 1964

End-use	*Low Density Polymer*	*High Density Polymer*
Film and sheet	42%	3%
Injection moulding	23%	27%
Wire and cable	9%	3%†
Blow moulding	11½%	48½%†
Pipe	5%	—
Monofil	—	12%†
Miscellaneous	9½%	3%

† Substantially Phillips-type copolymer (see Chapter 8).

Although the bulk of bottles produced in Britain are made from glass there has been a steadily increasing number of bottles made from plastics materials, some 80% or more from polyethylene. Such products are tough and are available in varying degrees of flexibility. Although the market for polyethylene as a bottle material may be expected to increase there may also be expected to be increased competition from polypropylene, rigid p.v.c. and polystyrene. Many large containers are also blow moulded from polyethylene.

The use of polyethylene as an electrical insulator, particularly for wire and cable work, has increased continuously over the years. Over 20,000 tons of the polymer were used in Britain for this purpose during 1964. Although the development of satellite communications could reduce the amount of material being used for submarine cables there is little doubt that insulation applications will continue to be an important outlet.

Although not suitable for conveying hot water, polyethylene pipe has found widespread use in domestic plumbing. Long term experience, exhaustive testing, resistance to damage by frozen water and public confidence are expected to lead to much wider application of the polymer for cold-water piping.

Whereas most of the packaging applications of polyethylene result from the use of film, polyethylene-coated paper is also important, particularly in sacks and milk cartons. The desirable properties of a polythene coated paper are low moisture permeability, high grease-proofness, good flexibility even at low temperature, heat sealability and high chemical inertness.

Although the above applications consume over 90% of the polyethylene produced there are a number of other important end-uses. Filament for ropes, fishing nets and fabrics are an important outlet for high density polyethylene; about 1,000 tons of powdered polymers are used for dip coating, flame spraying, rotational moulding and other outlets, whilst fabricated sheet is important in chemical plant.

In spite of the development of newer thermoplastics, the growth rate for polyethylene remains one of the highest for plastics materials. Whilst polypropylene and the other crystalline olefin polymers may capture some of the traditional polyethylene markets the dominant position held by polyethylene in the statistics of plastics production is likely to be enhanced in the next few years.

REFERENCES

1. VON PECHMANN, H., *Chem. Ber.*, **31**, 2643 (1898)
2. BAMBERGER, E., and TSCHIRNER, F., *Chem. Ber.*, **33**, 955 (1900)
3. CAROTHERS, W. H., HILL, J. W., KIRBY, J. E., and JACOBSON, R. A., *J. Am. Chem. Soc.*, **52**, 5279 (1930)
4. KOCH, H., and IBING, G., *Brennstoff-Chem.*, **16**, 141 (1955)
5. FAWCETT, E. W., and GIBSON, R. O., *J. Chem. Soc.*, 386 (1934)
6. *British Patent* 471590
7. ZIEGLER, K. E., *Angew. Chem.*, **67**, 426, 541 (1955)
8. PETICOLAS, W. L., and WATKINS, J. M., *Paper presented at 129th meeting of Am. Chem. Soc.*, Dallas, Texas (1956)
9. *Technical trade literature*, I.C.I. Plastics Ltd., Welwyn Garden City
10. SANDIFORD, D. J. H., and WILLBOURN, A. H., *Polythene* (Eds. RENFREW, D., and MORGAN, P.), Iliffe, London (2nd edn), Chapter 8 (1960)

11. *Technical trade literature*, British Resin Products Ltd., London
12. FERGUSON, J., WRIGHT, B., and HAWARD, R. N., *J. Appl. Chem.*, **14**, 53 (1964)
13. EXLEY, P. A., and STABLER, H. J., *A.P.I. Project Theses*
14. CLEGG, P. L., and HUCK, N. D., *Plastics*, **26** (282), 114 (1961); (283) 107 (1961)
15. *Brit. Plastics*, **38**, 2 (1965)

BIBLIOGRAPHY

Polyethylene

KRESSER, T. O. J., *Polyethylene*, Reinhold, New York (1960)
Crystalline Olefin Polymers (Eds. RAFF, R. A. V., and DOAK, K. W.), Interscience, New York (1964)
Polythene—the Technology and Uses of Ethylene Polymers (Eds. RENFREW, A., and MORGAN, P.), Iliffe Books Ltd., London, 2nd Edn. (1960)
SITTIG, M., *Polyolefin Resin Processes*, Gulf Publishing Co., Houston (1961)
TOPCHIEV, A. V., and KRENTSEL, B. A., *Polyolefines*, Pergamon Press, Oxford (1962)

Processing of Thermoplastics (General)

BERNHARDT, E. C., *Processing of Thermoplastic Materials*, Reinhold, New York (1959)
FISHER, E. G., *Extrusion of Plastics*, Iliffe Books Ltd., London, 2nd Edn. (1964)
MCKELVEY, J. A., *Polymer Processing*, John Wiley, New York (1962)
MUNNS, M. G., *Plastics Moulding Plant*, Vol. 2—*Injection Moulding Equipment*, Iliffe Books Ltd., London (1964)
SHARPE, D., *Trans. Plastics Inst.*, **29**, 54 (1961). (Discussion of fluidised bed coating technique)

8

Polyolefins other than Polyethylene, and Diene Rubbers

8.1 POLYPROPYLENE

Until the mid-1950's the only polyolefins of commercial importance were polyethylene, polyisobutylene and isobutylene–isoprene copolymers (butyl rubber). Attempts to produce polymers from other olefins had, at best, only resulted in the preparation of low molecular weight materials of no apparent commercial value.

In 1954 G. Natta of Milan following on the work of K. Ziegler in Germany found that certain 'Ziegler-type' catalysts were capable of producing high molecular weight polymers from propylene and many other olefins. By variations on the form of the catalysts used Natta was able to produce a number of different types of high molecular weight polypropylenes which differed extensively in their properties. One form, now known as isotactic polypropylene, was in many ways similar to high density polyethylene but with a higher softening point, rigidity and hardness, whilst another form, the atactic polymer, was amorphous and had little strength.

Although first reported only in 1954, isotactic polypropylene became commercially available from Montecatini (Moplen) in 1957 and by 1960 was being marketed by a number of companies including I.C.I. (Propathene) and Shell (Carlona P). It is now finding steadily increasing use in injection and blow moulding, packaging film, fibres and filaments and for a variety of other applications. None of the other new polyolefins have yet achieved commercial importance in the field of plastics but many exhibit behaviour which indicates that they have some potential value for specialist applications. Several copolymers containing ethylene have also recently become available and have already found a number of uses.

8.1.1 Preparation of polypropylene

There are many points of resemblance between the production of polypropylene and polyethylene using Ziegler-type catalysts. In both cases the monomers are produced by the cracking of petroleum products such as natural gas or light oils. For the preparation of polypropylene the C_3 fraction (propylene and propane) is the basic intermediate and this may be separated from the other gases without undue difficulty by fractional distillation. The separation of propylene from propane is rather more difficult and involves careful attention to the design of the distillation plant. For polymer preparation impurities such as water and methyl acetylene must be carefully removed. A typical catalyst system may be prepared by reacting titanium trichloride with aluminium triethyl, aluminium tributyl or aluminium diethyl monochloride in naphtha under nitrogen to form a slurry consisting of about 10% catalyst and 90% naphtha. The properties of the polymer are strongly dependent on the catalyst composition and its particle shape and size.

Propylene is charged into the polymerisation vessel under pressure whilst the catalyst solution and the reaction diluent (usually naphtha) are metered in separately. Reaction is carried out at temperatures of about 60°C for approximately 8 hours. In a typical process an 80–85% conversion to polymer is obtained. Since the reaction is carried out well below the polymer melting point the process involves a form of suspension rather than solution polymerisation. The polymer molecular weight can be controlled by the amount of hydrogen present, this gas acting as a chain transfer agent.

At this stage of the process the following materials are present in the polymerisation vessel:

1. Isotactic polymer.
2. Atactic polymer.
3. Solvent.
4. Monomer.
5. Catalyst.

The first step in separating these ingredients involves the transfer of the reaction mixture to a flash drum to remove the unreacted monomer which is purified (where necessary) and recycled. The residual slurry is centrifuged to remove the bulk of the solvent together with most of the atactic material which is soluble in the naphtha. The remaining material is then treated with an agent which decomposes the catalyst and dissolves the residue. A typical agent is methanol containing a trace of hydrochloric acid. The solution of residues in the methanol is removed by a centrifuging operation and the polymer is washed and dried at about 80°C. At this stage the polymer may be blended with antioxidants, extruded and cut into pellets. There are a number of variations in this basic process, many of which involve extra processes to reduce the atactic content of the polymer. A typical flow sheet for the manufacture of polypropylene is given in Fig. 8.1.

8.1.2   Structure and properties of polypropylene

Polypropylene is a linear hydrocarbon polymer containing little or no unsaturation. It is therefore not surprising that polypropylene and

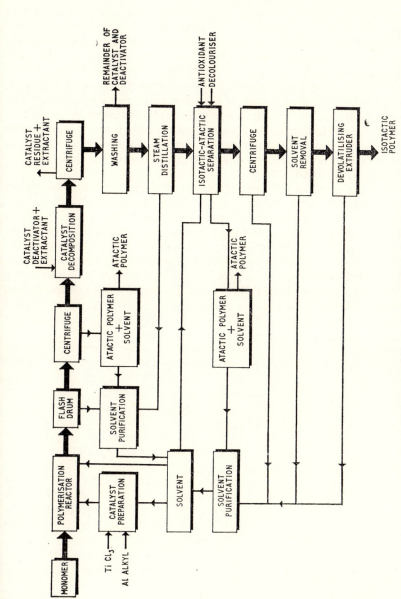

Fig. 8.1. Typical flow sheet for polypropylene manufacture

polyethylene have many similarities in their properties, particularly in their swelling and solution behaviour and in their electrical properties. In spite of the many similarities the presence of a methyl group attached to alternate carbon atoms on the chain backbone can alter the properties of the polymer in a number of ways. For example it can cause a slight stiffening of the chain and it can interfere with the molecular symmetry. The first effect leads to an increase in the crystalline melting point whereas the interference with molecular symmetry would tend to depress it. In the case of the most regular polypropylenes the net effect is a melting point some 50 degC higher than that of the most regular polyethylenes. The methyl side groups can also influence some aspects of chemical behaviour. For example the tertiary carbon atom provides a site for oxidation so that the polymer is less stable than polyethylene to the influence of oxygen. In addition thermal and high energy treatment leads to chain scission rather than cross linking.

The most significant influence of the methyl group is that it can lead to products of different tacticity, ranging from completely isotactic and syndiotactic structures to atactic molecules (see Chapter 4). The iso-tactic form is the most regular since the methyl groups are all disposed on one side of the molecule. Such molecules cannot crystallise in a planar zigzag form as do those of polyethylene because of the steric hindrance of the methyl groups but crystallise in a helix with three molecules being required for one turn of the helix. Both right-hand and left-hand helices occur but both forms can fit into the same crystal structure. Commercial polymers are usually about 90–95% isotactic. In these products, atactic and syndiotactic structures may be present either as complete molecules or as blocks of varying length in chains of otherwise isotactic molecules. Stereo-block polymers may also be formed in which a block of monomer residues with a right-handed helix is succeeded by a block with a left-handed helix. The frequency with which such changes in the helix direction occur can have an important influence on the crystallisation and hence the bulk properties of the polymer. In practice it is difficult to give a full descrip-tion of a specific propylene polymer although there has been marked progress in recent years. Many manufacturers simply state that their products are highly isotactic, others quote the polymer crystallinity obtained after some specified annealing treatment, whilst others quote the so-called 'isotactic index', the percentage of polymer insoluble in *n*-heptane. Both of these last two properties provide some rough measure of the isotacticity but are both subject to error. For example the iso-tactic index is affected by high molecular weight atactic polymer which is insoluble in *n*-heptane and by the presence of block copolymers of isotactic and atactic structures which may or may not dissolve according to the proportion of each type present.

In spite of these problems the general effects of varying the degree of isotacticity are well known. Whereas the atactic polymer is an amor-phous somewhat rubbery material of little value, the isotactic polymer is stiff, highly crystalline and with a high melting point. Within the range of commercial polymers the greater the amount of isotactic material the greater the crystallinity and hence the greater the softening point, stiffness, tensile strength, modulus and hardness, all other structural features being equal (Fig. 8.2).

The influence of molecular weight on the bulk properties of polypropylene is often opposite to that experienced with most of other well-known polymers. Although an increase in molecular weight leads to an increase

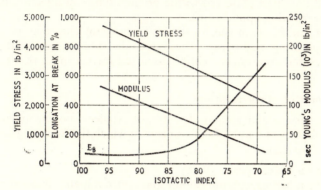

Fig. 8.2. Effect of isotacticity on tensile properties. (Reproduced by permission of ICI Plastics Division)

in melt viscosity and impact strength, in accord with most other polymers, it also leads to a lower yield strength, lower hardness, lower stiffness and softening point. This effect is believed to be due to the fact that high molecular weight polymer does not crystallise so easily as lower molecular weight material and it is the differences in the degree of crystallisation which affect the bulk properties. It may also be mentioned that an increase in molecular weight leads to a reduction in brittle point (see Table 8.1).

Only a limited amount of information is available concerning the effects of molecular weight distribution. There is however evidence that the narrower the distribution the more Newtonian are the melt flow properties. It has been noted that with polymers of molecular weights suitable for moulding and extrusion, polymers of wide distribution (M_w/M_n e.g. about 6) are stiffer and more brittle than those with a M_w/M_n ratio of about 2.

The morphological structure of polypropylene is rather complex and at least four different types of spherulite have been observed. The properties of the polymer will depend on the size and type of crystal structure formed and this will in turn be dependent on the relative rates of nucleation to crystal growth. The ratio of these two rates can be controlled by varying the rate of cooling and by the incorporation of nucleating agents. In general the smaller the crystal structures the greater the transparency and flex resistance and the less the rigidity and heat resistance.

One unfortunate characteristic property of polypropylene is the dominating transition point which occurs at about 0°C with the result that the polymer becomes brittle as this temperature is approached. Even at room temperature the impact strength of some grades leaves something to be desired. Products of improved strength and lower brittle points may be obtained by block copolymerisation of propylene with small amounts (2–3%) of ethylene. Such materials are now commercially available (known variously as polyallomers or just as propylene copolymers) and are often preferred to the homopolymer in injection moulding and bottle blowing applications.

Further variations in the properties of polyethylenes may be achieved by incorporating additives. These include rubbers, antioxidants and glass fibres and their effects will be discussed further in Section 8.1.4.

8.1.3   Properties of polypropylene

Although very similar to high density polyethylene, polypropylene differs from the former in a number of respects of which the following are among the most important:

1. It has a lower density (0·90).
2. It has a higher softening point and hence a higher maximum service temperature. Articles can withstand boiling water and be subject to many steam sterilising operations. For example mouldings have been sterilised in hospitals for over 1,000 hours at 135°C in both wet and dry conditions without severe damage.
3. Polypropylene appears to be free from environmental stress cracking problems.
4. It has a higher brittle point.
5. It is more susceptible to oxidation.

As shown in the previous section the mechanical and thermal properties of polypropylene are dependent on the isotacticity, the molecular weight and on other structure features. The properties of five commercial materials (all made by the same manufacturer and subjected to the same test methods) which are of approximately the same isotactic content but which differ in molecular weight and in being either homopolymers or block copolymers are compared in Table 8.1.

The figures in Table 8.1 show quite clearly how an increase in molecular weight (decrease in melt flow index) causes a reduction in tensile strength, stiffness, hardness and brittle point but an increase in impact strength. The general effects of isotactic index and melt flow index on some mechanical and thermal properties are also shown graphically in Figs. 8.3–8.6.[1]

Many features of the processing behaviour of polypropylene may be predicted from consideration of thermal properties. The specific heat of polypropylene is lower than that of polyethylene but higher than that of polystyrene. Therefore the plasticising capacity of an injection moulding machine using polypropylene is lower than when polystyrene is used but generally higher than with a high density polyethylene.

Studies of melt flow properties of polypropylene indicate that it is more non-Newtonian than polyethylene in that the apparent viscosity declines

Table 8.1 SOME MECHANICAL AND THERMAL PROPERTIES OF COMMERCIAL POLYPROPYLENES

Property	Test method	Homopolymers			Copolymers	
Melt flow index	(a)	3·0	0·7	0·2	3·0	0·2
Tensile strength (lb in⁻²)	(b)	5,000	4,400	4,200	4,200	3,700
Elongation at break (%)	(b)	350	115	175	40	240
Flexural modulus (lb in⁻²)	—	190,000	170,000	160,000	187,000	150,000
Brittleness temperature (°C)	I.C.I./ASTM D.746	+15	0	0	−15	−20
Vicat softening point (°C)	BS 2782	145–150	148	148	148	147
Rockwell hardness (R-scale)	—	95	90	90	95	88·5
Impact strength (ft lb)	(c)	10	25	34	34	42·5

(a) Standard polyethylene grader: load 2·16 kg at 230°C.
(b) Straining rate 18 in/min.
(c) Falling weight test on 14 in diameter moulded bowls at 20°C.

more rapidly with increase in shear rate. The melt viscosity is also more sensitive to temperature. Van der Wegt[2] has shown that if the log (apparent viscosity) is plotted against log (shear stress) for a number of polypropylene grades differing in molecular weight, molecular weight distribution and measured at different temperatures the curves obtained have practically the same shape and differ only in position.

The standard melt flow index machine is often used for characterising the flow properties of polypropylene and to provide a rough measure of molecular weight. Under the conditions normally employed for polyethylene (2·16 kg load at 190°C) the flow rate is too low for accurate measurement and in practice higher loads, e.g. 10 kg, and/or higher temperatures are used. It has been found[3] that a considerable pressure drop exists in the barrel so that the flow towards the end of a test run is higher than at the beginning.

The moulding shrinkage of polypropylene is less than that experienced with polyethylenes but is dependent on such processing factors as mould temperature, melt temperature and plunger dwell time. In general conditions which tend to reduce the growth of crystal structures will tend to reduce shrinkage, for example low mould temperatures will encourage quenching of the melt. It is also found that low shrinkage values are obtained with high melt temperatures. This is probably due to the fact that high melt temperatures lead to a highly disordered melt whereas some molecular order may be present in melts which have not been heated much above the crystalline melting point. Such regions of order would provide sites for crystal nucleation and hence crystallisation would be more rapid when cooling was carried out.

The electrical properties of polypropylene are very similar to those of high density polyethylenes. In particular the power factor is critically dependent on the amount of catalyst residues in the polymer. Some typical properties are given in Table 8.2 but it should be noted that these

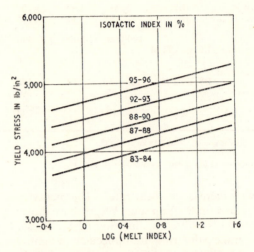

Fig. 8.3. *Variation of tensile yield stress with melt flow index (10 kg load at 190°C) and isotactic index. (After Crespi and Ranelli[1])*

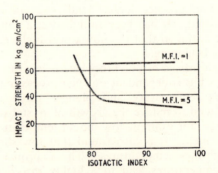

Fig. 8.4. *Variation of impact strength with melt flow index (10 kg at 190°C) and isotactic index. (After Crespi and Ranelli[1])*

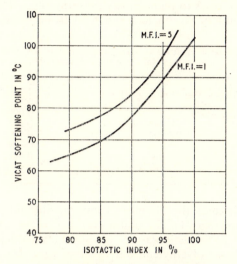

Fig. 8.5. *Variation of vicat softening point (5 kg load) with isotactic index and melt flow index. (After Crespi and Ranelli[1])*

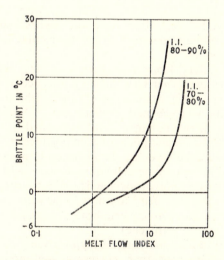

Fig. 8.6. *Variation of ASTM D746 brittle point with melt flow index and isotactic index. (After Crespi and Ranelli[1])*

properties are dependent on the antioxidant system employed as well as on the catalyst residues.

As with electrical properties the chemical resistance of polypropylene shows many similarities to high density polyethylene. The two polymers have similar solubility parameters and tend to be swollen by the same

Table 8.2 SOME TYPICAL ELECTRICAL PROPERTIES OF A HIGH HEAT STABILITY GRADE OF POLYPROPYLENE[4]

Dielectric constant at 5×10^6 c/s	2·25
Volume resistivity (Ω cm)	$>10^{17}$
Power factor at 10^2 c/s	0·0009
10^3 c/s	0·001
10^4 c/s	0·0009
10^5 c/s	0·006
10^6 c/s	0·0004
5×10^6 c/s	0·0005

liquids.   In both cases the absence of any possible interaction between the crystalline polymer and the liquid prevents solution of the polymers in any liquids at room temperature.   In some instances polypropylene is more affected than polyethylene but in other cases the reverse is true. Similar remarks may be made concerning the permeability of the two polymers to liquids and gases.   With many permeants polypropylene shows the lowest permeability but not, for example with hexane.   It may be mentioned in this context that although high density polyethylene is usually intermediate between low density polyethylene and polypropylene, where the permeant causes stress cracking (as with a silicone oil), the high density polyethylenes often have the highest permeability.   The fact that polypropylene is resistant to environmental stress cracking has already been mentioned.

Polypropylene differs from polyethylene in its chemical reactivity because of the presence of tertiary carbon atoms occurring alternately on the chain backbone.   Of particular significance is the susceptibility of the polymer to oxidation at elevated temperatures.   Some estimate of the difference between the two polymers can be obtained from Fig. 8.7[5] which compares the rates of oxygen uptake of each polymer at 200°F.   Substantial improvements can be made by the inclusion of antioxidants and such additives are used in all commercial compounds.   Whereas polyethylene cross links on oxidation, polypropylene degrades to form lower molecular weight products.   Similar effects are noted when the polymer is exposed to high energy radiation and when heated with peroxides (conditions which will cross link polyethylene).

Although a crystalline polymer, polypropylene mouldings are less opaque when unpigmented than corresponding mouldings from high density polyethylene.   This is largely due to the fact that the differences between the amorphous and crystal densities are less with polypropylene ($0 \cdot 85$ and $0 \cdot 94$ g cm^{-3} respectively) than with polyethylene (see Chapter 6). Biaxially stretched film has a high clarity since layering of the crystalline structures reduces the variations in refractive index across the thickness of the film and this in turn reduces the amount of light scattering.[6]

Biaxial stretching also leads to polymers of improved tensile strength. The effect of increasing the amount of stretching on the tensile strength and breaking elongation are given in Table 8.3.

There are other differences between cast, monoaxially oriented, and balanced biaxially oriented film. Typical figures illustrating these effects are given in Table 8.4.

When film is produced by air-cooled tubular blowing methods cooling rates are slower and larger degrees of crystallinity result. Hence tubular

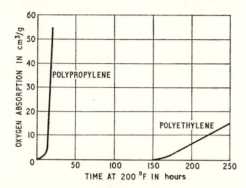

Fig. 8.7. Comparison of oxidation rates of unstablised polyethylene and polypropylene. (After Kresser[5])

film is slightly stronger in tension but has lower impact strength, tear strength and breaking elongation. The film also has more haze, and less gloss but somewhat better gas barrier properties and greater freedom from blocking.

8.1.4 Additives

Commercial grades of polypropylene may be blended with a number of other additives. Of these the most important are

1. Pigments.
2. Carbon black.
3. Glass fibres.
4. Rubbers.
5. Antioxidants.

The function and requirement of pigments and carbon black are almost identical to those with polyethylene. Glass fibres are included in special grades to improve the stiffness. The presence of up to 30% glass fibre has little effect on flow, tensile strength and hardness. Butyl and other rubbers have been blended with polypropylene to reduce the brittleness of

Table 8.3   EFFECT OF PERCENTAGE STRETCH ON TENSILE
PROPERTIES OF POLYPROPYLENE FILM[7]

Stretch (%)	Tensile strength (lb in^{-2})	E_B (%)
None	·5,600	500
200	8,400	250
400	14,000	115
600	22,400	40
900	23,800	40

the latter but with the advent of the block polymers their use has diminished.

Antioxidants are necessary components of all polypropylene compounds and the selection of such ingredients is an important factor in
determining the success of a given commercial material. A number of

Table 8.4   COMPARISON OF CAST, MONOAXIALLY ORIENTED AND BIAXIALLY ORIENTED
POLYPROPYLENE FILM[8]

Property	Cast polymer	Monoaxially oriented	Balanced oriented
Tensile strength (lb in^{-2})			
Machine direction	5,700	8,000	26,000
Transverse direction	3,200	40,000	22,000
Elongation at break (%)			
Machine direction	425	300	80
Transverse direction	300	40	65
ASTM D.523 gloss			
(45° head)	75–80	>80	>80
Low temperature brittleness	brittle at °C	excellent	excellent
Coefficient of friction	0·4	0·4–0·5	0·8

antioxidants used in rubber compounds, generally amines, phenols or
thio-compounds have proved of value as also have some polyethylene
antioxidants and, rather surprisingly, some well known p.v.c. stabilisers
such as calcium stearate. Best results are often obtained where more than
one stabiliser is used. For example one stabiliser may be used to stabilise
the material to processing conditions and another to stabilise it against
long-term aging. In some cases the long-term antioxidant has been
found to be susceptible to volatilisation or de-activation during processing
but this can be reduced by careful selection of the second antioxidant. Examples of antioxidants include dilauryl thiodipropionate, and
tris(2-methyl-4-hydroxy-5t-butyl phenyl) butane.

8.1.5   Processing characteristics

Polypropylene may generally be processed by methods very similar to
those used with the polyethylenes, particularly high density polyethylene.
The main differences are the lower specific heat and the greater sensitivity
of flow properties to temperature and shear rate. The moulding shrinkage

is lower than with polyethylene but higher than with polystyrene. The effect of injection moulding variables on the shrinkage has already been discussed. Most processing operations involve the use of melt temperatures in the range 210–250°C. Because of the tendency of polypropylene to oxidise heating times should be kept down to a minimum.

8.1.6 Applications

Although it is inevitable that polypropylene will be compared more frequently with polyethylene than with any other polymer its use as an injection moulding, blow moulding and extrusion material also necessitates comparison with polystyrene and related products, cellulose acetate and cellulose acetate-butyrate, each of which has a similar rigidity. When comparisons are made it is also necessary to distinguish between conventional homopolymers and the special copolymers now being marketed. A somewhat crude comparison between these different polymers is attempted in Table 8.5 but further details should be sought out from the appropriate chapters dealing with the other materials.

Such a table must be subject to certain generalisations and in special instances there may be some variations in the above ratings. There are also many other factors which must also be taken into account and the choice of a particular polymer for a given application will depend on a careful study of the product requirements and the properties of potential materials. In spite of their recent introduction polypropylene homopolymers and copolymers have found applications for mouldings where

Table 8.5 COMPARATIVE RATINGS OF VARIOUS POLYMERS IN RESPECT OF SOME SELECTED PROPERTIES

Polymer	Clarity of moulding	Temperature resistance	Toughness
Polypropylene (homopolymer)	opaque	A	F–G
Polypropylene (copolymer)	opaque	B	D–F
Polyethylene (high density)	opaque	D	D–F
Polystyrene	clear	E	H
High impact polystyrene	opaque	F	D–G
ABS polymers	opaque	C	A
Cellulose acetate	clear	G	B–C
C.A.B.	clear	H	B–C

A indicates the most favourable material, H the least favourable.

such properties as good appearance, sterilisability, environmental stress cracking resistance and good heat resistance are of importance. One further particularly useful property of polypropylene is the excellent resistance of thin sections to continued flexing. This has led to the introduction of a number of one piece mouldings for boxes, cases and automobile accelerator pedals in which the hinge is an integral part of the moulding. It was estimated that in 1963 in both Britain and the United States about half of the polypropylene produced was processed by injection moulding. The special copolymer grades with their higher impact strength and lower brittle point are rapidly absorbing a large part of this market. Typical mouldings include hospital sterilisable equipment, luggage, stacking chairs, washing machine parts, toilet cisterns and various car parts such as dome lights, kick panels, door frame parts and accelerator pedals.

Film is another important outlet for polypropylene. When made by a chill-roll casting process a transparent material similar to regenerated cellulose (Cellophane) film may be obtained. Such material is also available in a number of oriented and laminated forms and is finding application as a packaging material where its optical properties, heat resistance and chemical resistance can be exploited to advantage.

Polypropylene monofilaments combine low density with a high tenacity and good abrasion resistance and are finding some application in ropes, brush tufting and netting. There is also steadily increasing interest in polypropylene fibres and a wide range of different types of fabric have been produced. It has been estimated that in 1963 30% of British production of polypropylene was used in fibre applications.

The polymer has found some small scale outlets in other directions such as sheet, pipe, wire coating and blow moulding. Consumption of the polymer in these directions is however dependent on finding applications for which polypropylene is the most suitable material.

8.2 OTHER POLYOLEFINS

A number of polymers have been produced from higher olefins using catalysts of the Ziegler–Natta type. In addition polyisobutylene which may be produced by cationic polymerisation has been commercially available for over thirty years.

Fig. 8.8 shows the effect of increasing the length of the side chain on the

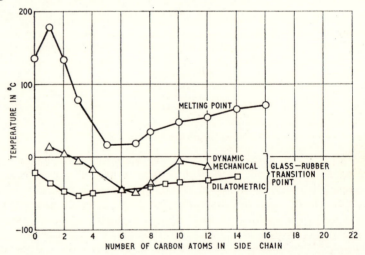

Fig. 8.8. Effect of side chain branching on the melting point and glass transition temperature of polyolefins ($-CHR-CH_2-)_n$ (R straight chain)[9]

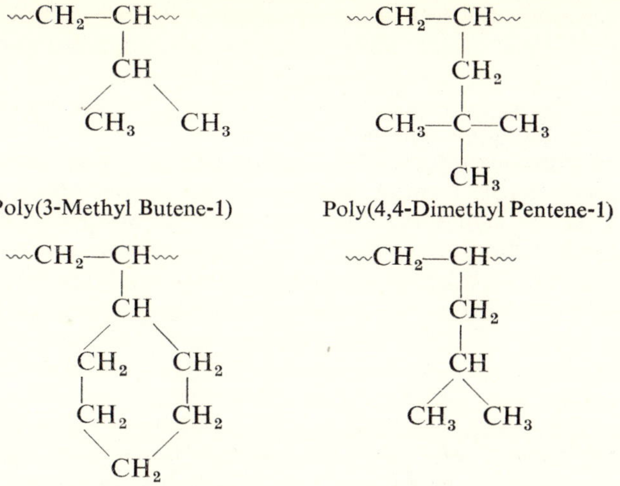

~~CH₂—CH~~   ~~CH₂—CH~~
       | |
      CH   CH₂
     ╱   ╲ |
  CH₃   CH₃   CH₃—C—CH₃
                                             |
                                            CH₃

Poly(3-Methyl Butene-1)   Poly(4,4-Dimethyl Pentene-1)

~~CH₂—CH~~   ~~CH₂—CH~~
       | |
      CH   CH₂
     ╱   ╲ |
  CH₂   CH₂   CH
   | | ╱   ╲
  CH₂   CH₂   CH₃   CH₃
     ╲   ╱
      CH₂

Poly(Vinyl Cyclohexane)   Poly(4-Methyl Pentene-1)

Fig. 8.9

melting point and glass transition temperature of a number of poly-α-olefins. As discussed previously the melting point of isotactic poly-propylene is higher than that of polyethylene because the chain stiffness of the polymer has a more dominating influence than the reduction in symmetry. With an increase in side chain length (polybutene-1 and polypentene-1) molecular packing becomes more difficult and with the increased flexibility of the side chain there is a reduction in the melting point. A lower limit is reached with polyoctene-1 and polynonene-1 and with polymers from higher α-olefins the melting point increases with increase in the length of the side chain. This effect has been attributed to side chain crystallisation. It is interesting to note that a polyolefin with n carbon atoms in the side chain frequently has a similar melting point to a paraffin with $2n$ carbon atoms. Published data[9] on glass transition temperatures show similar but less dramatic changes.

A number of polyolefins with branched side chains have also been pre-pared (see Fig. 8.9). Because of their increased cohesive energy, ability for the molecules to pack and because of the effect of increasing chain stiffness some of these polymers have a very high melting point. For example poly(3-methyl butene-1) melts at about 240°C and poly(4,4-dimethyl pentene-1) is reported to have a melting point of between 300°C and 350°C. Certain cyclic side chains can also lead to high melting polymers, for example, poly(vinyl cyclohexane) melts at 342°C.

None of the polymers from unbranched olefins, other than ethylene or propylene have yet become of importance as plastics materials although some of them are of interest as both adhesives and release agents. One limitation of a number of these materials is their tendency to undergo complex morphological changes on standing with the result that fissures and planes of weakness may develop.

Of the polymers from branched olefins *polyisobutylene*, (p.i.b.) is the most well known. This material has characteristics intermediate between those of a viscous liquid and a rubber. This means that the polymer is susceptible to extensive cold flow and therefore its use is restricted to non-self supporting applications such as adhesives, paper and fabric coatings and calking compounds. When copolymerised with about 2% of iso-prene, isobutylene forms a valuable vulcanisable rubber known as butyl rubber (see Section 8.4). The large scale preparation of this copolymer results in it being much cheaper than polyisobutylene and as a consequence is often used in place of p.i.b. even where rubbery properties are not specifically required.

The densities of the amorphous and crystalline zones of *poly(4-methyl pentene-1)* are similar ($\sim$0·83) so that the polymer has a high clarity even when large spherulitic structures are present. Its high clarity com-bined with a high crystalline melting point of 240°C leading to a Vicat softening point of 179°C has resulted in its introduction on a commercial development scale by I.C.I. in 1965 (TPX-R polymers). Whilst tensile yield strength and modulus are similar to those obtained with polypropy-lene, poly(4-methyl-pentene-1) is generally more liable to brittle failure (see also Table 8.7).

8.3   COPOLYMERS CONTAINING ETHYLENE

Many monomers have been copolymerised with ethylene using a variety of polymerisation systems, in some cases leading to commercial pro-ducts. Copolymerisation of ethylene with other olefins leads to hydro-carbon polymers with reduced regularity and hence lower density, inferior mechanical properties, lower softening point and lower brittle point.

Table 8.6   COMPARISON OF MAJOR PROPERTIES OF COPOLYMERS WITH HOMOPOLYMERS

	Copolymer		*Homopolymer*	
Density	0·95	0·95	0·96	0·96
Melt flow index	0·3	4·0	0·2	3·5
Tensile strength (lb in^{-2})	3,600	3,600	4,400	4,400
Elongation (%)	70	30	30	15
Vicat softening point (°C)	255	255	260	260
Environmental stress cracking (F_{50} hr)	400	20	60	2
Izod impact (ft/lb in. notch)	4	0·8	5	1·5

Two random copolymers of this type are of importance, ethylene–propy-lene copolymers and ethylene–butene-1 copolymers. The use and properties of polypropylene containing a small quantity of ethylene in stereo-blocks within the molecule has already been discussed. Although referred to commercially as ethylene–propylene copolymers these materials are essentially slightly modified polypropylene. The random ethylene-propylene polymers are rubbery and are discussed further in Section 8.4.

Of the all-olefin copolymers the ethylene–butene-1 products are of most interest to the plastics industry.[10] Such copolymers, prepared by the

Phillips process, were introduced commercially in September 1958 an immediately found applications in blown containers and injection mouldings. Table 8.6 compares some of the more important properties of two grades of such copolymers with two grades of Phillips-type homopolymer.

From this table it will be noted that in terms of the mechanical and thermal properties quoted the copolymers are marginally inferior to the homopolymers. They do however show a marked improvement in resistance to environmental stress cracking. It has also been shown that the resistance to thermal stress cracking and to creep are better than with the homopolymer.[11] This has led to widespread use in detergent bottles, pipes, monofilaments and cables.

Ethylene has also been copolymerised with a number of non-olefinic monomers and of the copolymers produced those with vinyl acetate have so far proved the most significant commercially. The presence of vinyl acetate residues in the chain reduces the polymer regularity and hence by the vinyl acetate content the amount of crystallinity may be controlled. Copolymers based on 45% vinyl acetate are rubbery and may be vulcanised with peroxides. These are not of commercial value as yet. Copolymers with about 30% vinyl acetate residues (Elvax-Du Pont) are flexible resins soluble in toluene and benzene at room temperature and with a tensile strength of about 1,000 lb in^{-2} and a density of about 0·95. Their main uses are as wax additives and as adhesive ingredients.

Copolymers with a somewhat lower vinyl acetate content have also been introduced, these materials having greater flexibility and toughness than low density polyethylene. They are also immune to environmental stress cracking. It is to be expected that these materials will find uses where flexibility approaching that found in plasticised p.v.c. is desired but where the problems associated with the use of plasticisers, such as leaching, are to be avoided. Rather similar products may be produced by copolymerisation of ethylene and ethyl acrylate. Commercial polymers of this type (Zetafin-Dow) are said to be similar to plasticised p.v.c. but with greater processing stability.

In September 1964 the Du Pont company announced that they were producing materials that had characteristics of both thermoplastics and thermosetting materials. These materials, known as *ionomers*, are prepared by copolymerising ethylene with a small amount (1–10% in the basic patent) of an unsaturated carboxylic acid such as methacrylic acid using the high pressure process. Such copolymers are then treated with the derivative of a metal such as sodium methoxide or magnesium acetate with the result that the carboxylic group appears to ionise. It would seem that this leads to some form of ionic cross-link which is stable at normal ambient temperatures but which reversibly breaks down on heating. In this way it is possible to obtain materials which possess the advantages of cross-linking at ambient temperatures, for example they have enhanced toughness and stiffness, but which behave as linear polymers at elevated temperatures and may be processed and even reprocessed without undue difficulty. In the case of the commercial materials already

Table 8.7 PROPERTIES OF THREE RECENTLY INTRODUCED OLEFIN POLYMERS. POLY(4-METHYL PENTENE-1), AN IONOMER (SURLYN A) AND AN ETHYLENE-VINYL ACETATE COPOLYMER (ALKATHENE VJG 501)

Property	Poly(4-methyl pentene-1)	Ionomer	EVA copolymer
Specific gravity	0·83	0·93	0·926
Yield strength (10^3 lb in^{-2})	4	2–2	1·05
Tension modulus (10^3 lb in^{-2})	160	28–40	11
Usual form of fracture	brittle	tough	tough
ASTM brittleness temp.	—	−100°C	−70°C
Vicat softening point	179°C	71°C	83°C
Clarity	transparent	transparent	opaque
Moisture vapour transmission g/m² per 24hr per 0·001 in at 38°C and 90% R.H.	100	~40	59
Power factor 10^3 c/s	0·00015	0·0015	0·0024

available (e.g. Surlyn A-Du Pont) copolymerisation has had the not unexpected effect of depressing crystallinity although not completely eliminating it so that the materials are also transparent. Other properties claimed for the ionomers are excellent oil and grease resistance, excellent resistance to stress cracking and a higher moisture vapour permeability (due to the lower crystallinity) than polyethylene.

It remains to be seen how the price disadvantage of the ionomers, compared to polyethylene and plasticised p.v.c. will hinder their development. At least they open up the prospect of a whole new range of polymeric materials although the concept of ionic bonds in polymers is not new. Polymers with ionic groups attached along the chain and showing the properties of both polymers and electrolytes have been known for some time. Known as polyelectrolytes these materials show ionic dissociation in water and find use for a variety of purposes such as thickening agents. Examples are sodium polyacrylate, ammonium polymethacrylate (both anionic polyelectrolytes) and poly(4-vinyl-N-butyl pyridinium bromide), a cationic polyelectrolyte. Also somewhat related are the ion-exchange resins, cross-linked polymers containing ionic groups which may be reversibly exchanged and which are used in water softening, in chromatography and for various industrial purposes. In general however the polyelectrolytes and ion-exchange resins are intractable materials and not processable on convential plastics machinery. The value of the ionomer is that the amount of ionic bonding has been limited and so yields useful and tractable plastics materials. It is also now possible to envisage a range of rubbers which vulcanise by ionic cross-linking simply as they cool on emergence from an extruder or in the mould of an injection moulding machine.

8.4 OLEFIN AND DIENE RUBBERS

Rubbers form one of the most important groups of commercial polymers and may be considered as a special class of plastics materials which possess recoverable high elasticity. Since however the use of these materials need not be, nor is, restricted to applications in which recoverable high elasticity is a requirement it is not considered out of place to deal very briefly with these materials here. Although the bulk of commercial rubbers are based on dienes or olefins other speciality rubbers are available and are briefly discussed elsewhere. These include the acrylate rubbers (Chapter 12), fluorine-containing rubbers (Chapter 10), polysulphides (Chapter 16), polyurethanes (Chapter 23) and the silicones (Chapter 25). Natural rubber and some of its chemical derivatives are also further considered in Chapter 26.

Some pertinent statistics concerning the usage of the more important rubbers are given in Table 8.8.

Natural rubber (see also Chapter 26) still has a higher global consumption than any synthetic rubber but the consumption growth rate is much lower than is the case with s.b.r. In the United States, which has in recent years set the pattern of usage of rubber, natural rubber is much the

Table 8.8 PRODUCTION AND CONSUMPTION STATISTICS FOR NATURAL SYNTHETIC RUBBERS[12] (long tons)

(a) Global Consumption 1964

Natural rubber	2,230,000
Synthetic rubber	2,805,000

(b) United States Production and Consumption 1964

Rubber	Total production	Consumption	
		Tyre products	*Non-tyre products*
Natural rubber	—	313,491	169,557
S.B.R.	1,255,362	694,846	347,842
Neoprene-type rubber	141,069	5,117	96,760
Butyl rubber	98,910	48,044	26,078
Nitrile rubber	52,430	329	41,187
Stereo regulars	185,981	133,387	19,583
Others	31,058	2,943	25,492
Total	1,764,810	1,198,157	726,499

(c) United Kingdom Consumption 1963

Products	Natural rubber	Synthetic rubber
Tyres and tyre products	77,400	85,800
Cellular products	10,190	4,750
Footwear	8,510	8,840
Belting	5,750	1,580
Cables	3,500	3,360
Miscellaneous	62,350	34,870
Total	167,700	139,200

less important of the two. The polymer is obtained by coagulating the latex of certain trees, in particular the *Hevea brasiliensis* and is chemically high molecular weight cis-1,4-polyisoprene. Since the raw rubber is elastic (but temperature sensitive and soluble in hydrocarbons) it is necessary to reduce the molecular weight to make the polymer plastic and processable. This is brought about by mastication (mechanical shearing) on a two-roll mill, in an internal mixer or in certain types of extruder. The polymer may then be compounded with fillers, antioxidants, pigments and other ingredients. In order to regenerate the rubbery state the polymer must be lightly cross linked, sulphur being almost universally used in conjunction with certain agents known as accelerators and also zinc oxide and stearic acid. Vulcanisation takes place after shaping at elevated temperatures (140–190°C) for lengths of time which can range from 2 minutes to well over an hour. Although unfilled natural rubber vulcanisates have a high tensile strength compared with those of other rubbers (largely as a result of crystallisation that is induced during stretching), where a high degree of abrasion and tear resistance is required finely divided fillers such as carbon black and certain silicas are used. These materials are known as reinforcing agents. Natural rubber is a low cost rubber and has a high resilience. It is attacked by agents which react with a double bond including oxygen, ozone, halogens and hydrochloric acid. The deteriorating influences of oxygen and ozone may be moderated by the use of antioxidants and antiozonants. Although the largest application of natural rubber is in tyres it is also widely used as a diversity of applications some of which do not require the property of high elasticity such as flooring, cables, belting and various mechanical goods.

In many of these instances natural rubber has been replaced by flexible thermoplastics where these have been found to be more suitable or cheaper. In other instances highly-filled rubber compounds may be quite suitable and more economic than the thermoplastics. With the development of antiozonants and the injection moulding process for rubber, examples of rubbers replacing plastics, may become less uncommon.

Styrene–butadiene rubber (s.b.r.) is the major general purpose synthetic rubber. It was first used on a large scale during World War II as a rubber substitute under the name GR–S. At that time it was inferior to the natural product but steady improvement over the years, batch-to-batch consistency and low cost have resulted in its widespread acceptance. It is processed by the same techniques as natural rubber and sulphur is commonly used for vulcanisation. The inclusion of a finely-divided filler such as carbon black is necessary if good tensile strength, tear resistance and abrasion resistance are to be achieved since the polymer does not crystallise on stretching. S.B.R. is affected by agents that attack the double bond and is similar to natural rubber in other respects. Compounds tend to have a lower resilience than corresponding natural rubber compounds so that in tyres and dynamic applications s.b.r. products tend to suffer a greater heat build-up.

Synthetic cis-1,4-polyisoprene has been available since the mid 1950's. These materials are somewhat less regular than the natural polymer with cis- contents ranging from about 92% to about 98% according to the method of production. Compounds from current commercial polymers are softer, have lower modulus and generally have lower viscosities during processing than their natural rubber counterparts. Although processing techniques are similar the softness of the polymer necessitates modifications to compounding techniques in order to achieve an adequate dispersion. Although available on a commercial scale, and at low cost, the synthetic polymer has not yet secured a substantial market. Synthetic *trans-1,4-polyisoprene* is now produced by Dunlop and is widely used instead of its natural counterparts (gutta percha and balata) for the manufacture of golf ball covers. It is preferred because of its lower cost and less variability of properties.

Cis-1,4-polybutadiene is another low cost rubber of recent development. A number of different polymers are available and most have a cis- content in excess of 90%. These materials give compounds with a higher resilience than those from s.b.r. and natural rubber at room temperature and with generally better low temperature properties. Tyres have a lower heat build up but differences in wear behaviour are dependent on the severity of the conditions and on temperature. Polymers of low cis- content have been introduced to give polymers with low resilience for use in tyres requiring good road grip in wet conditions. (The, perhaps, rather surprising relationship between low resilience and road grip is now well established. Such high hysteresis compounds do however suffer from greater heat-build up and whereas water is available as a coolant in wet conditions, 'high hysteresis tyres' can run hot in dry weather.)

Polychloroprene, well known under the trade name of Neoprene, is the most widely used of the special purpose synthetics. Unlike the rubbers described above it is not vulcanised by sulphur, zinc oxide being the preferred material. Commercial polymers are essentially trans-1,4-polymers and are easily crystallised. Vulcanisates have a high tensile strength in the absence of reinforcing fillers. Polychloroprenes produce rubber compounds superior to natural rubber and s.b.r. in oil resistance, heat resistance and resistance to sunlight. Polychloroprene is more expensive than s.b.r. and natural rubber and so is used only when the cheaper products are unsuitable. Applications include wire and cable coverings, belting, industrial hose and coated fabrics. The *nitrile rubbers* are copolymers obtained by emulsion polymerisation of about 70 parts of butadiene with about 30 parts acrylonitrile. They are processed by techniques similar to those used with natural rubber and are vulcanised by sulphur. They are used mainly in applications requiring oil and petrol resistance being superior in this respect to, and more expensive than, the polychloroprene rubbers. They also form the basis of many useful contact adhesives.

Butyl rubber is a copolymer produced by ionic polymerisation of about 98 parts of isobutylene and 2 parts of isoprene. The limited unsaturation of the polymer gives products with superior heat and ozone resistance to natural rubber. The polymer also has a very low air permeability for a rubber and so finds wide application in tyre inner tubes. Because of its good heat resistance it also finds use in cables, steam-hose linings and other specialised applications.   It may be vulcanised with sulphur.

Ethylene-propylene rubbers (e.p.r.) are new materials arising from the development of stereospecific polymerisation.   These copolymers have limited stereoregularity and hence do not crystallise.   They may be vulcanised by peroxides and require reinforcement with carbon black to obtain strong vulcanisates.   The rubber has a low price structure and is resistant to ozone and to many chemicals and non-hydrocarbon solvents. It has good resistance to high temperature aging.   At present in the development stage e.p.r. may be expected to find uses for medium high temperature cable applications, steam hose and various mechanical goods.   Because of its low cost it may also become strongly competitive with general purpose rubbers such as natural rubber and s.b.r.   Ter-polymers containing unsaturated groups have also appeared which may be vulcanised with sulphur.   They are currently the subject of intensive investigation.

Chlorosulphonated polyethylene (Hypalon) is produced by treatment of polyethylene with chlorine in the presence of a small quantity of sulphur dioxide.   The formula for the polymer may be represented as shown in Fig. 8.10.

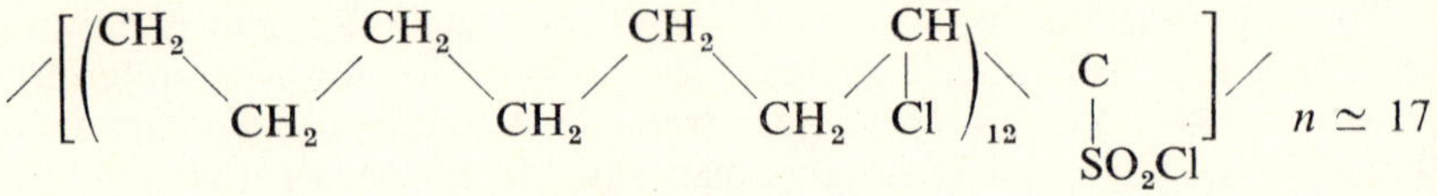

Fig. 8.10

The introduction of chlorine groups in a random manner reduces the ability of the polymer to crystallise but if more than one chlorine atom is introduced per seven carbon atoms the stiffness increases with an increase in chlorine content.   The SO_2Cl group provides a site for vulcanisation which can be carried out with litharge and water by the postulated mechanism shown in Fig. 8.11.

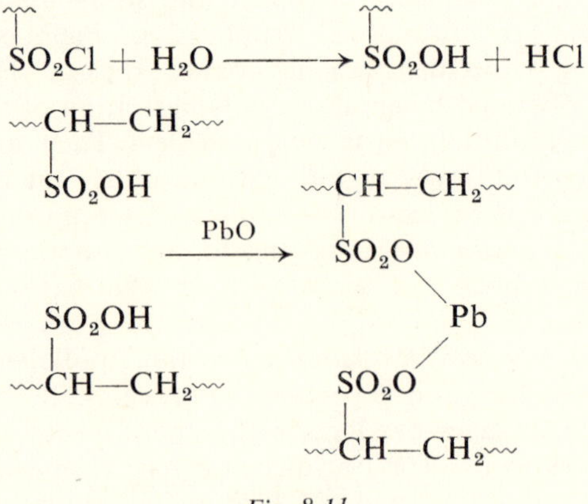

Fig. 8.11

These rubbers have good resistance to flexing heat, ozone and many chemicals. They are superior to butyl rubber in ozone resistance, compression set and abrasion resistance but their use is limited by their high cost. Applications include fabric coatings, protection coatings for other elastomers and white sidewalls for tyres.

REFERENCES

1. CRESPI, D., and RANALLI, F., *Trans. Plastics Inst.*, **27,** 55 (1959)
2. VAN DER WEGT, *Trans. Plastics Inst.*, **32,** 165 (1964)
3. CHARLEY, R. V., *Brit. Plastics*, **34,** 476 (1961)
4. *Technical trade literature*, I.C.I. Ltd., Welwyn Garden City
5. KRESSER, T. O. J., *Polypropylene*, Reinhold, New York (1960)
6. PRITCHARD, R., *Soc. Plastics Eng. J.*, **4,** 66 (1964)
7. TURNER, L. W., and YARSLEY, V. E., *Brit. Plastics.*, **34,** 32 (1962)
8. O'DONNELL, J. F., and SELLDORF, J. J., *Mod. Packaging*, **33,** 133 (1961)
9. RAINE, H. C., *Trans. Plastics Inst.*, **28,** 153 (1960)
10. PRITCHARD, J. E., MCGLAMERY, R. M., and BOEKE, P. J., *Mod. Plastics*, **37,** (2) 132 (1959)
11. HAYES, R., and WEBSTER, W., *Trans. Plastics Inst.*, **32,** 219 (1964)
12. *Rubber Stat. Bull.*, **19,** (6) (1965)

BIBLIOGRAPHY

KRESSER, T. O. J., *Polypropylene*, Reinhold, New York (1960)

Vinyl Chloride Polymers

9.1 INTRODUCTION

It is an interesting paradox that one of the least stable of commercially available polymers should also be, in terms of tonnage consumption at least, one of the two most important plastics materials available today. Yet this is the unusual position held by poly(vinyl chloride) (p.v.c.), a material whose commercial success has been to a large extent due to the discovery of suitable stabilisers and other additives which has enabled useful thermoplastic compounds to be produced.

The preparation of the monomer was first reported by Regnault[1] in 1835. The method used was to treat ethylene dichloride with an alcoholic solution of potassium hydroxide. Vinyl bromide was also obtained by a similar method using ethylene dibromide. It is reported that in 1872 vinyl bromide was also prepared by reacting acetylene and hydrogen bromide.[2] The analogous reaction with hydrogen chloride, discovered by F. Klatte[3] in 1912 subsequently became one of the two major routes for vinyl chloride production.

The polymerisation in sealed tubes of vinyl chloride and vinyl bromide, when exposed to sunlight, was reported in 1872 by Baumann.[4] Further work on these polymerisations was carried out by Ostromislensky in Moscow and this was duly reported in 1912.[5]

Commercial interest in poly(vinyl chloride) was revealed in a number of patents independently filed in 1928 by the Carbide and Carbon Chemical Corporation, Du Pont and I.G. Farben. In each case the patents dealt with vinyl chloride–vinyl acetate copolymers. This was because the homopolymer could only be processed in the melt state at temperatures where high decomposition rates occurred. In comparison the copolymers, which could be processed at much lower temperatures, were less affected by processing operations.

An alternative approach, which in due course became of great commercial significance, was made by W. L. Semon.[6] He found that if poly(vinyl chloride) was heated at 150°C with tritolyl phosphate, rubber-like masses which remained homogeneous at room temperature could be obtained. The blending of p.v.c. with this and other similar non-volatile liquids to give *plasticised p.v.c.* is now of great importance in the plastics industry.

During the next few years p.v.c. was steadily developed in the United States and in Germany. Both countries were producing the material commercially before the Second World War. In Great Britain I.C.I. in 1942 and the Distillers Company in 1943 also commenced pilot-plant production of p.v.c., a material then in demand as a rubber substitute for cable insulation. Paste-forming grades suitable for the production of leathercloth also became available soon afterwards.

After the war developments in Britain and the United States were concerned largely with plasticised p.v.c., handled mainly by extrusion, calendering and paste techniques. However on the continent of Europe, particularly in Germany, development work was also proceeding with unplasticised p.v.c. a rigid material which has only achieved recent significance in Britain and has yet to make its full impact felt. The use of copolymers has not grown in the same way as the homopolymer in spite of their early importance. Instead the former have become useful special purpose materials for the flooring, gramophone record and surface coating industries. Perhaps the greatest developments over the past few years have not been concerned with the molecular structure of the polymers but rather with the particles formed during polymerisation. Such factors as particle shape, size, size distribution and porosity vitally affect the processing characteristics of the polymer and a more complete knowledge of their influence has led to many useful new grades of polymer.

Today p.v.c. is being produced in many countries, in particular Britain, France, Germany, Japan, Italy and the United States. The major British p.v.c. polymer producers are I.C.I. (Corvic), British Geon (Breon), Bakelite (Vybak), and Shell Chemical (Carina).

9.2 PREPARATION OF VINYL CHLORIDE

There are three general methods of interest for the preparation of vinyl chloride, one for laboratory synthesis and the other two for commercial production.[7]

Vinyl chloride is most conveniently prepared in the laboratory by the addition of ethylene dichloride in drops on to a warm 10% solution of sodium hydroxide or potassium hydroxide in a 1:1 ethyl alcohol–water mixture (Fig. 9.1). At one time this method was of commercial interest. It does however suffer from the disadvantage that half the chlorine of the ethylene dichloride is consumed in the manufacture of common salt.

$$CH_2\!-\!CH_2 + NaOH \xrightarrow{\;60°C\;} CH_2\!=\!CH + NaCl + H_2O$$
$$\;\;|\quad\;\;| \qquad\qquad\qquad\qquad\quad |$$
$$\;Cl\quad Cl \qquad\qquad\qquad\qquad\;\; Cl$$

Fig. 9.1

For many years a major route to the production of vinyl chloride has been the addition of hydrochloric acid to acetylene (Fig. 9.2). The

$$CH\equiv CH + HCl \longrightarrow CH_2\!=\!\underset{\underset{Cl}{|}}{CH} + 22\!\cdot\!8 \text{ kcal/mole}$$

Fig. 9.2

acetylene is usually prepared by addition of water to calcium carbide which itself is prepared by heating together coke and lime. To remove impurities such as water, arsine and phosphine the acetylene may be compressed to 15 lb in^{-2}, passed through a scrubbing tower and chilled to $-10°C$ to remove some of the water present and then scrubbed with concentrated sulphuric acid.

Hydrochloric acid may conveniently be prepared by combustion of hydrogen with chlorine. In a typical process dry hydrogen chloride is passed into a vapour blender to be mixed with an equimolar proportion of dry acetylene. The presence of chlorine may cause an explosion and thus a device is used to detect any sudden rise in temperature. In such circumstances the hydrogen chloride is automatically diverted to the atmosphere. The mixture of gases is then led to a multi-tubular reactor, each tube of which is packed with a mercuric chloride catalyst on an activated carbon support. The reaction is initiated by heat but once started cooling has to be applied to control the highly exothermic reaction at about 90–100°C. In addition to the main reaction the side reactions shown in Fig. 9.3 may occur.

$$CH\equiv CH + H_2O \longrightarrow CH_3CHO$$

$$CH_2\!=\!\underset{\underset{Cl}{|}}{CH} + HCl \longrightarrow CH_3CH\cdot Cl_2$$

Fig. 9.3

The gases from the reactor are then cooled and subjected to a caustic wash to remove unreacted hydrogen chloride. This is then followed by a methanol wash to remove water introduced during the caustic wash. A final purification to remove aldehydes and ethylidene dichloride, formed during side reactions, is then carried out by low temperature fractionation. The resulting pure vinyl chloride is then stored under nitrogen in a stainless steel tank.

In recent years with the development of the petrochemical industry there has been a swing from acetylene to ethylene in the manufacture of vinyl chloride (a gas boiling at $-14°C$). The process commonly employed involves the formation of ethylene dichloride which is then dehydrochlorinated to give vinyl chloride. Ethylene dichloride is produced by reacting ethylene with chlorine in the liquid phase using an iron chloride catalyst at 30–50°C. Pure ethylene is not strictly necessary and it is possible to pass an unsplit ethylene–ethane stream through the reactor. The reaction is exothermic and it is necessary to carry out the reaction at low temperatures since higher temperatures favour substitution chlorination to yield tri- and tetra-chloroethanes.

The ethylene dichloride is converted to vinyl chloride by heating in tubes at 300–600°C (typically 500°C). The resulting vapour, a mixture of vinyl chloride, hydrogen chloride and ethylene dichloride (Fig. 9.4), is

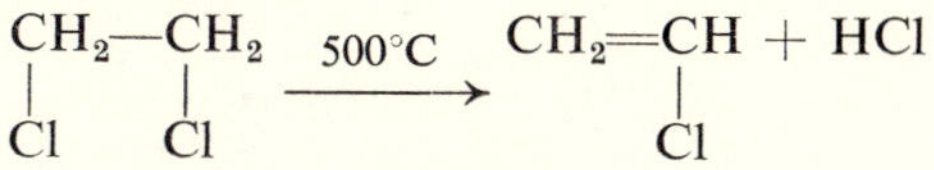

$$\underset{\substack{|\\Cl}}{CH_2}-\underset{\substack{|\\Cl}}{CH_2} \xrightarrow{500°C} \underset{\substack{|\\Cl}}{CH_2}=CH + HCl$$

Fig. 9.4

quenched downstream to freeze the equilibrium. The vinyl chloride is recovered by distillation.

Whereas the route from acetylene utilises all of the chlorine involved, in the ethylene route half of the chlorine is consumed in the formation of hydrogen chloride contaminated by aromatic hydrocarbons. The ethylene route is however normally the more economic as ethylene is cheaper than acetylene and also the hydrogen chloride may itself be used, for example in the production of vinyl chloride via the acetylene route.

In one process the ethylene and acetylene routes have been combined. A mixed feedstock of acetylene and ethylene together with diluents such as carbon monoxide, hydrogen and methane, all obtained by pyrolysis of a naphtha fraction, is employed. The only pyrolysis products that need to be removed are higher unsaturated hydrocarbons. The mixed stream is first sent to a hydrochlorination reactor where the acetylene present reacts to give vinyl chloride. The vinyl chloride is washed out of the stream and the remainder of the feed gas, consisting of ethylene and inert diluents, is passed to a dichloroethane reactor. The chlorine–ethylene reaction is carried out in the vapour phase over a fixed-bed catalyst. The resulting dichloroethane (ethylene dichloride) is then cracked to give vinyl chloride and hydrochloric acid, these being individually isolated by distillation processes.

Potential advantages of this route are the ability to use a mixed feedstock and the fact that hydrochloric acid formed in one part of the process may be used in the other part.

9.3 POLYMERISATION

In commercial practice vinyl chloride is polymerised by free radical mechanisms in bulk, in suspension and emulsion. Solution polymerisation methods are of less commercial importance, at least in Europe.

Bulk polymerisation is carried out commercially by Pechiney-St. Gobain (France). Details of the process are not available but in one patented example[8] vinyl chloride is polymerised with 0·8% of its own weight of benzoyl peroxide in a rotating cylinder containing stainless steel balls for 17 hours at 58°C. Bulk polymerisation is heterogeneous since the polymer is insoluble in the monomer. The reaction is autocatalysed by the presence of solid polymer whilst the concentration of catalyst has little effect on the polymerisation rate. This is believed to be due to the overriding effect of monomer transfer reactions on the chain length. As in all vinyl chloride polymerisations oxygen has a profound inhibiting effect.

Suspension polymerisations are generally easier to control and there is little loss in clarity or electrical insulation properties. Particle shape,

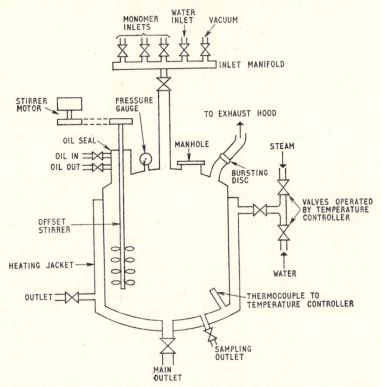

Fig. 9.5.   Typical polymerisation vessel suitable for suspension or emulsion polymerisation of vinyl chloride

size and size distribution may be controlled by varying the dispersing system and the rate of stirring. A typical polymerisation vessel for suspension polymerisation is shown in Fig. 9.5.

A typical charge would be:

Monomer	Vinyl chloride	30–50 parts
Dispersing agent	Gelatine	0·001 parts
Modifier	Trichloroethylene	0·1 parts
Initiator	Caprylyle peroxide	0·001 parts
	Demineralised water	90 parts

Poly(vinyl alcohol) is also used instead of, or as partial replacement for the gelatine. The trichloroethylene is a solvent transfer agent used to control molecular weight. In recent developments leading to the 'easy-processing' polymers porous particles are achieved with more complex suspending agents such as vinyl acetate–maleic anhydride copolymers and/ or fatty acid esters of glycerol, ethylene glycol or pentaerythrytol.

Using the above recipe the dispersing agent is first dissolved in a known weight of water, and added to the kettle. The rest of the water, the peroxide and the modifier are then added to the kettle which is sealed down and evacuated to 28 in. Hg. The vinyl chloride is then drawn in from the weighing vessel. In some cases pressurised oxygen-free nitrogen may be used to force all the monomer into the vessel. This is then closed and heated to about 50°C. Through the rise in temperature a pressure of about 100 lb in^{-2} will be developed in the reactor. As reaction proceeds the temperature is maintained until the pressure starts to fall, due to consumption of the monomer. When the pressure has dropped to 10–20 lb in^{-2} excess monomer is vented off and the batch cooled down discharged and dried, usually by a Venturi drying system. The product is checked for particle size by sieve analysis, for colour, contamination and for viscosity in a dilute solution. Other tests may be carried out on electrical insulation and paste grades.

Because of its low water solubility (0·09% at 20°C) vinyl chloride may be polymerised in emulsion. Using secondary alkyl sulphonates or alkali salts of alkyl sulphates as emulsifiers rapid polymerisation can occur in oxygen-free environments. The use of 'redox' initiating systems has made possible rapid reaction at temperatures as low as 20°C whilst in recent laboratory work sub-zero temperatures have been used. As explained in Chapter 2 water-soluble initiators are employed. Sodium persulphate, potassium persulphate and hydrogen peroxide are typical initiators whilst bisulphites and ferrous salts are useful reducing agents. Modifiers are often employed to control the molecular weight. Reaction times are commonly of the order of 1–2 hours. After polymerisation the particles are normally spray dried. There will thus be residual emulsifier which will adversely affect clarity and electrical insulation properties. Some improvement is however obtained using special washing operations.

Vinyl chloride is occasionally copolymerised with other monomers, notably with vinylidene chloride and vinyl acetate, the latter two in minor proportions. Where vinylidene chloride is used as comonomer there is a reduction in the overall polymerisation rate. In addition since vinylidene chloride radicals add preferentially to a vinylidene chloride molecule during chain growth a heterogeneous product is formed. This can be overcome by drip feeding the vinylidene chloride throughout the reaction at such a rate as to give a constant monomer composition. In the case of vinyl chloride–vinyl acetate copolymerisations the vinyl chloride is consumed preferentially and this will, without special steps being taken, also lead to some heterogeneity.

9.4 STRUCTURE OF POLY(VINYL CHLORIDE)

It is useful to compare the structures of p.v.c. and polyethylene (Fig. 9.6) since this enables predictions of the properties of the former to be made.

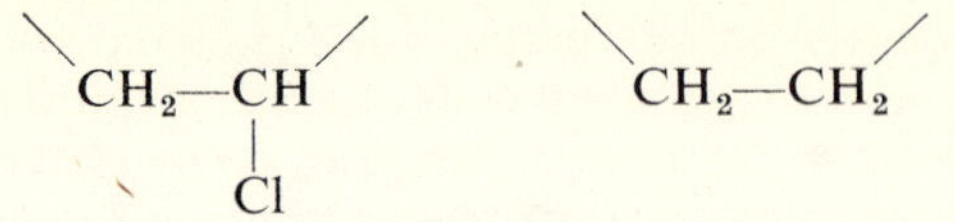

$$\overset{\diagdown}{CH_2}\!\!-\!\!\overset{\diagup}{CH} \qquad\qquad \overset{\diagdown}{CH_2}\!\!-\!\!\overset{\diagup}{CH_2}$$
$$\overset{|}{Cl}$$

Fig. 9.6

Both materials are linear polymers and substantially thermoplastic. The presence of the chlorine atom causes an increase in the inter chain attraction and hence an increase in the hardness and stiffness of the polymer. P.V.C. is also more polar than polyethylene because of the C—Cl dipole. Thus p.v.c. has a higher dielectric constant and power factor than polyethylene, although at temperatures below the glass transition temperature ($+80°C$) the power factor is still comparatively low (0·01–0·05 at 60 c/s) because of the immobility of the dipole.

The solubility parameter of p.v.c. is about 9·5 and the polymer is thus resistant to non-polar solvents which have a lower solubility parameter. In fact it has very limited solubility; the only solvents that are effective being those which appear to be capable of some form of interaction with the polymer. It is suggested[9] that p.v.c. is capable of acting as a weak proton donor and thus effective solvents are weak proton acceptors. These include cyclohexanone ($\delta = 9·9$) and tetrahydrofuran ($\delta = 9·5$). There are many materials that are suitable plasticisers for p.v.c. They have similar solubility parameters to p.v.c. and are also weak proton acceptors. These are of too high a molecular weight and too large a molecule size to dissolve the polymer at room temperature but they may be incorporated by mixing at elevated temperatures to give mixtures stable at room temperature. The presence of chlorine in large quantities in the polymer renders it flame retarding. The presence of plasticisers however reduces the resistance to burning.

Much work has been carried out in order to elucidate the molecular structure of poly(vinyl chloride). In 1939, Marvel, Sample and Roy[10] dechlorinated p.v.c. with zinc dust to give linked cyclic structures (Fig. 9.7).

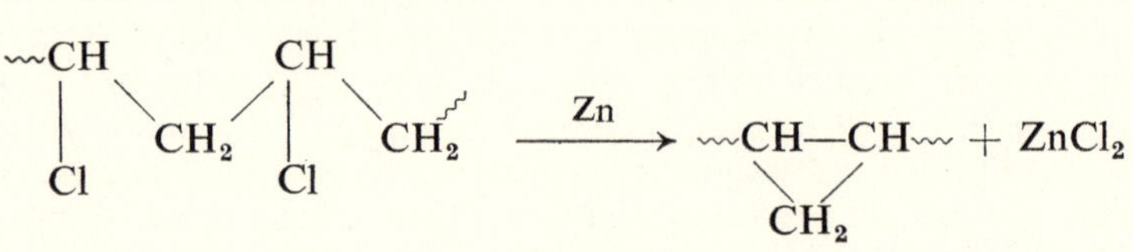

$$\sim\!\!CH \qquad CH$$
$$\overset{|}{Cl}\ \diagdown\!CH_2\ \diagup\ \overset{|}{Cl}\ \diagdown\!CH_2 \xrightarrow{\ Zn\ } \sim\!\!CH\!\!-\!\!CH\!\!\sim + ZnCl_2$$
$$\diagdown\!CH_2\!\diagup$$

Fig. 9.7

By noting the amount of chlorine that could be removed in this way they were able to determine whether the polymer was formed by head-to-tail linkage (Fig. 9.8 (a)) or head-head and tail-tail linkage (Fig. 9.8 (b)).

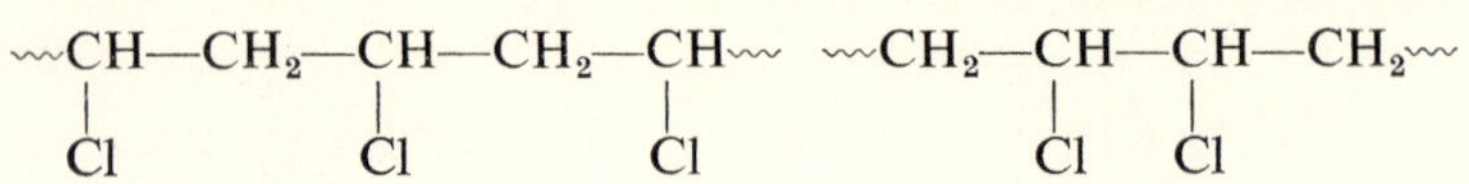

$$\sim\!\!CH\!\!-\!\!CH_2\!\!-\!\!CH\!\!-\!\!CH_2\!\!-\!\!CH\!\!\sim \qquad \sim\!\!CH_2\!\!-\!\!CH\!\!-\!\!CH\!\!-\!\!CH_2\!\!\sim$$
$$\overset{|}{Cl}\qquad\ \ \overset{|}{Cl}\qquad\ \ \overset{|}{Cl} \qquad\qquad \overset{|}{Cl}\quad\ \overset{|}{Cl}$$
$$(a) \qquad\qquad\qquad\qquad (b)$$

Fig. 9.8

It would be expected that if linkage was by the latter mechanism complete dechlorination would occur, the adjacent chlorine atoms being removed together. In the case of head-tail polymerisation it would be expected that because the reaction does not occur in steps along the chain, but at random, a number of unreacted chlorine atoms would become isolated and thus complete dechlorination could not occur (Fig. 9.9).

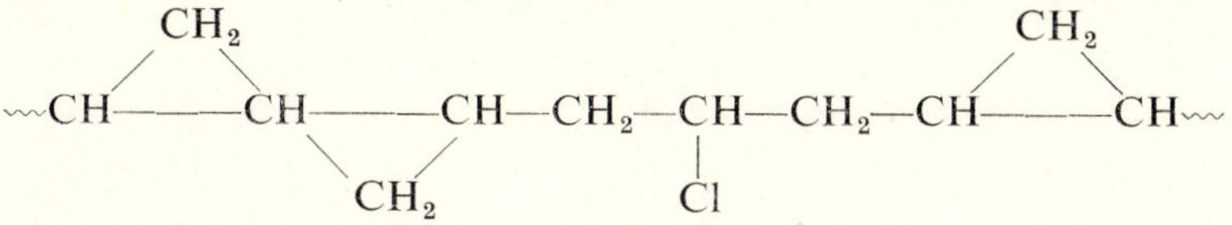

Fig. 9.9

It was found that the amount of chlorine that could be removed (84–87%) was in close agreement to that predicted by Flory[11] on statistical grounds for structure Fig. 9.8 (a). It is of interest to note that similar statistical calculations are of relevance in the cyclisation of natural rubber and in the formation of the poly(vinyl acetals) and ketals from poly(vinyl alcohol). Since the classical work of Marvel it has been shown by diverse techniques that head-to-tail structures are almost invariably formed in addition polymerisations.

X-ray studies indicate that the vinyl chloride polymer as normally prepared in commercial processes is substantially amorphous although some small amount of crystallinity (about 5% as measured by X-ray diffraction methods) is present. It has been reported by Fuller[12] in 1940 and Natta and Carradini[13] in 1956 that examination of the crystalline zones indicates a repeat distance of 5·1 Å which is consistent with a syndiotactic (i.e. alternating) structure. Thus we have a general picture of the conventional p.v.c. polymer as being largely atactic in structure but also possessing some short syndiotactic segments.

There is also evidence that commercial p.v.c., i.e. polymer produced by free radical initiation at about 50°C, contains about 16 branches per molecule. This evidence is largely obtained by converting the p.v.c. to a form of polyethylene by reaction with lithium aluminium hydride and examination of the product by infra-red techniques.

It is of interest to note that free radical polymerisation at lower temperatures, e.g. at −40°C using γ-radiation or by the use of highly active initiators such as the alkyl boranes, increasingly favours the formation of syndiotactic structures. These more regular polymers also show freedom from branching. The greater regularity and absence of branching result in crystalline polymers with decreased solubility in cyclohexanone. The polymers also have, as would be expected, greater density and a high softening point. Such polymers however are more difficult to process and the products generally more brittle. It has however been claimed that through careful control of molecular weight and crystallinity useful materials can be obtained.

P.V.C. has a rather limited thermal stability. This is rather surprising since it is known that low molecular weight materials containing similar structures are far more stable.   It would thus appear that this instability is due to imperfections or weak points in the structure at which degradation can commence.   The mechanism of degradation is far from being well understood and since the various theories do not as yet contribute to the technology of p.v.c. they will not be dealt with here.   The technological aspects of degradation are however considered in Section 9.5.1, dealing with stabilisers, and also in the processing sections.

9.4.1   Characterisation of commercial polymers

As indicated in the previous section, commercial p.v.c. polymers are largely amorphous, slightly branched molecules with the monomer residues arranged in a head-to-tail sequence.   Individual grades of material do however differ in average molecular weight, molecular weight distribution, particle shape, size and size distribution, and in the presence of impurities.   Some grades may also contain small quantities of comonomer residues.

For commercial purposes the molecular weight is usually characterised from measurements of the viscosity of dilute solutions.   It has been shown that, for dilute solutions, the relation between the viscosity and the molecular weight (in this case the 'viscosity-average' molecular weight) may be given by the relationship

$$[\eta] = KM^a$$

where K and a are constants

$\qquad$ M is the molecular weight

$\qquad$ $[\eta]$ is the intrinsic viscosity or limiting viscosity number

This is obtained by plotting $(\eta - \eta_0)/\eta_0 c$ against concentration c and noting its extrapolated value at infinite dilution.   In this case η is the viscosity of the polymer solution and η_0 the viscosity of the pure solvent. By correlating with results obtained by direct techniques (e.g. by osmosis) it is possible to give values for K for a given polymer and hence subsequently use the relationship to obtain the molecular weight.   In practice it is more common to characterise the molecular weight of a p.v.c. polymer by its Fikentscher K-value rather than to quote an actual figure for molecular weight.   This is *not* the same K as given in the above equation but is obtained from the following relationship and is a measure of the molecular weight, the lower the K-value the lower the molecular weight.

$$\log_{10} \eta_{\text{rel}} = \left[\left(\frac{75K^2 \times 10^{-6}}{1 + 1 \cdot 5Kc \times 10^{-3}} \right) + (K \times 10^{-3}) \right] c$$

where η_{rel} = relative viscosity = η/η_0, K = K-value and c = concentration in g/100 ml.

The K-value is however rather dependent on the method of measurement, and unfortunately different laboratories frequently use different solvents and different polymer concentrations. Comparisons of *K*-values quoted by different manufacturers must thus be treated with great care. Matthews and Pearson[14] have published a valuable correlation between different viscosity measurements. The data in Table 9.1 is abstracted from this correlation and illustrates the differences between different

Table 9.1 COMPARISON OF *K*-VALUES FOR P.V.C. USING DIFFERENT SOLVENTS[14]

K-values at 25°C				
0·5 g/100 ml ethylene dichloride	*0·5 g/100 ml cyclo-hexanone*	*0·4% nitro-benzene*	*Weight-average Mol. Wt.*	*Number-average Mol. Wt.*
45	48	52	54,000	26,000
50	53·9	57·5	70,000	36,000
55	59·5	62·5	100,000	45,500
60	65·2	68	140,000	55,000
65	70·8	73	200,000	64,000
70	76·5	78	260,000	73,000

K-values and also gives some indication of their relationship with weight-average and number-average molecular weights. In practice commercial polymers usually have *K*-values between 55 and 80. In this range the effect of molecular weight on mechanical properties is quite small. It does however have a greater effect on processing, the higher molecular weight grades being more difficult to process.

The properties of p.v.c. may also be expected to depend on the molecular weight distribution. There is however little published information on the nature of the molecular weight distribution of commercial polymers and on the effect of varying distribution. With the development of simpler techniques for measuring these distributions it is to be expected that information will eventually become available which will be of practical value.

With commercial polymers the major differences are, perhaps, not differences in molecular structure but in the characteristics of the particle, i.e. its shape, size, size distribution and porosity. Such differences will considerably affect the processing behaviour of a polymer.

A typical conventional suspension polymer is supplied in particles about 50–100 μm in diameter. The shape of the particle will depend on the dispersing agent system employed during polymerisation. Where a protective colloid such as gelatine is employed, smooth spherical particles are obtained which have a sufficiently low surface area–volume ratio to reduce the rate of plasticiser absorption at room temperature to a negligible figure. The use of certain other dispersing systems, such as a maleic acid–vinyl copolymer leads to an irregular type of particle with many voids and cavities. Such particles have a much higher surface area volume ratio, typically about 8 times greater and these materials will flux with plasticisers much more rapidly on heating. The advent of such porous suspension (granular) polymers, referred to as easy-processing resins, is one of the more important developments in p.v.c. technology that has taken place in recent years.

Emulsion polymerisation leads to much finer particles, of average diameter of the order of $0 \cdot 1$–$1 \cdot 0$ μm. In some commercial emulsion polymers these primary particles aggregate into hollow particles or 'cenospheres' with diameters of about 30–100 μm. These emulsion polymer

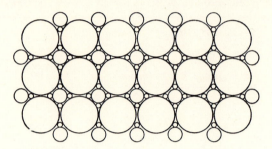

Fig. 9.10. p.v.c. paste polymer particles with distribution of size—efficient packing

particles have a high surface area–volume ratio and thus allow rapid fluxing with plasticisers. Their use is however restricted by the presence of large quantities of soaps, and other ingredients necessary for emulsion polymerisation, which adversely affect clarity and electrical insulation properties.

If p.v.c. polymer particles are mixed, at room temperature, with plasticisers the immediate product may take one of two forms. If there is insufficient plasticiser to fill all the gaps between the particle a 'mush' will

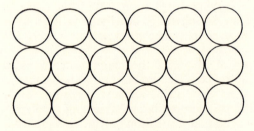

Fig. 9.11. p.v.c. paste polymer with homogeneous particle size—less efficient packing

be produced. If all the voids are filled then the particles will become suspended in the plasticiser and a paste will be formed. In the case of conventional granular polymer, or with emulsion polymer cenospheres, the particles are too large to remain in suspension and will settle out. Therefore compounds used in 'paste-processes' must use polymers with a small particle size. On the other hand there is a lower limit to this, since small particles will have a very high surface–volume ratio and measurable plasticiser absorption will occur at room temperature to give a paste whose viscosity will increase unduly with time. As a consequence paste polymers have an average particle size of about 0·2–1·5 μm.

It is found that the viscosity of a paste made from a fixed polymer–plasticiser ratio depends to a great extent on the particle size and size distribution. In essence, in order to obtain a low viscosity paste the less the amount of plasticiser required to fill the voids between particles the better. Any additional plasticiser present is then available to act as a lubricant for the particles, facilitating their general mobility in suspension. Thus in general a paste polymer in which the pastes have a wide particle size distribution (but within the limit set by problems of plasticiser absorption and settling-out) so that particles pack efficiently, will give lower viscosity pastes than those of constant particle size. The polymer particles

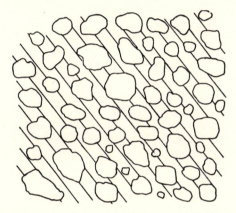

Fig. 9.12. Paste polymer suspended in plasticiser

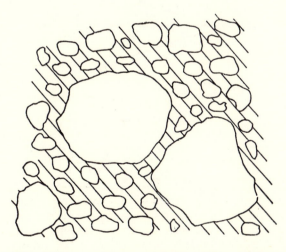

Fig. 9.13. p.v.c. paste containing filler polymer. Less plasticiser required to fill voids in unit volume

shown in Fig. 9.10 pack more closely and with less voids than those in Fig. 9.11 and hence give a lower viscosity polymer.

The success of 'filler' polymers used in increasing quantities in p.v.c. technology in recent years can be considered as an extension of this principle. These filler polymers are made by suspension (granular,

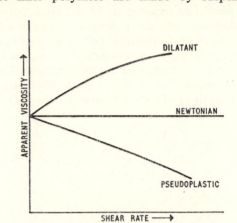

Fig. 9.14.   Classification of liquids according to dependence of apparent viscosity on shear rate

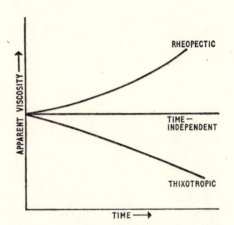

Fig. 9.15.   Classification of liquids according to dependence of apparent viscosity on time (constant shear rate)

dispersion) polymerisation and by themselves the particles are too large to make stable pastes.   However in the presence of paste-polymer particles they remain in stable suspension.   Being very much larger than paste-polymer particles they take up comparitively large volumes in which no plasticiser is required whatsoever.   This is shown in Figs. 9.12 and 9.13

where it is seen that the replacement in space of a mixture of paste-polymer particles and plasticiser by a large granular polymer particle releases plasticiser for use as lubricant, i.e. a viscosity depressant.

There is a further complication with paste polymers in that paste behaviour is non-Newtonian. The viscosities of paste are dependent on the shear rate and on the time of shear. The main possibilities are shown in Figs. 9.14 and 9.15. Thus a past viscosity may increase with shear rate (dilatancy) or decrease (shear thinning or pseudo-plasticity). Some pastes may show dilantant tendencies over one range of shear rates but be shear thinning over another range. Such behaviour has a profound effect on such processes as spreading. The viscosities may also decrease with time of stirring (thixotropy) or increase with it (rheopexy). Thus particular care must be taken in measuring rheological properties.

A certain amount of correlation between shear rate dependence and particle characteristics is now possible. It has been observed that spherical particles giving a high degree of packing are closest to Newtonian liquids in their behaviour. Very coarse lumpy uneven granules do not slide past each other easily in pastes and tend to become more entangled as shear rate increases. Such pastes commonly show dilatant behaviour. It has also been found that spherical particles of homogeneous size give shear thinning pastes. This may be due to the fact that these particles tend to aggregate (an observed fact) at rest whilst shearing causes dis-aggregation and hence easier movement of particles.

In addition to homopolymers of varying molecular and particle structure, copolymers are also available commercially in which vinyl chloride is the principal monomer. Comonomers used commercially include vinyl acetate, vinylidene chloride, acrylonitrile, vinyl isobutyl ether, and maleic, fumaric and acrylic esters. Of these the first two only are of importance to the plastics industry. The main function of introducing comonomers is to reduce the regularity of the polymer structure and thus lower the inter-chain forces. The polymers may therefore be processed at much lower temperatures and are useful in the manufacture of gramophone records and flooring compositions.

9.5 COMPOUNDING INGREDIENTS

In the massive form poly(vinyl chloride) is a colourless rigid material with limited heat stability and with a tendency to adhere to metallic surfaces when heated. For these, and other, reasons it is necessary to compound the polymer with other ingredients to make useful plastics materials. By such means it is possible to produce a wide range of products including rigid piping and soft elastic cellular materials.

A p.v.c. compound may contain the following ingredients:

1. Polymer.
2. Stabilisers.
3. Plasticisers.
4. Extenders.
5. Lubricants.
6. Fillers.
7. Pigment.

Other miscellaneous materials also used occasionally include fire retardants, optical bleaches and blowing agents.

9.5.1   Stabilisers

It is an observed fact that heating p.v.c. at temperatures above 70°C has a number of adverse effects on the properties of the polymer.   At processing temperatures used in practice (150–200°C) sufficient degradation may take place during standard processing operations to render the product useless.   It has been found that incorporation of certain materials known as stabilisers retards or moderates the degradation reaction so that useful processed materials may be obtained.

There is a great deal of uncertainty as to the mechanism of p.v.c. degradation but certain facts have emerged.   Firstly dehydrochlorination occurs at an early stage in the degradation process.   There is some infra-red evidence that as hydrogen chloride is removed polyene structures are formed (Fig. 9.16).

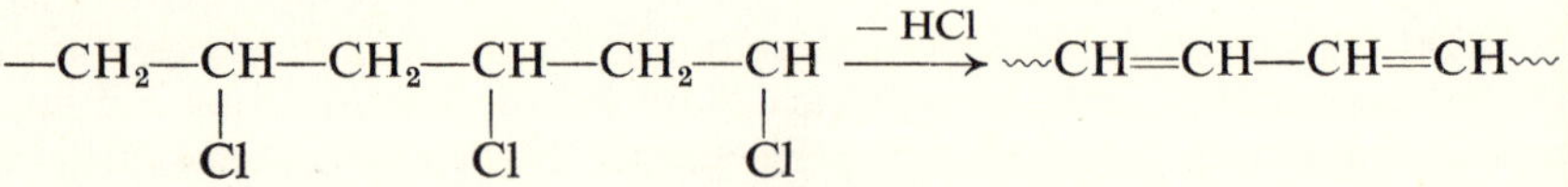

$$-CH_2-CH-CH_2-CH-CH_2-CH \xrightarrow{-HCl} \sim CH=CH-CH=CH\sim$$
$$\quad\quad\quad | \quad\quad\quad\quad\quad | \quad\quad\quad\quad\quad |$$
$$\quad\quad\quad Cl \quad\quad\quad\quad\quad Cl \quad\quad\quad\quad Cl$$

Fig. 9.16

At one time it was thought that the liberated hydrogen chloride caused autocatalytic liberation of further hydrogen chloride but there is now growing evidence that this is not the case.   A second fact that has emerged is that oxygen has an effect on the reaction.   It is believed that oxygen can cause both chain scission and cross-linking reactions whilst it has also been observed that the presence of oxygen accelerates colour formation.

The first physical manifestation of degradation, used in the widest sense, is a change in the colour of p.v.c.   Initially water-white, on heating it will turn, in sequence, pale yellow, orange, brown and black.   Further degradation causes adverse changes in mechanical and electrical properties.   For most commercial purposes however the 'end-point' is in fact the formation of colour.   With some applications some colour change can be acceptable, in other cases little or no change may be tolerated. Thus in practice changes in colour on heating provide a simple and readily obtainable criterion of degradation.

Many attempts have been made to correlate rate of colour formation with rate of hydrogen chloride evolution in polymer and compounded products but such attempts have not been successful.   Thus for practical purposes a stabiliser is most conveniently defined as a material that retards the formation of colour in unpigmented p.v.c. compounds. Beyond this the situation becomes complex.   Some stabilisers increase the rate of dehydrochlorination, some decrease it.   With some stabilisers the development of colour occurs slowly and steadily whilst with others formation of deep colours may take place quite suddenly after a prolonged induction time.   Some stabilisers are synergistic.   That is x parts of one stabiliser with y parts of a second stabiliser are often much more effective than $x + y$ parts of either used on its own. Some stabilisers are more effective in the presence of oxygen, some are less effective. Some systems are highly effective in one grade of p.v.c. polymer but only moderate in another. The presence of other additives such as plasticisers and fillers can also strongly influence the efficiency of a stabiliser. Thus phosphates and chlorinated extenders frequently reduce the stability of a compound.

The choice of a stabiliser thus becomes an empirical yet systematic process. The following factors are the most important which must be considered:

1. The grade of polymer used.
2. The nature of other ingredients present.
3. The cost of stabiliser required to give adequate stabilisation for the processing and anticipated service life of the compound.
4. The clarity required of the compound.
5. Toxicity.
6. The effect on lubrication, printing, heat sealing and plate-out.

Many tests have been devised for assessing the effect of a stabiliser in a given compound. That most successfully employed is to prepare a moulded sheet of a compound under strictly controlled conditions and to heat samples in a ventilated oven for various periods of time at various elevated temperatures. Small pieces of the samples are then bound on to cards giving data on the processing and heating conditions. These are then available for visible comparison of colour forming. For technical reports it is common to give the colour a numerical rating. Thus water-white samples can be given a number 0, a pale yellow 1, an orange 2, a brown 3 and a black sample 4 and these numbers can be tabulated in the report. It is however common experience that such a technique is far less effective in imparting information than the use of well-displayed samples.

In some laboratories samples are heated for prolonged periods in a press at a suitably elevated temperature. Such results may frequently fail to correlate with oven-heated samples since oxygen is largely excluded from the samples.

As already indicated, the measurement of dehydrochlorination rates is not a practical way of assessing the effect of a stabiliser. Thus the congo red test sometimes specified in standards, in which a piece of congo red paper is held in a test tube above a quantity of heated p.v.c. and the time taken for the paper to turn blue due to the evolution of a certain amount of hydrogen chloride, cannot be considered as being of much value.

Many stabilisers are also useful in improving the resistance of p.v.c. to weathering, particularly against degradation by ultraviolet radiation. This is an important consideration in building applications and other uses which involve outdoor exposure. The efficiency of stabilisers in improving resistance of p.v.c. compounds to degradation is best measured by lengthy outdoor weathering tests. Accelerated weathering tests using

Xenon lamps or carbon arcs have not proved to be reliable even for purposes of comparison.

The most important class of stabilisers are the lead compounds which form lead chloride on reaction with hydrogen chloride evolved during decomposition. As a class the lead compounds give rise to products of varying opacity, are toxic and turn black in the presence of certain sulphur containing compounds but are good heat stabilisers.

Of these materials *basic lead carbonate* has been, and probably still is, the most important stabiliser for p.v.c. It may be considered as typical of the lead compounds and has a low weight cost. It is appreciated that weight cost is not however the best criterion to be considered in assessing the economics of a stabiliser. Far more relevant is the cost required to stabilise the material to an acceptable level for the processing and service conditions involved. One additional disadvantage of the lead carbonate is that it may decompose with the evolution of carbon dioxide at the higher range of processing conditions thus leading to a porous product.

For this reason *tribasic lead sulphate*, a good heat stabiliser which gives polymer compounds with better electrical insulation properties than lead carbonate, has increased in popularity in recent years at the expense of white lead. Its weight cost is somewhat higher than that of lead carbonate but less than most other stabilisers. This material is used widely in rigid compounds, in electrical insulation compounds and to an even greater extent in general purpose formulations.

Other lead stabilisers are of much more specific applications. *Dibasic lead phosphite* gives compounds of good light stability but because of its higher cost compared with the sulphate and the carbonate its use is now restricted. In spite of its even greater weight cost *dibasic lead phthalate* finds a variety of specialised applications. Because it is an excellent heat stabiliser it is used in heat resistant insulation compounds (for example in 105° wire). It is also used in high fidelity gramophone records, in p.v.c. coatings for steel which contain polymerisable plasticisers and in expanded p.v.c. formulations which use azodicarbonamide as a blowing agent. In this latter application the phthalate stabiliser also acts as a 'kicker' to accelerate the decomposition of the blowing agent.

Normal and *dibasic lead stearate* have a stabilising effect but their main uses are as lubricants (see Section 9.5.4). *Lead silicate* is sometimes used in leathercloth formulations but is today of little importance. Other lead compounds now of negligible importance are coprecipitated *lead orthosilicates* and *lead salicylate*.

Whilst lead compounds have been, and still are, the most important class of stabiliser for p.v.c. the metallic soaps or salts have steadily increased in their importance and they are now widely used. At one time a wide range of *metal stearates, ricinoleates, palmitates* and *octoates* were offered as possible stabilisers and the efficiency of many of them has been examined. Today only the compounds of *cadmium, barium, calcium* and *zinc* are prominent as p.v.c. stabilisers.

The most important of these are *cadmium–barium* systems. These first became significant when it was discovered that stabilisers often behaved synergistically. Of the many stabilising systems investigated cadmium–barium systems gave considerable promise. The first of these systems to be used successfully were based on cadmium octoate in conjunction with barium ricinoleate. Alone the cadmium salt gave good initial colour but turned black after a relatively short heating period. The use of the barium soap in conjunction with the cadmium salt extended this period. The addition of antioxidants such as *trisnonyl phenyl phosphite* was found to greatly increase the heat stability whilst the further addition of *epoxidised oils* gave even better results. It was however found that on exposure to light an interaction took place between the ricinoleate and the epoxidised oil with the formation of products incompatible with the p.v.c. These products exuded and caused tackiness of the compound and problems in calendering. Replacement of the octoate and ricinoleate with *laurates* avoided the undesirable interaction but instead led to plate-out, difficulties in heat sealing and printing and compounds yellowish in colour and lacking in clarity. However the laurates continue to find some limited use in p.v.c. compounding.

Somewhat better results have been obtained with *octoates* and *benzoates* but these still lead to some plate-out. The use of liquid *cadmium–barium phenates* has today largely resolved the problem of plate-out whilst the addition of a trace of a zinc salt helps to improve the colour. Greater clarity may often be obtained by the addition of a trace of stearic acid or stearyl alcohol. Thus a modern so-called cadmium–barium stabilising system may contain a large number of components. A typical 'packaged' stabiliser could have the following composition:

Cadmium–barium phenate	2–3 parts
Epoxidised oils	3–5 parts
Stearic acid	0·5–1 part
Trisnonyl phenyl phosphite	1 part
Zinc octoate	0·5 parts

It appears that the zinc salt functions by preferentially reacting with sulphur to form white zinc sulphide rather than coloured cadmium sulphide and thus helps to reduce colour in the compound. Some metallic soaps also find some use for applications where freedom from toxicity is particularly important. Mixtures of calcium and zinc compounds together with epoxidised oils are commonly used for non-toxic purposes. These give products of limited long term stability but of excellent colour and clarity. For flooring compositions, magnesium–barium, calcium–barium and copper-barium compounds are sometimes used in conjunction with pentaerythritol. The latter material has the function of chelating iron present in the asbestos and thus reducing colour formation.

Another group of stabilisers are the *organo-tin compounds*. These materials found early application because of their resistance to sulphur and because they can yield crystal-clear compounds. The older organo-tin compounds such as *dibutyl tin dilaurate* however give only limited heat stability and problems may arise with high processing temperatures. *Dibutyl tin maleate* imparts somewhat greater heat resistance. The availability of a number of sulphur-containing organo-tin compounds, such as *dibutyl tin di-iso-octylthioglycollate*, which impart excellent heat resistance and clarity has to some extent increased the scope of organo-tin compounds. They are however more expensive in terms of weight cost and the development of improved systems based on cadmium–barium salts has so far limited the use of the organo-tin chemicals. Some of the organo-tin compounds also appear to possess some measure of toxicity. It should be noted that the sulphur-containing organo-tin compounds should not be used where lead derivatives are present in the p.v.c. compound since cross staining will occur to form black lead sulphide. Such lead compounds could be present as added stabiliser or even because the polymer on manufacture was washed with water fed through lead pipes.

Mention has already been made of *epoxide* stabilisers. They are of two classes and are rarely used alone. The first class are the epoxidised oils which are commonly employed in conjunction with the cadmium–barium systems. The second class are the conventional bis-phenol A epoxide resins (see Chapter 22). Although rarely employed alone, used in conjunction with a trace of zinc octoate (2 parts resin, 0·1 part octoate) compounds may be produced with very good heat stability.

A further class of stabilisers are the amines, such as *diphenyl urea* and *2-phenyl indole*. These materials are effective with certain emulsion polymers but rather ineffective with many other polymers.

9.5.2   Plasticisers

The tonnage of plasticisers consumed in Great Britain each year exceeds the annual tonnage consumption of most plastics materials. Only p.v.c. the polyolefins, the styrene polymers, the aminoplastics and, possibly, the phenolics are used in large quantity.

As explained in Chapter 5, these materials are essentially non-volatile solvents for p.v.c. Because of their molecular size they have a very low rate of diffusion into p.v.c. at room temperature but at temperatures of about 150°C molecular mixing can occur in a short period to give products of flexibility varying according to the type and amount of plasticiser added.

All p.v.c. plasticisers have a solubility parameter similar to that of p.v.c. It appears that differences between liquids in their plasticising behaviour is due to differences in the degree of interaction between polymer and plasticiser. Thus such phosphates as tritolyl phosphate, which have a high degree of interaction, gel rapidly with polymer, are more difficult to extract with solvents and give compounds with the highest brittle point. Liquids such as dioctyl adipate with the lowest interaction with polymer have the converse effect whilst the phthalates which are intermediate in their degree of interaction are the best all-round materials.

Phthalates prepared from alcohols with about eight carbon atoms are by far the most important class and probably constitute about 75% of plasticisers used. There are a number of materials which are very similar in their effect on p.v.c. compounds but for economic reasons *di-isooctyl phthalate* (d.i.o.p.) and the phthalate ester of the C7-C9 oxo-alcohol, often known unofficially as *dialphanyl phthalate* (d.a.p.), are most commonly used. (The term dialphanyl arises from the I.C.I. trade name for the C7-C9 alcohols–Alphanol 79.) As mentioned in the previous paragraph these materials give the best all-round plasticising properties. D.I.O.P. has somewhat less odour whilst d.a.p. has the greatest heat stability. *Diethyl hexyl phthalate* (d.e.h.p. or d.o.p.), for many years the most common plasticiser, is now rarely used in the United Kingdom, this fact being due to its somewhat higher cost. Because of its slightly lower plasticising efficiency, an economically desirable feature when the volume cost of a plasticiser is less than that of polymer, *dinonyl phthalate* (d.n.p.) may also be an economic proposition. Its gelation rate with p.v.c. is marginally less than with d.i.o.p., d.a.p. and d.e.h.p.

In spite of their high volatility and water extractability, *dibutyl phthalate* and *di-isobutyl* phthalate continue to be used in p.v.c. They are efficient plasticisers and their limitations are of greatest significance in thin sheet.

Certain higher phthalates have become available in recent years. For example *ditridecyl phthalate* and *di-isodecyl phthalate* are used in high temperature cable insulation, the former having the better high temperature properties. Because of its greater hydrocarbon nature than d.i.o.p., di-isodecyl phthalate has lower water extractability and is used for example with epoxidised oils in baby-pants. Whereas the bulk of phthalate plasticisers are based on branched alcohols, phthalates based on mixed normal alcohols have also recently become available. These materials give compounds with good low temperature resistance and high resilience.

The use of phosphates as plasticisers has not grown as rapidly as the phthalates in recent years because of their higher cost. Whereas the development of the petrochemical industry has resulted in the availability of d.i.o.p. and d.a.p. at lower cost than was experienced with d.e.h.p., parallel developments have not been possible with the phosphates. However because of its lower price structure the use of *trixylyl phosphate* (t.x.p.) has grown at the expense of *tritolyl phosphate* (t.t.p.). Both of the phosphates give compounds of greater flame resistance than corresponding phthalate compounds. They are also highly compatible, and have greater solvent resistance than the phthalates. On the debit side they are toxic and give products with a high cold flex temperature. Their main uses are in insulation and in mine belting.

For some applications it is important to have a compound with good low temperature resistance, i.e. with a low cold flex temperature. For these purposes alphatic esters are of great value. They have a lower interaction with p.v.c. and thus are incorporated with greater difficulty and extracted with greater facility. For many years the sebacates such as *dibutyl sebacate* (d.b.s.) and *dioctyl sebacate* (d.o.s.) were used where good low temperature properties are required. Today they have been largely replaced by cheaper esters of similar effect in p.v.c. devised from mixed acids produced by the petrochemical industry. The most important

Table 9.2 COMPARISON OF THE PRINCIPAL PLASTICISERS FOR P.V.C.

Plasticiser	Properties of plasticiser		Properties of plasticised p.v.c. sheet[a]					
	Specific gravity at 25°C	Refractive index n_D^{20}	Efficiency pro-portions[b]	Cold flex. temp. (°C)	Volatility (%)[c]	Water extraction (%)[d]	Iso-octane extraction (%)[e]	Volume resistivity at 20°C (Ω cm)
Di-n-butyl phthalate	1·045	1·49	54	−18	7·20	0·50	1·85	1×10^{13}
Diethyl hexyl phthalate	0·985	1·49	63	−26	0·35	0·10	20·3	$1·5 \times 10^{13}$
Di-iso-octyl phthalate	0·984	1·49	64	−25	0·40	0·05	20·2	1×10^{13}
Di-C$_7$–C$_9$ phthalate	0·995	1·49	58·7	−26	0·35	0·10	20·1	2×10^{13}
Dinonyl phthalate	0·970	1·48	69·5	−20	0·35	0·05	28·9	2×10^{13}
Tritolyl phosphate	1·165	1·56	71	−7	0·30	0·10	0·6	2×10^{13}
Trixylyl phosphate	1·137	1·55	75	−5	0·25	0·10	0·7	3×10^{13}
Di-iso-octyl adipate	0·927	1·45	52·7	−42	0·55	0·10	23·9	5×10^{11}
Iso-octyl ester of saturated C$_4$–C$_6$ dibasic acids	0·930	1·45	55	−42	1·70	0·20	22·7	$2·5 \times 10^{11}$
Epoxidised soya bean oil	0·993	1·47	66·5	−20	0·20	0·10	1·30	6×10^{12}
Polypropylene sebacate	1·060	1·47	—	—	very low	—	very low	—

(a) All properties are measured on sheets of modulus of 1,100 lb in^{-2} at 100% elongation.
(b) The efficiency proportion is the number of parts of plasticisers required per 100 parts of polymer to to give a modulus of 1,100 lb in^{-2} at 100% elongation.
(c) Weight loss of test-pieces 3 in × 2 in × 0·05 in in 24 hours in air circulated oven at 85°C.
(d) 240 hours at 23°C.
(e) 24 hours at 23°C.

of these mixed acids are the a.g.s. acids (a mixture of adipic, glutaric and succinic acids). These are esterified with octyl, nonyl and decyl alcohols to give plasticisers occasionally referred to as *sugludates*. The sebacate, adipate and sugludate-type plasticisers can also be used to give compounds of high resilience.

A number of other special-purpose plasticisers are also available. The *epoxidised oils* and related materials are good plasticisers and very good light stabilisers and are often used in small quantities in p.v.c. compounds. Polymeric plasticisers such as *polypropylene adipate*, *polypropylene sebacate* and similar products capped with lauric acid end groups are used where non-volatility and good hydrocarbon resistance is important. They are however rather expensive and are rather difficult to flux with p.v.c. Certain esters of citric acid find an outlet where minimum toxicity is of importance.

The development of p.v.c. as a metal-finishing material has led to the need for a good p.v.c.–metal adhesive. For some purposes it is found convenient to incorporate the adhesive component into the p.v.c. compound. Esters based on allyl alcohol, such as *diallyl phthalate* and various poly-unsaturated acrylates have proved useful in improving the adhesion and may be considered as polymerisable plasticisers. In p.v.c. pastes they can be made to cross link by the action of peroxides or perbenzoates simultaneously with the fluxing of the p.v.c. When the paste is spread on to metal the 'cured' coating can have a high degree of adhesion. The high adhesion of these rather complex compounds has led to their development as metal-to-metal adhesives used, for example in car manufacture. Metal coatings may also be produced from plasticised powders containing polymerisable plasticisers by means of fluidised bed or electrostatic spraying techniques.

Many other liquids have been found to be effective plasticisers for p.v.c. but are of limited commercial value, at least in Britain. The effect of plasticisers on the properties of p.v.c. is illustrated in Fig. 9.17 (a–e).

9.5.3 Extenders

A number of materials exist which are not in themselves plasticisers for p.v.c. because of their very limited compatibility with the polymer, but in conjunction with a true plasticiser a mixture is achieved which has a reasonable compatibility. Commercial *extenders*, as these materials are called, are cheaper than plasticisers and can often be used to replace up to a third of the plasticiser without serious adverse effects on the properties of the compound.

Three commonly employed types of extender are:

1. Chlorinated paraffin waxes.
2. Chlorinated liquid paraffinic fractions.
3. Oil extracts.

The solubility parameter of these extenders are generally somewhat lower than that of p.v.c. They are thus tolerated in only small amounts

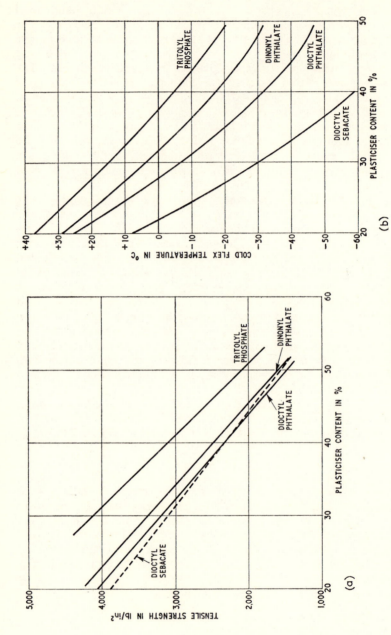

Fig. 9.17. See page 180

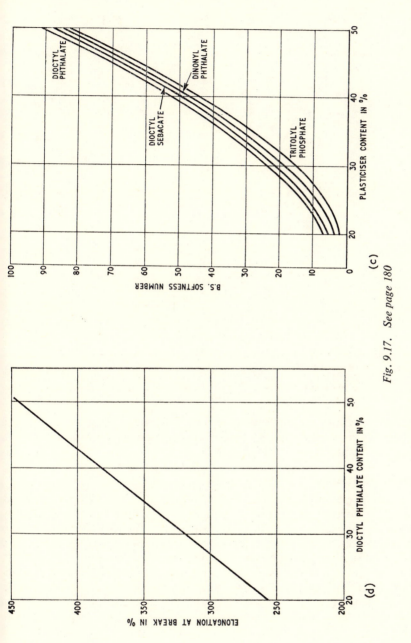

Fig. 9.17. See page 180

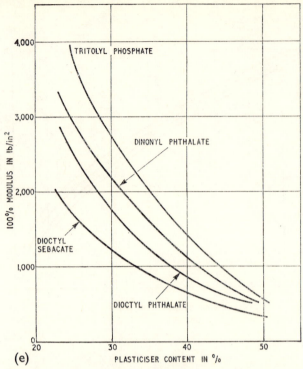

Fig. 9.17. *Effect of change of plasticiser on the properties of polyvinyl chloride compounds.*[15] *(a) Tensile strength. (b) Cold flex temperature. (c) BS softness number. (d) Elongation at break. (e) 100% modulus. (The Distillers Company Ltd.)*

when conventional plasticisers of low solubility parameter, e.g. the sebacates, are used but in greater amounts when phosphates such a tritolyl phosphate are employed.

9.5.4 Lubricants

In plasticised p.v.c. the main function of a lubricant is to prevent sticking of the compound to processing equipment. This is brought about by selecting a material of limited compatability which will thus sweat out during processing to form a film between the bulk of the compound and the metal surfaces of the processing equipment.

In Britain *calcium stearate* has been most commonly used with non-transparent products and *stearic acid* with transparent compounds. In the United States *normal lead stearate*, which melts during processing and lubricates like wax is commonly employed. *Dibasic lead stearate*, which does not melt, lubricates like graphite and which improves flow properties is also used.

Successful formulation of unplasticised p.v.c. compound necessitates a careful choice of lubricant and commercial compounds may contain a mixture of lubricants.   For example dibasic lead stearate, calcium stearate and either stearic acid or a wax may all be found in a single compound. In such rigid p.v.c. compounds the lubricant appears to function as a flow promoter as much as a lubricant in the sense defined above.

It has been pointed out[16] that for rigid p.v.c. extrusion compositions best results are obtained with a lubricant, or mixture of lubricants, which melt in the range 100–120°C, since these generally give a lubricating film of the correct viscosity at the processing temperature of about 165°C. For calendering operations it is suggested that lubricants should be chosen with higher melting points, i.e. in the range 140–160°C.   *Aluminium* and *magnesium stearate* fall within this melting point range.

9.5.5   Fillers

Fillers are commonly employed in opaque p.v.c. compounds in order to reduce cost.   They may also be employed for technical reasons such as to increase the hardness of a flooring compound, to reduce tackiness of highly plasticised compounds, to improve electrical insulation properties and to improve the hot deformation resistance of cables.

In evaluating the economics of a filler it is important to consider the volume of filler that can be added before bringing the processing and service properties below that which can be tolerated.   Thus in some cases it may be more economical to use a filler with a higher volume cost because more can be incorporated.   To judge the economics of a filler simply on its price per unit weight is of little merit.

For electrical insulation *china clay* is commonly employed whilst various *calcium carbonates* (whiting, ground limestone, precipitated calcium carbonate, and coated calcium carbonate) are used for general purpose work.   Also occasionally employed are *talc, light magnesium carbonate, barytes* (barium sulphate) and the *silicas* and *silicates*.   For flooring applications *asbestos* is an important filler.   The effect of fillers on some properties of plasticised p.v.c. are shown in Fig. 9.18 (a–d).

9.5.6   Pigments

A large number of pigments are now commercially available which are recommended for use with p.v.c.   Before selecting a pigment the following questions should be asked:

1. Will the pigment withstand processing conditions anticipated, i.e. will it decompose, fade or plate-out?
2. Will the pigment adversely affect the functioning of stabiliser and lubricant?
3. Will the pigment be stable to conditions of service, i.e. will it fade, be leached out or will it bleed?
4. Will the pigment adversely affect properties that are relevant to the end-usage?   (N.B. many pigments will reduce the volume resistivity of a compound.)

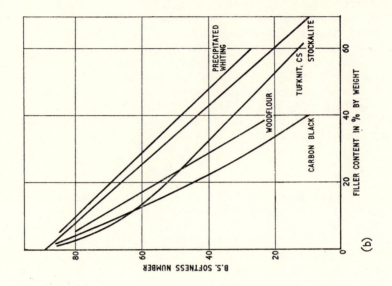

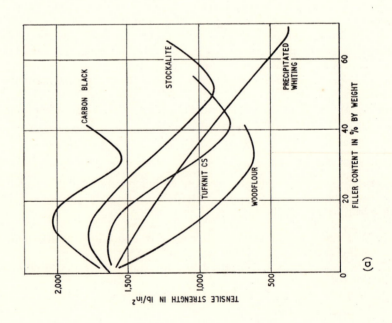

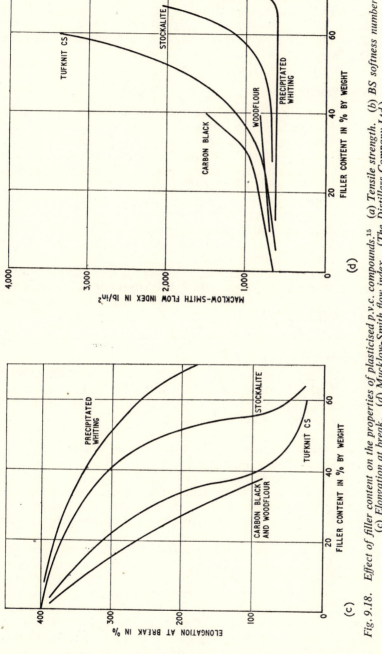

Fig. 9.18. Effect of filler content on the properties of plasticised p.v.c. compounds.[15] *(a) Tensile strength. (b) BS softness number. (c) Elongation at break. (d) Mucklow-Smith flow index. (The Distillers Company Ltd.)*

When there remains a choice of pigments which fulfil the above requirements then economic factors have to be taken into account.   The cost function relevant is again not the weight cost or volume cost but the cost of adding the amount of pigment required to give the right colour in the compound.   Thus a pigment with a high covering power may be more economic to use than a pigment of lower cost per pound but with a lower covering power.

9.5.7   Miscellaneous additives

A number of ingredients may be used from time to time in p.v.c. formulations.   For example blowing agents such as azodicarbonamide and azodiisobutyronitrile are frequently used in the manufacture of cellular p.v.c.

Antimony oxide[17] is useful in improving the fire retardance of p.v.c. compounds.   This is sometimes necessary since although p.v.c. itself has good flame retardance phthalate plasticisers will burn.

For some applications it is necessary that static charge should not accumulate on the product.   This is important in such diverse applications as mine belting and gramophone records.   The use of antistatic agents such as quaternary ammonium compounds has been of some limited value in solving this problem.

The viscosity of p.v.c. pastes may be reduced in many instances by the presence of certain polyethylene glycol derivatives and related materials. Because of their tendency to exude the use of these viscosity depressants should be restricted to levels of less than 1% of the total mix.

9.5.8   Formulations

It is obvious that the range of possible formulations based on poly(vinyl chloride) and related copolymers is very wide indeed.   For each end-use the requirements must be carefully considered and a formulation devised that will give a compound of adequate properties at the lowest cost.   In assessing cost it is not only important to consider the cost of the compound but also comparative processing costs, the possible cost of storing additional materials and many other cost factors.

The few formulations given below are intended as a general guide. They should not be taken as recommendations for a specific application where many factors, not considered in the brief discussion here, would need to be taken into account.   Formula 1 gives a typical general purpose insulation compound

Suspension polymer is chosen because its relative freedom from emulsifier and other surface active material gives polymers of better electrical insulation characteristics than emulsion polymers.   Di-isoctyl phthalate is a low cost good all-round plasticiser whilst some trixylyl phosphate is added to improve the fire-retarding properties.   China clay is a cheap filler with good insulation properties, whilst lead sulphate gives compounds of high heat stability, long term ageing stability and good insulation characteristics.   Formula 2 is a transparent calendering compound

Formula 1

Suspension polymer	100
D.I.O.P.	40
Trixylyl phosphate	20
China clay	20
Tribasic lead sulphate	7
Stearic acid	1
Pigment	2

Formula 2

Suspension polymer	100
D.I.O.P.	40
Ba–Cd phenate	3
Trisnonyl phenyl phosphite	1
Epoxidised oil	5
Stearic acid	1

Formula 3

Suspension polymer	100
D.I.O.P.	40
Epoxidised oils	10
Ca–Zn stabiliser	2·5

Each of the ingredients are chosen with a view to obtaining high clarity at a moderate cost. In Formula 3 the stabilising system has been replaced by a less powerful, but also, less toxic stabiliser.

Formula 4

Suspension polymer	100
D.A.P.	50
Extender	25
Whiting	30
Lead sulphate	6
Calcium stearate	1
Pigment	3

Formula 5

Vinyl chloride-vinyl acetate copolymer	100
D.A.P.	30
Extender	15
Whiting	150
Asbestos	150
Ba–Zn complex	3
Calcium stearate	1
Pigment	as required

Formula 6

Vinyl chloride–acetate copolymer	100
Dibasic lead stearate	0·75
Dibasic lead phthalate	0·75
Lamp black	2

The requirements for garden hose are somewhat less critical and both filler and extender may be incorporated to reduce cost (Formula 4).

The main requirements for a flooring composition are that it should be hard, durable and competitive in price with other materials. This calls for highly filled materials which are consequently harder to process than unfilled materials. The problem is alleviated by use of copolymers with their easier processing characteristics.   Formula 5 is an example of a flooring recipe.

Copolymers are also used in gramophone record formulations (Formula 6).   No filler can be tolerated and stabilisers and lubricants are chosen that give records of minimum surface noise.   Antistatic agents may also be incorporated into the compound.

Formula 7

Paste-making polymer	100
D.I.O.P.	50
Extender	20
Ba–Cd system	2
Epoxidised oil	3
Pigment	as required

Formula 7 is a leathercloth formulation for use in spreading techniques. There are many possible formulations and that given is for a product with a soft dry feel.

9.6 PROPERTIES OF P.V.C. COMPOUNDS

Because of the wide range of possible formulations it is difficult to make generalisations about the properties of p.v.c. compounds.   This problem is illustrated in Table 9.3 which shows some differences between three distinct types of compound.

Mechanical properties are considerably affected by the type and amount of plasticiser.   This was clearly shown in Fig. 9.17.   To a lesser extent fillers will affect the physical properties, as indicated in Fig. 9.18.

Unplasticised p.v.c. is a rigid material whilst the plasticised material is flexible and even rubbery at high plasticiser loadings.   It is of interest to note that the incorporation of small amounts of plasticiser, i.e. less than 20%, does not give compounds of impact strength higher than that of unplasticised grades, in fact the impact strength appears to go through a minimum at about 10% plasticiser concentration.   As a result of this behaviour, lightly plasticised grades are only used when ease of processing is more important than in achieving a compound with a good impact strength.

Poly(vinyl chloride) has a good resistance to hydrocarbons but some plasticisers, particularly the less polar ones such as dibutyl sebacate are extracted by materials such as iso-octane.   The polymer is also resistant to most aqueous solutions including those of alkalis and dilute mineral acids.   Below the second order transition temperature, poly(vinyl chloride) compounds are reasonably good electrical insulators over a wide range of frequencies but above the second order transition temperature their value as an insulator is limited to low frequency applications.   The more the plasticiser present the lower the volume resistivity.

Vinyl chloride–vinyl acetate copolymers have lower softening points than the homopolymers and compounds and may be processed at lower

Table 9.3 PROPERTIES OF THREE TYPES OF P.V.C. COMPOUND

	Unplasticised p.v.c.	*Vinyl chloride–vinyl acetate copolymer (sheet)*	*P.V.C.+ 50 p.h.r. d.i.o.p.*
Specific gravity	1·4	1·35	1·31
Tensile strength	8,500	7,000	2,700
Elongation at break %	5	5	300
B.S. softness No.	—	—	35
Vicat softening (°C)	80	70	flexible at room temperature

temperatures than those used for analogous homopolymer compounds. The copolymers have better vacuum-forming characteristics, are soluble in ketones, esters and certain chlorinated hydrocarbons but have generally an inferior long term heat stability. The effect of percentage comonomer on the properties of a copolymer are illustrated in Fig. 9.19.[18]

9.7 PROCESSING

Consideration of the methods of processing vinyl chloride polymers is most conveniently made under the following divisions:

1. Melt processing of plasticised p.v.c.
2. Melt processing of unplasticised p.v.c.
3. Processing of pastes.
4. Processing of latices.
5. Copolymers.

9.7.1 Plasticised p.v.c.

The melt processing of plasticised p.v.c. normally involves the following stages:

1. Pre-mixing polymer and other ingredients.
2. Fluxing the ingredients.
3. Converting the fluxed product into a suitable shape for further processing, e.g. granulating for injection moulding or extrusion.
4. Heating the product to such an extent that it can be formed by such processes as calendering, extruding, etc. and cooling the formed product before removal from the shaping zone.

The many possible variations and modifications to this sequence have been admirably summarised by Matthews[19] (see Fig. 9.20).

In most of these routes, premixing is carried out in a trough mixer at room temperature to give a damp powdery mass or 'mush'. This may then be fluxed on a two-roll mill, in an internal mixer, or in a continuous compounder such as the Werner and Pfleiderer ZSK machine. For many operations the compounded mass is then granulated or pelleted. This can be carried out as part of the continuous compounding process whereas

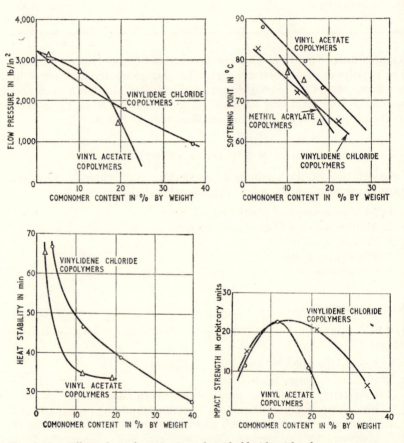

Fig. 9.19.   *Effect of copolymerisation of vinyl chloride with other monomers on the properties of unplasticised compounds.   (After Wheldon.*[18])

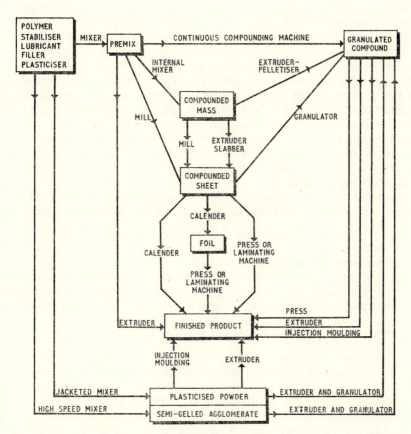

Fig. 9.20.   *Routes from raw materials to finished products illustrating different compounding techniques with p.v.c. compounds.*[19]

a mass mixed in an internal mixer may be either fed to an extruder–pelletiser, which extrudes strands which are then cut up into pellets, to an extruder–slabber which produces a sheet subsequently fed to a dicing machine, or to a two-roll mill which also provides a sheet for subsequent dicing.   Sheet may also be fed directly to a calender whilst still hot and it may also be used for pressing into sheet.

In recent years dry blending techniques have become more popular. In those processes the mixture of ingredients is either subjected to vigorous stirring, gentle heating or both.   As a result of such treatment the plasticiser is absorbed into the polymer particles to give dry free-flowing powders. This process is most easily worked with easy-processing polymers. Although the intensity of mixing is not so great as in an internal mixer, mixing is taken to a stage where it can be subsequently completed during the plasticising stage of an extrusion operation.   The important advantage of this process is that it frees the polymer from subjection to one high temperature process and thus reduces risk of decomposition and of deterioration in electrical insulation properties.   For successful use of dry blending processes it is important to ensure an adequate degree of mixing in the final product.   This involves not only care in the development ·of the mixing process but also care in the choice of extrusion machinery and conditions.

Extrusion operations are involved in making cables, garden hose, pipes and sections.   It is necessary to ensure the following conditions:

1. That the compound is dry.   (This is not normally a problem and special predrying operations are rarely necessary if the material has been properly stored.)
2. That the compound is not allowed to stagnate in heated zones of the extruder and thus decompose.   The 'life' of many compounds at processing temperatures is little more than the normal residence time of the material in the extruder.
3. That there is a good means of temperature control.

It is not possible to give detailed recommendations of processing conditions as these will depend on the nature of the product, the formulation of the compound used and the equipment available.   However the temperature is generally of the order of 130°C at the rear of the barrel and this increases to about 170–180°C at the die.   Screw speeds are of the order of 10–70 rev/min.   A typical screw would have a length–diameter ratio of about 15:1 and a compression ratio of 2:1.   It is however possible to extrude plasticised p.v.c. on extruders with lower length–diameter ratios and compression ratios, particularly with extruders of screw diameters in excess of 2 in.   In some cases it may however be necessary to improve the degree of homogenisation by use of stainless steel mesh screens behind the breaker plate and by the use of screw cooling water.   Higher length–diameter screws are often preferred for dry blends in order to ensure adequate mixing.

A large amount of plasticised p.v.c. is fabricated by calendering techniques using calenders of either the inverted–L or, preferably, of the inclined–Z type. A major problem is the control of gauge (thickness). Transverse variations due to bowl bending may be reduced or partially compensated for by the use of bowls of greater diameter–width ratio, by cambering the bowls, by bowl cross alignment or by deliberate bowl bending. Longitudinal variation may be largely eliminated by preloading the bowls to prevent the journals floating in their bearings. For technical reasons it is more convenient to preload on Z-type calenders than on L- and inverted L-type calenders. Bowl temperatures are in the range 140–200°C. Large scale leathercloth manufacture is today commonly carried out using calendering techniques.

Plasticised p.v.c. is not easily injection moulded in conventional plunger machines. This is because it is difficult to bring compound furthest from the cylinder walls to a temperature which gives the compound adequate flow without decomposing the material nearest to the cylinder wall. The introduction of preplasticising machines, which amongst other advantageous characteristics do not rely on heat transfer solely by conduction, has largely overcome this problem. A number of special low pressure machines have been especially developed for use in the manufacture of p.v.c. shoes and shoe soles. For these purposes, compounds based on lower molecular weight homopolymers (K-value 55–60) or copolymers are frequently employed.

Fluidised-bed techniques, pioneered with low density polythene, have been applied in recent years to p.v.c. powders. These powders can be produced by grinding of conventional granules, either at ambient or subzero temperatures or by the use of dry blends (plasticised powders). The fluidised-bed process is somewhat competitive with some well-established paste techniques, and has the advantage of a considerable flexibility in compound design.

9.7.2 Unplasticised p.v.c.

With unplasticised p.v.c. processing conditions are more critical because of the high melt viscosity of the compound. In order to achieve adequate flow properties it is necessary to work at temperatures at which decomposition can take place quite rapidly. Accurate temperature control of equipment is therefore necessary. The successful development of unplasticised p.v.c. in recent years is also due to a large extent to the advent of better stabilising and lubricating systems.

In the extrusion of p.v.c. it is essential not to overwork the polymer. Thus screws of low compression ratio (2:1 or less) are usually used. Dies are carefully tapered to ensure a slow but steady reduction in cross-sectional area from the breaker plate to the die tips and the die parallel is frequently 15–25 times the section thickness.

Injection moulding of unplasticised p.v.c. has been made possible by the advent of the in-line screw preplasticising machines. These machines may be operated under moderate processing conditions to give products of high quality. Although it is possible to extrude rigid p.v.c. sheet, it is commonly made by compression moulding techniques, either by laminating hide from a sheeting or mixing mill or by moulding granules. Such sheet may be welded using hot gas welding guns to produce chemical plant and other industrial equipment. The sheet may be shaped by heating and subjecting it to mechanical or air pressure. The methods used are similar to those originally developed to deal with poly(methyl methacrylate).

9.7.3   Pastes

As explained in Section 9.4 a paste is obtained when the voids between particles are completely filled with a plasticiser so that the polymer particles are suspended in it. It has also been pointed out that to ensure a stable paste there is an upper and a lower limit to the order of particle size. Finally it has been stressed that both the flow and fluxing characteristics of a paste are, to no small extent, dependent on the particle shape, size and size distribution.

A number of basic paste types may be distinguished. The most important classes are the plastisols, the organosols, plastisols incorporating filler polymers (including the rigisols), plastigels, hot melt compounds, and compounds for producing cellular products.

The first four types are most conveniently distinguished by reference to formulations A to D in Table 9.4. Formulation A is a conventional

Table 9.4

	A	*B*	*C*	*D*
Paste-making polymer	100	100	55	100
Filler polymer	—	—	45	—
Plasticiser (e.g. d.o.p.)	80	30	80	80
Filler (e.g. china clay)	10	10	10	10
Stabiliser (e.g. white lead)	4	4	4	4
Naphtha	—	50	—	—
Aluminium stearate	—	—	—	4

plastisol. The viscosity of the paste is largely controlled by the choice of type and amount of polymer and plasticiser. In order to achieve a sufficiently low viscosity for processing, large quantities of plasticiser must be added, thereby giving a product of lower hardness, modulus, tensile strength and other mechanical properties than would be the case if less plasticiser could be used. In many applications this is not a serious problem and plastisols are of some considerable importance commercially.

The influence of plasticiser content on viscosity is shown in Fig. 9.21. It is also to be noted that because of plasticiser absorption the viscosity of pastes does invariably increase on storage. The rate of increase is a function of the plasticiser used (Fig. 9.22).

Plastisols are converted into tough rubbery products by heating at about 160°C. At this temperature plasticiser diffusion occurs to give a molecular mixture of plasticiser and polymer. In some processes the

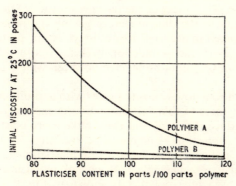

Fig. 9.21. *Effect of plasticiser (dialphanyl phthalate) on initial paste viscosity with two commercial paste polymers*

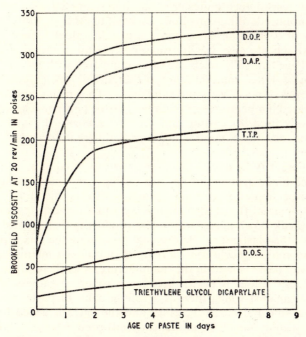

Fig. 9.22. *Effect of storage on the viscosity of p.v.c. pastes using different plasticisers.*[15] *(The Distillers Company Ltd.)*

paste is only partially fluxed initially to give a 'cheesy' pregel, full gelation being carried out at some later stage.

Organosols are characterised by the presence of a volatile organic diluent whose function is solely to reduce the paste viscosity. After application it is necessary to remove the diluent before gelling the paste. Organosols are therefore restricted in use to processes in which the paste is spread into a thin film, such as in the production of leathercloth. Because of the extra processes involved organosols have not been widely used, in Europe at least.

One way of obtaining low viscosity pastes with a minimum of plasticiser is to use filler polymers (Table 9.4, mix C). As explained in Section 9.4 these materials reduce the voids and thus make more plasticiser available for particle lubrication, thus reducing paste viscosity. It has been found that a good filler polymer must have a low plasticiser absorption and a high packing density.[20] The effect of partial replacement of a paste-making polymer by a filler-polymer is shown in Fig. 9.23. The use of filler polymers has increased considerably in recent years and has greatly increased the scope of p.v.c. pastes. The term *rigisol* is applied to pastes prepared using filler polymers and only small quantities (approx. 20 parts per 100 parts polymer) of plasticiser.

Mix D is a typical *plastigel*. The incorporation of such materials as fumed silicas, certain bentonites or aluminium stearate gives a paste which shows pronounced Bingham Body behaviour (i.e. it only flows on application of a shearing stress above a certain value). Such putty-like materials, which are also usually thixotropic may be hand shaped and subsequently gelled. Plastigels are often compared with *hot melt compositions*. These latter materials are prepared by fluxing polymer with large quantities of plasticisers and extenders. They melt and become very fluid at elevated temperatures so that they may be poured. These materials are extensively used for casting and prototype work. Formulations for making expanded p.v.c. compounds are of different types and are considered later.

Many different types of equipment are used for mixing p.v.c. pastes, of which sigma-blade trough mixers are perhaps the most popular. To facilitate dispersion it is common practice to mix the dry ingredients initially with part of the plasticiser in order to keep shearing stresses high enough to break down aggregates. The product is then diluted with the remainder of the plasticiser. Before final processing the mix is preferably deaerated to remove occlusions developed during the mixing operation.

Much p.v.c. paste is used in the manufacture of leathercloth by spreading techniques. Cloth is drawn between a roller or endless belt and a doctor blade against which there is a rolling bank of paste. A layer of paste is thus smeared on to the cloth and the paste is gelled by passing through a heated tunnel or underneath infra-red heaters. Whilst still hot, embossing operations may be carried out using patterned rollers. The cloth is then cooled and wound up. Where a flexible leathercloth is desired a dilatant paste compound is employed so that 'strike-through' is reduced to a minimum. Conversely where it is desired that the paste should enter the interstices of the cloth a shear-thinning paste is employed. In recent years there has been an increased interest in cellular leathercloth using modified spreading techniques. In some leathercloth formulations the surface has an undesirable tackiness. In such cases this can be overcome by applying a p.v.c.–acrylic lacquer.

Many useful products can be made by dipping heated formers into p.v.c. paste, allowing part of the paste to pregel on to the former, withdrawing the former (and pregel) from the paste and heating in an oven

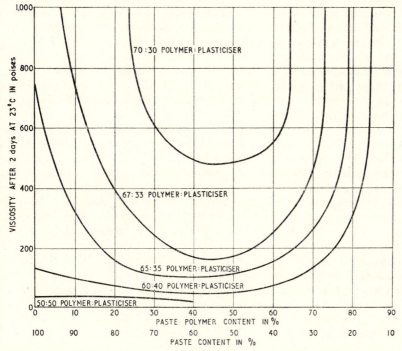

Fig. 9.23. *Effect on viscosity of partial replacement of a p.v.c. paste polymer (Geon 121) in a p.v.c. paste with a non-porous suspension polymer of high packing density.*[15]   (*The Distillers Company Ltd.*)

to complete gelation. It is not difficult to produce objects up to $\frac{1}{8}$ in. thick in this way, the thickness depending on the temperature of the former, its heat capacity (mass $\times$ specific heat), the time of immersion and the compound formulation. Metal objects may be permanently coated by the same technique.

The slush moulding process is in principle very similar. Paste is poured into a hot mould, some of which pregels and adheres to the walls of the mould cavity. Excess paste is poured out and the mould is then heated to complete the gelation of the adhering paste in an oven at about 155–170°C. Both the hot dipping and slush moulding processes involve repeated heating of paste and care must be taken to avoid formation of lumps. Uneven wall thicknesses can occur due to drainage but this may be reduced by the use of the so-called thixotropic fillers, such as fumed silicas, bentonites and aluminium stearate.

In recent years rotational casting methods have made the slush moulding process virtually obsolete. In these processes an amount of material equal to the weight of the finished product is poured into a mould. The mould is then closed and rotated slowly about two axes so that the paste flows easily over the cavity walls in an oven at about 200–250°C. When the compound has gelled, the moulds are cooled and the moulding removed. Compared with the slush moulding process there is no wastage of material, little flash, and a more even wall thickness. Completely enclosed hollow articles such as playballs are most conveniently made.

Paste injection moulding processes have also been developed in recent years. In one technique used for applying p.v.c. soles to shoe uppers the paste is injected by gas under pressure into a hot mould, the last and shoe upper forming the top half of the mould. The paste gels in the mould, adheres to the upper in the presence of a suitable adhesive and is stripped hot from the mould.

A number of methods have been devised for producing cellular products from p.v.c. pastes. One approach is to blend the paste with carbon dioxide, the latter either in the solid form or under pressure. The mixture is then heated to volatilise the carbon dioxide to produce a foam which is then gelled at a higher temperature. Flexible, substantially open-cell structures may be made in this way. Closed-cell products may be made if a blowing agent such as azo-di-isobutyronitrile is incorporated into the paste. The paste is then heated in a filled mould to cause the compound to gel and the blowing agent to decompose. Because the mould is full, expansion cannot take place at this stage. The mould is then thoroughly cooled and the as yet unexpanded block removed and transferred to an oven where it is heated at about 100°C and uniform expansion occurs.

9.7.4 Copolymers

Vinyl chloride-vinyl acetate copolymers may be processed at lower temperatures than those used for the homopolymer. Their main applications are in gramophone records and flooring. Gramophone record compositions are unfilled and contain only stabiliser, lubricant, pigment, and, optionally, an antistatic agent. Preheated compound is normally moulded in compression presses at about 130–140°C. Flooring compositions contain about 30–40 parts plasticiser per hundred parts copolymer and about 400 parts filler (usually a mixture of asbestos and chalk). Internal mixer discharge temperatures are typically about 130°C whilst calender roll temperatures are usually some 10–20 degC below this.

9.7.5 Latices

Poly(vinyl chloride) is commercially available in the form of aqueous colloidal dispersions (latices). They are the uncoagulated products of emulsion polymerisation processes and are used to coat or impregnate textiles and paper. The individual particles are somewhat less than 1 μm in diameter. The latex may be coagulated by concentrated acids, polyvalent cations and by dehydration with water-miscible liquids.

Compounding ingredients are of two types:

1. Surface active agents which modify the properties of the latex.
2. Ingredients that modify the properties of the end-product.

An example of the first type is the emulsion stabiliser as exemplified by sodium oleyl sulphate, cetyl pyridinium chloride and poly(ethylene oxide) derivatives. For a number of applications it is desirable that the latex be thickened before use in which case thickening agents such as water-soluble cellulose ethers or certain alginates or methacrylates may be employed. Antifoams such as silicone oils are occasionally required.

Fillers are often employed to reduce the surface tack of the final pro-duct. Examples are talc and china clay. If powdered materials are added directly to a latex they compete for the emulsion stabiliser present and tend to coagulate the latex. They are therefore added as an aqueous dispersion prepared by ball milling the filler with water and a dispersing agent, for example a naphthalene formaldehyde sulphonate at a con-centration of about 1% on the water content. Heat and light stabilisers which are solids must be added in the same way.

Liquid ingredients, such as plasticisers, are usually added as an emulsion using ammonium oleate as an emulsifying agent. The oleate is formed *in situ* by mixing together the plasticiser and oleic acid and adding this to water containing some ammonia. The emulsion may then be stirred into the latex.

Unplasticised latexes will not form a coherent film on drying and about 25% plasticiser is required before such a film is produced. As with pastes it is necessary to flux the compound at about 150–160°C in order to produce a useful product.

9.8 APPLICATIONS

P.V.C. compounds may reasonably be considered as the most versatile of plastics materials; its uses range from building applications to toys and baby-pants.

Plasticised p.v.c. has been used for electrical insulation for many years. Although of only limited value in high frequency work it is of great value as an insulator, for direct current and low frequency alternating current carriers. It has almost completely replaced rubber insulated wire for domestic flex and is widely used industrially. P.V.C. sheathing is widely used in cables where polythene is employed as the insulator.

P.V.C. compounds, particularly unplasticised grades, are extensively used in chemical plant. It is necessary, when considering p.v.c. compounds for applications in which contact with chemicals will occur, to check that all of the ingredients will be resistant to them and also that they will not be leached out by them. Ingredients of the p.v.c. compound must not affect the nature and properties of the chemicals.   Unplasticised p.v.c. is particularly useful in acid recovery plant and in plant for handling hydrocarbons, many of which adversely affect the polyolefins.

Unplasticised p.v.c., which when carefully compounded and processed has excellent resistance to weathering, has a great potential importance to the building industry.   It is now becoming used increasingly in place of traditional materials for many uses since, when the cost of installation is taken into account, it is frequently cheaper.   Important uses include guttering and waste piping, window frames that neither corrode nor rot, and translucent roof sheeting with good flame retarding properties. Plasticised p.v.c. sheet backed by such a variety of materials as plywood, chipboard, asbestos, concrete and aluminium is of value for wall cladding for both interior and exterior application.   Garden hose from plasticised p.v.c. has been in use for many years.

In transport, p.v.c. leathercloth is widely used for upholstery and trim. Compared with leather the p.v.c. material has greater abrasion resistance and flexing resistance and can be more easily washed.   Good appearance is more likely to be maintained with p.v.c. leathercloth than with real leather.   The continued preference of many people for leather is probably due largely to their preference for the pleasant mellow smell of new leather upholstery compared with the somewhat unpleasant acrid smell of new leathercloth.   Other transportation uses include the flexible rear windows of sports cars, car covers and tool bags.   P.V.C. adhesives, generally containing a polymerisable plasticiser are finding useful outlets in car manufacture.

In house and other furnishings p.v.c. compounds find many outlets. P.V.C. leathercloth is now generally accepted in kitchen upholstery whilst printed sheet is used in strictly utilitarian kitchen and bathroom cur-taining.   Developments in paste and powder techniques have led to important metal finishing applications for example in stacking chairs. Edging strip is widely used in office furniture.   Washable wallpapers are obtained by treating the paper with p.v.c. compounds.

The importance of p.v.c. in packaging has increased in recent years. All-p.v.c. sacks, like the all-polyethylene sacks, enable fertilisers and other products to be stored out of doors.   Unplasticised p.v.c. bottles have better clarity, oil resistance and barrier properties than those made from polyethylene and are now being used for fruit juices and many other liquids.   Shampoo packs are also frequently made from p.v.c. extrudates.

There are a number of examples of p.v.c. used in personal apparel. Ladies' handbags are frequently produced from p.v.c. leathercloth whilst 'plastic rainwear' and baby-pants are made by high frequency welding of calendered plasticised p.v.c. sheet.   All-p.v.c. shoes are of value in beachwear and as standard footwear in countries at present in a low state of economic development.   P.V.C. is now proving to be an excellent abrasion resistant shoe-soling material.   The sole may be applied by injection moulding, paste injection moulding and adhesive techniques. P.V.C. is also used in beachwear, in playballs and in babies' soft dolls.

At one time one of the largest applications for p.v.c. was in mine belting. Although still important in terms of the actual tonnage of material consumption this market has not grown as have other p.v.c. outlets in recent years.

As previously mentioned there are two important applications for vinyl chloride–vinyl acetate copolymers, gramophone records and flooring compositions.

From the above examples it can be seen that the range of usage of poly-(vinyl chloride) is wide indeed. Success in any one of these applications is however vitally dependent on careful attention to compound formulation, processing conditions and to product design. Whilst p.v.c. compounds have been and are being used successfully in all of the products mentioned in this section poor examples have also been marketed in which due care was not paid to these three factors.

9.9 MISCELLANEOUS PRODUCTS

In addition to the homopolymer and copolymer compounds already described there are two other products based on poly(vinyl chloride) which merit discussion. These are poly(vinyl chloride)–rubber blends and post-chlorinated p.v.c.

9.9.1   P.V.C.–rubber blends

A number of useful compositions may be produced by blending p.v.c. with other polymers in various ratios. Of these compositions the most important are those containing only a minor proportion of a second polymer, which is usually a rubber. The usual objective in producing these blends is to produce a material of greatly improved toughness. This has been achieved with certain blends, some of which find commercial application as 'high-impact' rigid p.v.c. compounds.

Poly(vinyl chloride) is not the only thermoplastic to be toughened with rubbers and in fact much of the theory on the toughening effect of rubbers has been evolved using polystyrene–rubber blends. The theory of toughening is thus discussed in more detail in Chapter 13 dealing with polystyrene. There is evidence that with polystyrene a two-phase system of rubber particles surrounded by a thermoplastic polymer matrix is necessary for toughening. One possible mechanism of toughening is that the rubber particles may 'bridge' a propagating crack and, because of its high extensibility, will absorb a much larger amount of the energy causing the crack than an equivalent volume of the thermoplastic which has a low extensibility. There is some evidence that a two phase structure is also necessary with p.v.c.

In order to achieve a good two-phase system there are certain requirements. Firstly, the rubber should be sufficiently incompatible with the p.v.c. to form a separate phase. Secondly, the rubber should be sufficiently compatible with the p.v.c. for good adhesion behaviour of the rubber and p.v.c. phases. Thirdly, the rubber must be rubbery at any service temperature envisaged for the polymer blend. From the first and second requirements it will be seen that a semi-incompatible rubber will be required for the best results. In practice it is observed that the greatest improvement in impact strength is obtained using a rubber of solubility parameter about 0·2–0·4 units below that of p.v.c. ($\delta = 9\cdot5$). Data by Bramfitt and Heaps[19] is largely in accord with this supposition.

Table 9.5 EFFECT OF SOLUBILITY PARAMETER OF RUBBER ON ITS EFFECT ON THE IMPACT STRENGTH OF A P.V.C.–RUBBER BLEND. (5 PARTS RUBBER PER 100 PARTS P.V.C.)[19]

Rubber	Solubility parameter (cal cm^{-3})$^{1/2}$	Izod impact strength (ft lb/in. of notch)
Butadiene–2 vinyl pyridine (30:70)	9·45	3
Chlorinated polyethylene (44% chlorine)	9·4	4·4
Butadiene–styrene–acrylonitrile (67:17:16)	9·2	17·4
Butadiene–2 vinyl pyridine (40:60)	9·2	10·0
Butadiene–methyl methacrylate (35:65)	8·9	2·8
Butadiene–methyl acrylate	8·8	3·3
Butadiene–methyl isopropenyl ketone	8·5	15·9
Butadiene–diethyl fumarate	8·4	15·0

The anomalous effect of the last two rubbers in the table with low solubility parameters is possibly explained by specific interaction of p.v.c. with carbonyl and carboxyl groups present respectively in the ketone- and fumarate-containing rubbers, which give a measure of increased compatibility.

It is important to note that variation in the monomer ratio of the copolymer, which will cause changes in the solubility parameter, will have a large effect on the impact strength of the blend.

Best results are obtained with the semi-incompatible rubbers where there is little cross linking of the rubbers (i.e. it has a minimum gel content). On the other hand it has been found that cross-linked compatible rubbers can have a marked toughening effect above a certain loading. P.V.C. polymers with a high molecular weight give better impact strengths than polymers of lower molecular weight. It is claimed that emulsion polymers are superior to granular polymers in this respect.

There is only limited information available concerning the rubbers

Table 9.6 COMPARISON OF RIGID P.V.C. WITH HIGH IMPACT

Property	Method	Rigid p.v.c.	Toughened p.v.c.
Izod impact strength at 23°C (ft lb in notch)	BS2782	1–5	15
Tensile strength (lb in^{-2})	ASTM D.638	7,500	6,000
Elongation at break (%)	ASTM D.638	20–100	150
Flexural strength (lb in^{-2})	ASTM D.790	14,500	11,500
Flexural modulus (lb in^{-2})	ASTM D.638	$4\cdot25 \times 10^5$	$3\cdot25 \times 10^5$
Softening point (°C)	BS2782	80	79
Specific gravity	BS2782	1·4	1·38

now being currently used commercially. Whereas at one time nitrile rubbers were used almost exclusively, more complex products are now often employed. Some typical properties of a commercial high impact p.v.c. are compared with conventional rigid p.v.c. in Table 9.6.[19]

Chlorinated polyethylenes have been used in Germany to produce toughened blends (Hostalit Z). These are claimed[21] to be superior to rubbers containing butadiene since the absence of double bonds gives better aging properties. It has also been claimed that ASTM Izod impact strengths in the range 13–24 have been obtained with these materials.

9.9.2 After-chlorinated products

The process of post-chlorinating p.v.c. was carried out in Germany during the Second World War in order to obtain poly(vinyl chloride) products soluble in low cost solvents and which could therefore be used for fibres and lacquers. The derivative was generally prepared by passing chlorine through a solution of p.v.c. in tetrachloroethane at between 50 and 100°C. Solvents for the product included methylene chloride, butyl acetate and acetone. These materials were of limited value because of their poor colour, poor light stability, shock brittleness and comparatively low softening point.

In recent developments useful products have been obtained by chlorination at lower temperatures. In one process the reaction is said to be carried out photochemically in aqueous dispersion in the presence of a swelling agent such as chloroform. At low temperatures and in the presence of excess chlorine the chlorine adds to the carbon atom that does not already have an attached chlorine. The product is thus in effect a copolymer of vinyl chloride and symmetrical dichloroethylene. The more the chlorination the greater the density and the higher the transition temperature. It is thus possible to produce rigid compounds of higher maximum service temperature than that of normal unplasticised p.v.c. and such materials are now under active development.

REFERENCES

1. REGNAULT, V., *Ann.*, **15**, 63 (1835); *Ann. Chim.*, II, **59**, 358 (1835)
2. REBOUD, N. E., *Jahresber.*, 304 (1872)
3. *German Patent*, 278,249 (F. Klatte)
4. BAUMANN, E., *Ann.*, **163**, 312 (1872)
5. OSTROMISLENSKY, I., *Chem. Zentral.*, I, 1980 (1912)
6. *U.S. Patent*, 2,188,396
7. *European Chem. News*, **4** (89), 35 (1963)
8. *British Patent*, 842,690
9. SMALL, P. A., *J. Appl. Chem.*, **3**, 71 (1953)
10. MARVEL, C. S., SAMPLE, J. H., and ROY, M. F., *J. Am. Chem. Soc.*, **61**, 3241 (1939)
11. FLORY, P. J., *J. Am. Chem. Soc.*, **61**, 1518 (1939)
12. FULLER, C. S., *Chem. Rev.*, **26**, 162 (1940)
13. NATTA, G., and CARRADINI, P., *J. Polymer Sci.*, **20**, 262 (1956)
14. MATTHEWS, G. A. R., and PEARSON, R. B., *Plastics*, **28** (307), 98 (1963)
15. Technical Trade Literature, British Geon Ltd., London
16. JACOBSON, U., *Brit. Plastics*, **34**, 328 (1961)
17. SKINNER, S. J., and BOLAM, S. E., *Rubber and Plastics Age*, **37**, 169 (1956)
18. WELDON, L. H. P., *Trans. Plastics Inst.*, **24**, 303 (1956)

19. MATTHEWS, G. A. R., *Advances in P.V.C. Compounding and Processing* (Ed. KAUFMAN, M.), Maclaren, London, p. 53 (1962)
20. STOCKER, J. H. J., A.P.I. Project Thesis (1960)
21. FREY, H. H., *Kunstoffe*, **49**, 50 (1959)

BIBLIOGRAPHY

CHEVASSUS, F., and DE BROUTELLES, R., *The Stabilization of Vinyl Chloride*, Edward Arnold, London (1963)
KREKELER, K., and WICK, G., *Kunstoff-Handbuch* (Band II)–*Polyvinylchlorid*, Hanser, Munich, Pts. I and II (1963)
MATTHEWS, G. A. R., *Advances in P.V.C. Compounding and Processing* (Ed. KAUFMAN, M.), Maclaren, London, p. 53 (1962)
MAYO SMITH, W., *Vinyl Resins*, Reinhold, New York (1958)
PENN, W. S., *P.V.C. Technology*, MacLaren, London (1962)

10

Fluorine-containing Polymers

10.1 INTRODUCTION

The high thermal stability of the carbon–fluorine bond has led to considerable interest in fluorine-containing polymers as heat resistant plastics and rubbers. The first patents, taken out by I.G. Farben in 1934, related to polychlorotrifluoroethylene[1] (p.c.t.f.e.) (Fig. 10.1 (a)), these materials being subsequently manufactured in Germany and the United States. P.C.T.F.E. has been of limited application and it was the discovery of polytetrafluoroethylene (p.t.f.e.) (Fig. 10.1 (b)) by Plunkett[2] in 1938 which gave an impetus to the study of fluorine-containing polymers. Today, p.t.f.e. probably accounts for at least 90% of the fluorinated polymers produced and, in spite of its high cost, has a great diversity of uses.

Other fluorine-containing materials which have become highly specialist plastics materials include poly(vinyl fluoride) (Fig. 10.1 (c)), poly(vinylidene fluoride) (Fig. 10.1 (b)) and the tetrafluoroethylene–hexafluoropropene copolymers, whilst others have become of interest as oil-resisting rubbers.

10.2 POLYTETRAFLUOROETHYLENE

In addition to the presence of stable C—F bonds, the p.t.f.e. molecule possesses other features which lead to materials of outstanding heat resistance, chemical resistance and electrical insulation characteristics and with a low coefficient of friction. It is today produced by a number of chemical manufacturers such as Du Pont (Teflon), I.C.I. (Fluon), Hoechst (Hostaflon TF), Rhône-Poulenc (Soreflon), Montecatini (Algoflan), Nitto Chemical-Japan (Tetraflon) and Daikin Kogyo-Japan (Polyflon).

10.2.1 Preparation of monomer

Tetrafluoroethylene was first prepared in 1933. The current commercial syntheses are based on fluorspar, sulphuric acid and chloroform.

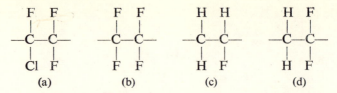

Fig. 10.1   (a) *Polychlorotrifluoroethylene (p.c.t.f.e.).*   (b) *Polytetrafluoroethylene (p.t.f.e.).*   (c) *Poly(vinyl fluoride).*   (d) *Poly(vinylidene fluoride)*

The reaction of fluorspar (CaF_2) and sulphuric acid yields hydrofluoric acid

$$CaF_2 + H_2SO_4 \longrightarrow CaSO_4 + 2HF$$

Treatment of chloroform, obtained by reacting methanol and chlorine, with the hydrofluoric acid yields monochlorodifluoromethane, also used as a refrigerant, which is a gas boiling at $-40 \cdot 8°C$.

$$CHCl_3 + 2HF \longrightarrow CHClF_2 + 2HCl$$

The monochlorodifluoromethane may be converted to tetrafluoroethylene by pyrolysis, for example, by passing through a platinum tube at 700°C.

$$2CHClF_2 \longrightarrow CF_2 = CF_2 + 2HCl$$

Other fluorine compounds are produced during pyrolysis including some highly toxic ring structures. Since very pure monomer is required for polymerisation, the gas is first scrubbed to remove any hydrochloric acid and then distilled to separate other impurities. Tetrafluoroethylene has a boiling point of $-76 \cdot 3°C$. For safe storage under pressure the oxygen content should be below 20 p.p.m. Traces of compounds which react preferentially with oxygen such as 0·5% dipentene, benzaldehyde or methyl methacrylate may be added as stabilisers.

10.2.2   Polymerisation

Pure uninhibited tetrafluoroethylene can polymerise with violence, even at temperatures initially below that of room temperature. There is little published information concerning details of commercial polymerisation. In one patent[3] example a silver-plated reactor was quarter-filled with a solution consisting of 0·2 parts ammonium persulphate, 1·5 parts borax and 100 parts of water, and with a pH of 9·2. The reactor was closed, evacuated and 30 parts of monomer were let in. The reactor was agitated for one hour at 80°C and after cooling gave an 86% yield of polymer.

P.T.F.E. is made commercially by two major processes, one leading to the so-called 'granular' polymer and the second leading to a dispersion of polymer of much finer particle size and lower molecular weight. One method of producing the latter[4] involved the use of a 0·1% aqueous disuccinic acid peroxide solution. The reactions were carried out at temperatures up to 90°C. It is understood that the Du Pont dispersion polymers, at least, are produced by methods based on the patent containing the above example.

10.2.3 Structure and properties

Polytetrafluoroethylene is a linear polymer free from any significant amount of branching (Fig. 10.2).

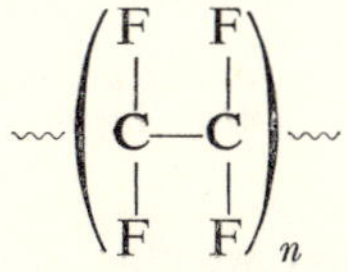

$$\left(\begin{array}{c} \text{F} \quad \text{F} \\ | \quad | \\ \text{C} - \text{C} \\ | \quad | \\ \text{F} \quad \text{F} \end{array} \right)_n$$

Fig. 10.2

Whereas the molecule of polyethylene is in the form of a planar zigzag in the crystalline zone this is sterically impossible with that of p.t.f.e. due to the fluorine atoms being larger than those of hydrogen. As a consequence the molecule takes up a twisted zigzag with the fluorine atoms packing tightly in a spiral around the carbon—carbon skeleton. A complete turn of the spiral will involve over 26 carbon atoms below 19°C and 30 above it, there being a transition point involving a 1% volume change at this temperature. The compact interlocking of the fluorine atoms leads to a molecule of great stiffness and it is this feature which leads to the high crystalline melting point and thermal form stability of the polymer.

The intermolecular attraction between p.t.f.e. molecules is very small, the computed solubility parameter being 6·2. The polymer in bulk does not thus have the high rigidity and tensile strength which is often associated with polymers with a high softening point.

The carbon—fluorine bond is very stable. Further, where two fluorine atoms are attached to a single carbon atom there is a reduction in the C—F bond distance from 1·42 Å to 1·35 Å. As a result bond strengths may be as high as 120 kcal/mole. Since the only other bond present is the stable C—C bond, p.t.f.e. has a very high heat stability, even when heated above its crystalline melting point of 327°C.

Because of its high crystallinity and incapability of specific interaction, there are no solvents at room temperature. At temperatures approaching the melting point certain fluorinated liquids such as perfluorinated kerosenes will dissolve the polymer.

The properties of p.t.f.e. are dependent on the type of polymer and the method of processing. The polymer may differ in particle size and/or molecular weight. The particle size will influence ease of processing and the quantity of voids in the finished product whilst the molecular weight will influence crystallinity and hence many physical properties. The processing techniques will also affect both crystallinity and void content.

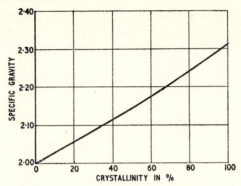

Fig. 10.3.   Density as a function of crystallinity
in p.t.f.e.   (After Thomas et al.[5])

The weight-average molecular weights of commercial polymers appear to be very high and are in the range 400,000 to 9,000,000.   I.C.I. report that their materials have a molecular weight in the range 500,000 to 5,000,000 and a percentage crystallinity greater than 94% as manufactured.   Fabricated parts are less crystalline.   The degree of crystallinity of the finished product will depend on the rate of cooling from the processing temperatures.   Slow cooling will lead to high crystallinity with fast cooling giving the opposite effect.   Low molecular weight materials will also be more crystalline.

Fig. 10.3 shows the relationship between percentage crystallinity and specific gravity at 23°C.   By measuring the specific gravity of mouldings

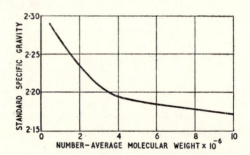

Fig.   10.4.   Standard   specific   gravity   of
p.t.f.e.   as   a   function   of   molecular   weight.
(After Thomas et al.[5])

prepared under rigorously controlled conditions Thomas[5] and co-workers were able to obtain a measure of molecular weight which they were able to calibrate with results obtained by end-group and infrared analysis (Fig. 10.4).

It is observed that the dispersion polymer, which is of finer particle size and lower molecular weight, gives products with a vastly improved resistance to flexing and also distinctly higher tensile strengths. These improvements appear to arise through the formation of fibre-like structures in the mass of polymer during processing.

10.2.4   General properties

P.T.F.E. is a tough, flexible, non-resilient material of moderate tensile strength but with excellent resistance to heat, chemicals and to the passage of an electric current.   It remains ductile in compression at temperatures as low as $4°K$ $(-269°C)$.

Table 10.1 lists some typical values for p.t.f.e. mouldings compared with other fluorine-containing thermoplastics.

As with other plastics materials, temperature has a considerable effect on mechanical properties.   This is clearly illustrated in Fig. 10.5 in the

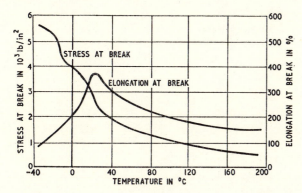

Fig. 10.5. *Effect of temperature on the stress at break and elongation at break of p.t.f.e.*[6]   (*Reproduced by permission of ICI Plastics Division*)

case of stress to break and elongation at break.   Even at 20°C unfilled p.t.f.e. has a measurable creep with compression loads as low as 300 lb in^{-2}.

The coefficient of friction is unusually low and stated to be lower than that of any other solid.   A number of different values have been quoted in the literature but are usually in the range 0·02–0·10 for polymer to polymer.

P.T.F.E. is an outstanding insulator over a wide range of temperature and frequency.   The volume resistivity (100 sec value) exceeds 10^{18} Ω cm and it appears that any current measured is a polarisation current rather than a conduction current.   The power factor is negligible in the temperature range $-60°C$ to $+250°C$ at frequencies up to 10^{10} c/s.   The polymer has a low dielectric constant similarly unaffected by frequency.   The only effect of temperature is to alter the density which has been found to influence the dielectric constant according to the relationship

$$\text{Dielectric constant} = \frac{1 + 0·238D}{1 - 0·119D}$$

where D = specific gravity

Table 10.1 PROPERTIES OF P.T.F.E. AND OTHER FLUORINE-CONTAINING THERMOPLASTICS

Property	ASTM test	P.T.F.E.	P.C.T.F.E.	P.V.F.c	P.V.D.F.	T.F.E.–H.F.P.
Specific gravity	D.792	2·1–2·3	2·1	1·38–1·57*	1·76	2·16
Tensile strength at 23°C (lb in^{-2})	D.638	2,500–3,800	4,300–5,700	9·6–19 $\times$ 10^3*	7,000	2,700–3,100
Elongation at break at 23°C (%)	D.638	200–300	100–200	110–260*	100–300	250–350
Izod impact strength at 23°C (ft lb/in.)	D.256	2·0	1·2–1·3	—	3·5	2·9
Deflection temp. under load of 66 lb in^{-2} (°C)	D.648	121	58	—	150	88
Water absorption (%)	D.570	0·005	negligible	<0·5	0·04	negligible
Coefficient of friction[a]	—	0·09–0·12	0·4	—	0·04	0·08–0·425
Power factor 60 c/s	D.150	<0·0003	0·010	0·01 (100 c/s)	0·049	<0·0003
Power factor 10^6 c/s	D.150	<0·0003	0·010	0·08 (10^4 c/s)	0·17	<0·0003
Dielectric constant 60 c/s	D.150	2·1	3·0	6·8–8·5 (10^3 c/s)	8·4	2·1
Dielectric constant 10^6 c/s	D.150	2·1	2·5	—	6·6	2·1
Volume resistivity (Ω cm)	D.257	>10^{18}	10^{18}	10^{13}–10^{14}	10^{14}	>10^{13}
Electric strength[b] (V/0·001 in)	D.149	400–500	530	3,000–6,000*	260	500–600

a Polymer to metal.   b Short time on 0·080 in thick sheet.   c Tests on Tedlar film.   Different test methods involved where marked*.

Fig. 10.6 shows the influence of temperature on specific volume (reciprocal specific gravity). The exact form of the curve is somewhat dependent on the crystallinity and the rate of temperature change. A small transition is observed at about 19°C and a first order transition (melting) at about 327°C. Above this temperature the material does not exhibit true flow but is rubbery. A melt viscosity of 10^{10}–10^{11} poises has been measured at about 350°C. A slow rate of decomposition may be

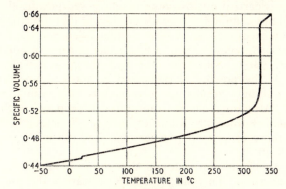

Fig. 10.6. *Variation of specific volume of p.t.f.e. with temperature.*[6] *(Reproduced by permission of ICI Plastics Division)*

detected at the melting point and this increases with a further increase in temperature. Processing temperatures, except possibly in the case of extrusion, are, however, rarely above 380°C.

The chemical resistance of p.t.f.e. is exceptional. There are no solvents and it is only attacked at room temperature by molten alkali metals and in some cases by fluorine. Treatment with a solution of sodium metal in liquid ammonia will sufficiently alter the surface of a p.t.f.e. sample to enable it to be cemented to other materials using epoxide resin adhesives.

Although it has good weathering resistance, p.t.f.e. is degraded by high energy radiation. Exposure to a dosage of 70 Mrad will halve the tensile strength of a given sample. The polymer is not wetted by water and does not measurably absorb it. The permeability to gases is low, the water vapour transmission rate being approximately half that of low density polyethylene and poly(ethylene terephthalate).

10.2.5 Processing

P.T.F.E. is normally available in three forms:

1. Granular polymers with median particle size of 300 or 600 μm.
2. Dispersion polymer obtained by coagulation of a dispersion. It consists of agglomerates with an average diameter of 450 μm made up of primary particles 0·1 μm in diameter.
3. Dispersions (latices) containing about 60% polymer in particles with an average diameter of about 0·16 μm.

The exceptionally high melt viscosity above the melting point (about 10^{10}–10^{11} poises at 350°C) prevents the use of the usual techniques for processing thermoplastics.   In the case of granular polymers, methods allied to those used with ceramics and in powder metallurgy are employed instead.   In principle this involves preforming the powder, usually at room temperature, sintering at a temperature above the melting point, typically at about 370°C, and then cooling.

Preforming is carried out by compressing sieved powder that has been evenly loaded into a mould.   It has been shown[7] that if the pressure is too low there is an excessively large void content whereas if the pressures are too high cleavage planes are produced by one portion of polymer sliding over another.   Best results are obtained using a pressure of about 3 ton in^{-2}.   If the powder is preheated at 100°C immediately before preforming, optimum results are obtained at preforming pressures of 1·5 ton in^{-2}.   Mouldings made from preforms prepared in this way, i.e. using preheated powder, are found to have tensile strengths appreciably higher than those using cold powder (e.g. 2,800 lb in^{-2} compared with about 2,000 lb in^{-2}).

For many applications it is found that the technique of free sintering is quite satisfactory.   This simply involves heating the preform in an oven at about 380°C for a time of 90 minutes plus a further 60 minutes for every 0·25 in. thickness.   For example a sample 0·5 in. thick will require sintering for 3·5 hours.   The ovens should be ventilated to the open air to prevent toxic decomposition products accumulating in the working area.

After sintering, the moulding is cooled.   Thin sections, i.e. less than $\frac{1}{16}$ in. thick, may be removed from the oven and allowed to cool naturally or may be quenched by placing between cold flat metal plates and light pressure applied.   Sections up to 1 in. thick are preferably cooled in an oven, cooling at a rate of 20 degC/hr when maximum dimensional stability is required.   Thicker sections are usually cooled under pressure.   In this case the preform can be sintered in the preform mould and the mould and moulding then transferred to a press in which they are cooled under a gradually increased pressure.   It is necessary that this should reach the preforming pressure as the sample goes through the transition temperature (327°C) and it should be maintained until the sample reaches room temperature.

Shrinkage of about 5–10% occurs at right angles to the direction of the preforming force.   The amount of shrinkage is mainly dependent on the rate of cooling, but also to a minor extent, on the preforming pressure.

The above process is limited to simple shapes whose principal dimension is not more than four times, and preferably less than twice, that of the next largest dimension.   More intricate shapes must be made by machining or in some instances by a coining operation which involves stamping a sintered moulding of the same weight and approximate dimensions as the finished part at 320°C.

Granular polymer may also be extruded, albeit at very low rates (1–6 in/min), by means of both screw and ram extruders.   In both machines the extruder serves to feed cold powder into a long, heated sintering die whose overall length is about 90 times its internal diameter. The polymer, preferably a presintered grade of 600 μm particle size, is compacted, sintered and partially cooled before leaving the die. Since compacting is still taking place as the polymer enters the sintering zone it is possible to obtain extrudates reasonably free from voids, a factor which is reflected in their high tensile strength and elongation.

P.T.F.E. mouldings and extrudates may be machined without difficulty. Film may be obtained by peeling from a pressure sintered ring and this may be welded to similar film by heat sealing under light pressure at about 350°C.

Dispersion polymer, which leads to products with improved tensile strength and flex life, is not easily fabricated by the above techniques. It has, however, been found possible to produce preforms by mixing with 15–25% of a lubricant, extruding and then removing the lubricant and sintering. Because of the need to remove the lubricant it is only possible to produce thin section extrudates by this method.

In a typical process a preform billet is produced by compacting a mixture of 83 parts p.t.f.e. dispersion polymer and 17 parts of petroleum ether (100–120°C fraction). This is then extruded using a vertical ram extruder. The extrudate is subsequently heated in an oven at about 105°C to remove the lubricant, this being followed by sintering at about 380°C. By this process it is possible to produce thin-walled tube with excellent flexing fatigue resistance and to coat wire with very thin coatings of polymer.

Tape may be made by a similar process. In this case the lubricant selected is a non-volatile oil. The preform is placed in the extruder and a rod extruded. The rod is then passed between a pair of calender rolls at about 60–80°C. The unsintered tape is often used for lapping wire and for making lapped tube. Sintering is carried out after fabrication. The current most important application of unsintered tape is in pipe-thread sealing.

If sintered tape is required, the product from the calender is first degreased by passing through boiling trichloroethylene and then sintered by passing through a salt bath. This tape is superior to that made by machining from granular polymer mouldings.

P.T.F.E. dispersions[8] may be used in a variety of ways. Filled p.t.f.e. moulding material may be made by stirring fillers into the dispersion, coagulating with acetone, drying at 280–290°C and disintegrating the resulting cake of material. Asbestos and glass cloth fabrics may be impregnated with p.t.f.e. by passing through the dispersion, drying and sintering the polymer. Glass-cloth p.t.f.e. laminates may be produced by plying-up layers of impregnated cloth and pressing at about 330°C. The dispersions are also used for coating p.t.f.e. on to metal to produce surfaces which are non-adhesive and which have a very low coefficient of friction.

Whenever p.t.f.e. is used in a sintered form there are two points that should always be borne in mind. Firstly, at sintering temperatures toxic cyclic fluorinated compounds are formed and it is thus necessary to ventilate ovens and to use fume hoods whenever fumes of such toxic compounds are produced. It is particularly important that p.t.f.e. dust should not contaminate cigarettes or tobacco since the smoker will inhale the decomposition products. Secondly, scrupulous standards of cleanliness are necessary to prevent dust, which is frequently organic in nature, contaminating p.t.f.e. products before sintering. If this did happen organic dust would carbonise on sintering leaving a product both unsightly and with inferior electrical properties.

10.2.6 Additives

Because of the high processing temperatures there are few pigments suitable for use with p.t.f.e. A number of inorganic pigments particularly the cadmium compounds, iron oxides and ultramarines may however be used.

The resistance of p.t.f.e. to creep can be improved by blending in up to 25% of glass or asbestos fibre using p.t.f.e. dispersions as mentioned in the previous section. By the same technique alumina, silica and lithia may be incorporated to give compounds of improved dimensional stability coupled with good electrical insulation properties. Molybdenum disulphide and graphite improve dimensional stability without losing the low coefficient of friction whilst the use of barium ferrite will produce a material that can be magnetised. The incorporation of titanium dioxide serves to increase the dielectric constant whilst certain compounds of boron increase the resistance to neutron bombardment.

10.2.7 Applications

The use of p.t.f.e. in a great diversity of applications may be ascribed to the following properties:

1. Chemical inertness.
2. Exceptional weathering resistance. Samples exposed in Florida for 10 years showed little change in physical properties.
3. The excellent electrical insulation characteristics.
4. The excellent heat resistance.
5. The non-adhesive properties.
6. The very low coefficient of friction.

However, world production is probably still below 6,000 tons per annum and this is a reflection of the high volume cost, the rather specialised techniques involving lengthy processing times and to a smaller extent the high creep rate under load.

Because of its chemical inertness over a wide temperature range it is used in a variety of seals, gaskets, packings, valve and pump parts and in laboratory equipment.

Its excellent electrical insulation properties lead to its use in wire insulation, valve holders, insulated transformers, hermetic seals for condensers, in laminates for printed circuitry and for many other miscellaneous electrical applications.

P.T.F.E. is used for lining chutes and coating other metal objects where a low coefficient of friction or non-adhesive characteristic are required. Because of its excellent flexing resistance, inner linings made from dispersion polymer are used in flexible steam hose. A variety of mouldings are used in aircraft and missiles and also in other applications where use at elevated temperatures is required.

Because of its high volume cost p.t.f.e. is not generally used to produce large objects. In many cases however it is possible to coat a metal object with a layer of p.t.f.e. and hence meet the particular requirement.

10.3 TETRAFLUOROETHYLENE–HEXAFLUOROPROPYLENE COPOLYMERS

These materials were first introduced by Du Pont in 1956 and are now known as Teflon FEP resins. (FEP = fluorinated ethylene–propylene.) They are essentially a melt processable form of p.t.f.e.

The commercial polymers are mechanically similar to p.t.f.e. but with a somewhat greater impact strength. They also have the same excellent electrical insulation properties and chemical inertness. Weathering tests in Florida showed no change in properties after four years. The material also shows exceptional non-adhesiveness. The coefficient of friction of the resin is low but somewhat higher than that of p.t.f.e. Films up to 0·010 in thick show good transparency.

The maximum service temperature is about 60 degC lower than that of p.t.f.e. for use under equivalent conditions. Continuous service at 200°C is possible for a number of applications. The polymer melts at about 290°C.

Injection moulding and extrusion may be carried out at temperatures in the range of 300–380°C. The polymer has a high melt viscosity and melt fracture occurs at a lower shear rate (about 10^2 sec^{-1}) than with low density polythene (about 10^3 sec^{-1}) or nylon 66 (about 10^5 sec^{-1}). Extruders should thus be designed to operate at low shear rates whilst large runners and gates are employed in injection moulds.

The advantage of being able to injection mould and extrude these copolymers has perhaps had a less marked effect than might have been expected. This is because the fabrication of p.t.f.e. has been developed by firms closely related to the engineering industries rather than by the conventional plastics fabricators. The p.t.f.e. fabricators, because they do not normally possess conventional injection moulding and extrusion machines, would see no obvious advantage in melt processability. At the same time the conventional plastics fabricators, if they wished to enter the field of fabricated fluorine-containing thermoplastics, would have to modify their existing machinery in order to cope with the high processing temperatures and high melt viscosity. In spite of these retarding influences the use of FEP copolymers may be expected to grow.

At the present time they are used for a variety of electrical and chemically resistant mouldings, for coatings, for flexible printed circuits and for wire insulation.

10.4 POLYCHLOROTRIFLUOROETHYLENE (P.C.T.F.E.)

Polychlorotrifluoroethylene was the first fluorinated polymer to be produced on an experimental scale and polymers were used in Germany and in the United States early in the Second World War. P.C.T.F.E. was used, in particular, in connection with the atomic bomb project in the handling of corrosive materials such as uranium hexafluoride.

The monomer may conveniently be produced from hexachloroethane via trifluorotrichloroethane

$$CCl_3 \cdot CCl_3 \xrightarrow{\text{HF}} CClF_2 \cdot CCl_2F \xrightarrow{\text{Zn}} CF_2{=}CFCl$$

Polymerisation may be carried out by techniques akin to those used in the manufacture of p.t.f.e. The preparation of polymers in yields of up to 88% are described in one patent.[9] Water was used as a diluent in concentrations of from one to five times the weight of the monomer, a gas with a boiling point of $-27 \cdot 9°C$. Solid polymers were formed with reaction temperatures of 0–40°C; at higher reaction temperatures liquid polymers are formed.

Pressures varied from 20 to 1,500 lb in^{-2} and reaction times were of the order of 5–35 hours. Reaction promoters included peroxides and salts of persulphuric and perphosphoric acids. 'Activators', 'accelerators' and buffering agents were also discussed in the patent. The process of manufacture of Kel-F is understood to be based on this patent.

The major differences in properties between p.t.f.e. and p.c.t.f.e. can be related to chemical structure. The introduction of a chlorine atom, which is larger than the fluorine atom, breaks up the very neat symmetry which is shown by p.t.f.e. and thus reduces the close chain packing. It is still however possible for the molecules to crystallise, albeit to a lower extent than p.t.f.e. The introduction of the chlorine atom in breaking up the molecular symmetry appears to increase the chain flexibility and this leads to a lower softening point. On the other hand the higher interchain attraction results in a harder polymer with a higher tensile strength. The unbalanced electrical structure adversely affects the electrical insulation properties of the material and limits its use in high frequency applications.

Because of the lower tendency to crystallisation it is possible to produce thin transparent films.

The chemical resistance of p.c.t.f.e. is good but not as good as p.t.f.e. Under certain circumstances substances such as chlorosulphonic acid, molten caustic alkalis and molten alkali metal will adversely affect the material. Alcohols, acids, phenols and aliphatic hydrocarbons have little effect but certain aromatic hydrocarbons, esters, halogenated hydrocarbons and ethers may cause swelling at elevated temperatures.

The polymer melts at 216°C and above this temperature shows better cohesion of the melt than p.t.f.e. It may be processed by conventional thermoplastics processing methods at temperatures in the range 230–290°C. Because of the high melt viscosity high injection moulding pressures are required.

P.C.T.F.E. is more expensive than p.t.f.e. and its use is comparatively limited. With the advent of FEP copolymers the advantage of melt processability is no longer, alone, a sufficient justification for its use. The particular advantages of the material are its transparency in thin films and its greater hardness and tensile strength as compared to p.t.f.e. and FEP copolymers. Examples of its use include transparent windows for chemical and other apparatus where glass or other materials cannot be used, seals, gaskets and O-rings and some electrical applications such as hook-up wire and terminal insulators.

P.C.T.F.E. is marketed by Hoechst as Hostaflon C2 and in the United States by Minnesota Mining and Manufacturing (Kel-F) and Allied Chemical (Halon). Typical values for various physical properties are given in Table 10.1.

10.5 POLY(VINYL FLUORIDE) (P.V.F.)

Poly(vinyl fluoride) was first introduced in the early 1960's, in film form, by Du Pont under the trade name Tedlar. Details of the commercial method of preparing the monomer have not been disclosed but it may be prepared by addition of hydrogen fluoride to acetylene at about 40°C.

$$CH{\equiv}CH + HF \xrightarrow[\text{on Charcoal}]{\text{HgCl}_2} CH_2{=}CHF$$

It may also be prepared by pyrolysis of 1,1-difluoroethane at 725°C over a chromium fluoride catalyst in a platinum tube or by the action of zinc dust on difluorobromoethane at 50°C.

The polymers were first described by Newkirk.[10] Polymerisation may be brought about by subjecting acetylene-free vinyl fluoride to pressures of up to 1,000 atm at 80°C in the presence of water and a trace of benzoyl peroxide.

Although poly(vinyl fluoride) resembles p.v.c. in its low water absorption, resistance to hydrolysis, insolubility in common solvents at room temperature and a tendency to split off hydrogen halides at elevated temperatures, it has a much greater tendency to crystallise. This is because the fluorine atom (cf. the chlorine atom) is sufficiently small to allow molecules to pack in the same way as polythene.

P.V.F. has better heat resistance than p.v.c. and exceptionally good weather resistance. It will burn slowly. Instability at processing temperatures makes handling difficult but this problem has been sufficiently overcome for Du Pont to be able to market their Tedlar film.

P.V.F. film is now being used in the manufacture of weather-resisting laminates, for agricultural glazing and in electrical applications.

10.6 POLY(VINYLIDENE FLUORIDE)

This material first became available on a semi-commercial basis in the United States in 1961 under the name Kynar (Pennsalt Chemicals Corp.). Two grades were offered, one with a molecular weight of approximately 600,000 and the other with a molecular weight of about 300,000. The crystalline melting point is about 170°C and this enables the polymer to be injection moulded and extruded in the usual way.

Poly(vinylidene fluoride) has good tensile and impact strengths and is flexible in thin sections. Although it has generally good chemical resistance, strongly polar solvents such as dimethyl acetamide tend to dissolve the polymer, whilst some strongly basic primary amines such as *n*-butylamine lead to discoloration and embrittlement. The material has very good weather resistance and may be used continuously at temperatures up to 150°C.

10.7 FLUORINE-CONTAINING RUBBERS

The fluorinated elastomers[11] were originally developed for military purposes, particularly in applications where oil resistance coupled with good resistance to low temperatures was required. Commercial materials are now available which partly satisfy these requirements and which also satisfy, to some extent, the requirements of radiation resistance, weather resistance and heat resistance. These elastomers have been developed largely in the United States, in many cases as a result of contracts for the U.S. Army or Air Force. Many hundreds of polymers and copolymers have been investigated and some have become commercially available on a small scale.

The commercial materials are generally incapable of crystallisation, this being ensured either by the use of copolymers or atactic homopolymers. The polymers also invariably contain C—H bonds which help to increase the chain flexibility, an essential feature of a rubber, and which also provide a site for cross linking via polyamines, peroxides and polyisocyanates.

The first commercial material was Fluoroprene introduced by Du Pont

Table 10.2 COMMERCIAL FLUORINATED RUBBERS

Type	Designation	Manufacturer	Remarks
Chlorotrifluoroethylene–vinylidene fluoride 50:50	Kel F Elastomer 5500	before 1957—M. W. Kellogg since 1957—M.M.M.	high compression set at elevated temps. Unsuitable at sub-zero temps.
30:70	Kel F Elastomer 3700		
Hexafluoropropylene–vinylidene fluoride	Fluorel Viton	M.M.M. Du Pont	30:70 ratio
Fluoracrylate	Poly 1F4 Poly 2F4	M.M.M. M.M.M.	ex-Poly FBA ex-Poly FMFPA

M.M.M. = Minnesota Mining and Manufacturing Co.

in 1948 and a polymer of 2-fluorobutadiene-1,3. However its properties did not justify its high cost and production was discontinued. In the 1950's a number of further materials were introduced superior in various respects to Fluoroprene and those which have been marketed are indicated in Table 10.2. The formulae of Poly 1F4 and Poly 2F4 are as shown in Fig. 10.7.

These two materials have an oil resistance inferior to the other two classes of rubber and their use is now very limited. The h.f.p.–v.d.f.

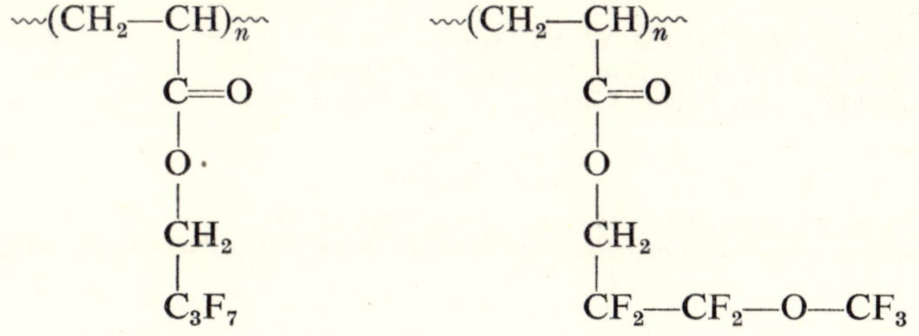

Fig. 10.7 (a) Poly 1F4. (b) Poly 2F4

copolymers (Viton and Fluorel) are produced by emulsion polymerisation using potassium persulphate as initiator. They are generally superior in heat aging and oil resistance to the c.t.f.e.–v.d.f. elastomers but have similarly poor low temperature properties. They are currently the most important class of fluorine-containing rubbers.

Fluorine-containing silicone rubbers show low temperature properties better than other oil-resistant rubbers but are not as good as the ordinary

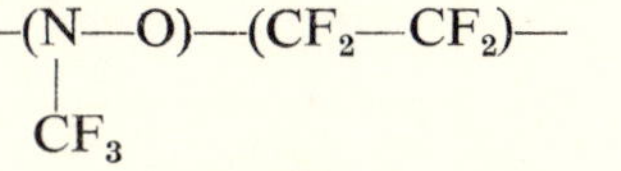

$$-(N-O)-(CF_2-CF_2)-$$
$$|$$
$$CF_3$$

Fig. 10.8

silicone rubbers in this respect. They have an oil resistance better than the nitrile and polychloroprene rubbers but inferior to the h.f.p.–v.d.f. material. They show excellent heat aging resistance.

Of the many other rubbers which have progressed beyond the test-tube scale of operation the fluorinated nitroso compounds are of particular interest. They are typically of structure in Fig. 10.8 being prepared by copolymerisation of trifluoronitrosomethane with tetrafluoroethylene. Polymers have been prepared of high molecular weight which will not burn in a Bunsen flame. They have a glass transition temperature of $-50°C$ and a very good chemical resistance, even to chlorine trifluoride. The material may be vulcanised by polyamines such as a combination of triethylene tetramine and hexamethylene diamine carbamate. Vulcanisates have a low tensile strength but this can be improved by reinforcement with fine fillers. Other nitroso-type copolymers have been prepared and it is to be expected that varieties of this class of material will be marketed.

REFERENCES

1. *German Patent*, 677,071;   *French Patent*, 796,026;   *British Patent*, 465,520 (I.G. Farben)
2. *U.S. Patent*, 2,230,654 (Kinetic Chemicals Inc.)
3. *U.S. Patent*, 2,393,967 (Du Pont)
4. *U.S. Patent*, 2,534,058 (Du Pont)
5. THOMAS, P. E., LONTZ, J. F., SPERATI, C. A., and MCPHERSON, J. L., *Soc. Plastics Engrs J.*, **12**, (5) 89 (1956)
6. Technical Trade Literature, I.C.I. Plastics Ltd., Welwyn Garden City
7. BOWLEY, G. W., *Plastics Progress 1957* (Ed. P. Morgan), Iliffe Books Ltd., London (1958)
8. WHITCUT, H. M., *Plastics Progress 1955* (Ed. P. Morgan), Iliffe Books Ltd., p. 103 (1956)
9. *U.S. Patent*, 2,689,241 (M. W. Kellogg)
10. NEWKIRK, A. E., *J. Am. Chem. Soc.*, **68**, 2467 (1946)
11. MONTERMOSO, J. C., *Rubber Chem. Tech.*, **34**, 1521 (1961)

BIBLIOGRAPHY

RUDNER, M. A., *Fluorocarbons*, Reinhold, New York (1958)
SCHILDKNECHT, C. E., *Vinyl and Related Polymers*, John Wiley, New York (1952)

11

Poly(vinyl acetate) and its Derivatives

11.1 INTRODUCTION

Because of its high cold flow, poly(vinyl acetate) is of little value in the form of mouldings and extrusions. However, because of its good adhesion to a number of substrates, and to some extent because of its cold flow a large quantity is produced for use in emulsion paints, adhesives and various textile finishing operations. A minor proportion of the material is also converted into poly(vinyl alcohol) and the poly(vinyl acetal)s which are of some interest to the plastics industry.

11.2 POLY(VINYL ACETATE)

11.2.1 Preparation of the monomer

Vinyl acetate is commonly prepared industrially[1] by the reaction of acetylene with acetic acid.

This reaction may be carried out either in the liquid or vapour phase. In a typical liquid-phase preparation, acetylene is passed through an agitated solution of glacial acetic acid and acetic anhydride containing mercuric sulphate, preferably formed *in situ*, in a finely divided state as catalyst.

Owing to the tendency for ethylidene diacetate to be formed at elevated temperatures, care is taken for the rapid removal of vinyl acetate from the reaction vessel as soon as it is formed (Fig. 11.1).

In a typical system the reaction vessel is at 75–80°C and the vinyl acetate formed is swept out into a condenser at 72–74°C by means of circulating excess acetylene. This prevents distillation of higher boiling components but allows the vinyl acetate and acetylene through. The former is separated out by cooling and the acetylene recycled.

Vapour-phase synthesis may be carried out by passing a mixture of acetylene and acetic acid through a reaction tube at 210–250°C. Typical catalysts for this reaction are cadmium acetate, zinc acetate and zinc

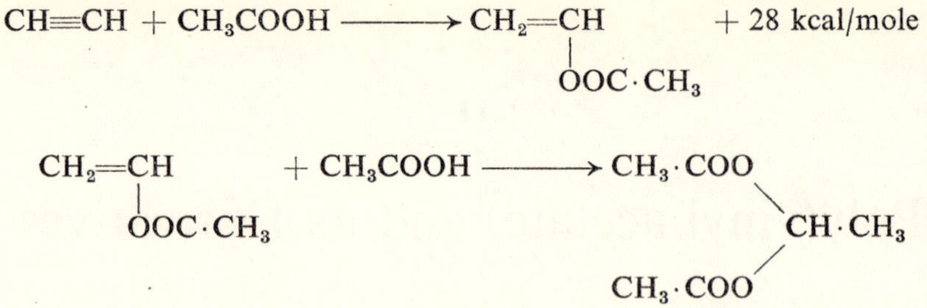

$$CH\equiv CH + CH_3COOH \longrightarrow CH_2=CH \qquad + 28 \text{ kcal/mole}$$
$$\overset{|}{OOC \cdot CH_3}$$

$$CH_2=CH \qquad + CH_3COOH \longrightarrow CH_3 \cdot COO$$
$$\overset{|}{OOC \cdot CH_3} \qquad\qquad\qquad\qquad\qquad CH \cdot CH_3$$
$$CH_3 \cdot COO$$

Fig. 11.1

silicate. The monomer in each of the above mentioned processes is purified by distillation.

Purified monomer is usually inhibited before shipment by such materials as copper resinate, diphenylamine or hydroquinone, which are generally removed before polymerisation. The monomer is a sweet-smelling liquid partially miscible with water and with the following properties: boiling point at 760 mmHg, 72·5°C; specific gravity at 20°C, 0·934; refractive index n_D^{20}, 1·395; vapour pressure at 20°C, 90 mmHg.

In 1953 the Celanese Corporation of America introduced a route for the production of vinyl acetate from light petroleum gases. This involved the oxidation of butane which yields such products as acetic acid and acetone. Two derivatives of these products are acetic anhydride and acetaldehyde which then react together to give ethylidene diacetate (Fig. 11.2).

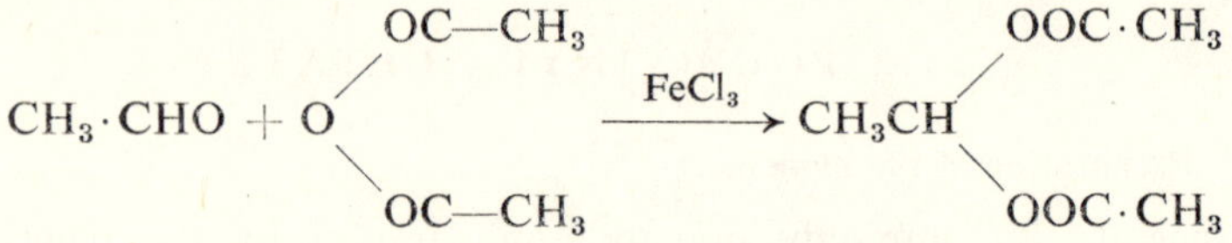

$$CH_3 \cdot CHO + O \underset{OC-CH_3}{\overset{OC-CH_3}{\Big\langle}} \xrightarrow{FeCl_3} CH_3CH \underset{OOC \cdot CH_3}{\overset{OOC \cdot CH_3}{\Big\langle}}$$

Fig. 11.2

Exposure of the ethylidene diacetate to an aromatic sulphonic acid in the presence of five times its weight of acetic anhydride as diluent at 136°C will yield the following mixture: 40% vinyl acetate; 28% acetic acid; 20% acetic anhydride; 4% ethylidene diacetate; 8% acetaldehyde.

The latter four products may all be reused after separation.

Other routes to vinyl acetate from petroleum, particularly via ethylene, are under active development. In one process ethylene is oxidised to acetaldehyde by passing through a solution containing palladium and cupric chlorides. It is believed that a palladium chloride–ethylene complex is formed that is subsequently hydrolysed to acetaldehyde. Vinyl acetate may be obtained from acetaldehyde as described above for the Celanese route.

A one-stage process for producing vinyl acetate directly from ethylene has also been disclosed. In this process ethylene is passed through a substantially anhydrous suspension or solution of acetic acid containing cupric chloride and copper or sodium acetate together with a palladium catalyst to yield vinyl acetate.

11.2.2 Polymerisation

Vinyl acetate may be easily polymerised in bulk, solution, emulsion and suspension. At conversions above 30%, chain transfer to polymer or monomer may occur. In the case of both polymer and monomer transfer two mechanisms are possible, one at the tertiary carbon, the other (illustrated in Fig. 11.3) at the acetate group.

The radical formed at either the tertiary carbon atom or at the acetate group will then initiate polymerisation and form branched structures.

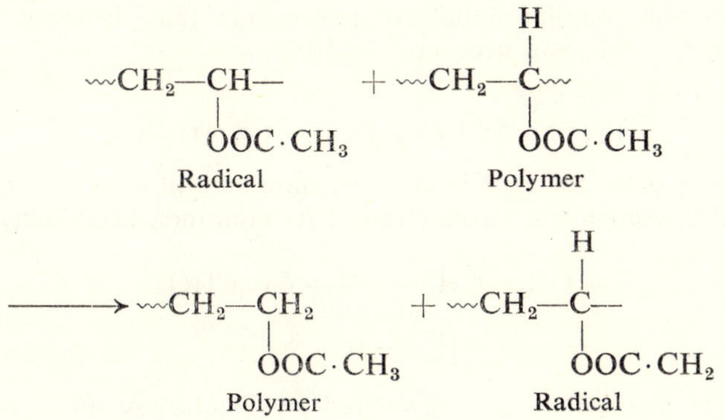

$$\sim\!\!CH_2\!-\!CH\!- \quad + \sim\!\!CH_2\!-\!\overset{\displaystyle H}{\underset{\displaystyle OOC\cdot CH_3}{C}}\!\!\sim$$

$$\underset{\text{Radical}}{\overset{\displaystyle |}{OOC\cdot CH_3}} \qquad \underset{\text{Polymer}}{}$$

$$\longrightarrow \sim\!\!CH_2\!-\!CH_2 \quad + \sim\!\!CH_2\!-\!\overset{\displaystyle H}{\underset{\displaystyle OOC\cdot CH_2}{C}}\!\!-$$

$$\underset{\text{Polymer}}{\overset{\displaystyle |}{OOC\cdot CH_3}} \qquad \underset{\text{Radical}}{}$$

Fig. 11.3

Since poly(vinyl acetate) is usually used in an emulsion form, the emulsion polymerisation process is commonly used. In a typical system, approximately equal quantities of vinyl acetate and water are stirred together in the presence of a suitable colloid–emulsifier system, such as poly(vinyl alcohol) and sodium lauryl sulphate, and a water-soluble initiator such as potassium persulphate.

Polymerisation takes place over a period of about 4 hours at 70°C. The reaction is exothermic and provision must be made for cooling when the batch size exceeds a few litres. In order to achieve better control of the process and to obtain particles with a smaller particle size, part of the monomer is first polymerised and the rest, with some of the initiator, is then steadily added over a period of 3–4 hours. To minimise the hydrolysis of vinyl acetate or possible comonomers during polymerisation, it is necessary to control the pH throughout reaction. For this purpose a buffer such as sodium acetate is commonly employed.

11.2.3 Properties and uses

Poly(vinyl acetate) is too soft and shows excessive 'cold flow' for use in moulded plastics. This is no doubt associated with the fact that the glass

transition temperature of 28°C is little above the usual ambient temperatures and in fact in many places at various times the glass temperature
may be the lower.   It has a density of 1·19 g cm^{-3} and a refractive index
of 1·47.   Commercial polymers are atactic and, since they do not crystallise, transparent (if free from emulsifier).   They are successfully used in
emulsion paints, as adhesives for textiles, paper and wood, as a sizing
material and as a 'permanent starch'.   A number of grades are supplied
by manufacturers which differ in molecular weight and in the nature of
comonomers (e.g. vinyl maleate) which are occasionally used.

The polymers are usually supplied as emulsions which also differ in the
particle size, the sign of the charge on the particle, the pH of the aqueous
phase and in other details.

An amorphous polymer with a solubility parameter of 9·5, it dissolves
in solvents with similar solubility parameters (e.g. benzene $\delta = 9\cdot2$,
chloroform $\delta = 9\cdot3$, and acetone $\delta = 10\cdot0$).

11.3 POLY(VINYL ALCOHOL)

Vinyl alcohol does not exist in the free state and all attempts to prepare
it have led instead to the production of its tautomer, acetaldehyde.

$$CH_2{=}CH \underset{\longleftarrow}{\overset{\hspace{2em}}{\longrightarrow}} CH_3CHO$$
$$\underset{OH}{|}$$

Poly(vinyl alcohol) is thus prepared by alcoholysis of a poly(vinyl
ester) and in practice poly(vinyl acetate) is used (Fig. 11.4).

$$\sim\!\!\sim\!CH_2{-}CH\!\sim\!\!\sim \quad + CH_3OH \longrightarrow \sim\!\!\sim\!CH_2{-}CH\!\sim\!\!\sim$$
$$\underset{OOC\cdot CH_3}{|} \qquad\qquad\qquad\qquad\qquad \underset{OH}{|}$$
$$+ CH_3COOCH_3$$

Fig. 11.4

The term hydrolysis is sometimes incorrectly used to describe this process.   In fact water does not react readily to yield poly(vinyl alcohol)s
and may actually retard reaction where certain catalysts are used.

Either methanol or ethanol may be used to effect alcoholysis but the
former is often preferred because of its miscibility with poly(vinyl acetate)
at room temperature and its ability to give products of better colour.
Where methanol is employed, methyl acetate may be incorporated as a
second solvent.   It is also formed during reaction.   The concentration of
poly(vinyl acetate) in the alcohol is usually between 10 and 20%.

Either acid or base catalysis may be employed.   Alkaline catalysts such
as caustic soda or sodium methoxide give more rapid alcoholysis.   With
alkaline catalysts, increasing catalyst concentration, usually less than 1%
in the case of sodium methoxide, will result in decreasing residual acetate
content and this phenomenon is used as a method of controlling the degree
of alcoholysis.   Variations in reaction time provide only a secondary

means of controlling the reaction. At 60°C the reaction may take less than an hour but at 20°C complete 'hydrolysis' may take up to 8 hours.

The use of acid catalysts such as dry hydrochloric acid has been described in the literature but are less suitable when incompletely 'hydrolysed' products are desired as it is difficult to obtain reproducible results.

Commercial poly(vinyl alcohol) (e.g. Gelvatol, Elvanol, Mowiol and Rhodoviol) is available in a number of grades which differ in molecular weight and in the residual acetate content. Because alcoholysis will cause scission of branched polymers at the points where branching has proceeded via the acetate group, poly(vinyl alcohol) polymer will have a lower molecular weight than the poly(vinyl acetate) from which it is made.

11.3.1 Structure and properties

Poly(vinyl acetate) is an atactic material and is amorphous. Whilst the structure of poly(vinyl alcohol) is also atactic the polymer exhibits crystallinity and has essentially the same crystal lattice as polythene. This is because the hydroxyl groups are small enough to fit into the lattice without disrupting it.

The presence of hydroxyl groups attached to the main chain has a number of significant effects. The first effect is that the polymer is hydrophilic and will dissolve in water to a greater or lesser extent according to the degree of 'hydrolysis' and the temperature. Polymers with a degree of 'hydrolysis' in the range of 87–89% are readily soluble in cold water. An increase in the degree of 'hydrolysis' will result in a reduction in the ease of solubility and fully 'hydrolysed' polymers are only dissolved by heating to temperatures above 85°C.

This anomolous effect is due to the greater extent of hydrogen bonding in the completely 'hydrolysed' polymers. Hydrogen bonding also leads to a number of other effects, for example unplasticised poly(vinyl alcohol) decomposes below its flow temperature. The polymer also has a very high tensile strength and is very tough. Films cast from high molecular weight grades, conditioned to 35% humidity, are claimed[2] to have tensile strengths as high as 18,000 lb in^{-2}.

The properties will be greatly dependent on humidity; the higher the humidity, the more the water absorbed. Since water acts as a plasticiser there will be a reduction in tensile strength but an increase in elongation and tear strength. Fig. 11.5 shows the relationship between tensile strength, percentage 'hydrolysis' and humidity.

Because of its high polarity, poly(vinyl alcohol) is very resistant to hydrocarbons such as petrol. Although the polymer will dissolve in lower alcohol–water mixtures, it does not dissolve in pure alcohols. As it is crystalline as well as highly polar only a few organic solvents, such as diethylene triamine and triethylene tetramine are effective at room temperature. As might be expected the hydroxyl group is very reactive and many derivatives have been prepared.

The polymer may be plasticised by polar liquids capable of forming

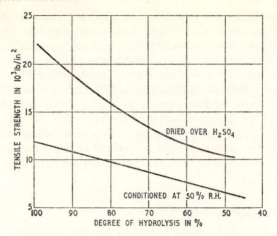

Fig. 11.5. Relation between tensile strength and degree of 'hydrolysis' for unplasticised polyvinyl alcohol film. (After Davidson and Sittig[2])

hydrogen bonds with the hydroxyl groups. Glycerin is commonly used for this purpose.

11.3.2 Applications

Poly(vinyl alcohol) is employed for a variety of purposes. Film cast from aqueous alcohol solution is an important release agent in the manufacture of reinforced plastics. Incompletely 'hydrolysed' grades are currently being developed for water-soluble packages for bath salts, bleaches, insecticides and disinfectants. Techniques for making tubular blown film, similar to that used with polythene, have been developed for this purpose. Moulded and extruded products which combine oil resistance with toughness and flexibility are produced in the United States but have never become popular in Europe.

Poly(vinyl alcohol) will function as a non-ionic surface-active agent and is used in suspension polymerisation as a protective colloid. In many applications it serves as a binder and thickener in addition to an emulsifying agent. The polymer is also employed in adhesives, binders, paper sizing, paper coatings, textile sizing, ceramics, cosmetics and as a steel quenchant.

Japanese workers have developed fibres from poly(vinyl alcohol). The polymer is wet spun from warm water into a concentrated aqueous solution of sodium sulphate containing sulphuric acid and formaldehyde, the latter insolubilising the alcohol by formation of formal groups.

11.4 THE POLY(VINYL ACETALS)

Treatment of poly(vinyl alcohol) with aldehydes and ketones leads to the formation of poly(vinyl acetals) and poly(vinyl ketals), of which only the former products are of any commercial significance (Fig. 11.6).

The products are amorphous resins whose rigidity and softening point depend on the aldehyde used. Poly(vinyl butyral) with the larger side chain is softer than poly(vinyl formal). Since the reaction between the aldehyde and the hydroxyl groups occurs at random, some hydroxyl groups become isolated and are incapable of reaction. A poly(vinyl acetal) molecule will thus contain:

1. Acetal groups.
2. Residual hydroxyl groups.
3. Residual acetate groups, due to incomplete 'hydrolysis' of poly(vinyl acetate) to poly(vinyl alcohol).

11.4.1 Poly(vinyl formal)

The poly(vinyl acetals) may be made either from poly(vinyl alcohol) or directly from poly(vinyl acetate) without separating the alcohol. In the case of poly(vinyl formal) the direct process is normally used.

In a typical process, 100 parts of poly(vinyl acetate) are added to a mixture of 200 parts acetic acid and 70 parts water, which has been warmed to about 70°C, and stirred to complete solution. 60 parts of 40% formalin and 4 parts sulphuric acid (catalyst) are added and reaction is carried out for 24 hours at 70°C. Water is added to the mixture with rapid agitation to precipitate the granules which are then washed free from acid and dried.

A number of grades of poly(vinyl formal) are commercially available

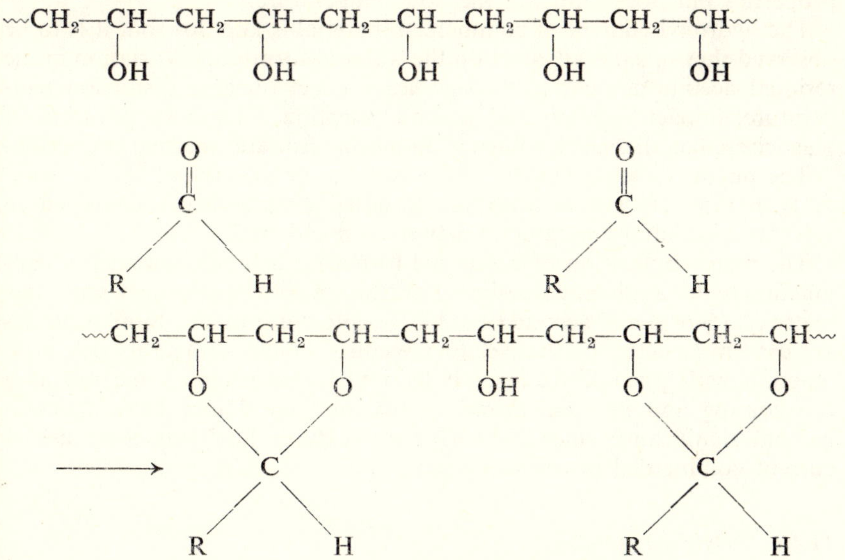

A Poly(Vinyl Acetal)

Fig. 11.6

(Formvar, Mowital) which vary in degree of polymerisation, hydroxyl content and residual acetate content.

Table 11.1[3] shows the influence of these variables on some properties. The residual hydroxyl content is expressed in terms of poly(vinyl alcohol) content and residual acetate in terms of poly(vinyl acetate) content.

Table 11.1 INFLUENCE OF STRUCTURE VARIABLES ON THE PROPERTIES OF POLY(VINYL FORMAL)

	ASTM test	*Various grades of poly(vinyl formal)*				
Av. D. of P.	—	500	500	350	430	350
Poly(vinyl alcohol) (%)	—	5–6	7–9	7–9	5–7	5–7
Poly(vinyl acetate) (%)	—	9·5–13	9·5–13	9·5–13	20–27	40–50
Flow temperature (°C)	D.569–48T	160–170	160–170	140–145	145–150	—
Heat distortion temperature (°C)	D.648–49T	88–93	88–93	88–93	75–80	50–60
Tensile strength (10^{-3} lb in^{-2})	D.638–41T	10	10	10	10	10
Elongation (%)	D.638–41T	7–20	10–50	10–50	4–5	3–4
Impact strength (Izod $\frac{1}{2}$ in × $\frac{1}{2}$ in) (ft lb/in)	D.256–43T	1·2–2·0	1·2–2·0	1·0–1·4	0·5–0·7	0·4–0·6
Water absorption (%)	D.570–40T	0·75	1·1	1·1	1·5	1–5

It will be observed that molecular weight has little effect on mechanical properties but does influence the flow temperature.

The hydroxyl content of commercial material is kept low but it is to be observed that this has an effect on the water absorption. Variation in the residual acetate content has a significant effect on heat distortion temperature, impact strength and water absorption. The incorporation of plasticisers has the usual influence on mechanical and thermal properties.

The polymer, being amorphous, is soluble in solvents of similar solubility parameter, grades with low residual acetate being dissolved in solvents of solubility parameter between 9·8 and 10·7.

The main application of poly(vinyl formal) is as a wire enamel in conjunction with a phenolic resin. For this purpose, polymers with low hydroxyl (5–6%) and acetate (9·5–13%) content are used. Similar grades are used in structural adhesives (e.g. Redux) which are also used in conjunction with phenolic resins. Poly(vinyl formal) finds some use as a can coating and in wash primers. Injection mouldings have no commercial significance since they have no features justifying their use at current commercial prices.

11.4.2 Poly(vinyl acetal)

Poly(vinyl acetal) itself is now of little commercial importance. The material may be injection moulded but has no particular properties which merit its use. It is occasionally used in conjunction with nitrocellulose in lacquers, as a vehicle for wash primers and as a stiffener for fabrics.

11.4.3 Poly(vinyl butyral)

As a safety glass interleaver, poly(vinyl butyral) (Butacite, Saflex) is extensively used because of its high adhesion to glass, toughness, light stability, clarity and moisture insensitivity.

It also finds miscellaneous applications in textile and metal coatings and in adhesive formulations. Where it is to be used as a safety glass interleaver, a very pure product is required and this is most conveniently prepared from poly(vinyl alcohol) rather than by the direct process from poly(vinyl acetate).

In a typical process 140 parts of fully 'hydrolysed' poly(vinyl alcohol) is suspended in 800 parts of ethanol; 80 parts of butyraldehyde and 8 parts of sulphuric acid are added and the reaction is carried out at about 80°C for 5–6 hours.

The solution of poly(vinyl butyral) is diluted with methanol and the polymer precipitated by the addition of water during vigorous agitation. The polymer is then stabilised, washed and dried.

Highly 'hydrolysed' poly(vinyl alcohol) is normally used as a starting point. For safety glass applications about 25% of the hydroxyl groups are left unreacted. In this application the polymer is plasticised with an ester such as dibutyl sebacate or triethylene glycol di-2-ethyl butyrate, about 30 parts of plasticiser being used per 100 parts of polymer. The compound is then calendered to a thickness of 0·015 in and coated with a layer of sodium bicarbonate to prevent blocking. To produce safety glass the film is washed and dried and then placed between two pieces of glass which are then subjected to mild heat and pressure. Bulletproof glass is made by laminating together several layers of glass and poly(vinyl butyral) film.

In the United States the use of laminated safety glass is obligatory on automobile windscreens. In Great Britain, where the cheaper heat-toughened glass is allowed, such laminated glass is only used in sports cars and in cars for export. It is also used in aircraft glazing.

11.4.4 Other organic vinyl ester polymers

Whilst poly(vinyl acetate) is an important commercial material, polymers from higher esters such as vinyl propionate have made no commercial impact. As is to be expected the longer side chains give softer polymers similar in many respects to the higher acrylates and methacrylates (Chapter 12). Poly(vinyl chloroacetate) was once introduced as a non-inflammable competitor for celluloid but is no longer of any value.

Whilst vinyl acetate is reluctant to copolymerise with many other common monomers a number of important copolymers exist. Two of particular interest to the plastics industry are the ethylene–vinyl acetate (Chapter 8) and vinyl chloride–vinyl acetate copolymers (Chapter 9).

REFERENCES

1. HORN, O., *Chem. Ind.*, 1749 (1955)
2. DAVIDSON, R. L., and SITTIG, M., *Water-soluble Resins*, Reinhold, New York (1962)
3. FITZHUGH, A. F., and LAVIN, E., *J. Electrochem. Soc.*, **100** (8), 351 (1953)

BIBLIOGRAPHY

General

KAINER, F., *Polyvinylalkohole*, Enka, Stuttgart (1949)
SCHILDKNECHT, C. E., *Vinyl and Related Polymers*, John Wiley, New York (1952)

Polyvinyl Acetate

WHEELER, O. L., LAVIN, E., and CROZIER, R. N., *J. Polymer Sci.*, **9,** 157 (1952)

Polyvinyl Alcohol

Brit. Plastics., **16,** 77, 84, 122 (1944)
DAVIDSON, R. L., and SITTIG, M., *Water-soluble Resins*, Reinhold, New York (1962)

Polyvinyl Acetals

FITZHUGH, A. F., and LAVIN, E., *J. Electrochem. Soc.*, **100** (8), 351 (1953)
PLATZER, N., *Mod. Plastics*, **28,** 142 (1951)

12

Acrylic Plastics

12.1 INTRODUCTION

Poly(methyl methacrylate) (Fig. 12.1, I) is, commercially, the most important member of a range of *acrylic* polymers which may be considered structurally as derivatives of acrylic acid (II)

This family includes a range of polyacrylates (III), polymethacrylates (IV) and the important fibre-forming polymer, polyacrylonitrile (V).

Methyl, ethyl and allyl acrylate were first prepared in 1873 by Caspary and Tollens,[1] and of these materials the last was observed to polymerise. In 1880 Kahlbaum[2] reported the polymerisation of methyl acrylate and at approximately the same time Fittig[3,4] found that methacrylic acid and some of its derivatives readily polymerised.

In 1901 Otto Röhm reported on his studies of acrylic polymers for his doctoral dissertation. His interest in these materials however did not cease at this stage and eventually in 1927 the Röhm and Haas concern at Darmstadt, Germany commenced limited production of poly(methyl acrylate) under the trade names Acryloid and Plexigum. These were soft gummy products of interest as surface coatings rather than as mouldable plastics materials. About 1930 R. Hill prepared poly(methyl methacrylate) and found it to be a rigid, transparent polymer, potentially useful as an aircraft glazing material.[5]

The first methacrylic esters were prepared by dehydration of hydroxyiso-butyric esters, prohibitively expensive starting points for commercial synthesis. In 1932 J. W. C. Crawford[6] discovered a new route to the monomer using cheap and readily available chemicals—acetone, hydro-cyanic acid, methanol and sulphuric acid—and it is his process which is today used, with minor modifications, throughout the world. Sheet poly(methyl methacrylate) became prominent during the Second World War for aircraft glazing, a use predicted by Hill in his early patents, and since then has found other applications in many fields.

Examples of commercial poly(methyl methacrylate) sheet are Perspex (I.C.I.), Oroglas (Rohm and Haas, U.S.A.) and Plexiglas (Röhm and Haas

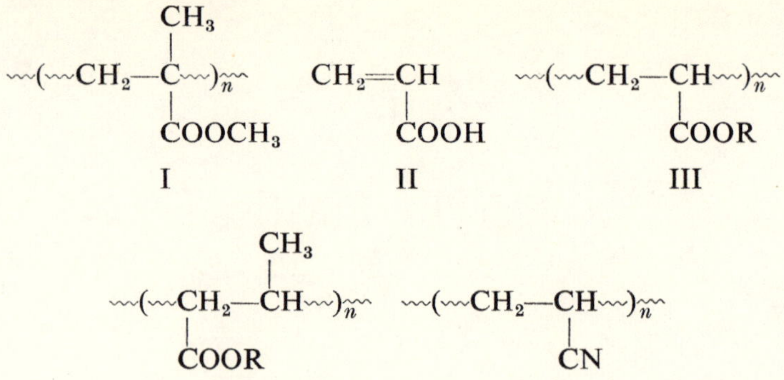

$$\text{\textasciitilde(\textasciitilde CH}_2\text{—}\underset{\underset{\text{COOCH}_3}{|}}{\overset{\overset{\text{CH}_3}{|}}{\text{C}}}\text{\textasciitilde)}_n\text{\textasciitilde} \qquad \underset{\underset{\text{COOH}}{|}}{\text{CH}_2\text{=CH}} \qquad \text{\textasciitilde(\textasciitilde CH}_2\text{—}\underset{\underset{\text{COOR}}{|}}{\text{CH\textasciitilde)}}_n\text{\textasciitilde}$$

I   II   III

$$\text{\textasciitilde(\textasciitilde CH}_2\text{—}\underset{\underset{\text{COOR}}{|}}{\overset{\overset{\text{CH}_3}{|}}{\text{CH}}}\text{\textasciitilde)}_n\text{\textasciitilde} \qquad \text{\textasciitilde(\textasciitilde CH}_2\text{—}\underset{\underset{\text{CN}}{|}}{\text{CH\textasciitilde)}}_n\text{\textasciitilde}$$

(IV)   (V)

Fig. 12.1

GmbH, Germany).   Poly(methyl methacrylate) moulding powders include Diakon (I.C.I.), Acry-ace (Fudow Chemical Co., Japan), Lucite (Du Pont) and Vedril (Montecatini).

12.2 POLY(METHYL METHACRYLATE)

12.2.1   Preparation of monomer

The successful commercial utilisation of poly(methyl methacrylate) is due in no small measure to the process of producing the monomer from acetone developed by Crawford of I.C.I. which enabled the polymer to be produced at a competitive price.   Some details of the process as operated by the Rohm and Haas Company of Philadelphia have been disclosed.[7]

Acetone is first reacted with hydrogen cyanide to give acetone cyano-hydrin (Fig. 12.1).

$$\underset{\text{CH}_3}{\overset{\text{CH}_3}{\diagdown\diagup}}\text{C=O} + \text{HCN} \longrightarrow \text{CH}_3\text{—}\underset{\underset{\text{CN}}{|}}{\overset{\overset{\text{CH}_3}{|}}{\text{C}}}\text{—OH}$$

Fig. 12.2

The cyanohydrin is then treated with 98% sulphuric acid in a cooled hydrolysis kettle to yield methacrylamide sulphate (Fig. 12.3).

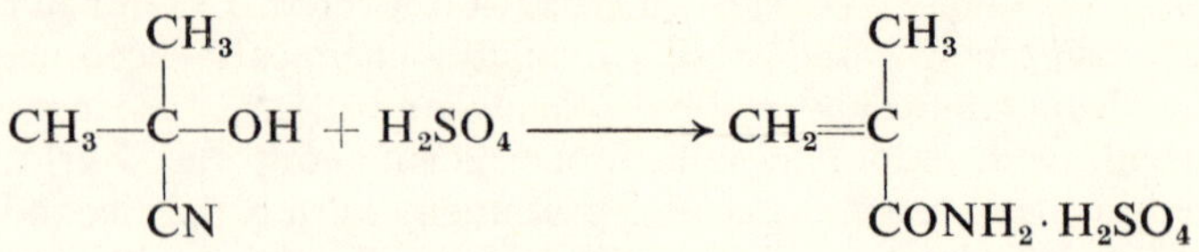

$$\text{CH}_3\text{—}\underset{\underset{\text{CN}}{|}}{\overset{\overset{\text{CH}_3}{|}}{\text{C}}}\text{—OH} + \text{H}_2\text{SO}_4 \longrightarrow \text{CH}_2\text{=}\underset{\underset{\text{CONH}_2 \cdot \text{H}_2\text{SO}_4}{|}}{\overset{\overset{\text{CH}_3}{|}}{\text{C}}}$$

Fig. 12.3

The sulphate is not isolated from the reaction mixture which passes into an esterification kettle and reacts continuously with methanol (Fig. 12.4).

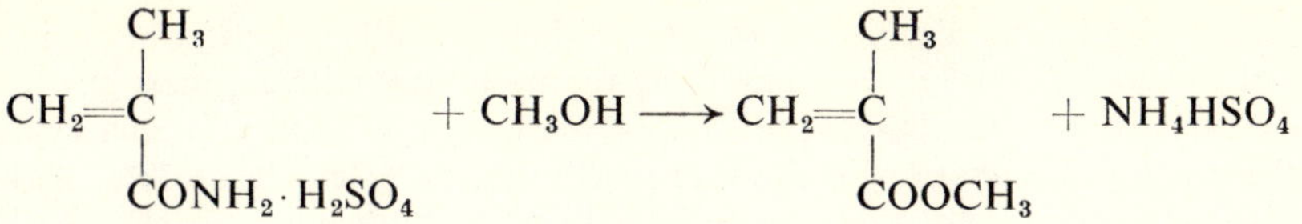

$$\underset{\underset{CONH_2 \cdot H_2SO_4}{|}}{\overset{\overset{CH_3}{|}}{CH_2\!=\!C}} \;+\; CH_3OH \;\longrightarrow\; \underset{\underset{COOCH_3}{|}}{\overset{\overset{CH_3}{|}}{CH_2\!=\!C}} \;+\; NH_4HSO_4$$

Fig. 12.4

The esterified stream, which may contain inhibitors to prevent premature polymerisation, is then passed to a stripping column which separates the methyl methacrylate, methanol and some water from the residue made up of sulphuric acid, ammonium bisulphate and the remainder of the water. The methyl methacrylate is subsequently separated and purified by further distillation.

An alternative route to the monomer is stated to be under development by the Escambia Chemical Company starting from isobutylene (Fig. 12.5).

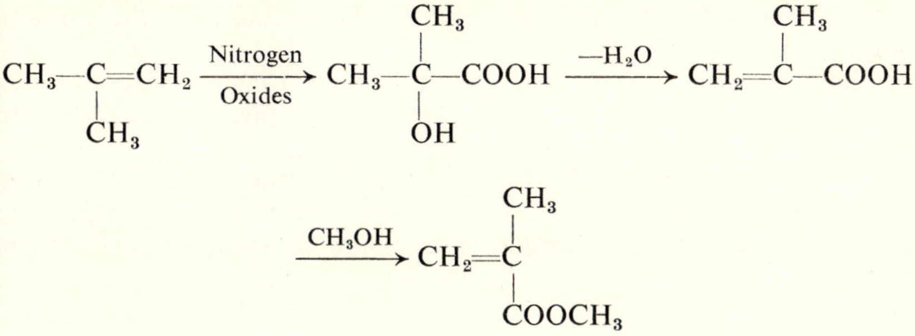

$$\underset{\underset{CH_3}{|}}{CH_3\!-\!C\!=\!CH_2} \xrightarrow[\text{Oxides}]{\text{Nitrogen}} \underset{\underset{OH}{|}}{\overset{\overset{CH_3}{|}}{CH_3\!-\!C\!-\!COOH}} \xrightarrow{-H_2O} \overset{\overset{CH_3}{|}}{CH_2\!=\!C\!-\!COOH}$$

$$\xrightarrow{CH_3OH} \underset{\underset{COOCH_3}{|}}{\overset{\overset{CH_3}{|}}{CH_2\!=\!C}}$$

Fig. 12.5

The monomer is a mobile liquid with a characteristic sweet smelling odour and with the following properties:

Boiling point (760 mmHg)	100·5°C
Density D_4^{20}	0·936–0·940
Refractive index n_D^{20}	1·413–1·416
Heat of polymerisation	11·6 kcal/mole

12.2.2 Polymerisation

Methyl methacrylate will polymerise readily and the effect may be observed with uninhibited samples of monomers during storage. In commercial practice the monomer is supplied with up to 0·10% of an inhibitor such as hydroquinone which is removed before polymerisation, either by distillation under reduced pressure or, in some cases, by washing with an alkaline solution.

Free-radical polymerisation techniques involving peroxides or azo-diisobutyronitrile at temperatures up to about 100°C are employed commercially. The presence of oxygen in the system will affect the rate of reaction and the nature of the products, owing to the formation of methacrylate peroxides in a side reaction. It is therefore common practice to polymerise in the absence of oxygen, either by bulk polymerisation in a full cell or chamber or by blanketing the monomer with an inert gas.

It has been observed that in the polymerisation of methyl methacrylate there is an acceleration in the rate of conversion after about 20% of the monomer has been converted. The average molecular weight of the polymer also increases during polymerisation. It has been shown that these results are obtained even under conditions where there is a negligible rise in the temperature (< 1 degC) of the reaction mixture.

The explanation for these effects is that the chain termination reaction slows down during conversion. This is thought to be due to the fact that the monomer is a relatively poor solvent for the polymer and so the natural tendency of polymers and polymer radicals is to coil up in the monomer. Such coiling often means that the growing radical end is inside the coil and although reaction with small monomer molecules is not affected there is some difficulty in two growing radical ends coming together and this becomes considerably more difficult as the viscosity of the reaction mixture increases.

The effect can in fact be induced in poly(methyl methacrylate) through increasing the viscosity by adding poly(methyl methacrylate) or even another polymer such as cellulose tripropionate.

On the industrial scale there are two general methods of polymerisation of importance, bulk polymerisation to give sheet, rod and tube, and suspension polymerisation leading to moulding and extrusion composition.

Polymerisation in bulk

Bulk polymerisation is extensively used in the manufacture of sheet and to a lesser extent rod and tube. In order to produce a marketable material it is important to take the following factors into account:

1. The exotherm developed during cure.
2. The acceleration in conversion rate due to increasing viscosity.
3. The effect of oxygen.
4. The extensive shrinkage in conversion from monomer to polymer ($\sim 20\%$).
5. The need to produce sheet of even thickness.
6. The need to produce sheet of constant quality.
7. The need to produce sheet free from impurities and imperfections.

In order to reduce the shrinkage in the casting cell, and also to reduce problems of leakage from the cell, it is normal practice to prepare a 'prepolymer'. In a typical process monomer freed from inhibitor is heated with agitation for about 8 minutes at 90°C with 0·5% benzoyl peroxide and then cooled to room temperature. Plasticiser, colouring agents and ultra-violet light absorbers may be incorporated at this stage if required. The resulting syrup, consisting of a solution of polymer in monomer is then filtered and stored in a refrigerator if it is not required for immediate use. The heating involved in making the prepolymer may also be of assistance in removing oxygen dissolved in the monomer.

The preparation of a prepolymer requires careful control and can be somewhat difficult in large scale operations. An alternative approach is to prepare a syrup by dissolving some polymer in the monomer and adding some peroxide to the mixture. As in the case of a prepolymer syrup, such a syrup will cause less shrinkage on polymerisation and fewer leakage problems.

Acrylic sheet is prepared by pouring the syrup into a casting cell. This consists of two plates of heat resistant polished glass provided with a separating gasket round the edges. The gasket commonly consists of a hollow flexible tube made from a rubber, or from plasticised poly(vinyl alcohol). The cell is filled by opening up the gasket at a corner or edge and metering in the syrup, care being taken to completely fill the cell before closing up the gasket. The cell is held together by spring loaded clamps or spring clips so the plates will come closer together as the reacting mixture shrinks during polymerisation. This technique will enable the sheet to be free of sink marks and voids.

It is important to use rigid glass sheet and to apply pressure to the plates in such a manner that they do not bow out as this would lead to sheet of uneven thickness.

The filled cells are then led through a heating tunnel. In a typical system the time to pass through the tunnel is about 16 hours. For the first fourteen hours the cell passes through heating zones at about 40°C. Under these conditions polymerisation occurs slowly. Any acceleration of the rate due to either the rise in temperature through the exothermic reaction or due to the viscosity–chain termination effect will be small. It is particularly important that the temperature of any part of the syrup is not more than 100°C since this would cause the monomer to boil. By the end of this period the bulk of the monomer has reacted and the cell passes through the hotter zones. After 15 hours (total time) the cell is at about 97°C at which temperature it is held for a further half-hour. The sheet is then cooled and removed from the cell. In order to reduce any internal stresses the sheet may be annealed by heating to about 140°C and, before being dispatched to the customer, the sheet is masked with some protective paper using gelatine or, preferably, with a pressure-sensitive adhesive.

When casting large blocks, the exotherm problem is more severe and it may be necessary to polymerise inside a pressure vessel and thus raise the boiling point of the monomer.

In order to compensate for shrinkage, special techniques are required in the manufacture of rod. In one process, vertical aluminium tubes are filled with syrup and slowly lowered into a water bath at 40°C. As the lowest level of syrup polymerises, it contracts and the higher levels of syrup thus sink down the tube, often under pressure from a reservoir of syrup feeding into the tubes.

Acrylic tubes may be prepared by adding a calculated amount of syrup to an aluminium tube, sealing both ends, purging the air with nitrogen and then rotating horizontally at a constant rate. The whole assembly is heated and the syrup polymerises on the wall of the rotating tube. The natural shrinkage of the material enables the casting to be removed quite easily.

Suspension polymerisation

The average molecular weight of most bulk polymerised poly(methyl methacrylates) is too high to give a material which has adequate flow properties for injection moulding and extrusion.

By rolling on a two-roll mill the molecular weight of the polymer can be greatly reduced by mechanical scission, analogous to that involved in the mastication of natural rubber, and so mouldable materials may be obtained. However bulk polymerisation is expensive and the additional milling and grinding processes necessary make this process uneconomic in addition to increasing the risk of contamination.

As a result the suspension polymerisation of methyl methacrylate has been developed to produce commercial materials such as Diakon made by I.C.I. Such a polymerisation can be carried out rapidly, usually in less than an hour because there is no serious exotherm problem.

There is, however, a problem in controlling the particle size of the beads formed and further in preventing their agglomeration, problems common to all suspension type polymerisations. The particle size of the beads is determined by the shape and size of the reactor, the type and rate of agitation and also the nature of suspending agents and protective colloids present. Suspending agents used include talc, magnesium carbonate and aluminium oxide whilst poly(vinyl alcohol) and sodium polymethacrylate are among materials used as protective colloids.

In one process described in the literature[8] one part of methyl methacrylate was agitated with 2 parts of water and 0·2% benzoyl peroxide was employed as the catalyst. Eight to 18 g of magnesium carbonate per litre of reactants were added, the lower amount being used for larger beads, the larger for small beads. The reaction temperature was 80°C initially but this rose to 120°C because of the exothermic reaction. Polymerisation was complete in about an hour. The magnesium carbonate was removed by adding sulphuric acid to the mixture. The beads were then filtered off, carefully washed and dried.

Other additives that may be incorporated include sodium hydrogen phosphates as buffering agents to stabilise the pH of the reaction medium, lauryl mercaptan or trichlorethylene as chain transfer agents to control molecular weight, a lubricant such as stearic acid and small amounts of an emulsifier such as sodium lauryl sulphate.

The dried beads may be supplied as injection moulding material without further treatment or they may be compounded with additives and granulated.

12.2.3 Structure and properties

Commercial poly(methyl methacrylate) is an atactic polymer and since the methyl and the ester groups are incapable of being interchanged in a crystal lattice these polymers are therefore amorphous and transparent.

Since the substituents on the α-carbon atom restrict chain flexibility and since the side groups are polar and relatively small and there is thus fairly substantial inter-chain attraction, the polymer is hard and rigid with a glass transition temperature of 110°C.

The solubility of poly(methyl methacrylate) is consistent with that expected of an amorphous thermoplastic with a solubility parameter of about 9·2. Solvents include ethyl acetate ($\delta = 9\cdot1$), ethylene dichloride ($\delta = 9\cdot8$), trichlorethylene ($\delta = 9\cdot3$), chloroform ($\delta = 9\cdot3$) and toluene ($\delta = 8\cdot8$). Poly(methyl methacrylate) sheet however usually has a high average molecular weight ($\sim 10^6$) and contains some very large molecules. Because of this it may be difficult to completely dissolve such material.

Poly(methyl methacrylate) is a polar material, the polar groups being on a side-chain rather than on the skeletal backbone of the polymer. Because of this the polymer has a rather high dielectric constant and power factor at room temperature even though this is well below the glass transition temperature.

12.2.4   General properties of poly(methyl methacrylate)

As indicated in the previous section poly(methyl methacrylate) is a hard, rigid, transparent material. Commercial grades have extremely good weathering resistance compared with other thermoplastics.

The properties of three types of poly(methyl methacrylate) (sheet based on high molecular weight polymer, lower molecular weight injection moulding material and a commercial copolymer) are given in Table 12.1.

Table 12.1   SOME PROPERTIES OF METHYL METHACRYLATE POLYMERS

Property	Units	ASTM test method	Acrylic sheet*	Moulding composition†	Copolymer‡
Molecular weight	—	—	$\sim 10^6$	$\sim 60{,}000$	—
Specific gravity	—	D.792	1·19	1·18	1·17
Tensile strength	10^3 lb in^{-2}	D.638		10·5	
Tensile modulus	10^3 lb in^{-2}	—	~ 430	~ 350	~ 400
Flexural strength	10^3 lb in^{-2}	—	~ 20	~ 18	~ 18
Flexural modulus	10^3 lb in^{-2}	—	~ 400	~ 400	
Rockwell hardness	—	D.785	M.100	M.103	
Scratch hardness (Moh's scale)	—			2–3	
Water absorption [% in 24 hr (20°C)]	%	D.570	0·2	0·3	0·25
Izod impact strength	ft lb	(B.S.) 2782		0·40	
Vicat softening pt.	°C			109–112	
Heat distortion Temperature	°C (264 lb in^{-2}	D.648	100	85–95	80
Refractive index	$n_D{}^{20}$		1·49	1·49	1·49
Volume resistivity (20°C)	ohm cm		$>10^{14}$	$>10^{15}$	—
Dielectric constant at 10^3 c/s 60% R.H.	(20°C)		3·0	3·1	

* Perspex (I.C.I.).   † Diakon M (I.C.I.).   ‡ Asterite (I.C.I.).

As might be expected of a somewhat polar thermoplastics material, mechanical, electrical and other properties are strongly dependent on temperature, testing 'rate' and humidity.   Detailed data on the influence of these variables have been made available by at least one manufacturer and the following remarks are intended only as an illustration of the effects rather than as an attempt at providing complete data.

Fig. 12.6 shows the considerable temperature sensitivity of the tensile strength of acrylic sheet whilst Fig. 12.7 shows how the fracturing stress

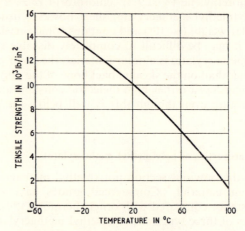

Fig. 12.6.   Effect of temperature on tensile strength of acrylic sheet (Perspex) at constant rate of strain (0·044% per second).   (Reproduced by permission of ICI Plastics Division)

decreases with the period of loading.   Mouldings from acrylic polymers usually show considerable molecular orientation.   It is observed that a moulding with a high degree of frozen-in orientation is stronger and tougher in the direction parallel to the orientation than in the transverse direction.

Poly(methyl methacrylate) is recognised to be somewhat tougher than polystyrene (after consideration of both laboratory tests and common experience) but is less tough than cellulose acetate or the ABS polymers. It is superior to untreated glass in terms of impact resistance and although it cracks, any fragments formed are less sharp and jagged than those of glass and, normally, consequently less harmful.   However, oriented acrylic sheet such as may result from double curvature shaping shatters with a conchoidal fracture and fragments and broken edges can be quite sharp.   Although harder than most other thermoplastics the scratch resistance does leave something to be desired.   Shallow scratches may however be removed by polishing.

The optical properties of poly(methyl methacrylate) are particularly important.   Poly(methyl methacrylate) absorbs very little light but there is about 4% reflection at each polymer–air interface for normal incident light.   Thus the light transmission of normal incident light through a

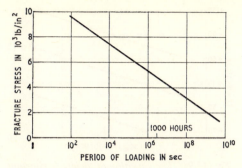

Fig. 12.7. *Effect of period of loading on fracturing stress at 25°C of acrylic sheet (Perspex). (Reproduced by permission of ICI Plastics Division)*

parallel sheet of acrylic material free from blemishes is about 92%. The influence of the wavelength of light on transmission is shown in Fig. 12.8.

The interesting property of total internal reflection may be conveniently exploited in poly(methyl methacrylate). Since the critical angle for the polymer–air boundary is 42° a wide light beam may be transmitted through long lengths of solid polymer. Light may thus be 'piped' round curves and there is little loss where the radius of curvature is greater than three times the thickness of the sheet or rod. Scratched and roughened surface will reduce the internal reflection. This is normally undesirable but a roughened or cut area can also be deliberately incorporated to 'let out' the light at that point.

Poly(methyl methacrylate) is a good electrical insulator for low frequency work, but is inferior to such polymers as polyethylene and polystyrene, particularly at high frequencies. The influence of temperature and frequency on the dielectric constant is shown in Fig. 12.9.

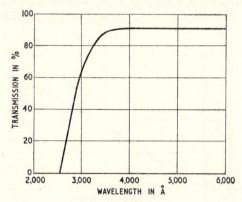

Fig. 12.8. *Light transmission of acrylic polymer ($\frac{1}{4}$ in. thick moulded Diakon. Parallel light beam normally incident on surface). (Reproduced by permission of ICI Plastics Division)*

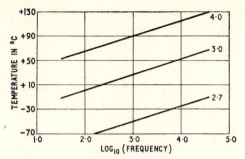

Fig. 12.9. *The variation of dielectric constant with temperature and frequency (Perspex) (the lines join points of equal dielectric constant). (Reproduced by permission of ICI Plastics Division)*

The apparent volume resistivity is dependent on the polarisation time (Fig. 12.10). The initial polarisation current is effective for some time and if only a short time is allowed before taking measurements low values for volume resistivity will be obtained.

As may be expected of an amorphous polymer in the middle range of the solubility parameter table, poly(methyl methacrylate) is soluble in a number of solvents with similar solubility parameters. Some examples were given in the previous section. The polymer is attacked by mineral acids but is resistant to alkalies, water and most aqueous inorganic salt solutions. A number of organic materials although not solvents may cause crazing and cracking, e.g. aliphatic alcohols.

12.2.5   Additives

Poly(methyl methacrylate) may be blended with a number of additives. Of these the most important are dyes and pigments and these should be

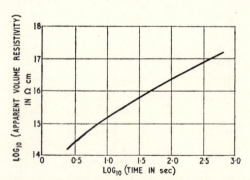

Fig. 12.10. *The dependence of apparent volume resistivity on time of polarisation of acrylic polymer (Perspex). (Reproduced by permission of ICI Plastics Division)*

stable to both processing and service conditions. Two particular requirements are, firstly, that when used in castings they should not affect the
polymerisation reaction and, secondly, that they should have good
weathering resistance.

Plasticisers are sometimes added to the polymer, dibutyl phthalate being
commonly employed in quantities of the order of 5%. Use in moulding
powders will enhance the melt flow but somewhat reduce the mechanical
properties of the finished product.

Further improvement in light stability may be achieved by addition of
small quantities of ultraviolet absorbers. Typical examples include
phenyl salicylate, 2:4-dihydroxy benzophenone, resorcinol monobenzoate,
methyl salicylate and stilbene.

12.2.6 Processing

In commercial practice three lines of approach are employed in order to
produce articles from poly(methyl methacrylate). They are:

1. Processing in the melt state such as by injection moulding and
   extrusion.
2. Manipulation of sheet, rod and tube.
3. The use of monomer–polymer doughs.

There are a number of general points to be borne in mind when processing the polymer in the molten state which may be summarised as
follows:

1. The polymer granules tend to pick up moisture (up to 0·3%).
   Although most commercial grades are supplied in dry condition
   subsequent exposure before use to atmospheric conditions will lead
   to frothy mouldings and extrudates, owing to volatilisation of the
   water in the heating cylinders. Particular care should be taken
   with reground scrap.
2. The melt viscosities at the processing temperatures employed are
   considerably higher than those of polystyrene, polyethylene and
   plasticised p.v.c. This means that the equipment used must be
   robust and capable of generating high extrusion and injection
   pressures. The use of reciprocating screw type injection moulding
   machines is particularly recommended.

   The melt viscosity is more sensitive to temperature than that of
   most thermoplastics (Fig. 12.11) and this means that for accurate,
   consistent and reproducible results, good temperature control is
   required on all equipment.
3. Since the material is amorphous the moulding shrinkage is low and
   normally less than 0·008 in/in.

A great number of poly(methyl methacrylate) products are produced by
manipulation of sheet, rod and tube. Such forms may easily be machined
using drills, circular saws and bandsaws providing care is taken not to
overheat the polymer. It is very difficult to weld the sheet satisfactorily

but cementing techniques have been highly developed. Acrylic parts may be joined using solvents such as chloroform or by use of solutions of polymer in a suitable solvent. Generally however the best results are obtained, particularly where there is a gap filling requirement, by use of a monomer–polymer solution. Commercial cements of this type either contain a photocatalyst to allow hardening by ultraviolet light polymerisation or contain a promoter so that on addition of a peroxide, polymerisation of the monomer is sufficiently rapid at room temperature to harden the cement in less than one hour.

When heated above the glass transition temperature ($\sim 100°C$), acrylic sheet from high molecular weight polymer becomes rubbery. The

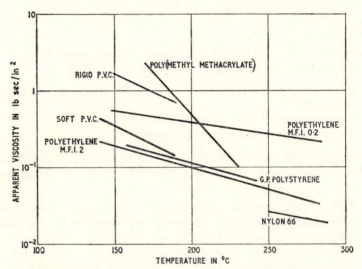

Fig. 12.11. *Viscosity–temperature curves for poly(methyl methacrylate) and other thermoplastics. (Reproduced by permission of ICI Plastics Division)*

rubbery range extends for some 60 degC. Further raising of the temperature causes decomposition rather than melting. The reasonably wide rubbery range, cf. cellulose acetate, high-impact polystyrene and polyethylene, enables the sheet to be heated in ovens rather than having to be heated while clamped to the shaping apparatus. Poly(methyl methacrylate) is not however suitable for normal vacuum forming operations since the modulus of the material in the rubbery state is too great to allow shaping simply by atmospheric pressure. As a result a large number of techniques have been devised using air pressure, mechanical pressure, or both in combination, and sometimes also involving vacuum assistance.

In 1962 the copolymer sheet material Asterite was introduced by I.C.I. as an acrylic material suitable for vacuum forming.

The use of monomer–polymer doughs has been largely confined to the production of dentures. A plaster of Paris mould is first prepared from a supplied impression of the mouth. Polymer powder containing a suitable polymerisation initiator is then mixed with some monomer to form a dough. A portion of the dough is then placed in the mould which is closed, clamped and heated in boiling water. After polymerisation, which usually takes less than half an hour, the mould is cooled and opened. This technique could also be usefully employed for other applications where only a few numbers-off are required but does not seem to have been exploited.

12.2.7 Applications

The major uses of poly(methyl methacrylate) arise from its high light transmission and good outdoor weathering properties. It is also a useful moulding material for applications where good appearance, reasonable toughness and rigidity are requirements which are considered to justify the extra cost of the polymer as compared with the larger tonnage plastics.

The material is eminently suitable for display signs, illuminated and non-illuminated, and for both internal and external use. The properties of importance here are weatherability, the variety of techniques possible which enable a wide range of signs to be produced and, in some cases, transparency.

In lighting fittings poly(methyl methacrylate) finds an important outlet. Street lamp housings originally shaped from sheet are now injection moulded. Ceiling lighting for railway stations, school rooms, factories and offices frequently incorporate poly(methyl methacrylate) housings. In many of these applications opalescent material is used which is effective in diffusing the light source. Poly(methyl methacrylate) is the standard material for automobile rear lamp housings.

The methacrylic polymer remains a useful glazing material. In aircraft applications it is used extensively on aircraft which fly at speeds less than Mach 1·0. They form the familiar 'bubble' body of many helicopters. On land, acrylic sheet is useful for coach roof lights, motor cycle windscreens and in do-it-yourself 'cabins' for tractors and earth-moving equipment. Injection mouldings are frequently used for plaques on the centre of steering wheels and on some fascia panelling.

Transparent guards for foodstuffs, machines and even baby-incubators may be fabricated simply from acrylic sheet. It should however be pointed out that due to the rather rapid surface deterioration and the lack of 'sparkle' the material is not ideally suited as a cover for displayed goods.

Acrylic sheet is also employed for many other diverse applications, including baths and wash-basins, which have considerable design versatility, are available in a wide range of colours, and are cheaper and much lighter than similar products from conventional materials.

The best-known application of moulded poly(methyl methacrylate) in Great Britain is the G.P.O. telephone. These instruments have however had a rather mixed reception and have been criticised in that they are easily damaged. The tougher ABS materials offer a strong challenge to the acrylics in this outlet and, in fact, are used variously for this purpose.

Decorative plaques are produced by injection moulding poly(methyl methacrylate) and then coating the back of the transparent moulding with a thin coat of metal by the vacuum deposition technique or with a paint by spraying.   By suitable masking, more than one metal and more than one colour paint may be used to enhance the appearance.   These plaques are frequently used in the centre of car steering wheels, refrigerators and other equipment where an eye-catching motif is considered desirable.

If the surface of an acrylic sheet, rod or tube is roughened or carved, less light is internally reflected and the material is often rather brighter at these non-polished surfaces.   The use of this effect enables highly attractive carvings to be produced.   Similarly, lettering cut into sheet, particularly fluorescent sheet becomes 'lit-up' and this effect is useful in display signs.

The use of acrylic materials for dentures has already been mentioned.

12.3 MISCELLANEOUS METHACRYLATE AND CHLOROACRYLATE POLYMERS AND COPOLYMERS

A large number of methacrylate polymers have been prepared in addition to poly(methyl methacrylate).   In many respects the properties of these materials are analogous to those of the polyolefins described in Chapter 8.

As with other linear polymers the mechanical and thermal properties are dependent on the intermolecular attraction, the spatial symmetry and the chain stiffness.   If the poly(*n*-alkyl methacrylate)s are compared it is seen that as the side chain length increases the molecules become spaced apart and the inter-molecular attraction is reduced.   Thus as the chain length increases the softening point decreases and the polymers become rubbery at progressively lower temperatures (Fig. 12.12).[9]   However, where the number of carbon atoms in the side chain is 12 or more, the softening point, brittle point and other properties closely related to the glass transition temperature rise with increase in chain length.   As with the polyolefins this effect is due to side chain crystallisation.   It is to be noted that in the case of the polyolefins the side chain crystallisation has a much greater effect on melting point than on the glass temperature.   In studies on the methacrylates the property measured was the brittle point, a property generally more associated with the glass temperature.

Table 12.2   VICAT SOFTENING POINT OF METHACRYLATE POLYMERS OF TYPE
$CH_2=CH(R)COOCH_3$

R =			
$-CH_3$	119		
$-CH_2 \cdot CH_3$	81		
$-CH_2 \cdot CH_2 \cdot CH_3$	55	$-CH \cdot (CH_3)_2$	88
$-CH_2 \cdot CH_2 \cdot CH_2 \cdot CH_3$	30	$-CH_2 \cdot CH \cdot (CH_3)_2$	67
		$-C \cdot (CH_3)_3$	104
$-CH_2 \cdot CH_2 \cdot CH_2 \cdot CH_2 \cdot CH_3$	*	$-CH_2 \cdot CH_2 \cdot CH(CH_3)_2$	46
		$-CH_2 \cdot C \cdot (CH_3)_3$	115
		$-CH \cdot C(CH_3)_3$	119
		$\quad \mid$	
		CH_3	

* Too rubbery for testing

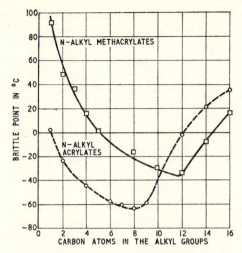

Fig. 12.12.   Brittle points of n-alkyl acrylate and methacrylate ester polymers. (After Rehberg and Fisher,[9] copyright 1948 by The American Chemical Society and reprinted by permission of the copyright owner)

A number of higher *n*-alkyl methacrylate polymers have found commercial usage.   The poly(*n*-butyl-), poly(*n*-octyl-) and poly(*n*-nonyl methacrylate)s have found use as leather finishes whilst poly(lauryl methacrylate) has become useful as a pour-point depressant and improver of viscosity–temperature characteristics of lubricating oils.

As is the case in the polyolefins, the polymethacrylates with branched side chains have higher softening points and are harder than their unbranched isomers.   The effect of branching on Vicat Softening point is shown in Table 12.2.[10]

This effect is not simply due to the better packing possible with the branched isomers.   It is now generally considered that the lumpy branched structures impede rotation about the carbon–carbon bond on the main chain thus giving a stiffer molecule with consequently higher transition temperature.

A number of copolymers based on methyl methacrylate have become commercially available.   Butadiene-methyl methacrylate latices (the butadiene being the major component) are used in paper and board finishing applications (Butakon M, I.C.I.).   In 1962 I.C.I. introduced Asterite, a copolymer of methyl methacrylate with a second acrylic monomer, for vacuum forming.   Some properties of this material were given alongside those of poly(methyl methacrylate) homopolymers in Table 12.1.

A number of other copolymers have been used in the manufacture of acrylic sheet in attempts to increase hardness, heat resistance and scratch resistance but the products tend to have reduced formability and impact strength.   A number of polyfunctional compounds such as allyl methacrylate and glycol dimethyl acrylate have been used to give cross-linked sheet. Some manufacturers supply the sheet in an incompletely cross-linked state which allows a limited amount of forming, after which the sheet may be further heated to complete the 'cure'.

Sheet from poly(methyl α-chloroacrylate) has also been available. This material has a higher softening point than poly(methyl methacrylate) It is, however, expensive, difficult to obtain in a water-white form and the monomer is most unpleasant to handle. It is because of these disadvantageous features that the polymer is believed to be no longer commercially available.

12.4 OTHER ACRYLIC POLYMERS

A number of acrylic polymers other than the poly methacrylates have been produced but these are not generally of interest as plastics materials.

Poly(acrylic acid) is insoluble in its monomer but soluble in water. It does not become thermoplastic when heated. The sodium and ammonium salts have been used as emulsion thickening agents in particular for rubber latex. Poly(methacrylic acid) (Fig. 12.13 (VI)) is similar in properties.

Polymers containing acrylonitrile (VII) are extensively used as fibres (Orlon, Acrilan, Courtelle). The monomer (boiling point 77·1°C) may be

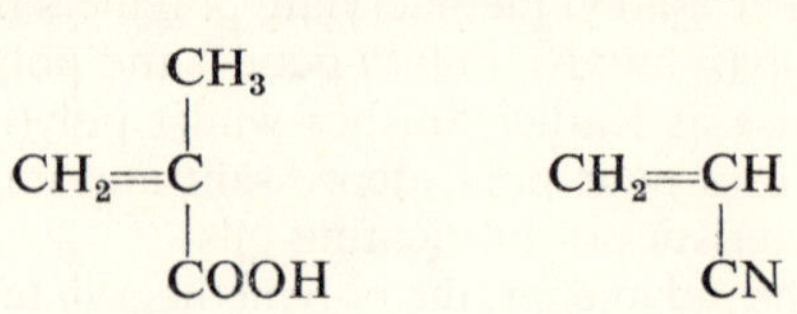

$$CH_2{=}\underset{\underset{COOH}{|}}{\overset{\overset{CH_3}{|}}{C}} \qquad CH_2{=}\underset{\underset{CN}{|}}{CH}$$

VI   VII

Fig. 12.13

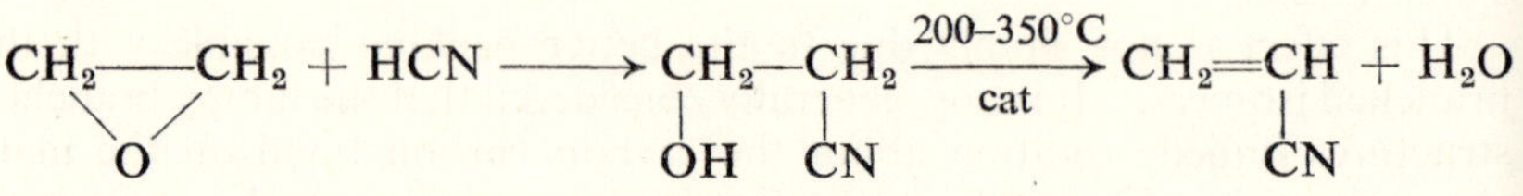

$$\underset{O}{CH_2{-}CH_2} + HCN \longrightarrow \underset{OH \quad CN}{CH_2{-}CH_2} \xrightarrow[cat]{200{-}350°C} \underset{CN}{CH_2{=}CH} + H_2O$$

Fig. 12.14

$$CH{\equiv}CH + HCN \longrightarrow \underset{CN}{CH_2{=}CH}$$

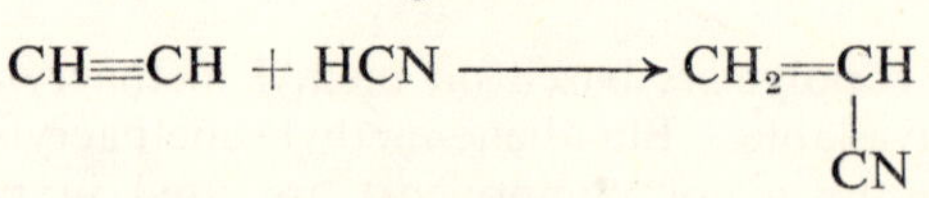

Fig. 12.15

prepared by dehydration of ethylene cyanohydrin, which is itself obtained from ethylene oxide (Fig. 12.14).

It may also be produced by reaction of acetylene with hydrogen cyanide (Fig. 12.15)

Catalyst systems based on cuprous chloride are frequently cited in the patent literature.

In the absence of inhibitors, acrylonitrile polymerises rapidly but the polymer is insoluble in the monomer and precipitates out. Because of the highly exothermic reaction, bulk polymerisations are difficult to control on the technical scale.

Polyacrylonitrile is only slightly thermoplastic even when heated to 200–300°C and unplasticised materials are very difficult to mould. The unmodified polymer is thus not used for plastics, but important thermoplastics containing acrylonitrile include the styrene–acrylonitrile copolymers, the ABS polymers (Chapter 13) and the vinylidene chloride–acrylonitrile copolymers (Chapter 14). Mention may also be made here of nitrile rubbers which are copolymers of butadiene and acrylonitrile (Chapter 8).

A large number of organic acrylic ester polymers have been prepared in the laboratory. Poly(methyl acrylate) is tough, leathery and flexible. With increase in chain length there is a drop in the brittle point but this reaches a minimum with poly-(*n*-octyl acrylate) (see Fig. 12.12). The increase in brittle point with the higher acrylates, which is similar to that observed with the poly-α-olefins and the poly(alkyl methacrylate)s is due to side chain crystallisation.

Poly(methyl acrylate) is water-sensitive and, unlike the corresponding methacrylate, is attacked by alkalis. This polymer and some of the lower acrylate polymers are used in leather finishing and as a textile size.

The alkyl-2-cyanoacrylates[11] have recently become of interest as adhesives. If, for example, methyl-2-cyanoacrylate is thinly spread between two layers to be bonded and the two layers brought into contact, the presence of moisture causes the monomer to polymerise anionically in a few seconds to yield a very strong bond.

Copolymers of ethyl acrylate with 2-chloroethyl vinyl ether in the ratio 95:5 are rubbery materials and are commercially available as Hycar 4021. Vulcanisation may be brought about by amines such as triethylene tetramine.

The vulcanisates exhibit good heat ageing resistance, flex life, ozone resistance and oil resistance but are somewhat lacking in resilience and flexibility at low temperatures. One application for these materials is as a transmission seal for public transport vehicle gear-boxes.

A number of thermosetting acrylic resins for use as surface coatings have appeared during recent years. These are generally complex copolymers and terpolymers such as a styrene–ethyl acrylate–alkoxy methyl acrylamide polymer. Coating resins have also been produced by blending methyl methacrylate with a non-drying alkyd.

The ease with which acrylic monomers may polymerise with each other and with other monomers had led to a host of compositions, frequently of undisclosed nature, being offered for use as moulding materials, casting resins, coating resins, finishing agents and in other applications. Let it suffice to say that many vastly differing compositions may be based on acrylic and methacrylic polymers.

REFERENCES

1. CASPARY, W., and TOLLENS, B., *Ann*, **167**, 241 (1873)
2. KAHLBAUM, G. W. A., *Ber*, **13**, 2348 (1880)
3. FITTIG, R., *Ber*, **12**, 1739 (1879)

4. FITTIG, R., and ENGELHORN, E., *Ann*, **200,** 65 (1880)
5. *U.S. Patent*, 1,980,483.   *British Patent* 395,687 (I.C.I.)
6. *U.S. Patent*, 2,042,458, *British Patent* 405,699 (I.C.I.)
7. SALKIND, M., RIDDLE, E. H., and KEEFER, R. W., *Ind Eng Chem*, **51,** 1232, 1328 (1959)
8. HORN, M. B., *Acrylic Resins*, Reinhold, New York (1960)
9. REHBERG, C. E., and FISHER, C. H., *Ind Eng Chem*, **40,** 1431 (1948)
10. CRAWFORD, J. W. C., *J Soc Chem Ind* **68,** 201 (1949)
11. COOVER, H. W., JOYNER, F. B., SHEARER, N. H., and WICKER, T. H., *Soc. Plastics Engrs J.*, **15,** 413 (1959)

BIBLIOGRAPHY

HORN, M. B., *Acrylic Resins*, Reinhold, New York (1960)
RIDDLE, E. H., *Monomeric Acrylic Esters*, Reinhold, New York (1954)
SCHILDKNECHT, C. E., *Vinyl and Related Polymers*, John Wiley, New York (1952)

13

Plastics Based on Styrene

13.1 INTRODUCTION

It may well be argued that the history of polystyrene is more closely bound up with the history of the twentieth century than is the case with any other plastics material.

In 1934, at the Dow Chemical Co., semi-plant scale work on styrene showed promise of commercial success. At the same time, I.G. Farben, in Germany were carrying out similar work on styrene and polystyrene. As a consequence there became available shortly before the Second World War a material of particular interest because of its good electrical insulation characteristics but otherwise considerably inferior to the polystyrenes available today. Because of these excellent electrical characteristics prices were paid of the order of several dollars per pound for these polymers.

In 1942 the Japanese overran Malaya and the then Dutch East Indies to cut off the main sources of natural rubber from the United States and the British Commonwealth. Because of this the U.S. Government initiated a crash programme for the installation of plants for the manufacture of a rubber from butadiene and styrene. This product, then known as GR–S (Government Rubber–Styrene) provided at that time an inferior substitute for natural rubber but, with a renewed availability of natural rubber at the end of the war the demand for GR–S slumped considerably. (Today the demand for s.b.r. (as GR–S is now known) has increased with the great improvements in quality that have been made and in 1964 U.S.A. production alone was of the order of 1,260,000 tons compared with the world natural rubber production of about 2,200,000 tons.)

After the war however there was a large surplus capacity of plant for the manufacture of styrene and polystyrene together with a great deal of knowledge and experience that had been collected over the war years. It was therefore found possible to produce polystyrene, not as an expensive electrical insulator, but as a cheap general purpose thermoplastic.

247

Because of such desirable properties as low cost, good mouldability, excellent colour range, transparency, rigidity and low moisture absorption, polystyrene is now the third most important thermosplastics material. In Great Britain the material is sold by a number of leading manufacturers such as Shell Chemical Ltd. (Carinex), Monsanto Chemicals Ltd. (Lustrex), British Resin Products Ltd. (Distrene), BX Plastics Ltd. (Bextrene), Sterling Moulding Materials Ltd. (Sternite) and Mobil Chemicals Ltd. (Erinoid Polystyrene).

13.2 PREPARATION OF THE MONOMER

In 1786 William Nicholson wrote *A Dictionary of Practical and Theoretical Chemistry*. In this work Nicholson mentions that a chemist named Neuman, on distillation of storax (a balsam derived from the tree *Liquambar orientalis*), had produced a fragrant 'empyreumatic oil'. In 1839 E. Simon[1] carried out some similar experiments, apparently quite independently, and again obtained this essential oil which he called styrol. In 1845 M. Glenard and R. Boudault[2] reported on the production of styrol (now known as styrene) by dry distillation of dragons blood, a resin obtained from the fruit of the Malayan rattan palm.

In 1869 Berthelot[3] reported the production of styrene by dehydrogenation of ethyl benzene. This method is the basis of present day commercial methods. Over the years many other methods were developed such as the decarboxylation of acids, dehydration of alcohols, pyrolysis of acetylene, pyrolysis of hydrocarbons and the chlorination and dehydrogenation of ethyl benzene.[4]

There are today two methods of interest, (*a*) the laboratory preparation, and (*b*) commercial preparation.

13.2.1   Laboratory preparation

The principal constituent of storax is cinnamic acid and for laboratory purposes styrene is still most easily obtained in high purity by dry distillation of cinnamic acid and its salts under atmospheric pressure (Fig. 13.1).

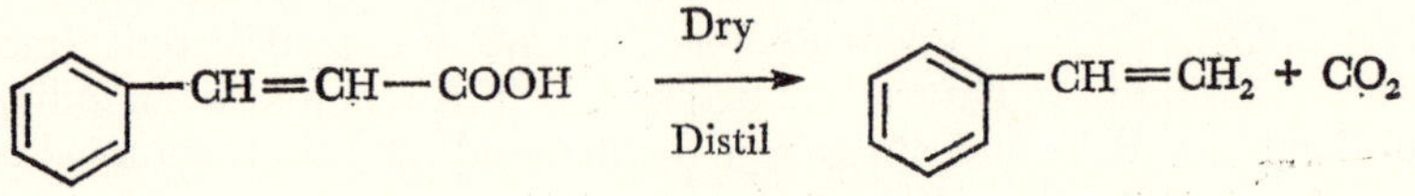

Fig. 13.1

The cinnamic acid is readily prepared by heating benzaldehyde with acetic anhydride and sodium acetate (the Perkin Reaction) (Fig. 13.2).

13.2.2   Commercial preparation

The bulk of commercial styrene is prepared by the Dow process or some similar system. The method involves the reaction of benzene and ethylene to ethyl benzene, its dehydrogenation to styrene and a final

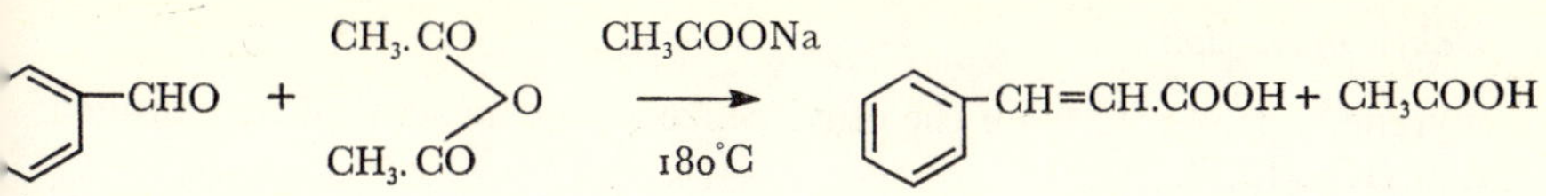

$$\text{C}_6\text{H}_5\text{-CHO} + \underset{\text{CH}_3.\text{CO}}{\overset{\text{CH}_3.\text{CO}}{>}}\text{O} \xrightarrow[\text{180°C}]{\text{CH}_3\text{COONa}} \text{C}_6\text{H}_5\text{-CH=CH.COOH} + \text{CH}_3\text{COOH}$$

Fig. 13.2

finishing stage. It is therefore useful to consider this process in each of the three stages.

Preparation of ethyl benzene

Ethyl benzene is prepared by reaction of ethylene and benzene in the presence of a Friedel–Crafts catalyst such as aluminium chloride at about 95°C (Fig. 13.3).

$$\text{C}_6\text{H}_6 + \text{C}_2\text{H}_4 \rightleftharpoons \text{C}_6\text{H}_5\text{-C}_2\text{H}_5$$

Fig. 13.3

To improve the catalyst efficiency some ethyl chloride is added which produces hydrochloric acid at the reaction temperatures.

The purity of the ethylene is not critical providing that acetylene is not present. The normal purity of ethylene used is about 95%. The purity of the benzene is somewhat higher at about 99% and it is important here that sulphur, as impurity, should be below 0·10%.

In order that the amount of side reaction should be reduced and to minimise the production of polyethyl benzenes, the molar ratios of feedstock and products are approximately as indicated in the following equation (Fig. 13.4).

$$0\cdot58\ \text{CH}_2\text{=CH}_2 + 1\cdot00\ \text{C}_6\text{H}_6 \longrightarrow \begin{array}{l} 0\cdot41\ \text{Ethyl benzene} \\ 0\cdot51\ \text{Benzene} \\ 0\cdot08\ \text{Polyethyl benzenes} \end{array}$$

Fig. 13.4

After passing through the reaction chamber the products are cooled and the aluminium chloride, which is in the form of a complex with the hydrocarbons, settles out. The ethyl benzene, benzene and polyethyl benzenes are separated by fractional distillation, the ethyl benzene having a purity of over 99%. The polyethyl benzenes are dehydrated by heating at 200°C in the presence of aluminium chloride and these products together with the unreacted benzene are recycled.

Recently plants have been installed by some manufacturers to produce ethyl benzene via catalytic reforming processes. The reforming process is one which converts aliphatic hydrocarbons into a mixture of aromatic hydrocarbons. This may be subsequently fractionated to give benzene, toluene and a 'xylene fraction' from which ethyl benzene may be obtained.

Dehydrogenation

Styrene is produced from the ethyl benzene by a process of dehydrogenation (Fig. 13.5).

This is an endothermic reaction in which a volume increase accompanies dehydrogenation. The reaction is therefore favoured by operation at reduced pressure. In practice steam is passed through with the ethyl

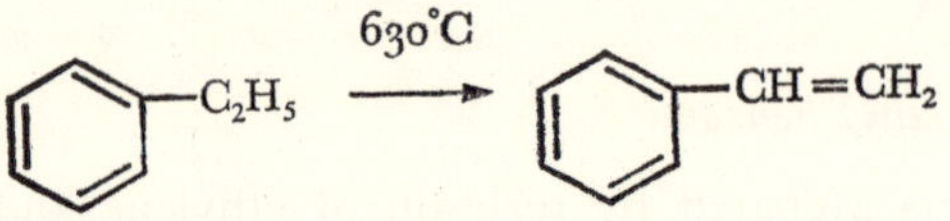

Fig. 13.5

benzene in order to reduce the partial pressure of the latter rather than carrying out a high temperature reaction under partial vacuum. By the use of selected catalysts such as magnesium oxide and iron oxide a conversion of 35–40% per pass with ultimate yields of 90–92% may be obtained.

Styrene purification

The dehydrogenation reaction produces 'crude styrene' which consists of approximately 37·0% styrene, 61% ethyl benzene and about 2% of aromatic hydrocarbon such as benzene and toluene with some tarry matter. The purification of the styrene is made rather difficult by the fact that the boiling point of styrene (145·2°C) is only 9 deg C higher than that of ethyl benzene and because of the strong tendency of styrene to polymerise at elevated temperatures. To achieve a successful distillation it is therefore necessary to provide suitable inhibitors for the styrene, to distil under a partial vacuum and to use specially designed columns.

In one process the crude styrene is first passed through a pot containing elemental sulphur, enough of which dissolves to become a polymerisation inhibitor. The benzene and toluene are then removed by distillation.

Table 13.1

Molecular weight	104·14
Density at 25°C	0·9019 g/ml
Refractive index at 25°C	1·5439
Boiling point	145·2°C
Vol. shrinkage on polymerisation	17%

The ethyl benzene is then separated from the styrene and tar by passing this through two distillation columns each with top temperatures of about 50°C and bottom temperatures of 90°C under a vacuum of about 35 mmHg. The tar and sulphur are removed by a final distillation column and the styrene is permanently inhibited by addition of 10 p.p.m. of t-butyl catechol which has less adverse effects on the final polymer than sulphur.

Styrene is a colourless mobile liquid with a pleasant smell when pure but with a disagreeable odour due to traces of aldehydes and ketones if allowed to oxidise by exposure to air. It is a solvent for polystyrene and many synthetic rubbers including s.b.r. but has only a very limited mutual solubility in water. Table 13.1 shows some of the principal properties of pure styrene.

Styrene takes part in a very large number of chemical reactions. In particular it has a strong tendency to polymerise on heating or on exposure to ultraviolet light.

13.2 POLYMERISATION

Polystyrene was first made by E. Simon in 1839 who at the time believed he had produced an oxidation product which he called styrol oxide. Since that time the polymerisation of styrene has been extensively studied. In fact a great deal of the work which now enables us to understand the fundamentals of polymerisation was carried out on styrene.

The polymer is today prepared by mass, solution, suspension and emulsion methods, the first three being the most important. Mass polymerisation has the advantage of apparent simplicity and gives a polymer of high clarity and very good electrical insulation characteristics. There are, however, severe problems due to the exothermic reaction and the product has a broad molecular wehigt distribution. Polymerisation in solution reduces the exotherm but may lead to problems of solvent recovery and solvent hazards. The solvent may also act as a chain transfer agent and cause a reduction in molecular weight. Suspension polymerisation avoids most of these problems but there is some contamination of the polymer by water and the suspension agent. Furthermore the polymer must be dried and aggregated before being sold as pellets suitable for injection moulding and extrusion. Emulsion polymerisation techniques are seldom used with polystyrene since the large quantities of soap used seriously affects clarity and electrical insulation characteristics. This process is therefore only used for the production of polystyrene latex.

13.3.1 Mass polymerisation

Continuous mass polymerisation units are extensively used for making polystyrene. Great care is necessary to prevent the heat of reaction accelerating the polymerisation to such an extent that the reaction gets out of control. The problem is made particularly difficult by the fact that heat can only be taken away from the points of higher temperature by conduction because of the very high viscosity of the reacting material, and also the low thermal conductivities of both styrene and polystyrene.

Most mass processes used today are a variation of that developed by Wolff in Germany before World War II. In this process the styrene is prepolymerised by heating (without initiators) in a prepolymerisation kettle at 80°C for 2 days until a 33–35% conversion to polymer is reached (see Fig. 13.6). The monomer–polymer mixture is then run into a tower about 25 ft high. The tower is fitted with heating and cooling jackets and internally with a number of heating and cooling coils. The top of the tower is maintained at a temperature of about 100°C, the centre at about 150°C and the bottom of the tower at about 180°C. The high bottom temperature not only ensures a higher conversion but boils off

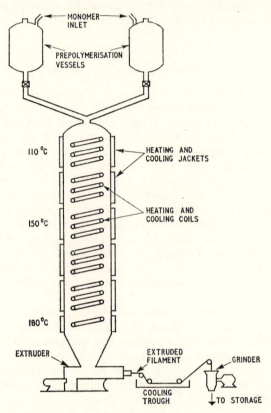

Fig. 13.6. *Tower process for mass polymerisation of styrene*

the residual styrene from the polymer. The base of the tower forms the hopper of an extruder from which the melt emerges as filaments which are cooled, disintegrated and packed.

That such a process is today commercially important is a measure of the success of chemical engineers in overcoming heat transfer problems involved with masses incapable of being stirred. An idea of the extent of the problem can be gauged from the fact that it takes 6 hours to cool a sample of polystyrene from 160°C using a cooling medium at 15°C when the heat transfer distance is 2 inches.

13.3.2 Solution polymerisation

By polymerising styrene in solution many problems associated with heat transfer and the physical movement of viscous masses are reduced, these advantages being offset by problems of solvent recovery and the possibility of chain transfer reactions. In 1955 Distrene Ltd started a plant at Barry in South Wales for the production of styrene by such a solution polymerisation process and some details have been made available.[5,6] The essential details of this process are indicated by Fig. 13.7.

Styrene and solvent are blended together and then pumped to the top of the first reactor which is divided into three heating zones. In the first

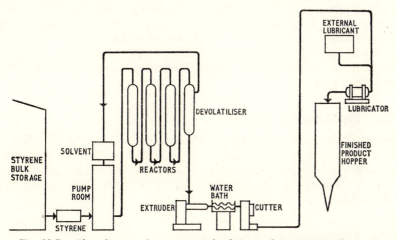

Fig. 13.7. *Flow diagram for commercial solution polymerisation of styrene*

zone the solution is heated to start up the polymerisation reaction but because of the exothermic reaction in the second and third zones of the first reactor and the three zones of the second reactor Dowtherm cooling coils are used to take heat out of the system. By the time the reaction mixture reaches the third reactor the polymerisation reaction has started to slow down and so the reaction mixture is reheated.

From the third reactor the polymer is then run into a devolatilising ('stripping') vessel in the form of thin strands. At a temperature of 225°C the solvent, residual monomer and some very low molecular weight polymers are removed, condensed and recycled. The polymer is then fed to extruder units, extruded as filaments, granulated, lubricated and stored to await dispatch.

13.3.3 Suspension polymerisation

Suspension polymerisation of styrene is widely practised commercially.[7] In this process the monomer is suspended in droplets $\frac{1}{32} - \frac{1}{64}$ in. in diameter in a fluid, usually water. The heat transfer distances for the dissipation of the exotherm are thus reduced to values in the range

$\frac{1}{64}-\frac{1}{128}$ in. Removal of heat from the low viscosity fluid medium presents little problem. The reaction is initiated by monomer-soluble initiators such as benzoyl peroxide.

It is necessary to coat the droplets effectively with some suspension agent, e.g. poly(vinyl alcohol), talc etc., to prevent them cohering. Control of the type and quantity of suspension agent and of the agitation has a pronounced effect on the resulting particles. It is not unknown for the whole of the polymerising mass to aggregate and settle to the bottom of the reaction vessel because of such conditions being incorrect. Following polymerisation, unreacted monomer may be removed by steam distillation and the polymer is washed and dried.

The disadvantages of the suspension process are that about 70% of the volume of the kettle is taken up by water, the need for a drying stage which could cause discoloration by degradation and the need to convert the small spheres formed into a larger shape suitable for handling. Furthermore, the suspension method cannot easily be converted into a continuous process.

13.3.4 Emulsion polymerisation

Because of the large quantities of soap left in the polymer, which adversely affects clarity, electrical insulation characteristics and problems in agitation and densification, this process is used only for making latices.

The techniques used are in many respects similar to those for emulsion polymerised p.v.c.

13.3.5 Characterisation and routine control

After polymerisation, or before purchase by a fabricator, it is necessary to check the quality of a batch of polymer. It is neither economical nor necessary to check each batch for every important property but there are a number of tests that may be made to indicate the quality of the polymer. Because of the exothermic reaction there is a tendency for the polymer to vary from granule to granule and it is therefore necessary to take care in sampling.

B.S.1493:1958 has specified acceptance limits for four grades of polystyrene. These limits are given in Table 13.2.

These properties are measured by means of tests laid down in the

Table 13.2

Property	Type A	Type B	Type C	Type D
Softening point (°C) min.	100	90	85	75
Methanol soluble matter (%) max.	1·5	3·5	5·0	7·0
Volatile matter (%) max.	0·5	1·5	2·5	3·0
Viscosity of 2% solution in toluene (cP) min.	2·0	2·0	1·9	1·7
Impact strength (ft lb) min.	0·12	0·12	0·12	0·12

standard. Electrical grades have additional requirements in respect of power factor and permittivity.

The solution viscosity is a measure of molecular weight. Commercial polymers normally have a viscosity average molecular weight in the range of 50,000–200,000. In addition to polystyrene the polymer will contain (i) monomer, (ii) impurities such as ethyl benzene and aldehydes and (iii) low molecular weight polymers (e.g. dimers and trimers). The methanol solubles content is a measure of the total of these low molecular weight products. The volatile content per cent is a measure of (i) and (ii) alone with perhaps a very small amount of the dimer and trimer. It is not simply a measure of the residual monomer for this when measured by ultraviolet absorption techniques accounts for only 50–60% of the volatiles.

The impact strength and softening point will depend on the molecular weight, residual impurities from polymerisation and or additives such as internal and external lubricants which may have been incorporated into the polymers.

13.3.6 Grades available

In addition to the high impact (toughened) polystyrenes dealt with later in this chapter, polystyrene is available in a number of grades. These may conveniently be grouped as follows:

1. General purposes grades. In these grades a balance is attempted to obtain good heat resistance, reasonably high setting up temperature, good flow properties and reasonable impact strength.
2. High molecular weight grades. Polystyrene has little strength if its molecular weight is below 50,000 but increases rapidly, with molecular weight up to 100,000 (Fig. 13.8). An increase in molecular weight above 100,000 has little further effect on tensile strength but continues to have an adverse effect on the ease of flow. Such higher molecular weight grades are sometimes used where improved impact strength is required without the loss of clarity that occurs with the toughened polystyrenes.

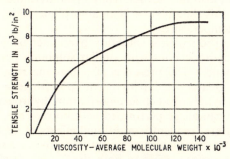

Fig. 13.8. Influence of molecular weight on the tensile strength of polystyrene

3. Heat resistant grades.   By reducing the amount of volatile matter the softening point of the polystyrene can be raised.   For example, by reducing the monomer content from 5% to 0% the softening point may be raised from 70°C to 100°C.   Commercial heat resisting grades usually have a softening point about 7 deg C above the softening point of general purpose polystyrene.

4. Easy flow grades.   By incorporating an internal lubricant such as butyl stearate or liquid paraffin, by using a polymer of lower molecular weight, by careful control of granule shape and size and by lubrication of the granules with an external lubricant such as zinc stearate, the flow properties of polystyrene may be improved with little effect on other properties apart from reduction of up to 10 deg C in the softening point.   These materials are very useful for thin-wall mouldings, for moulding with minimum frozen-in strains or other products where the moulding is rather intricate.   They have not however replaced general purpose polystyrene because of their lower setting up temperature which causes a prolongation of the injection moulding cycle.

13.4   PROPERTIES AND STRUCTURE OF POLYSTYRENE

Polystyrene has the simple repeating structure (Fig. 13.9).

As might be expected from such a substantially linear polymer it is thermoplastic.   Commercial polystyrene is an atactic polymer and therefore because of the random spatial position of the benzene ring it is incapable of crystallisation.   The benzene ring does however stiffen the

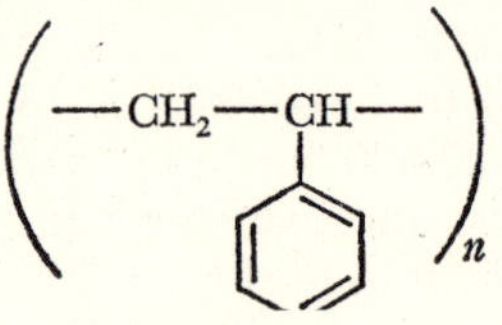

$$\left(-\text{CH}_2-\text{CH}-\right)_n$$

Fig. 13.9

polymer chain and also induces relatively high intermolecular forces.   As a consequence of this the polymer is an amorphous resin, hard and transparent at room temperature.

Being a hydrocarbon with a solubility parameter of 9·1 it is dissolved by a number of hydrocarbons with similar solubility parameters such as benzene and toluene.   The presence of a benzene ring results in polystyrene having greater reactivity than polyethylene.   Characteristic reactions of a phenyl group such as chlorination, hydrogenation, nitration and sulphonation can all be performed with polystyrene.   Chain rupture and discoloration are frequently additional effects of such reactions.

The pure hydrocarbon nature of polystyrene gives it excellent electrical insulation characteristics, owing to both the fundamentally good characteristics of the material and to the low water absorption of such a hydrocarbon polymer.   The insulation characteristics are therefore well main-tained in humid conditions.

13.5 GENERAL PROPERTIES

Polystyrene is a hard, rigid, transparent thermoplastic which emits a characteristic metallic ring when dropped.   It is free from odour and taste, burns with a sooty flame and has a low specific gravity of 1·054. Because of its low cost, good mouldability, low moisture absorption, good dimensional stability, good electric insulation properties, colourability and reasonable chemical resistance it is widely used as an injection moulding and vacuum forming material.   Additionally the low thermal conductivity has been made use of in polystyrene foam used for thermal insulation.   The principal limitations of the polymer are its brittleness, inability to withstand the temperature of boiling water and its mediocre oil resistance.

The mechanical properties of polystyrene depend to some extent on the nature of the polymer (e.g. its molecular weight), on the method of preparing the sample for testing and on the method of test, as is the case with all plastics materials.

Some typical mechanical properties of polystyrene are indicated in Table 13.3.   It will be observed that there is little real difference in the

Table 13.3

| Property | Test method | Grade of polystyrene | | | |
		General purpose	High mol. wt	Heat resistant	Easy flow
Tensile strength (10³ lb in⁻²)	ASTM D.638–58T	6–7	6·5–7·5	6·5–7·5	6–7
Elongation (%)	ASTM D.638–58T	1·0–2·5	1·0–2·5	1·0–2·5	1·0–2·5
Modulus in tension (10⁵ lb in⁻²)	ASTM D.638–58T	5·0	5·0	5·5	5·0
Flexural strength (10³ lb in⁻²)	ASTM D.790–58T	9–11	10–12	11–14	9–11
Impact strength (notched Izod ft lb in notch)	BS.1493	0·25–0·35	0·25–0·35	0·25–0·35	0·25–0·35

mechanical properties of the four types of straight polystyrene considered in the table.

Amongst the optical properties of polystyrene of importance are its high transmission of all wavelengths of visible light and its high refractive index (1·592) which gives it a particularly high 'brilliance'.   Certain factors may however mar the good optical characteristics such as haze and yellowing.   Haze is believed to be due to the presence of dust and also to variations in the refractive index caused by localised molecular orientation.   Yellowness may be due to coloured impurities in the monomer, or to impurities which cause reaction during polymerisation to contribute to the yellowness, or to aging. Yellowing on aging has been shown to be an oxidation reaction, the rate of yellowing increasing with decrease in wavelength, monomer content and in the presence of traces of sulphur. In one example radiation from a 15 W germicidal lamp (2537 Å) one foot from a polystyrene sample caused yellowing in a few hours. In another experiment it was found that the presence of 0·040% sulphur caused a sevenfold increase in the amount of yellowing after a given time in an artificial light aging cabinet. This has meant that sulphur cannot be tolerated as a polymerisation inhibitor during storage of the monomer. The incorporation of about 1% of saturated aliphatic amines, cyclic amines or aminoalcohols has been found to improve greatly the resistance to weathering. Examples of materials quoted in the literature include diethylamino ethanol and piperazine. Many proprietary materials are now marketed.

The electrical insulation characteristics of polystyrene are extremely good. Typical figures are given in Table 13.4.

The negligible effect of frequency on dielectric constant and power factor from 60 to 10^6 c/s is of particular interest. It should however be noted that at 10^7 c/s the power factor may increase about four-fold.

The chemical resistance of polystyrene is not generally as good as that of polyethylene. It is dissolved by a number of hydrocarbons such as

Table 13.4

Property	Value	ASTM test
Dielectric strength	500–700 volts per 0·001 in.	D.149–44
Volume resistivity	10^{17}–10^{19} ohm cm	D.257–46
Dielectric constant		
60–10^6 c/s	2·45–2·65	D.150–46T
Power factor		
60–10^6 c/s	1–2 × 10^{-4}	D.150–46T

benzene, toluene and ethyl benzene, by chlorinated hydrocarbons such as carbon tetrachloride, chloroform and o-dichlorobenzene, by a number of ketones (but not acetone), and esters and by a few oils (e.g. oil of verbena and ylang ylang oil). Many other materials, in particular acids, alcohols, oils, cosmetic creams and foodstuffs will cause crazing and cracking and in some cases chemical decomposition. The extent of attack will also depend on such factors as the grade of polystyrene, internal stresses in the polystyrene product, external stresses to which the part is subjected, to the time and temperature of exposure and to the concentration of the reagent. Furthermore many materials do not attack polystyrene individually but do so in combination. Cosmetic creams and patent medicines provide many examples of this synergistic type behaviour. Technical service bulletins supplied by the manufacturer provide useful information on the subject of chemical resistance.

Particular mention should be made of the influence of styrene monomer (Fig. 13.10). An increase of the residual monomer from 0 to 5% can cause a 30 deg C reduction in softening point. On the other hand there is

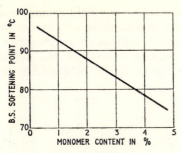
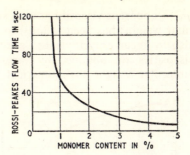

Fig. 13.10. Influence of styrene monomer content on the softening point and flow characteristics of polystyrene. (After Haward and Crabtree)[8]

a marked increase in the ease of flow. It is not however good practice to increase the flow properties in this way as the monomer will volatilise in the processing machine and the bubbles formed will be distorted to produce such faults as 'silver streaks' and 'mica marks' in the finished product.

The thermal properties are of interest to both the user of the end product and to the processor. From the user's point of view the principal features are the very low thermal conductivity (0·0003 c.g.s. units) and the comparatively low softening point. Standard tests give softening points of about 90°C, that is below the boiling point of water. In addition many properties are affected by temperature (Fig. 13.11).

In common with other thermoplastic melts polystyrene exhibits pseudoplastic behaviour. At shearing stresses below $6 \times 10^8/M$ dyn cm^{-2} (where M = molecular weight), the ratio of shear stress to shear rate is almost constant and the melt is substantially Newtonian. Above this shear stress non-Newtonian behaviour becomes pronounced. The much stronger dependence on shear rate of polystyrene compared with low

Table 13.5 EFFECT OF SHEAR STRESS ON MELT VISCOSITY[9]

	Apparent viscosity (poises) at 232°C	
Shearing stress	*G.P. polystyrene*	*Polyethylene D = 0·916 MFI = 2*
Zero shear	24,000	9,500
70,000 dyn cm^{-2}	1,500	5,300

density polyethylene is shown in Table 13.5. The shear stress selected is considered as typical during injection moulding of polystyrene.

The melt viscosity is also strongly dependent on temperature and molecular weight. Whilst it has been found[10] that the viscosity of some thermoplastics depends exponentially on the square root of the weight average molecular weight, the equation

$$\log \mu_a = K + 3·4 \log_{10} M_w$$

(μ_a = apparent viscosity at zero shear rate; M_w = weight average molecular weight) has been found[11] to be more applicable to polystyrene (and also to linear polyethylenes—see Chapter 7).   Although a wide range of viscosities are obtained by varying processing conditions and polymer molecular weight the numerical values generally lie between the high

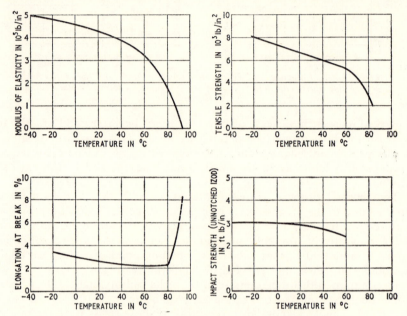

Fig. 13.11.   *Influence of temperature on some mechanical properties of polystyrene.   (After Boundy and Boyer[4])*

values experienced with unplasticised p.v.c. and the low values observed with the nylons (see also Fig. 12.11).

The specific heat of polystyrene is dependent on temperature and at 200°C the value is approximately double that at room temperature (Fig. 13.12).

13.6 RUBBER-MODIFIED POLYSTYRENES

For many applications polystyrene might be considered to be too brittle a polymer.   Because of this, polystyrene manufacturers have made a number of attempts to modify their products.

The methods of approaching this problem include:

1. Use of higher molecular weight polymers.
2. Use of plasticisers.
3. Incorporation of fillers such as glass fibre, wood flour, etc.
4. Deliberate orientation of the polymer molecules.
5. Copolymerisation.
6. The use of rubbery additives.

Of these methods the first gives only marginal improvements whilst the second approach has far too severe an effect on the softening point to be of any commercial value. The use of fillers has been practised to some extent in the United States but is not of importance in Europe. Deliberate orientation is limited to filament and sheet.

Very many copolymers with styrene as the principal constituent have been prepared and a number have been marketed. In some instances there is an appreciable increase in toughness but usually in such cases the softening point of the copolymer is much lower than that of the homopolymer. For example copolymers from 70 parts of styrene and 30 parts

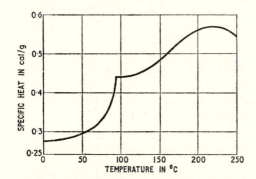

Fig. 13.12. Specific heat–temperature curve for polystyrene. (After Boundy and Boyer[4])

butadiene, although tough, are leather-like. Using lower quantities of butadiene the improvement in toughness is insufficient to tolerate the losses in temperature resistance.

The addition of rubbery materials to polystyrene is extensively used commercially. Rubber-modified styrenes are today used to about the same extent as straight polystyrene. A number of rubbers have been proposed but today either s.b.r. (styrene–butadiene rubbers) with a styrene content of about 25–30% or polybutadiene are generally used. Not only is s.b.r. extremely effective but as the world's most used synthetic rubber it is quite cheap. The use of polybutadiene as an alternative is a recent development.

The polystyrene and the rubber may be blended in a number of ways. Originally the ingredients were compounded in a 2-roll mill, in an internal mixer or in an extruder. The impact strength of the products was however little better than the unmodified polymer. Blending of s.b.r. latex and polystyrene latex, which is then followed by co-coagulation and drying has also been employed in the past but once again the improvement is only marginal.

Today the common practice is first to dissolve the rubber in the styrene monomer and then to polymerise the styrene in the usual way. By this process the resultant blend will contain not only s.b.r. and polystyrene but also a graft polymer where short styrene side chains have been attached to the s.b.r. molecules. This gives a marked improvement in the impact strengths that can be obtained.

It has been demonstrated that with s.b.r.–polystyrene blends the rubber should exist in discrete droplets, less than 50 μm in diameter where a good finish is required, within the polystyrene matrix. It is believed that in such a form the rubber can reduce crack propagation and hence fracture in two ways.[12] In the first case consider a crack propagating through a resin–rubber blend encountering a rubber particle and assume that the crack passes through the particle. Assuming a good rubber–polystyrene adhesion and a reasonably high elongation at break of the rubber, a rubber 'bridge' will tend to form across the crack and retard further cleavage. Energy will also be absorbed in stretching the rubber so that propagation will be slowed down. Where a crack on reaching a particle divides and passes round the particle there is a dissipation in the concentration of the cracking force. The strip of polymer between the two advancing cracks also becomes subject to a number of shearing and tensile stresses thus adding to the work which must be done in propagating the crack.

From the foregoing comments it is seen that the following features are necessary for a suitable blend:

1. The rubber and the polystyrene should not be compatible. If they are there will be molecular mixing and no improvement in toughness.
2. The rubber should not be too incompatible if good rubber–polystyrene adhesion is to be obtained.

In effect this means that, to achieve reasonable toughness, semicompatible rubbers should be used. Semicompatibility may be achieved by (a) selecting mixtures of slightly different solubility parameter to the polystyrene, (b) by judicious amounts of cross linking or (c) by judicious use of selected graft polymers. In current commercial grades it is probable that all three features are involved.

In commercial polymers there are three main variables to be considered:

1. The amount of s.b.r. added, usually 5–20% (see Fig. 13.13). An increase in the s.b.r. will increase the toughness but there will be an attendant reduction in softening point.
2. The size of the rubber particles. In a typical blend these would be in the range 1–10 μm.
3. The gel content (toluene insoluble per cent) of the rubber and the swelling index of the gel (the ratio of the volume of a swollen gel to its unswollen volume). The former is a measure of the amount of cross-linked material and the second a measure of the intensity of cross linking. It has been found[12] that a sample of medium gel content (5–20%) and a medium swelling index (10–20) gives the best impact strength in the blend.

A high impact polystyrene (polystyrene–s.b.r. blend) may have seven times the impact strength of ordinary polystyrene, but about half the tensile strength, a lower hardness and a softening point some 15°C lower. Because of the rubber content there may be a reduction in light and heat stability and stabilisers are normally incorporated.

The use of stabilisers (antioxidants) may however have adverse effects in that they inhibit cross linking of the rubber. C. B. Bucknall[13] has

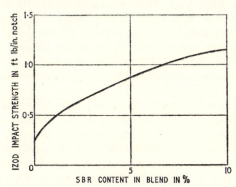

Fig. 13.13. *Effect of adding s.b.r. on the impact strength of polystyrene*[11]

studied the influence of phenolic antioxidants on polystyrene–s.b.r. alloys blended in an internal mixer at 180°C. He found that alloys containing 1% of certain phenolic antioxidants were gel-deficient in the rubber phase. The gel-deficient blends were blotchy in appearance, had lower flow rates compared with the normal materials, and mouldings were somewhat brittle. Substantial improvement in the impact properties was achieved when the anti-oxidant was added later in the mixing cycle after the rubber had reached a moderate degree of cross linking.

Specifications for three types of toughened polystyrene are given in B.S.3126:1959. The quantitative requirements of this specification are summarised in Table 13.6.

The methods of test used are similar to those specified in B.S.1493 but it should be noted that B.S.3126 states that impact strength results are to be expressed in terms of foot pounds per inch of notch instead of foot pounds.

Experiments on the toughening of polystyrene by the use of rubbers has not been restricted to s.b.r. The advent of stereoregular polybutadiene

Table 13.6

	Type 1	Type 2	Type 3
Impact strength (ft lb per in of notch) (min.)	1·0	0·5	0·5
Softening point (°C) (min.)	80	85	95
Tensile strength (10^3 lb in^{-2}) (min.)	3	3·5	4
Elongation at break (%) (min.)	15	10	7·5
Volatile matter (%) (max.)	3·0	2·5	2·5
Water absorption (mg) (max.)	20	15	20

P.M.—18

in particular has led to new developments. High cis content 1:4 poly-butadienes[14],[15] are claimed to be more effective in toughening polystyrene than s.b.r. Reasons given for this greater effectiveness include a better balance of the compatibility–incompatibility factors, the lower glass transition temperature of the rubber ($-100°C$ instead of $-55°C$ for s.b.r.), greater resilience than s.b.r. and the higher reactivity towards grafting. High impact materials with a degree of transparency have been prepared from 1:2-polybutadiene. In this case the double bond is present in a pendant vinyl group rather than in the main chain. It is thus more reactive and unless the grafting process is carried out at temperatures below 100°C and under carefully controlled condition excess cross linking may occur.

A number of polyisoprenes[16] have also been investigated as potential toughening agents. By careful control of grafting conditions high impact blends have been made from 3:4 polyisoprenes. Neither natural rubber (essentially cis-1.4-polyisoprene) nor the other synthetic polyiso-prenes give a significant reinforcing effect. Ethylene–propylene rubbers which have been peroxidised by bubbling oxygen through a solution of the polymer whilst being irradiated with ultraviolet light may also toughen polystyrene. The peroxidised rubber is dissolved in styrene monomer in the concentration range of 5–20% and the solution reacted at above 70°C. The peroxide or hydroperoxide groups present then decompose with the formation of free radicals which initiated growth of a polystyrene branch on the chain of the rubber molecule.

13.7 STYRENE–ACRYLONITRILE COPOLYMERS AND RELATED ALLOYS

Styrene–acrylonitrile copolymers ($\sim$20–30% acrylonitrile content) have been commercially available for a number of years. Until recently however the price of these materials was too high for them to find more than a few specialised outlets. Because of the polar nature of the acry-lonitrile molecule these copolymers have better resistance to hydrocarbons, oils and greases than polystyrene. They also have a higher softening point, a much better resistance to stress cracking and crazing and an enhanced impact strength yet retain the transparency of the homopolymer. The higher the acrylonitrile content the greater the toughness and chemical resistance but the greater the difficulty in moulding and the greater the yellowness of the resin. Typical resins have a water absorption about that of poly(methyl methacrylate), i.e. about ten times that of polystyrene but about one-tenth that of cellulose acetate.

The important features of rigidity and transparency make the material competitive with polystyrene, cellulose acetate and poly(methyl meth-acrylate) for a number of applications. In general the copolymer is cheaper than poly(methyl methacrylate) and cellulose acetate, tougher than poly(methyl methacrylate) and polystyrene and superior in chemical and most physical properties to polystyrene and cellulose acetate. It does not have such a high transparency or such good weathering properties as poly(methyl methacrylate). As a result of these considerations the styrene–acrylonitrile copolymers have found applications for dials, knobs and covers for domestic appliances, electrical equipment and car equipment; for picnic-ware and housewares and a number of other industrial and domestic applications with requirements somewhat more stringent than can be met by polystyrene.

Although tough enough for many uses styrene–acrylonitrile copolymers are inadequate in this respect for other purposes. As a consequence, a range of materials popularly referred to as ABS polymers first became available in the early 1950's. There are many ways of producing these materials, the two most important types of which are:

1. Blends of acrylonitrile–styrene copolymers with butadiene–acrylonitrile rubber (referred to below as Type 1).
2. Interpolymers of polybutadiene with styrene and acrylonitrile (referred to below as Type 2).

The Type 1 materials may be produced by blending on a 2-roll mill or in an internal mixer or blending the latices followed by coagulation or spray drying. In these circumstances the two materials are compatible and there is little improvement in the impact strength. If however the rubber is lightly cross linked by the use of small quantities of peroxides the resultant reduction in compatibility leads to considerable improvements in impact strength (see Figs. 13.4 and 13.15). A wide range of polymers may be made according to the nature of each copolymer and the proportion of each employed.[12] A typical blend would consist of

70 parts (70:30 styrene–acrylonitrile copolymer)

40 parts (63:35) butadiene–acrylonitrile rubber).

By altering these variables blends may be produced to give products varying in processability, toughness, low temperature toughness and heat resistance. The Kralastic and Royalite polymers are believed to be of this type.

Although the nitrile rubbers employed normally contain about 35% acrylonitrile the inclusion of nitrile rubber with a higher butadiene content will increase the toughness at low temperatures. For example whereas the typical blend cited above has an impact strength of only 0·9 ft lb/in. notch at 0°F, a blend of 70 parts styrene–acrylonitrile, 30 parts of nitrile rubber (35% acrylonitrile) and 10 parts nitrile rubber (26% acrylonitrile) will have an impact value of 4·5 ft lb/in. notch at that temperature.[12]

From Fig. 13.4 it will be seen that a minimum of about 20% cross-linked nitrile rubber is required in order to obtain tough products. For high impact material the acrylonitrile–styrene copolymer should have a high molecular weight. The commercial copolymers containing 20–30% acrylonitrile are suitable for the preparation of Type 1 ABS polymers. The amount of gelled rubber also has a profound effect on the strength as shown in Fig. 13.15. The samples used were prepared by milling the rubber and peroxide for 15 minutes, the styrene–acrylonitrile resin then being added and blended for a further 15 minutes. The physical nature of the blend is more complex than with the s.b.r. modified polystyrene as it appears that rubber molecule networks may exist in the resin phase.

To produce the Type 2 polymers, styrene and acrylonitrile are added to polybutadiene latex and the mixture warmed to about 50°C to allow absorption of the monomers. A water soluble initiator such as potassium persulphate is then added to polymerise the styrene and acrylonitrile. The

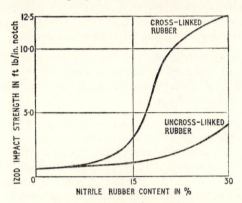

Fig. 13.14. Variation in impact strength against concentration of rubber in ABS Type 1 blends[12]

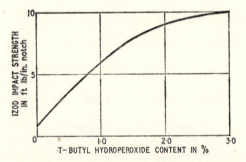

Fig. 13.15. Effect of cross-linking agent on impact strength of 75/25 styrene–acrylonitrile/ nitrile rubber blend

resultant material will be a mixture of polybutadiene, polybutadiene grafted with acrylonitrile and styrene, and styrene–acrylonitrile copolymer. The presence of graft polymer is essential since straightforward mixtures of polybutadiene and styrene–acrylonitrile copolymers are weak.

It is obvious that the range of possible ABS-type polymers is very large. Not only may the ratios of the three monomers be varied but the way in which they can be assembled into the final polymer can also be the subject of considerable modifications. Neither is it necessary to be restricted to the use of acrylonitrile, butadiene, and styrene. For example in 1964 Mazzucchelli Celluloide introduced Sicoflex MBS a transparent material (unlike conventional ABS materials which are opaque) which is said to be an alloy of resins and rubbery polymers in which the basic constituents are methacrylic esters, butadiene and styrene. Because of the wide range of products available and because the chemical nature of these materials is rarely divulged it is not possible to give detailed properties of these materials unless one resorts to lists of properties of named proprietary materials. In general however these materials have a high impact strength, have softening points as high as, and sometimes higher, than general purpose polystyrene, and moulded specimens generally have a very good surface appearance. This last property is particularly marked

Table 13.7

	G.P. poly-styrene	*Med. impact s.b.r. poly-styrene blend*	*High impact s.b.r. poly styrene blend*	*A–S co-polymer*	*ABS polymer*
Specific gravity	1·05	1·05	1·05	1·06	1·01–1·05
Izod impact strength B.S.2782 (ft lb per in. notch)	0·25	0·7	1·6	0·4–0·8	5–9
Tensile strength (10^3 lb in^{-2} ASTM D.638–58T)	7·5	5	3	6–10	4–5
Softening point (°C) B.S.2782	88–95	85	80	100	100
Volume resistivity (ohm cm)	$\sim 10^{18}$	$\sim 10^{15}–10^{16}$	$\sim 10^{15}–10^{16}$	—	$\sim 10^{13}$

with the Type 2 materials. Some typical properties of ABS polymers compared with other styrene-containing polymers are given in Table 13.7.

Although some 50–70% more expensive than general purpose poly-styrene, ABS polymers have found uses as a result of their toughness and good appearance. Their use has been somewhat greater in the United States than in Europe since the development of rigid p.v.c. has been rather slower in America. They will also meet serious competition from the recently developed polypropylene copolymers. Moulded applications include telephone housings, vacuum cleaner covers, and pump impellers. Calendered sheet is extensively used in the manufacture of suit cases, in car upholstery and for car crash pad covers. An expanded ABS laminate sheet material has recently been introduced capable of thermoforming.

13.8 MISCELLANEOUS POLYMERS AND COPOLYMERS

In addition to the polymers, copolymers and alloys already discussed styrene and its derivatives have been used for the polymerisation of a wide range of polymers and copolymers. Two of the more important applications of styrene, in s.b.r. and in polyester laminating resins, are dealt with in other chapters.

The influence of nuclear substituents on the properties of a homopolymer depends on the nature, size and shape of the substituent, the number of the substituents and the position of entry into the benzene ring.

Table 13.8 shows how some of these factors influence the softening point of the polymers of the lower *p*-alkyl styrenes.

It will be seen that increasing the length of a *n*-alkyl side group will cause a reduction in the interchain forces and a consequent reduction in

Table 13.8

Polymer	B.S.1524 softening point
Poly(*p*-methyl styrene)	88°C
Poly(*p-n*-propyl styrene)	R.T.
Poly(*p*-isopropyl styrene)	87°C
Poly(*p-n*-butyl styrene)	rubber
Poly(*p*-sec-butyl styrene)	86°C
Poly(*p*-tert-butyl styrene)	130°C

the transition temperature, and hence the softening point. Branched alkyl groups impede free rotation and may more than offset the chain separation effect to give higher softening points. Analogous effects have already been noted with the polyolefins and polyacrylates.

Polar substituents such as chlorine increase the interchain forces and hinder free rotation of the polymer chain. Hence polydichlorstyrenes have softening points above 100°C. One polydichlorstyrene has

Table 13.9

Polymer	V.S.P.(°C)
Poly(*m*-methyl styrene)	92
Poly(*o*-methyl styrene)	128
Poly(*p*-methyl styrene)	105
Poly(2,4-methyl styrene)	135
Poly(2,5-dimethyl styrene)	139
Poly(3,4-dimethyl styrene)	99
Poly(2,4,5,trimethyl styrene)	147
Poly(2,4,6 trimethyl styrene)	164
Poly(2,3,5,6 tetramethyl styrene)	150

been marketed commercially as Styramic HT. Such polymers are essentially self-extinguishing, have heat distortion temperatures of about 120°C and a specific gravity of about 1·40.

The nuclear substituted methyl styrenes have been the subject of much study and of these poly(vinyl toluene) (i.e. polymers of *m*- and *p*-methyl styrenes) has found use in surface coatings. The Vicat softening point of some nuclear substituted methyl styrenes is given in Table 13.9.

Catalytic dehydrogenation of cumene, obtained by alkylation of benzene with propylene will give α-methyl styrene (Fig. 13.16).

Both the alkylation and dehydrogenation may be carried out using equipment designed for the production of styrene.

It has not been found possible to prepare high polymers from α-methyl styrene by free radical methods and ionic catalysts are used. The reaction may be carried out at about −60°C in solution.

Polymers of α-methyl styrene have been marketed for various purposes but have not become of importance for mouldings and extrusions. On the other hand copolymers containing α-methyl styrene are currently marketed. Styrene–α-methyl styrene polymers are transparent, water-white materials with B.S. softening points of 104–106°C (cf. 100°C for

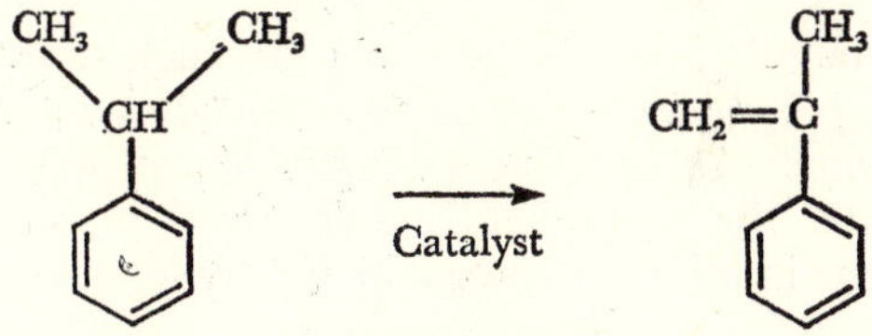

Fig. 13.16

normal polystyrenes). These materials have melt viscosities slightly higher than heat resistant polystyrene homopolymer.

Many other copolymers are mentioned in the literature and some of these have reached commercial status in the plastics or some related industry. The reason for the activity usually lies in the hope of finding a polymer which is of low cost, water white and rigid but which has a greater heat resistance and toughness than polystyrene. This hope has yet to be fulfilled.

13.9 STEREOREGULAR POLYSTYRENE

Polystyrene produced by free radical polymerisation techniques is atactic in structure and therefore amorphous. In 1955 Natta[17] and his co-workers reported the preparation of substantially isotactic polystyrene using aluminium alkyl–titanium halide catalyst complexes. Similar systems were also patented by Ziegler[18] at about the same time. The use of *n*-butyl lithium as a catalyst has been described by R. C. P. Cubban and D. Margerison.[19] Whereas, at room temperature, atactic polymers are produced, polymerisation at −30°C leads to isotactic polymer, with a narrow molecular weight distribution.

In the crystalline region isotactic polystyrene molecules take a helical form with three monomer residues per turn and an identity period of 6·65 Å. 100% crystalline polymer has a density of 1·12 compared with 1·05 for amorphous polymer and is also translucent. Because of its more regular nature the melting point of the polymer is as high as 230°C. Below the glass transition temperature of 97°C the polymer is rather brittle.

Because of the high melting point and high molecular weight it is difficult to process isotactic polystyrenes.   Various techniques have been suggested for injection moulding in the literature but whatever method is employed it is necessary that the moulding be heated to about 180°C, either within or outside of the mould to allow the material to develop a stable degree of crystallinity.

The brittleness of isotactic polystyrenes has hindered their commercial development.   Quoted Izod impact strengths are only 20% that of conventional atactic polymer.   Impact strength double that of the atactic material has however been claimed when isotactic polymer is blended with a synthetic rubber or a polyolefin.

13.10 PROCESSING OF POLYSTYRENE

Polystyrene and closely related thermoplastics such as the ABS polymers may be processed by such techniques as injection moulding, extrusion and blow moulding.   Of less importance is the processing in latex and solution form and the process of polymerisation casting.   The main factors to be borne in mind when considering polystyrene processing are:

1. The negligible water absorption avoids the needs for predrying granules.
2. The low specific heat (compared with polyethylene) enables the polymer to be rapidly heated in injection cylinders which therefore have a higher plasticising capacity with polystyrene than with polyethylene.   The setting-up rates in the injection moulds are also faster than with the polyolefins so that faster cycles are also possible.
3. The strong dependence of apparent viscosity on shear rate.   This necessitates particular care in the design of complex extrusion dies.
4. The absence of crystallisation gives polymers with low mould shrinkage.
5. Molecular orientation.

Although it is not difficult to make injection mouldings from polystyrene which appear to be satisfactory on visual examination it is another matter to produce mouldings free from internal stresses.   This problem is common to injection moulding of all polymers but is particularly serious with such rigid amorphous thermoplastics as polystyrene.

Internal stresses occur because when the melt is sheared as it enters the mould cavity the molecules tend to be distorted from the favoured coiled state.   If such molecules are allowed to freeze before they can re-coil ('relax') then they will set up a stress in the mass of the polymer as they attempt to regain the coiled form.   Stressed mouldings will be more brittle than unstressed mouldings and are liable to crack and craze, particularly in media such as white spirit.   They also show a characteristic pattern when viewed through crossed Polaroids.   It is because compression mouldings exhibit less frozen-in stresses that they are preferred for comparative testing.

To produce mouldings from polystyrene with minimum strain it is desirable to inject a melt, homogenous in its melt viscosity, at a high rate into a hot mould at an injection pressure such that the cavity pressure drops to zero as the melt solidifies. Limitations in the machines available or economic factors may however lead to less ideal conditions being employed.

A further source of stress may arise from incorrect mould design. For example if the ejector pins are designed in such a way to cause distortion of the mouldings, internal stresses may develop. This will happen if the mould is distorted while the centre is still molten, but cooling, since some molecules will freeze in the distorted position. On recovery by the moulding of its natural shape these molecules will be under stress.

A measure of the degree of frozen-in stresses may be obtained by comparing the properties of mouldings with known, preferably unstressed, samples, by immersion in white spirit and noting the degree of crazing, by alternately plunging samples in hot and cold water and noting the number of cycles to failure or by examination under polarised light. Annealing at temperatures just below the heat distortion temperature followed by slow cooling will in many cases give a useful reduction in the frozen-in stresses.

The main reason for extruding polystyrene is to prepare high-impact polystyrene sheet. Such sheet can be formed without difficulty by vacuum forming techniques. In principle the process consists of clamping the sheet above the mould, heating it so that it softens and becomes rubbery and then applying a vacuum to draw out the air between the mould and the sheet so that the sheet takes up the contours of the mould.

13.11 EXPANDED POLYSTYRENE[20,21,22]

Polystyrene is now available in certain forms in which the properties of the product are distinctly different from those of the parent polymer. Of these by far the most important is expanded polystyrene, an extremely valuable insulating material now available in densities as low as 1 lb ft^{-3}. A number of processes have been described in the literature for the manufacture of the cellular product of which four are of particular interest.

1. Polymerisation in bulk of styrene with azo-di-isobutyronitrile as initiator. This initiator evolves nitrogen as it decomposes so that expansion and polymerisation occur simultaneously. This method was amongst the earliest suggested but has not been of commercial importance. There has however been recent resurgence of interest in this process.
2. The Dow 'Log' Process. Polystyrene is blended with a low boiling chlorinated hydrocarbon and extruded. The solvent volatilises as the blend emerges from the die and the mass expands. This process is still used to some extent.
3. The B.A.S.F. Process. Styrene is blended with a low boiling hydrocarbon and then polymerised. The product is chipped. The chips

are then converted into expanded polymer as in method (4) described in detail below.

4. Bead Processes. These processes have generally replaced the above techniques. The styrene is polymerised by bead (suspension) polymerisation techniques. The blowing agent, typically 6% of low boiling petroleum ether fraction such as *n*-pentane, may be incorporated before polymerisation or used to impregnate the bead under heat and pressure in a post-polymerisation operation.

The impregnated beads may then be processed by two basically different techniques (*a*) the steam-moulding process, the most important industrially and (*b*) direct injection moulding or extrusion. In the steam-moulding process the beads are first 'prefoamed' by heating them in a steam bath. This causes the beads to expand to about forty times their previous size. At this stage the beads should not fuse or stick together in any way. It has been shown that expansion is due not only to volatilisation of the low boiling liquid (sometimes known as a pneumatogen) but also to an osmotic-type effect in which steam diffuses into the cells within the bead as they are formed by the expanding pneumatogen. The entry of steam into the cells causes a further increase in the internal pressure and causes further expansion. It has been estimated[20] that about half of the expansion is due to the effect of steam which can diffuse into the cells at a much greater rate than the pneumatogen can diffuse out. The expansion of the beads is critically dependent on both temperature and time of heating. At low steaming pressures the temperature obtained is about that of the softening point of polystyrene and it is important to balance the influences of polymer modulus, volatilisation rates and diffusion rates of steam and pneumatogen. In practice preforming temperatures of about 100°C are used. Initially the amount of bead expansion increases with the time of prefoaming. If however the beads are heated for too long the pneumatogen diffuses out of the cells and the residual gas cannot withstand the natural tendency of the bead to collapse. (This natural tendency is due to beads consisting largely of membranes of highly oriented polymers in a rubbery state at prefoaming temperatures. The natural tendency of molecules to disorient above the glass transition temperature, the reason why rubbers are elastic, was discussed in the early chapters of this book.)

The second stage of the process is to condition the beads, necessary because on cooling after prefoaming pneumatogen and steam within the cells condense and cause a partial vacuum within the cell. By allowing the beads to stand in air for at least 24 hours air can diffuse into the cells in order that at room temperature the pressure within the cell equilibrates with that outside.

The third stage of the process is the steam moulding operation itself. Here the prefoamed beads are charged into a chest or mould with perforated top, bottom and sides through which steam can be blown. Steam is blown through the prefoam to sweep air away and the pressure then allowed to increase to about 15 lb in^{-2}. The beads soften, air in the cells expands on heating, pneumatogen volatilises and steam once again permeates into the cells. In consequence the beads expand and being enclosed in the fixed volume of the mould consolidate into a solid block, the density of which is largely decided by the amount of expansion in the initial prefoaming process. Heating and cooling cycles are selected to give the best balance of economic operation, homogeneity in density through the block, good granule consolidation, good block external appearance and freedom from warping. This process may be used to give slabs which may be subsequently sliced to the appropriate size or alternatively to produce directly such objects as containers and flower pots. The steam moulding process, although lengthy, has the advantages of being able to make very large low density blocks and being very economic in the use of polymer.

Whilst it is possible to purchase standard equipment for the steam moulding process attempts continue to be made to make sweeping modifications to the process. These include the use of dielectric and microwave heating and the development of semicontinuous and continuous processes.

The alternative approach to the two-stage steam moulding process is that in which impregnated beads are fed directly to an injection moulding machine or extruder so that expansion and consolidation occur simultaneously. This approach has been used to produce expanded polystyrene sheet and paper by a tubular process reminiscent of that used with polyethylene. Bubble nucleating agents such as sodium bicarbonate and citric

Table 13.10

	Density (lb ft^{-3})	Thermal conductivity (Btu in ft^{-2} hr^{-1} degF^{-1})
Expanded polystyrene	1·0	0·22
Polyurethane foam (with chloro-fluorocarbon gas)	2·0	0·16
Expanded ebonite	3·75	0·21
Cork (expanded)	6·25	0·27
Wood	25·0	0·65
Glass wool	4·0	0·26
Expanded p.v.c.	2·5	0·22

acid which evolve carbon dioxide during processing are often incorporated to prevent the formation of a coarse pore structure. Typical film has a density of about 3 lb ft^{-3}. Injection moulding of impregnated beads gives an expanded product with densities of about 12–13 lb ft^{-3}. This cannot compare economically with steam moulding and is best considered as a low-cost polystyrene (in terms of volume) in which air and pneumatogen act as a filler. Such products generally have an inferior appearance to normal polystyrene mouldings.

The outstanding features of steam moulded polystyrene foam are its low density and low thermal conductivity. These are compared with other important insulating materials in Table 13.10.

Table 13.11

Mean temperature (°F)	Thermal conductivity (Btu in. ft^{-2} hr^{-1} degF^{-1})
50	0·24
0	0·21
−27	0·19
−40	0·18
−126	0·14

It is important to note that the thermal conductivity is dependent on the mean temperature involved in the test. The relationship may be illustrated by quoting results obtained from a commercial material of density 1 lb ft^{-3} (Table 13.11).

Other typical properties for a 1 lb ft^{-3} expanded polystyrene material are

Tensile strength	15–20 lb in^{-2}
Flexural strength	20–30 lb in^{-2}
Compression strength	10–15 lb in^{-2}
Water absorption	2% g/100 ml

13.12 ORIENTED POLYSTYRENE

Deliberately oriented polystyrene is available in two forms; filament (monoaxially oriented) and film (biaxially oriented). In both cases the increase in tensile strength in the direction of stretching is offset by a reduction in softening point because of the inherent instability of oriented molecules.

Filament is prepared by extrusion followed by hot stretching. It may be used for brush bristles or for decorative purposes such as in the manufacture of 'woven' lampshades.

Biaxially stretched film has proved of value as a packaging material. Specific uses include blister packaging, snap-on lids, overwrapping, 'windows' and de luxe packaging.

It may be produced by extrusion either by a tubular process or by a flat film extrusion.[23] The latter process appears to be preferred commercially as it allows greater flexibility of operation. The polystyrene is first extruded through a slit die at about 190°C and cooled to about 120°C by passing between rolls. The moving sheet then passes above a heater and is rewarmed to 130°C, the optimum stretching temperature. The sheet is then stretched laterally by means of driven edge rollers and longitudinally by using a haul-off rate greater than the extrusion rate. Lateral and longitudinal stretching is thus independently variable. In commercial processes stretch ratios of 3:1–4:1 in both directions are commonly employed (see Fig. 13.14).

Commercial oriented film has a tensile strength of 10–12,000 lb in^{-2} (cf. 6,000–8,000 for unstretched material) and an elongation of break of 10–20% (cf. 2–5%). The impact strength of bars laminated from biaxially stretched film have impact strengths of the order of 15 times greater than the basic polymer. The heat distortion temperature is negligibly affected. Whereas toughness and clarity are the principal desirable features of oriented polystyrene film the main disadvantages are the high moisture vapour transmission rate compared with polyethylene and the somewhat poor abrasion resistance.

Although it is possible to vacuum form these films the material has such a high modulus at its shaping temperatures that an exceptionally good vacuum is required for shaping. As a consequence of this the pressure

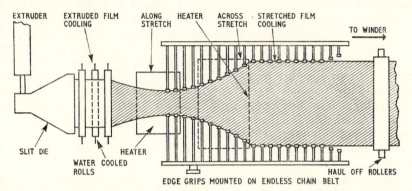

Fig. 13.17. Plax process for manufacture of biaxially stretched polystyrene film

forming technique has now been developed. In this process the sheet is clamped between the mould and a heated plate. Air is blown through the mould pressing the sheet against the hot plate. After a very short heating period the air supply is switched so that compressed air passes through holes in the heater plate and blows the sheet into the mould.

13.13 APPLICATIONS

Polystyrene and closely related copolymers and blends find application where rigidity and low cost are important prerequisites. Other properties which may also be relevant to the selection of these materials include transparency and high refractive index (of the homopolymer), freedom from taste, odour and toxicity, good electrical insulation characteristics, low water absorption and processability. In most countries tonnage consumption of the straight polystyrene (i.e. the homopolymer) is very similar to the consumption of the rubber–polystyrene blends. Consumption of ABS polymers in Great Britain and the United States is about 10% of that for polystyrene (including the blends) whilst the usage of styrene–acrylonitrile copolymers is about one-third that of ABS.

The largest outlet for polystyrene is in packaging applications. Specific uses include bottle caps, small jars and other injection moulded containers, blown containers (a somewhat recent development but which has found rapid acceptance for talcum powder), vacuum formed toughened polystyrene as liners for boxed goods and oriented polystyrene film for foodstuffs such as creamed cheese. Vacuum formed cigarette packets were introduced in the United States in the early 1960's and were claimed to be as economical to produce as those from cardboard.

A second important outlet is in refrigeration equipment where the low thermal conductivity and improved impact properties of polystyrene at low temperatures are an asset. Specific uses in this area include door liners and inner liners made from toughened polystyrene sheet, mouldings for flip lids, trays and other refrigerator 'furnishings' and expanded polystyrene for insulation. Although in the past most liners have been fabricated from sheet there is a current interest in injection moulding these parts since these will give greater design flexibility. It is also claimed that with sufficiently high production rates the injection process will be cheaper. For the insulation, expanded polystyrene faces competition from rigid polyurethane foams but at the time of writing has maintained its dominance for this outlet. The expanded polystyrene has a low density, a low weight cost, is less brittle and can be made fire retarding. Polyurethane foams produced by systems using auxiliary blowing agents such as fluorochloromethanes are claimed to have a lower thermal conductivity but this value will increase if the fluorocarbon gas is able to diffuse out of the system. The polyurethanes may also be formed *in situ*.

Polystyrene and high impact polystyrene mouldings are widely used for housewares, for example storage containers, for toys, games and sports equipment, radio and electrical equipment (largely as housings, knobs and switches), for bathroom and toilet fittings (such as cistern ball-cock floats) and for shoe heels. A small-quantity of light stabilised polymer is used for light fittings but because of the tendency of polystyrene to yellow, poly(methyl methacrylate) is usually preferred. Polystyrene monofilament finds limited use for brushes and for handicraft work.

Expanded polystyrene accounts for over 10% of the weight consumption of polystyrene and high impact polystyrene in Britain. The volume of expanded material produced annually in this country exceeds even the volume production of the polyolefins. About half of this class of material goes into building applications for thermal insulation and for between-floor sound insulation, about 30% is used for low temperature insulation and the rest for packaging and buoyancy applications and such miscellaneous products as flower pots and *jardinières*. It is to be expected that in future the building industry will use ever-increasing quantities of expanded polystyrene.

ABS polymers, because of their toughness and good appearance, have found steadily increasing usage for housings for domestic and industrial electrical equipment. In automobiles they have been used for instrument cluster boards and door handles whilst sheet has been used for the covers of crash pads. The use of injection moulded boot lids, bonnet lids and door panels has received and is receiving active consideration by car manufacturers. Whilst ABS polymers have in recent years provided a severe challenge to cellulosic plastics they are meeting competition from the propylene–ethylene block copolymers. Styrene–acrylonitrile copolymers which are somewhat tougher than polystyrene and which are also transparent are now used for drinking tumblers, housewares and housing for electrical equipment.

REFERENCES

1. SIMON, E., *Ann.*, **31**, 265–277 (1839)
2. GLENARD, M., and BOUDALT, R., *Ann.*, **53**, 325 (1845)
3. BERTHELOT, M., *Ann. Chim. Phys.* (4), **16**, 153–162 (1869)
4. BOUNDY, R. H., and BOYER, R. F., *Styrene, its Polymers, Copolymers and Derivatives*, Reinhold, New York (1952)
5. *Brit. Plastics*, **30**, 26 (1957)
6. *Plastics (London)*, **22**, 3 (1957)
7. SAMARAS, N. N. T., and PARRY, E., *J. Appl. Chem. (London)*, **1**, 243 (1951)
8. HAWARD, R. N., and CRABTREE, D. R., *Trans. Plastics Inst.*, **23**, 61 (1955)
9. GOGGIN, W. C., CHENEY, G. W., THAYER, G. B., *Plastics Technol.*, **2**, 85 (1956)
10. FLORY, P. J., *J. Am. Chem. Soc.*, **62**, 1057–70 (1940)
11. FOX, T. G., and FLORY, P. J., *J. Am. Chem. Soc.*, **70**, 2384–2395 (1948)
12. DAVENPORT, N. E., HUBBARD, L. W., and PETTIT, M. R., *Brit. Plastics*, **32**, 549 (1959)
13. BUCKNALL, C. B., *Trans. Instn. Rubber Ind.*, **39**, 221 (1963)
14. *British Patent*, 892, 910
15. *British Patent*, 897, 625
16. *British Patent*, 880, 928
17. NATTA, G., *J. Polymer Sci.*, **16**, 143 (1955)
18. *Belgian Patent*, 533, 362
19. CUBBAN, R. C. P., and MARGERISON, D., *Proc. Chem. Soc.*, 146 (1960)
20. SKINNER, S. J., BAXTER, S., and GREY, P. J., *Trans. Plastics Inst.*, **32**, 180 (1964)
21. SKINNER, S. J., *Trans. Plastics Inst.*, **32**, 212 (1964)
22. FERRIGNO, T. H., *Rigid Plastics Foams*, Reinhold, New York (1963)
23. JACK, J., *Brit. Plastics*, **34**, 312, 391 (1961)

BIBLIOGRAPHY

BOUNDY, R. H., and BOYER, R. F., *Styrene, its Polymers, Copolymers and Derivatives*, Reinhold, New York (1952)
GIBELLO, H., *Le Styrène et ses Polymères*, Dunod, Paris (1956)

14

Miscellaneous Ethenoid Thermoplastics

14.1 INTRODUCTION

In addition to the various ethenoid polymers discussed in the preceding seven chapters a large number of other polymers of this type have been described in the literature.[1] Some of these have achieved commercial significance and those which have interest as plastics or closely related materials are the subject of this chapter.

14.2 VINYLIDENE CHLORIDE POLYMERS AND COPOLYMERS

Vinylidene chloride polymerises spontaneously into poly(vinylidene chloride), a polymer sufficiently thermally unstable to be unable to withstand melt processing (Fig. 14.1).

By copolymerising the vinylidene chloride with about 10–15% of vinyl chloride, processable polymers may be obtained which are used in the manufacture of filaments and films. These copolymers have been marketed by the Dow Company since 1940 under the trade name Saran. Vinylidene chloride–acrylonitrile copolymers for use as coatings of low moisture permeability are also marketed (Saran, Viclan). Vinylidene chloride–vinyl chloride copolymers in which the vinylidene chloride is the minor component (2–20%) were mentioned in Chapter 9.

The monomer is produced from trichloroethane by dehydrochlorination (Fig. 14.2). This may be effected by pyrolysis at 400°C, by heating with lime or treatment with caustic soda. The trichlorethane itself may be obtained from ethylene, vinyl chloride or acetylene.

Vinylidene chloride is a clear mobile liquid which is highly inflammable and with the following physical properties

Boiling point	31·9°C at 760 mmHg
Specific gravity	1·233 at 15·5°C
Refractive index	1·4246 at 20°C

Specific heat 0.27 cal g^{-1} degC^{-1}
Heat of polymerisation 14.5 kcal/mole

Although miscible with many organic solvents it has a very low solubility in water (0.04%).

The handling of the monomer presents a number of problems. The monomer will polymerise on storage even under an inert gas. Polymer deposition may be observed after standing for less than a day. Exposure

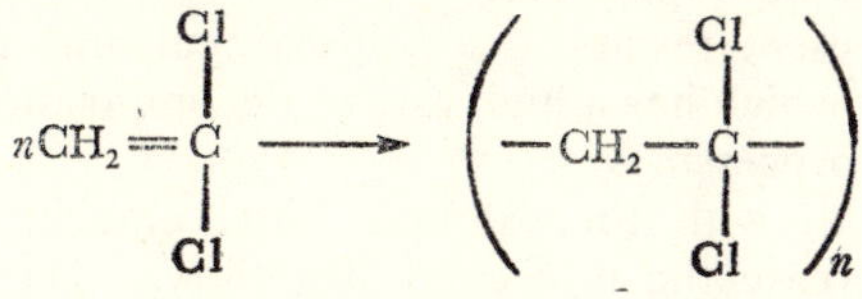

Fig. 14.1.

to air, to water or to light will accelerate polymerisation. A number of phenolic materials are effective inhibitors, a typical example being 0.02% p-methoxyphenol. Exposure to light, air and water must however still be avoided. The monomer has an anaesthetic action and chronic toxic properties and care must therefore be taken in its handling.

The polymer may be prepared readily in bulk, emulsion and suspension, the latter technique apparently being preferred on an industrial scale.

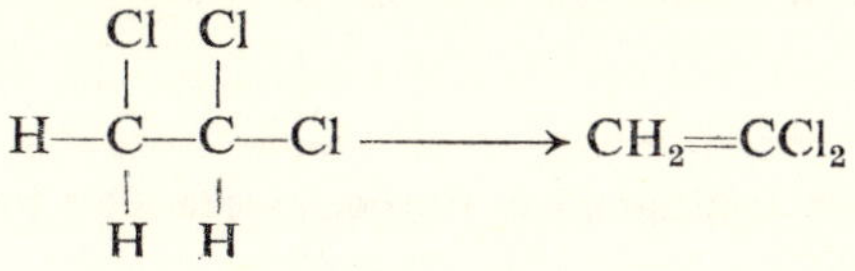

Fig. 14.2

The monomer must be free from oxygen and metallic impurities. Peroxides such as benzoyl peroxide are used in suspension polymerisations which may be carried out at room temperature or at slightly elevated temperatures. Persulphate initiators and the conventional emulsifying soaps may be used in emulsion polymerisation. The polymerisation rate for vinylidene chloride–vinyl chloride copolymers is markedly less than for either monomer polymerised alone.

Consideration of the structure of polyvinylidene chloride (Fig. 14.3) enables certain predictions to be made about its properties.

It will be seen that the molecule has an extremely regular structure and that questions of tacticity do not arise. The polymer is thus capable of

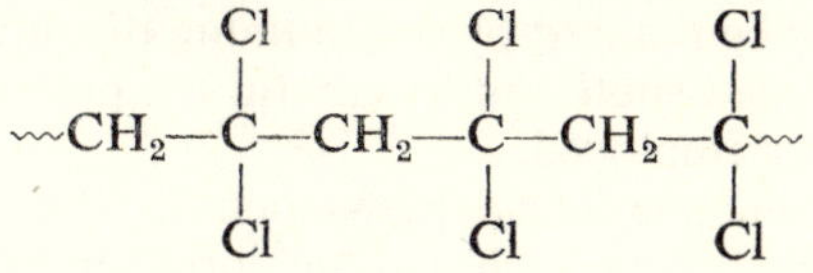

Fig. 14.3

crystallisation.   The resultant close packing and the heavy chlorine atom result in the polymer having a high density (1·875) and a low permeability to vapours and gases.

The solubility parameter is calculated at 9·8 and therefore the polymer is swollen by liquids of similar cohesive forces.   Since crystallisation is thermodynamically favoured even in the presence of liquids of similar solubility parameter and since there is little scope of specific interaction between polymer and liquid there are no effective solvents at room temperature for the homopolymer.

The chlorine present results in a self-extinguishing polymer.   It also leads to a polymer which has a high rate of decomposition at the temperatures required for processing.

Copolymerisation, with for example vinyl chloride will reduce the regularity and increase the molecular flexibility.   The copolymers may thus be processed at temperatures where the decomposition rates are less catastrophic.

Vinylidene chloride–vinyl chloride polymers are also self-extinguishing and possess very good resistance to a wide range of chemicals including acids and alkalies.   They are dissolved by some cyclic ethers and ketones.

Because of the extensive crystallisation, even in the copolymers, high strengths are achieved even though the molecular weights are quite low ($\sim$ 20,000–50,000).   A typical 85 : 15 copolymer plasticised with diphenyl ethyl ether has a melting point of about 170°C, a glass temperature of about -17°C and a maximum rate of crystallisation at approximately 90°C.[2]

14.2.1   Properties and applications of vinylidene chloride–vinyl chloride copolymers[3]

Since some properties of the vinylidene chloride–vinyl chloride copolymers are greatly dependent on crystallisation and orientation it is convenient to consider the applications of these copolymers and then to discuss the properties of the products.

The copolymers have been used in the manufacture of extruded pipe, moulded fittings and for other items of chemical plant.   They are however rarely used in Europe for this purpose because of cost and the low maximum service temperature.   Processing conditions are adjusted to give a high amount of crystallinity, for example by the use of moulds at about 90°C.   Heated parts of injection cylinders and extruder barrels which come into contact with the molten polymer should be made of special materials which do not cause decomposition of the polymer. Iron, steel and copper must be avoided.   The danger of thermal decomposition may be reduced by streamlining the interior of the cylinder or barrel to avoid dead-spots and by careful temperature control.   Steam heating is frequently employed.

Additives used include plasticisers such as diphenyl diethyl ether, ultraviolet light absorbers such as 5-chloro-2-hydroxy benzophenone (1–2% on the polymer) and stabilisers such as phenoxy propylene oxide.

The main application of the copolymers is in the manufacture of filaments.[4]   These may be extruded from steam-heated extruders with a screw compression ratio of 5:1 and a length:diameter of 10:1.   The filaments are extruded downwards (about 40 at a time) into a quench bath and then round drawing rollers which cause a three- to fourfold extension of the filaments and an increase in strength from about 10,000 to 36,000 lb in⁻².   The filaments are used for deck chair fabrics, car upholstery, decorative radio grilles, dolls' hair, filter presses and for sundry other applications where their toughness, flexibility, durability and chemical resistance are of importance.

Biaxially-stretched copolymer film is a useful though expensive packaging material (Saran Wrap-Dow) possessing exceptional clarity,

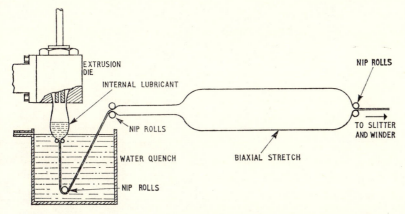

Fig. 14.4.   Extrusion process for the manufacture of biaxially oriented Saran film[2]

brilliance, toughness and water and gas impermeability.   A number of grades are available differing in transparency, surface composition and shrinkage characteristics.   It is produced by water quenching a molten tubular extrudate at 20°C and then stretching by air inflation at 20–50°C. Machine direction orientation of 2–4 : 1 and transverse orientation of 3–5 : 1 occurs and crystallisation is induced during orientation.[2]   The process is shown schematically in Fig. 14.4.   Some general properties of vinylidene chloride–vinyl chloride copolymers containing about 85% vinylidene chloride are given in Table 14.1.   Gas transmission date of typical films is given in Table 14.2.   The water vapour transmission is about 0·05–0·15 g/100 in²/24 hr at 70°F for 0·001 in. thick film.   The large variation in gas transmission values quoted are due to differences in formulation, films having the higher transmission having a softer feel.

14.2.2   Vinylidene chloride–acrylonitrile copolymers

Copolymers of vinylidene chloride with 5–50% acrylonitrile were investigated by I.G. Farben during the Second World War and found to be

Table 14.1 GENERAL PROPERTIES OF VINYLIDENE CHLORIDE–VINYL
CHLORIDE (85 : 15) COPOLYMER

Specific gravity	1·67–1·7
Refractive index	1·60–1·61
Specific heat	0·32 cal g^{-1} degC^{-1}
Max. service temperature	60°C (continuous)
Dielectric constant 10^2 c/s	4·9–5·3 (ASTM D.150)
10^5 c/s	3·4–4·0 (ASTM D.150)
Power factor 10^2 c/s	0·03–0·05 (ASTM D.150)
10^5 c/s	0·04–0·05 (ASTM D.150)
Volume resistivity	10^{12}–10^{16} ohm cm (ASTM D.257)
Tensile strength (unoriented)	8,000 lb in^{-2}
Tensile strength (filaments)	20,000–40,000 lb in^{-2}
Tensile strength (film)	8,000–20,000 lb in^{-2}

Table 14.2 GAS TRANSMISSION
(CM^3/100 IN^2/24 HR ATM) AT 73·4°F
(TABULATED TO A 1 MIL THICKNESS)
ASTM D.1434–56T

O$_2$	1·0–5·9
CO$_2$	3·8–45·7
N$_2$	0·16–1·6
Air	0·21–2·6
Freon 12	<0·03–4·0

Source: Dow Co. Literature

promising for cast films.   Early patents by I.C.I.[5] and Dow[6] indicated that
the copolymers were rigid, transparent and with a high impact strength.

The principal commercial outlet for these copolymers (Saran, Viclan)
has however been as coatings for cellophane, polyethylene, paper and other
materials.   Such coatings are of value because of their high moisture and
gas impermeability, chemical resistance, clarity, toughness and heat-
sealability.   The percentage of acrylonitrile used is normally in the range
5–15%.   Higher quantities facilitate solubility in ketone solvents whereas
lower amounts, i.e. higher vinylidene chloride contents, increase the
barrier properties.   The barrier properties of these copolymers are of the
same order as the vinylidene chloride–vinyl chloride copolymers and are
claimed in the trade literature to be between 100 and 1,000 times more
impermeable than low density polythene in respect of CO$_2$, nitrogen and
oxygen transmission.

In 1962 Courtaulds announced a flame-resisting fibre BHS said to be a
50:50 vinylidene chloride–acrylonitrile copolymer.   This product has
subsequently been renamed 'Teklan'.

A number of other copolymers with vinylidene chloride as the major
component have been marketed.   Prominent in the patent literature are
methyl methacrylate, methyl acrylate and ethyl acrylate.

14.3 COUMARONE–INDENE RESINS

Fractionation of coal tar naphtha (b.p. 150–200°C) yields a portion boiling at 168–172°C consisting mainly of coumarone and indene (Fig. 14.5).

The products bear a strong formal resemblance to styrene and may be polymerised. For commercial purposes the monomers are not separated but are polymerised *in situ* in the crude naphtha, sulphuric acid acting as an ionic catalyst to give polymers with a degree of polymerisation of 20–25.

In one process the naphtha fraction boiling between 160 and 180°C is washed with caustic soda to remove the acids and then with sulphuric acid to remove basic constituents such as pyridine and quinoline. The naphtha

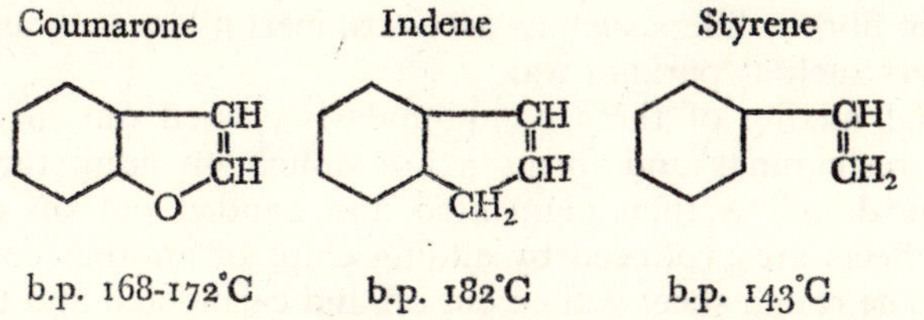

Coumarone Indene Styrene

b.p. 168-172°C b.p. 182°C b.p. 143°C

Fig. 14.5

is then frozen to remove naphthalene, agitated with sulphuric acid, then with caustic soda and finally with water. Concentrated sulphuric acid is then run into the purified naphtha at a temperature below 0°C. The reaction is stopped by addition of water after 5–10 minutes, any sediment is removed, the solution is neutralised and then washed with water. Residual naphtha is distilled off under vacuum leaving behind the resin which is run into trays for cooling.

By varying the coumarone–indene ratio and also the polymerisation conditions it is possible to obtain a range of products varying from hard and brittle to soft and sticky resins.

Being either brittle or soft these resins do not have the properties for moulding or extrusion compounds. There are however a number of

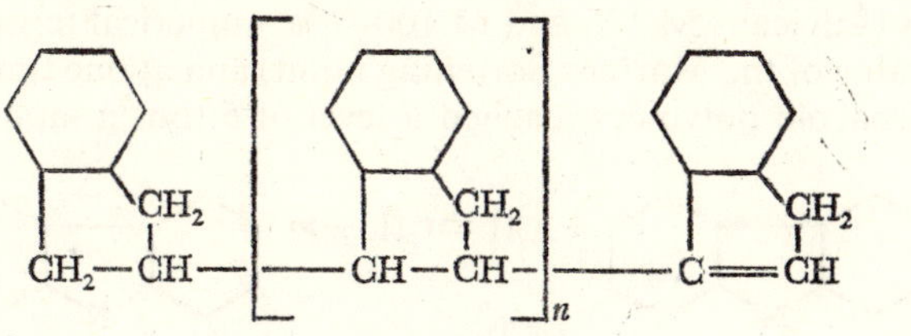

Structure of Polyindene

Fig. 14.6

properties which lead to these resins being used in large quantities. The resins are chemically inert and have good electrical insulation properties. They are compatible with a wide range of other plastics, rubbers, waxes, drying oils and bitumens and are soluble in hydrocarbons, ketones and esters.

The resins tend to be dark in colour and it has been suggested that this is due to a fulvenation process involving the unsaturated end group of a polymer molecule. Hydrogenation of the polymer molecule, thus eliminating unsaturation, helps to reduce discoloration.

Because of their wide compatibility and solubility, coumarone resins are used considerably in the paint and varnish industry. The resins also find application as softeners for plastics and rubbers such as p.v.c., bitumens and natural rubber.

Since the war the hard thermoplastic floor tile has been developed. These tiles use coumarone resins as a binder for the other ingredients which may contain fibrous fillers such as asbestos, inert fillers such as china clay and softeners such as paraffin wax.

The initial mixing of these compounds is carried out in an internal mixer, the resin melts and forms a hot dough on admixture with the fillers. The dough is then pigmented and banded out on a hot mill. Marbling effects are produced by adding chips of another colour to the mill nip. The rough sheet is then cut off and calendered and the product cut into tiles. These tiles may easily be cut when warmed thus making laying a simple operation. Because of the low cost of the raw materials and the relatively simple method of manufacture, coumarone tiles are cheaper than the vinyl tile based on vinyl chloride–vinyl acetate copolymers and are extensively used for both industrial and domestic flooring.

14.4 POLY(VINYL CARBAZOLE)[7,8]

Early in the Second World War there was a shortage of mica in Germany and in the United States. A need therefore arose for a material with good electrical insulation characteristics coupled with good heat resistance. In an attempt to meet this need poly(vinyl carbazole) was produced in both Germany (Luvican–I.G. Farben) and the United States (Polectron–General Aniline & Film Corporation). In addition to the homopolymer (Luvican M.150) the I.G. Farben complex also produced styrene copolymers (Luvicans M.125 and M.100—the numerical term corresponding to the value of the Martens Softening point) and at one time production of vinyl carbazole polymers reached a level of 5 tons a month. Because

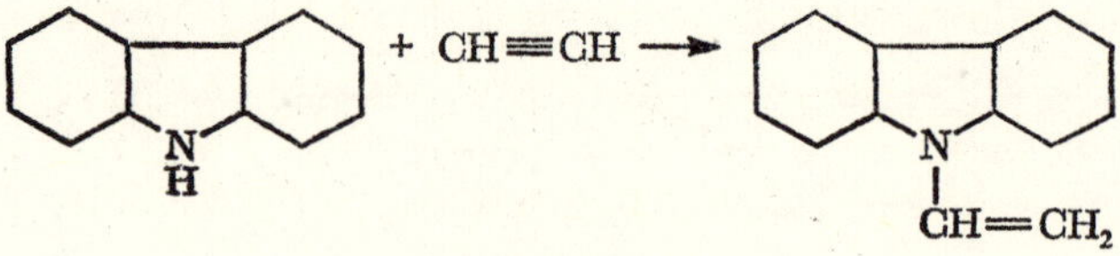

Fig. 14.7

of its brittleness and its tendency to cause an eczema-type of rash on people handling the material, production of these polymers is now very small.

Vinyl carbazole is obtained by reacting carbazole, readily available as a by-product of coal tar distillation, with acetylene in the presence of a catalyst and solvent under pressure (Fig. 14.7).

Typically, the reaction would be carried out at 140°C in white spirit with potassium carbazole as a catalyst. Davidge[9] has reported problems in polymerisation of *N*-vinyl carbazole prepared from carbazole obtained from coal tar, attributing this to the presence of sulphur. To overcome these problems carbazole has been prepared synthetically by reactions of cyclohexanone with phenyl hydrazine to give tetrahydrocarbazole which is then dehydrogenated with Raney nickel. *N*-vinyl carbazole is a solid with a melting point of 64–67°C.

High molecular weight polymers are produced by an adiabatic bulk polymerisation process[9,10] using di-tert butyl peroxide (0·02%) and 2:2′azobisdi-isobutyronitrile (0·01%) as initiators and pressurised with N_2. Heating to 80–90°C causes an onset of polymerisation and a rapid increase in temperature. After the maximum temperature has been reached the mass is allowed to cool under pressure. A typical current commercial material (Luvican M.170) has a *K*-value of about 70 (as assessed in a 1% tetrahydrofuran solution).

The polymerisation *in situ* of monomer impregnated into rolled and stacked condensers was at one time of commercial importance.[11]

The most important properties of poly(vinyl carbazole) are the electrical insulation characteristics (see Table 14.3) which are very good over a range of temperature and frequency.

The polymers are however more brittle than polystyrene and not suitable for applications which are to be subject to mechanical shock.

Poly(vinyl carbazole) is insoluble in alcohols, esters, ethers, ketones, carbon tetrachloride, aliphatic hydrocarbons and castor oil. It is swollen or dissolved by such agents as aromatic and chlorinated hydrocarbons and tetrahydrofuran.

The polymer is not easy to process and in injection moulding melt temperatures of 300°C are employed. In order to prevent excess embrittlement by shock cooling of the melt, mould temperatures as high as 150°C are employed. The polymer may also be compression moulded at temperatures of 250–260°C. The largest single application of poly(vinyl carbazole) is as a capacitor dielectric. It is also used in other electrical applications such as switch parts, cable connectors and coaxial cable spacers. In spite of its excellent electrical properties over a wide range of temperatures consumption of poly(vinyl carbazole) remains very limited. This is due to its brittle nature and to a lesser extent the health hazards involved in handling.

14.5 POLY(VINYL PYRROLIDONE)[12]

Poly(vinyl pyrrolidone) (p.v.p.) was introduced by the Germans in the Second World War as a blood plasma substitute.[13] A water-soluble polymer its main value is due to its ability of forming loose addition compounds with many substances.

The monomer is prepared from acetylene, formaldehyde and ammonia via 2:butyne-1:4-diol, butane 1:4 diol, γ-butyrolactone and α-pyrrolidone (Fig. 14.8).

Polymerisation is carried out in aqueous solution to produce a solution containing 30% polymer. The material may be marketed in this form or

Table 14.3

Volume resistivity	$>10^{16}$ ohm cm
Dielectric constant	2·9–3·0 (1 kc/s–300 Mc/s)
Power factor	0·0004–0·001 (1 kc/s–100 Mc/s)
Water absorption after 96 hours	0·1 (measured by DIN 53472 method)
Vicat softening point	200°C (DIN 57302)
Specific heat	0·3 cal g^{-1} $degC^{-1}$

spray dried to give a fine powder. Polymers may be produced with molecular weights in the range 10,000–100,000 (*K* values 20–100) of which products with a *K* value of 30–35 are the most important.[14,15]

In addition its water solubility poly(vinyl pyrrolidone) is soluble in a very wide range of materials including aliphatic halogenated hydrocarbons (methylene chloride, chloroform), many monohydric and polyhydric alcohols (methanol, ethanol, ethylene glycol), some ketones (acetyl acetone) and lactones (α-butyrolactone), lower aliphatic acids (glacial acetic acid) and the nitroparaffins. The polymer is also compatible with a wide range of other synthetic polymers, with gums and with plasticisers.

P.V.P. has found several applications in the textile industry because of its affinity for dyestuffs. Uses include dye stripping, removal of identification tints, in the formulation of sizes and finishes and to assist in dye-levelling operations. In the field of cosmetics p.v.p. is used because of its unique property of forming loose addition compounds with skin and hair. Hair lacquers may be formulated based on 4–6% p.v.p. in ethyl alcohol,

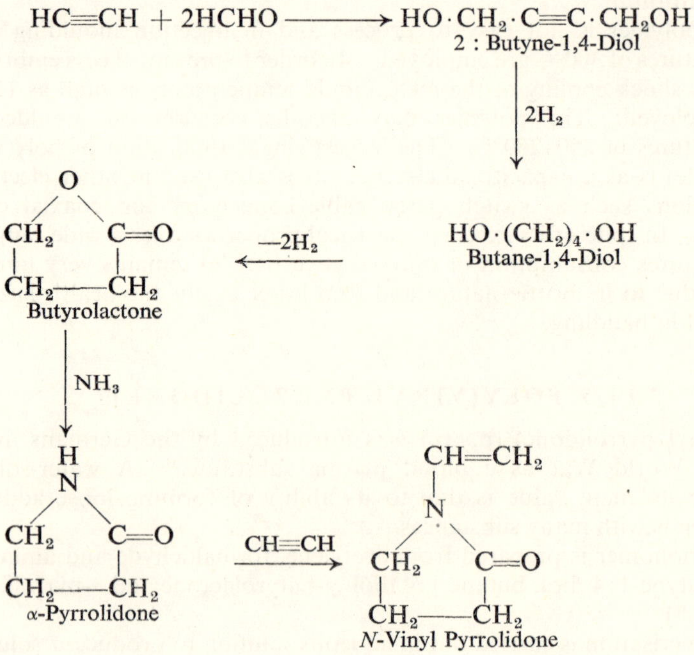
HC≡CH + 2HCHO
HO·CH2·C≡C·CH2OH
2 : Butyne-1,4-Diol
2H2
HO·(CH2)4·OH
Butane-1,4-Diol
—2H2
Butyrolactone
NH3
α-Pyrrolidone
CH≡CH
N-Vinyl Pyrrolidone

Fig. 14.8

whilst wave sets use about 1–2% of polymer. The polymer is also said to reduce the sting of after-shave lotion and is used in hand cream, lotions and liquid make-up.

On the Continent of Europe p.v.p. is still used as a blood plasma substitute, the original application, and is stockpiled for emergency use in the United States. It is not used for this purpose in Britain.

Because of its complexing action it finds miscellaneous uses in the pharmaceutical, brewing, soap and paper industries.

Copolymers of vinyl pyrrolidone with vinyl acetate, styrene and ethyl acrylate have been marketed by the General Aniline and Film Corporation.

14.6 POLY(VINYL ETHERS)[1,16]

It is not possible to polymerise vinyl ethers by free radically initiated methods but, as with isobutylene polymers, it is possible to make polymers using Friedel–Crafts type catalysts.

The poly(vinyl ethers), which were first made available in Germany before 1940, are not of importance in the plastics industry but have applications in adhesives, surface coatings and rubber technology. Of the many vinyl ether polymers prepared only those from the vinyl alkyl ethers and some halogenated variants are of interest. Two methods of monomer preparations may be used.

1. The direct vinylation of alcohols by acetylene diluted with nitrogen or methane (Reppe method) (Fig. 14.9).

$$ROH + CH{\equiv}CH \xrightarrow[\text{Diluted} \quad 130-180°C]{\text{KOR}} CH_2{=}CH + 30{\cdot}7 \text{ kcal/mole}$$
$$\underset{OR}{|}$$

Fig. 14.9

High pressure autoclaves may be used fitted with remote control behind safety barricades which are necessary because of the danger of explosions. In a typical process the autoclave is half filled with alcohol containing 15% potassium hydroxide or potassium alcoholate. The free space is then thoroughly purged with oxygen-free nitrogen and the temperature raised to 140°C. Acetylene and nitrogen are run in under pressures of about 100 $lbin^{-2}$. Conversions are usually taken to 70–80%.

2. Preparation via acetals (Carbide and Chemicals Corporation) (Fig. 14.10).

A typical catalyst for the final stage would be 10% palladium deposited on finely divided asbestos.

The vinyl alkyl ethers polymerise violently in the presence of small quantities of inorganic acids.

The following details for the commercial manufacture of poly(vinyl methyl ether) have been made available.[1] Agitated vinyl methyl ether at 5°C is treated over a period of 30 minutes with 0·2% of catalyst solution

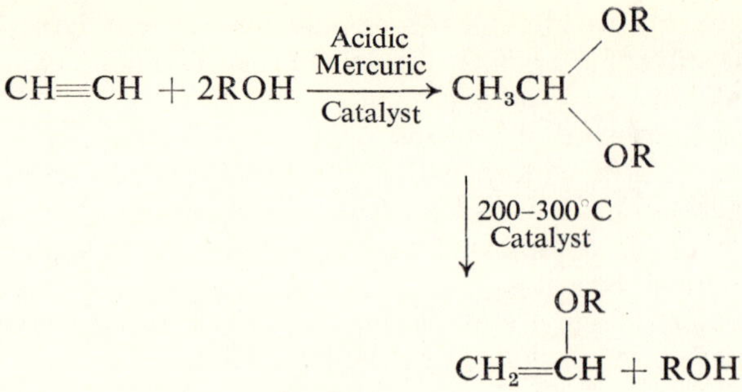

$$CH{\equiv}CH + 2ROH \xrightarrow[\text{Catalyst}]{\substack{\text{Acidic} \\ \text{Mercuric}}} CH_3CH\begin{smallmatrix} \diagup OR \\ \\ \diagdown OR \end{smallmatrix}$$

$$\Big\downarrow \substack{\text{200–300°C} \\ \text{Catalyst}}$$

$$CH_2{=}\overset{\displaystyle OR}{\overset{|}{C}}H + ROH$$

Fig. 14.10

consisting of 3% $BF_3 \cdot 2H_2O$ in dioxane.   When the reaction rises to 12°C the reaction is moderated by brine cooling.   Over the next 3–4 hours further monomer and catalyst is added.   The autoclave is then closed and the temperature allowed to rise slowly to 100°C.

The end of the reaction is indicated by the pressure and temperature observations.   The total reaction time is of the order of 16–18 hours.

The polymer is a water-soluble viscous liquid which has found application in the adhesive and rubber industries.   One particular use has been a heat sensitiser used in the manufacture of rubber latex dipped goods.

A number of higher poly(vinyl ether)s, in particular the ethyl and butyl polymers have found use as adhesives.   When antioxidants are incorporated pressure-sensitive adhesive tapes from poly(vinyl ethyl ether) are said to have twice the shelf life as similar tapes from natural rubber. Copolymers of vinyl isobutyl ether with methyl acrylate and ethyl acrylate (Acronal series) and with vinyl chloride have been commercially marketed. The first two products have been used as adhesives and impregnating agents for textile, paper and leather whilst the latter (Vinoflex MP 400) has found use in surface coatings.

14.7   OTHER ETHENOID POLYMERS[1]

In addition to the ethenoid polymers reviewed in this and the previous seven chapters many others have been prepared.   Few have however reached the pilot plant stage of manufacture and none appear, at present, to be of interest as plastics.   Typical of these materials are the poly(vinyl thioethers), the poly(vinyl isocyanates), the poly(vinyl ureas) and the poly(alkyl vinyl ketones).   Methyl-isopropenyl ketone and certain vinyl pyridine derivatives have been copolymerised with butadiene to give special purpose rubbers.

REFERENCES

1. SCHILDKNECHT, C. E., *Vinyl and Related Polymers*, John Wiley, New York (1952)

Vinylidene chloride polymers and copolymers

2. JACK, J., *Brit. Plastics*, **34**, 312 (1961)
3. GOGGIN, W. C., and LOWRY, R. D., *Ind. Eng. Chem.*, **34**, 327 (1942)
4. JACK, J., and HORSLEY, R. A., *J. Appl. Chem.*, **4**, 178 (1954)

5. *British Patent*, 570,711 (I.C.I.)
6. *U.S. Patent* 2,238,020 (Dow)

Poly(vinyl carbazole)

7. KLINE, G. M., *Mod. Plastics* **24**, 157 (1946)
8. CORNISH, E. H., *Plastics*, **28** (305), 61, March (1963)
9. DAVIDGE, H., *J. Appl. Chem.*, **9**, 241–246, 553–560 (1959)
10. *German Patent* 931,731; *Brit. Patent* 739,438 (BASF)
11. SHINE, W. M., *Mod. Plastics*, **25**, 130 (1947)

Poly(vinyl pyrrolidone)

12. DAVIDSON, R. L., and SITTIG, M. (Eds.) *Water-soluble Resins*, Reinhold, New York (1962)
13. B.I.O.S. Report 354, Item 22
14. *Ind. Chemist*, **29**, 122 (1953)
15. GREENFIELD, I., *Ind. Chemist*, **32**, 11 (1956)

Poly(vinyl ethers)

16. FIAT 856: BIOS 742: BIOS 1292

BIBLIOGRAPHY

SCHILDKNECHT, C. E., *Vinyl and Related Polymers*, John Wiley, New York (1952)

15

Polyamides and Related Materials

15.1 INTRODUCTION

Whilst by far the bulk of polyamide materials are used in the form of fibres, they have also become of some importance as speciality thermoplastics of particular use in engineering applications. The fibre-forming polyamides and their immediate chemical derivatives and copolymers are referred to as nylons. There are also available polyamides of more complex composition which are not fibre-forming and are structurally quite different. These materials are not considered as nylons (see Section 15.9).

The early development of the nylons is largely due to the work of W. H. Carothers and his colleagues who first synthesised nylon 66 in 1935 after extensive and classical researches into condensation polymerisation. Commercial production of this polymer for subsequent conversion into fibres was commenced by the Du Pont Company in December 1939. The first nylon mouldings were produced in 1941 but the polymer did not become well known in this form until about 1950.

In an attempt to circumvent the Du Pont patents, German chemists investigated a wide range of synthetic fibre-forming polymers in the late 1930's. This work resulted in the successful introduction of nylon 6 (and incidentally in the evolution of the polyurethanes) and today nylons 66 and 6 account for nearly all of the polyamides produced for fibre applications. Mention may however be made of nylons 7 (Enanth) and 9 (Pelargone) which are under investigation as fibres in the Soviet Union. Very many other polyamides have been prepared in the laboratory and a few have become of specialised interest as plastics materials including nylons 11, 12, 610, 66/610 and 66/610/6.

Of the many possible methods for preparing linear polyamides three are of commercial importance

1. The reaction of diamines with dicarboxylic acid, via a 'nylon salt' (Fig. 15.1).
2. Self-condensation of an ω-amino acid (Fig. 15.2).

290

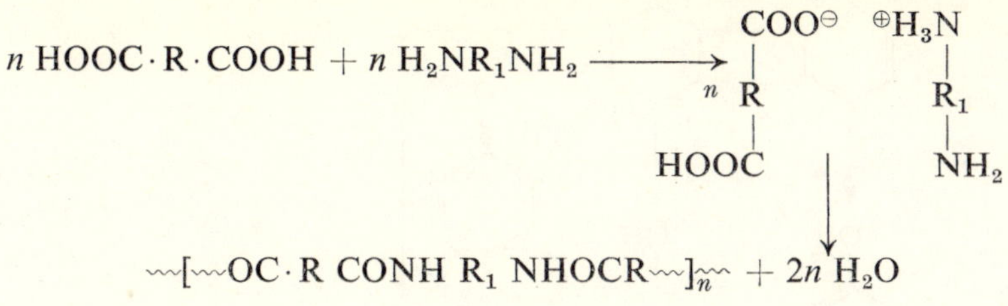

$$n \text{ HOOC} \cdot \text{R} \cdot \text{COOH} + n \text{ H}_2\text{NR}_1\text{NH}_2 \longrightarrow$$

$$\underset{n}{} \begin{array}{cc} \text{COO}^{\ominus} & {}^{\oplus}\text{H}_3\text{N} \\ | & | \\ \text{R} & \text{R}_1 \\ | & | \\ \text{HOOC} & \text{NH}_2 \end{array}$$

$$\sim\!\![\sim\!\!\text{OC} \cdot \text{R CONH R}_1 \text{ NHOCR}\!\!\sim]_n^{\sim} + 2n \text{ H}_2\text{O}$$

Fig. 15.1

$$n \text{ NH}_2\text{R COOH} \longrightarrow \sim(\sim\!\!\text{NHRCO})_n^{\sim} + n \text{ H}_2\text{O}$$

Fig. 15.2

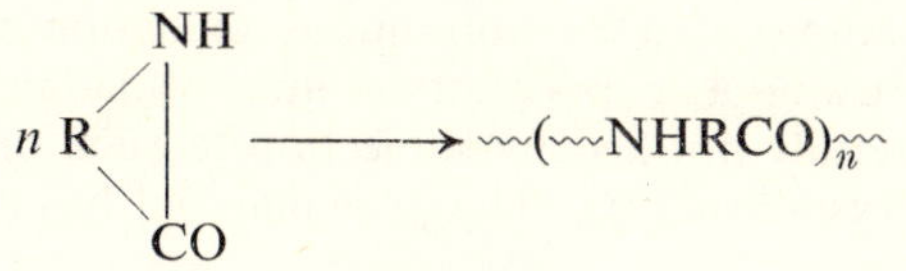

$$n \text{ R}\begin{array}{c} \text{NH} \\ \diagup | \\ | \\ \diagdown | \\ \text{CO} \end{array} \longrightarrow \sim(\sim\!\!\text{NHRCO})_n^{\sim}$$

Fig. 15.3

3. Opening of a lactam ring (Fig. 15.3).

An example of the first route is given in the preparation of nylon 66 which is made by reaction of hexamethylene diamine with adipic acid. The first '6' indicates the number of carbon atoms in the diamine and the second the number of carbon atoms in the acid. Thus, as a further example, nylon 6.10 is made by reacting hexamethylene diamine with sebacic acid ($\text{HOOC} \cdot (\text{CH}_2)_8 \cdot \text{COOH}$).

Where the material is denoted by a single number, viz nylon 6 and nylon 11, preparation from either a lactam or an ω-amino-acid is indicated. The polymer nylon 66/6.10 (60:40) indicates a copolymer using 60 parts of nylon 6.6 salt with 40 parts of nylon 6.10 salt.

15.2 PREPARATION OF INTERMEDIATES

15.2.1 Adipic acid

It is possible to produce adipic acid by a variety of methods from such diverse starting points as benzene, acetylene and waste agricultural products. In practice benzene is the favoured starting point and some of the more important routes for this material are shown (Fig. 15.4).

A typical route is that via cyclohexane, and cyclohexanol. To produce cyclohexane, benzene is subjected to continuous liquid phase hydrogenation at 340 lb in^{-2} pressure and a temperature of 210°C using a Raney nickel catalyst. After cooling and separation of the catalyst the product is fed to the cyclohexane store. In the next stage of the operation the cyclohexane is preheated and continuously oxidised in the liquid phase by air using a trace of cobalt naphthenate as catalyst. This gives an approximately 70% yield of a mixture of cyclohexanol and cyclohexanone with a

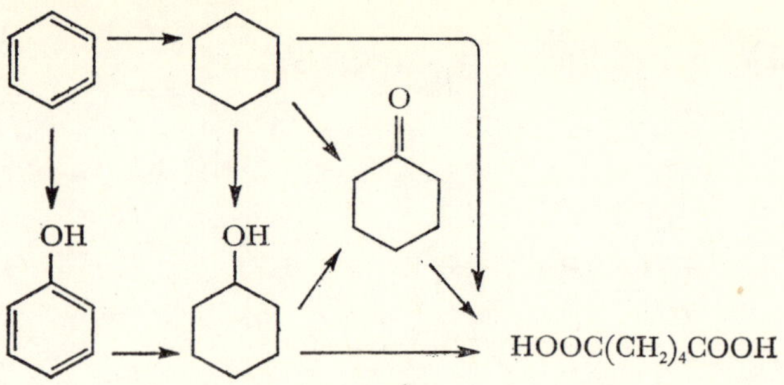

Fig. 15.4

small quantity of adipic acid.   The cyclohexanol–cyclohexanone mixture is converted to adipic acid by continuous oxidation with 50% HNO_3 at about 75°C using a copper–ammonium vanadate catalyst.   The adipic acid is carefully purified by subjection to such processes as steam distillation and crystallisation.   The pure material has a melting point of 151°C.

15.2.2   Hexamethylene diamine

Hexamethylene diamine may be conveniently prepared from adipic acid via adiponitrile

$$HOOC \cdot (CH_2)_4 \cdot COOH \longrightarrow NC \cdot (CH_2)_4 \cdot CN$$

$$\longrightarrow H_2N \cdot (CH_2)_6 \cdot NH_2$$

In a typical process adiponitrile is formed by the interaction of adipic acid and gaseous ammonia in the presence of a boron phosphate catalyst at 305–350°C.   The adiponitrile is purified and then subjected to continuous hydrogenation at 130°C and 4000 lb in^{-2} pressure in the presence of excess ammonia and a cobalt catalyst.   By-products such as hexamethylene imine are formed but the quantity produced is minimised by the use of excess ammonia.   Pure hexamethylene diamine (boiling point 90–92°C at 14 mmHg pressure, melting point 39°C) is obtained by distillation.   This diamine may also be prepared from furfural and from butadiene.[1]

15.2.3   Sebacic acid

This material is normally made from castor oil which is essentially glyceryl ricinoleate.   The castor oil is treated with caustic soda at high temperature, e.g. 250°C, so that saponification, leading to the formation of ricinoleic acid, is followed by a reaction giving sebacic acid and octanol-2 (Fig. 15.5).

Because of the by-products formed, the yield of sebacic acid is necessarily

Castor Oil

| NaOH
↓

Glycerol + $CH_3 \cdot (CH_2)_5 \cdot CH \cdot CH_2 \cdot CH = CH \cdot (CH_2)_7 \, COOH$

  $\overset{|}{OH}$

| NaOH + H_2O
↓

$CH_3 \cdot (CH_2)_5 \cdot CH \cdot CH_3 + HOOC \cdot (CH_2)_8 COOH + H_2$

  $\overset{|}{OH}$

Fig. 15.5

low and in practice yields of 50–55% (based on the castor oil) are considered to be good.

In a process said to be operated in Britain[1] castor oil is subjected to alkaline fusion under critically controlled conditions to produce a mixture of methyl hexyl ketone and ω-hydroxy decanoic acid. Interaction of these two materials at higher temperatures leads to the formation of sebacic acid, as the sodium salt, and capryl alcohol. Heating must be rapid and even and any tendency to preheating must be avoided. The sebacic acid is formed from the sodium salt by precipitation with sulphuric acid.

15.2.4 Caprolactam

Because it is easier to make and to purify caprolactam is preferred to ω-aminocaproic acid for the manufacture of nylon 6. Commercial production is based on cyclohexanol which may be produced, as already indicated, by a number of routes from benzene. The cyclohexanol is oxidised to cyclohexanone, typically by contact with a copper catalyst.

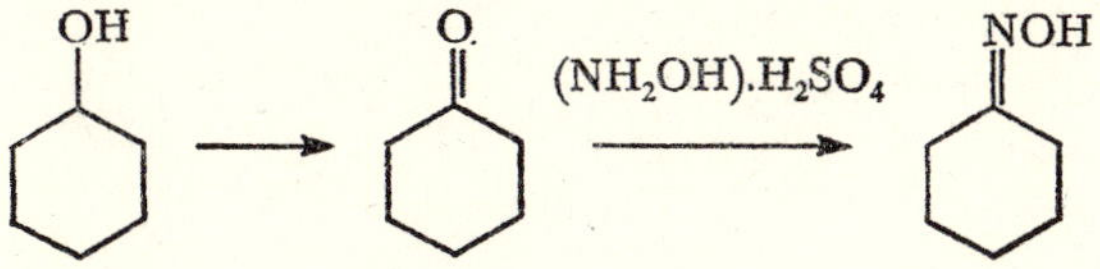

Cyclohexanol   Cyclohexanone

$(NH_2OH).H_2SO_4$

Fig. 15.6

The cyclohexanone is then treated with hydroxylamine sulphate, typically at 20°C, to give cyclohexanone oxime (Fig. 15.6).

After oximation the products are neutralised with ammonia and the oxime allowed to settle out.

Treatment of the oxime at elevated temperatures, e.g. 140°C, with sulphuric acid containing free sulphur trioxide leads to caprolactam through reactions which include a Beckmann rearrangement (Fig. 15.7).

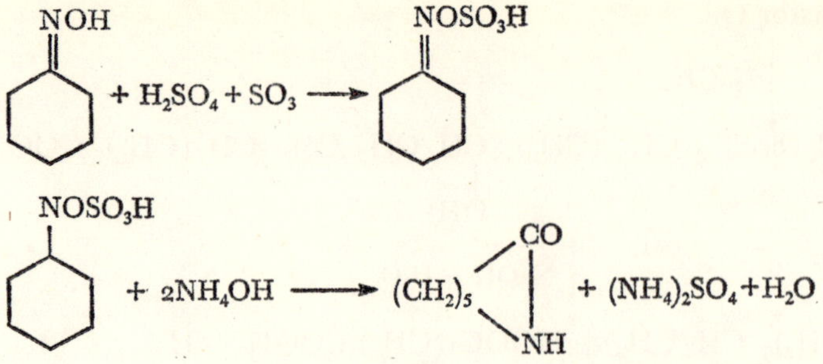

$$NOH \qquad \qquad NOSO_3H$$

$$\bigcirc + H_2SO_4 + SO_3 \longrightarrow \bigcirc$$

$$NOSO_3H$$

$$\bigcirc + 2NH_4OH \longrightarrow (CH_2)_5 \Big\langle \begin{array}{c} CO \\ | \\ NH \end{array} + (NH_4)_2SO_4 + H_2O$$

Fig. 15.7

In one process the resulting solution is continuously withdrawn and cooled rapidly to below 75°C to prevent hydrolysis and then further cooled before being neutralised with ammonia. After phase separation, the oil phase is then treated with trichloroethylene to extract the caprolactam which is then steam distilled. Pure caprolactam has a boiling point of 120°C at 10 mmHg pressure.

In the above process 5·1 tons of ammonium sulphate are produced as a by-product per ton of caprolactam.

15.2.5   ω–Aminoundecanoic acid

The starting point for this amino-acid, from which nylon 11 is obtained, is the vegetable product castor oil, composed largely of the triglyceride of ricinoleic acid. This is first subjected to treatment with methanol or ethanol to form the appropriate ricinoleic acid ester.

Cracking of the ester at about 500°C leads to the formation of the undecylenic acid ester together with such products as heptyl alcohol,

$$\text{Castor Oil} \longrightarrow CH_3 \cdot (CH_2)_5 \cdot \underset{\underset{OH}{|}}{CH} \cdot CH_2 \cdot CH = CH(CH_2)_7COOR$$

$$\xrightarrow[\sim 500°C]{Pyrolysis} CH_2{=}CH \cdot (CH_2)_8 \cdot COOR + C_6H_{13}CHO$$

Hydrolysis

$$CH_2{=}CH \cdot (CH_2)_8 \cdot COOH$$

Bromination

$$\underset{\underset{Br}{|}}{CH_2}{-}CH_2{-}(CH_2)_8COOH + CH_3 \cdot \underset{\underset{Br}{|}}{CH} \cdot (CH_2)_8COOH$$

NH_3

$$NH_2 \cdot (CH_2)_{10} \cdot COOH$$

Fig. 15.8

heptanoic acid and heptaldehyde. Undecylenic acid may then be obtained by hydrolysis of the ester. Treatment of the acid by HBr in the presence of a peroxide leads to ω-bromoundecanoic acid together with the 10-isomer which is removed. Treatment of the ω-bromo derivative with ammonia leads to ω-aminoundecanoic acid which has a melting point of 50°C (Fig. 15.8).

This amino-acid may also be produced via telomerisation reactions (see below).

15.2.6 ω–Aminoenanthic acid

Interest in this material as an intermediate for nylon 7 arises largely from the development by Russian scientists of the process of *telomerisation*,[2] a process yielding low molecular weight polymers of simple unsaturated compounds, the polymers possessing useful reactive end-groups. Of greatest interest to date is the reaction of ethylene with carbon tetrachloride initiated by a peroxide such as benzoyl peroxide. The reaction proceeds by the following stages

$$\underset{\substack{\text{Initiator} \\ \text{Radical}}}{\text{I}-} + CCl_4 \longrightarrow \underset{\substack{\text{Inert} \\ \text{Compound}}}{ICl} + \underset{\text{Radical}}{CCl_3-}$$

$$C \cdot Cl_3- + CH_2{=}CH_2 \longrightarrow Cl_3 \cdot C \cdot CH_2{-}CH_2{-}$$

$$Cl_3 \cdot C \cdot CH_2 \cdot CH_2{-} + CH_2{=}CH_2 \longrightarrow Cl_3C \cdot CH_2 \cdot CH_2 \cdot CH_2 \cdot CH_2{-} \text{ etc.}$$

$$Cl_3 \cdot C(CH_2 \cdot CH_2)_n{-} + CCl_4 \longrightarrow Cl_3C \cdot (CH_2 \cdot CH_2)_n Cl + CCl_3{-}$$

Because of the random nature of the occurrence of the chain transfer reaction which terminates molecular growth, polymers varying in molecular weight will be formed. For reaction at about 100°C, 100 atmospheres pressure and with an ethylene–carbon tetrachloride ratio of about 4:1 about 60% of the telomers have 7, 9 or 11 carbon atoms in the molecule. The individual telomers may be fractionated at reduced pressures. ω-amino acids may be obtained from the isolated telomers by hydrolysing the —CCl_3 group by heating the telomer with concentrated sulphuric acid for one hour at 90–100°C and then treating the product with an aqueous solution of ammonia under pressure. For example for ω-amino enanthic acid the following reactions occur

$$Cl(CH_2)_6CCl_3 \xrightarrow{H_2SO_4} Cl \cdot (CH_2)_6COOH$$

$$\xrightarrow{NH_3} NH_2 \cdot (CH_2)_6COOH + NH_4Cl$$

The amino acid and the ammonium chloride may conveniently be separated by passing through a column of ion-exchange resins. The amino-acid melts at 195°C.

ω-amino pelargonic acid (for nylon 9) and ω-amino undecanoic acid may also be prepared by this route.

15.2.7 Dodecyl lactam

Nylon 12 first became available on a semicommercial scale in 1963. The monomer, dodecyl lactam, is prepared from butadiene by a multistaged reaction. In one process butadiene is treated with a Ziegler-type catalyst system to yield the cyclic trimer, 1,5,9-cyclododecatriene. This may then be hydrogenated to give cyclododecane which is then subjected to direct

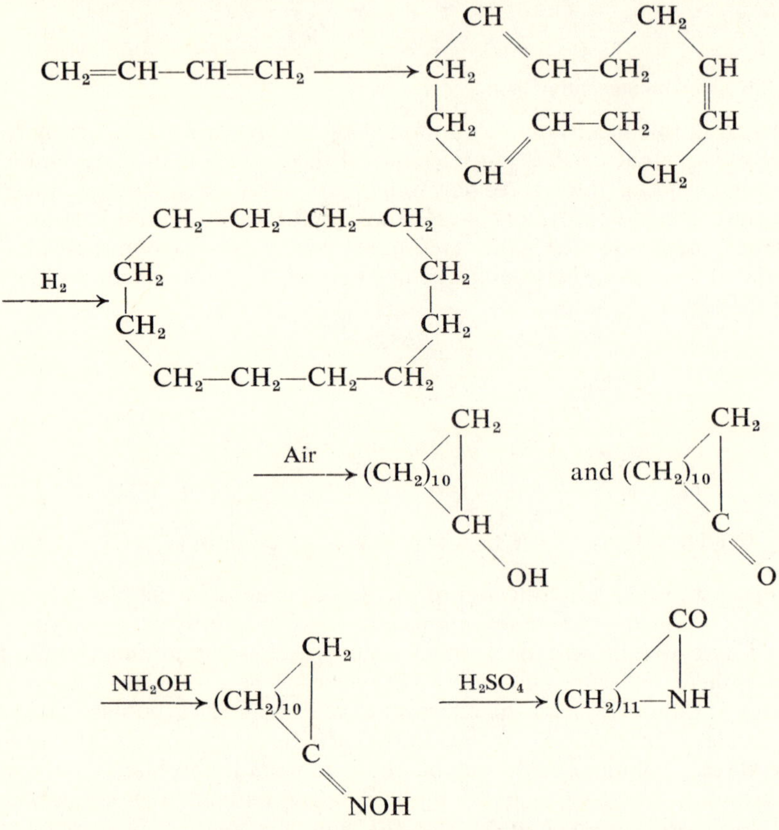

Fig. 15.9

air oxidation to give a mixture of cyclodecanol and cyclododecanone. Treatment of the mixture with hydroxylamine yields the corresponding oxime which on treatment with sulphuric acid re-arranges to form the lactam (Fig. 15.9).

A number of variations in the process are also being investigated including the direct photo-oximation of cyclododecane.

15.3 POLYMERISATION

As already indicated, the fibre-forming polyamides are produced commercially by reacting diamines with dibasic acids, by self-condensation of an amino-acid or by opening of a lactam ring.   Whatever method is chosen it is important that there should be an equivalence in the number of amine and acid groups for polymers of the highest molecular weight to be obtained.   In the case of the amino-acids and lactams this is ensured by the use of pure monomer but when diamines and dibasic acids are used it is necessary to form a salt to ensure such an equivalence.   Small quantities of monofunctional compounds are often used to regulate molecular weight.

15.3.1   Nylon 66 and nylon 610

The nylon 66 salt is prepared by reacting the hexamethylene diamine and the adipic acid in boiling methanol, the comparatively insoluble salt (melting point 190–191°C) precipitating out.

A 60% aqueous solution of the salt is then run into a stainless steel autoclave together with a trace of acetic acid to limit the molecular weight (9,000–15,000).   The vessel is sealed and the temperature raised to about 220°C.   The steam formed purges the air, and a pressure of 250 lb in^{-2} is developed.   After 1–2 hours the temperature is raised to 270–280°C and steam bled off to maintain the pressure at 250 lb in^{-2}.   The pressure is then reduced to atmospheric for one hour, after which the polymer is extruded by oxygen-free nitrogen on to a water-cooled casting wheel, to form a ribbon which is subsequently disintegrated.   Nylon 610 is prepared from the appropriate salt (melting point 170°C) by a similar technique.

15.3.2   Nylon 6

Both batch and continuous processes have been used for the manufacture of nylon 6.   In a typical batch process the caprolactam, water (which acts as a catalyst) and a molecular weight regulator, e.g. acetic acid, are charged into the vessel and reacted under a nitrogen blanket at 250°C for about 12 hours.   The product consists of about 90% high polymer and 10% low molecular weight material such as the monomer.   In order to achieve the best physical properties the low molecular weight materials may be removed by leaching and/or vacuum distillation.   In the continuous process the reactants are maintained in reservoirs which continuously feed reaction columns kept at a temperature of about 250°C.

The polymerisation casting of nylon 6 *in situ* in the mould has been developed in recent years.   Anionic polymerisation is normally employed; a typical system[3] uses as a catalyst 0·1–1 mol. % of acetic caprolactam and 0·15–0·50 mol. % of the sodium salt of caprolactam.   The reaction temperature initially is normally between 140 and 180°C but during polymerisation this rises by about 50°C.   Mouldings up to one ton in weight are claimed to have been produced by these casting techniques.

15.3.3   Nylon 11

This polymer may be prepared by stirring the molten ω-amino undecanoic acid at about 220°C.   The reaction may be followed by measurements of the electrical conductivity of the melt and the intrinsic viscosity of solutions in *m*-cresol.[4]   During condensation 0·4–0·6% of a 12-membered ring lactam may be formed by intramolecular condensation but this is not normally removed since its presence has little effect on the properties of the monomer.

15.3.4 Nylon 12

The opening of the caprolactam ring for nylon 6 involves an equilibrium reaction which is easily catalysed by water.   In the case of nylon 12 from dodecyl lactam higher temperatures, i.e. above 260°C, are necessary for opening the ring structures but since in this case the condensation is not an equilibrium reaction the process will yield almost 100% of high polymer.[5]

15.3.5 Nylon 7

The ω-amino-enanthic acid is polymerised in an aqueous solution under nitrogen at 14–15 atm pressure at 250–260°C.   The process takes several hours.[2]

15.4   RELATION OF STRUCTURE AND PROPERTIES

Polyamides such as nylons 66, 6, 610 and 11 are linear polymers and thus thermoplastic.   They contain polar —CONH— groups spaced out at regular intervals so that the polymers crystallise with a high intermolecular attraction.   These polymer chains also have aliphatic chain segments which give a measure of flexibility in the amorphous region.   Thus the combination of high interchain attraction in the crystalline zones and flexibility in the amorphous zones leads to polymers which are tough above their apparent glass transition temperatures.

The high intermolecular attraction leads to polymers of high melting point.   However above the melting point the melt viscosity is low because of the polymer flexibility at such high temperatures and the relatively low molecular weight.

Because of the high cohesive energy density and their crystalline state the polymers are only soluble in a few liquids of similar high solubility parameter and which are capable of specific interaction with the polymers.

The electrical insulation properties are quite good at room temperature in dry conditions and at low frequencies.   Because of the polar structure they are not good insulators for high frequency work and since they absorb water they are also generally unsuitable under humid conditions (see Section 15.8 for exceptions).

There are a number of structural variables which can considerably affect the properties of the aliphatic polyamides:

1. The distance between the repeating —CONH— group.   In the case of nylon 11 this is approximately twice that of nylon 6.   Thus nylon 11 has a lower interchain attraction, is softer, has a lower melting point and lower water absorption.   Nylon 11 may, in fact, be

considered to be intermediate in structure and properties between nylon 6 and polyethylene.

2. The number of methylene groups in the intermediates. It has been observed that polymers from intermediates with an even number of methylene groups have higher melting points than similar polymers with an odd number of methylene groups. This is seen clearly in Fig. 15.10[6] where it is seen that nylon 66 has a higher melting point

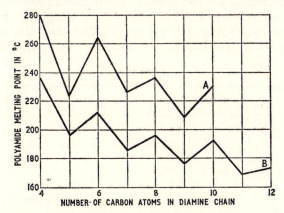

Fig. 15.10. Melting points of polyamides from aliphatic diamines: A, with adipic acid, B, with sebacic acid. (After Coffman et al[6])

than either nylon 56 or nylon 76. With polymers from amino-acids or lactams the same rule applies, nylon 7 having a higher melting point ($\sim$227°C) than either nylon 6 ($\sim$215°C) or nylon 8 ($\sim$180°C). These differences are due to the differences in the crystal structure of polymers with odd and even methylene groups which develop in order that oxygen atoms in one molecule are adjacent to amino groups of a second molecule. Hydrogen bonds with an NH–0 distance of 2·8 Å are produced and are the reason for the high strength and the high melting points of polyamides such as nylons 6,66 and 7. The crystal structures of the polyamides differ according to the type of polymer and in some cases such as with nylon 66 two crystal forms co-exist in the same mass of polymer. These structures have been discussed in detail elsewhere.[7,8,9,10]

3. The molecular weight. Specific types of nylon, e.g. 66, are frequently available in forms differing in molecular weight. The main difference between such grades is in melt viscosity, the more viscous grades being more suitable for processing by extrusion techniques.

4. *N*-substitution. Replacement of the hydrogen atom in the —CONH— group by such groups as $\sim$CH$_3$ and —CH$_2$OCH$_3$ will cause a reduction in the interchain attraction and a consequent decrease in softening point. Rubbery products may be obtained from methoxy methyl nylons. These materials are considered in more detail in Section 15.9.

The properties of the nylons are considerably affected by the amount of crystallisation. Whereas in some polymers, e.g. the polyacetals and p.c.t.f.e., processing conditions have only a minor influence on crystallinity, in the case of the nylons the crystallinity of a given polymer may vary by as much as 40%. Thus a moulding of nylon 6, slowly cooled and subsequently annealed may be 50–60% crystalline, while rapidly cooled thin-wall mouldings may be only 10% crystalline.

The glass transition temperatures of the commercial nylons are about or below room temperature and are reduced by absorption of water. Thus on storage the nylons may continue to crystallise, this being accompanied by shrinkage and an increase in density. This after-shrinkage is

Table 15.1

Property	Unit	ASTM test	66	6	610	11	12	66/610 (35:65)	66/610/6 (40:30:30)
Tensile stress at yield	10^3 lb in^{-2}	D.638	11·5	11	8·5	5·5	6·6	5·5	—
At break	10^3 lb in^{-2}	D.638	—	—	—	7·6	7·8	—	7·5
E_B	%	D.638	80–100	100–200	100–150	30–300	200	>200	~300
Tension modulus	10^5 lb in^{-2}		4·3	4	3	2	2	2	2
Impact strength	ft lb/½ in notch	(B.S.771)	1·0–1·5	1·5–3·0	1·6–2·0	—	—	2·0	—
S.G.	—		1·14	1·13	1·09	1·04	1·02	1·08	1·09
Rockwell hardness	—	D.785	118R	112R	111R	108R	107R	—	83R

particularly marked with nylon 66. The after-shrinkage process may be accelerated by annealing the samples at an elevated temperature, typically that which corresponds to the maximum crystallisation rate for that polymer (see also Section 15.7).

The greater the degree of crystallinity the less the water absorption and hence the less will be the effect of humidity on the properties of the polymer. The degree of crystallinity also has an effect on electrical and mechanical properties. In particular high crystallinity leads to high abrasion resistance.

According to the method of processing different morphological structures will be produced.[11] Slowly cooled melts may form spherulites, rapidly-cooled polymers may form only fine aggregates. It follows that in an injection moulding the morphological form of rapidly cooled surface layers may be quite different from that of the slower cooled centres. In recent years polymer manufactures have introduced nylons which give mouldings with an even structure right up to the surface and which have as a result an increased surface hardness and abrasion resistance.

15.5 GENERAL PROPERTIES OF THE NYLONS

Typical mechanical properties of some commercial grades of nylon are given in Table 15.1.

The figures given in the table are obtained on mouldings relatively free from orientation and tested under closely controlled conditions of temperature and humidity. Changes in these conditions or the use of additives may profoundly affect these properties. Details of the influence of these factors on mechanical properties have been published in the trade literature but Figs. 15.11–15.14 have been included to illustrate some salient features.

Fig. 15.11 shows the influence of temperature on the tension modulus of nylons 66 and 6 and Fig. 15.12 the effect of temperature on impact

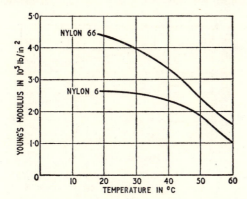

Fig. 15.11. *Effect of temperature on the Young's modulus of nylon 66 and nylon 6*

strength of nylon 66. Fig. 15.13 shows the profound plasticising influence of moisture on the modulus of nylons 6 and 66, while Fig. 15.14 illustrates the influence of moisture content on impact strength.

Laboratory tests and experience during use have demonstrated that the nylons have extremely good abrasion resistance. This may be further

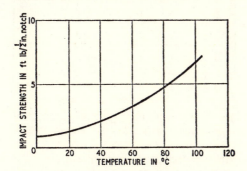

Fig. 15.12. *Effect of temperature on the impact strength of nylon 66. (Reproduced by permission of ICI Plastics Division)*

improved by addition of external lubricants and by processing under conditions which develop a highly crystalline hard surface, e.g. by use of hot injection moulds and by annealing in a non-oxidising fluid at an elevated temperature (150–200°C for nylon 66).

The coefficients of friction of the nylons are somewhat higher than the acetal resins (Chapter 16). Results obtainable will depend on the method of measurement but typical properties are given in Table 15.2.[12]

The effect of lubricants on the kinetic coefficient of friction of nylon 66 (like surfaces) is shown in Table 15.3.[12]

For bearing applications the upper working limits are determined by frictional heat build-up, this being related to the coefficient of friction

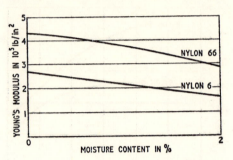

Fig. 15.13.   *Effect of moisture content on the Young's modulus of nylon 66 and nylon 6*

under working conditions. A measure of the upper working limits of a material for this application is the maximum *PV* value (the product of load in lb in^{-2} on the projected bearing area and the peripheral speed in ft min^{-1}) which can be tolerated. Maximum *PV* values of 500–1000 are suggested for continuous operation of unlubricated nylon 66 bearings whilst initially

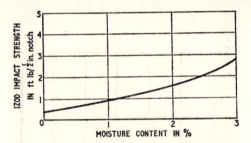

Fig. 15.14.   *Effect of moisture content on the impact strength of nylon 66.   (After Riley[12])*

oiled nylon bearings can be used intermittently at *PV* values of 8,000. Higher maximum *PV* values can be employed with continuously lubricated bearings (see also Chapter 16 for data on polyacetals).

The glass transition temperatures of the nylons are generally below room temperature so that the materials have a measure of flexibility in spite of their high crystallinity under general conditions of service. The polymers have fairly sharply defined melting points and above this temperature the homopolymers have low melt viscosities. Some thermal properties of the nylons are given in Table 15.4.

The nylons are reasonably good electrical insulators at low temperatures

Table 15.2   KINETIC COEFFICIENT OF FRICTION OF NYLON 66[12]

Moving surface	Stationary surface		
	Nylon (moulded)	Nylon (machined surface)	Mild steel
Nylon (moulded)	0·63	0·52	0·31
Nylon (machined surface)	0·45	0·46	0·33
Mild steel	0·41	0·41	0·6–1·0

Table 15.3   EFFECT OF LUBRICANTS ON THE KINETIC COEFFICIENT OF FRICTION OF NYLON 66 (LIKE SURFACES)[12]

Lubricant	Coefficient of friction
None	0·46
Water	0·24
Liquid paraffin	0·13
Graphite	0·28

Table 15.4   THERMAL PROPERTIES OF THE NYLONS

Property	Units	ASTM test	66	6	610	11	12	66/610/6 (40:30:30)
Melting point	°C	D.569	264	215	215	185	175	160
Deflection temp. (heat distortion temp.)	°C	D.648						
(264 lb in^{-2})			75	60	55	55	51	30
(66 lb in^{-2})			200	155	160	150	140	—
Coefficient of linear expansion	10^{-5} cm/ cm degC	—	10	9·5	15	15	12	—

Table 15.5   ELECTRICAL PROPERTIES OF THE NYLONS*

Property		Nylon 66	Nylon 6	Nylon 11	Nylon 610	Nylon 66/610/6 (40:30:30)	Nylon 66/610 (35:65)
Vol. resistivity	ohm cm (dry)	>10^{15}	>10^{15}	—	>10^{15}	—	—
	ohm cm 50%RH	10^{13}	—	—	10^{14}	10^{13}	10^{13}
	ohm cm 65%RH	10^{12}	—	10^{13}–10^{14}	—	—	—
Dielectric constant	10^3 c/s dry	3·6–6·0	3·6–6·0	—	3·6–6·0	—	—
	10^3 c/s 65%RH	—	—	3·7	—	—	—
	10^6 c/s 50%RH	3·4	—	—	—	—	—
Power factor	10^3 c/s dry	0·04	0·02– 0·06	—	0·02	—	—
	10^6 c/s 65%RH	—	—	0·06	—	—	—
Dielectric strength	V/0·001 in. 50%RH 25°C	>250	>250	—	>250	—	—

* The data on nylon 11 are from trade literature on Rilsan, those on the other polymers from information supplied by I.C.I.

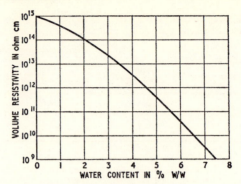

Fig. 15.15. *Effect of moisture content on the volume resistivity of nylon 66*

and under conditions of low humidity but the insulation properties deteriorate as humidity and temperature increase. The effects of the amount of absorbed water on the volume resistivity of nylon 66 is shown in Fig. 15.15. This effect is even greater with nylon 6 but markedly less with nylon 11. Some typical electrical properties of the nylons are given in Table 15.5.

Nylon 6, 66, 610 and 11 are polar crystalline materials with exceptionally good resistance to hydrocarbons. Esters, alkyl halides, and glycols have little effect. Alcohols generally have some swelling action and may in fact dissolve some copolymers (e.g. nylon 66/610/6). There are few solvents for the nylons of which the most common are formic acid, glacial acetic acid, phenols and cresols.

Mineral acids attack the nylons but the rate of attack depends on the type of nylon and the nature and concentration of the acid. Nitric acid is generally active at all concentrations. The nylons have very good resistance to alkalis at room temperature. Resistance to all chemicals is more limited at elevated temperatures.

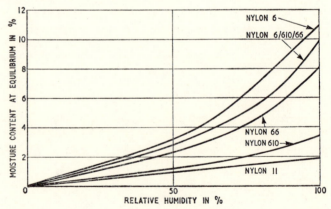

Fig. 15.16. *Effect of relative humidity on the water absorption of the nylons*

The nylons are hygroscopic. Fig. 15.16 shows how the equilibrium water absorption of different nylons varies with humidity at room temperature. Fig. 15.17 shows how the rate of moisture absorption is affected by the environmental conditions.

The absorbed water has a plasticising effect and thus will cause a reduction in tensile strength and modulus, and an increase in impact

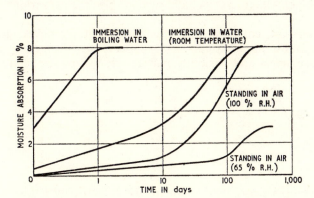

Fig. 15.17. *Effect of environmental conditions on rate of moisture absorption of nylon 66 ($\frac{1}{8}$ in. thick specimens)*

strength. As has already been mentioned the presence of absorbed water also results in a deterioration of electrical properties.

When in service indoors or otherwise protected from sunlight the nylons show no appreciable change of properties on aging at room temperature. Care should be taken in the use of the polymers when exposed to direct sunlight particularly in film and filament applications where embrittlement is liable to occur. Some improvement may be achieved if stabilised compounds are used (see Section 15.6). Continuous exposure to air at temperatures above 60°C will also cause surface discoloration and a lower impact strength of mouldings. The useful life of a moulding in service at 100°C will only be of the order of four to six weeks. If the moulding is immersed in oil, or otherwise shielded from oxygen a considerably longer life time may be expected. Heat-stabilised grades have markedly improved resistance.

15.6 ADDITIVES

The major nylon moulding materials are each available in a number of grades. These may differ in molecular weight but they may also differ in the nature of additives which may be present.

Additives used in nylon can be grouped as follows:

1. Heat stabilisers.
2. Light stabilisers.
3. Plasticisers.
4. Lubricants.

5. Reinforcing fillers.
6. Pigments.
7. Fungicides.

With the possible exception of pigments, which may be dry-blended by the processor, additives are incorporated by the manufacturers and only a limited amount of information about them is normally made available.

Amongst heat stabilisers quoted in the literature are syringic acid, phenyl-β-naphthylamine, mercaptobenzthiazole and mercaptobenzimidazole. Light stabilisers include carbon black and various phenolic materials.

Plasticisers are comparatively uncommon but plasticised grades are supplied by most manufacturers. Plasticisers lower the melting point and improve toughness and flexibility particularly at low temperatures. An

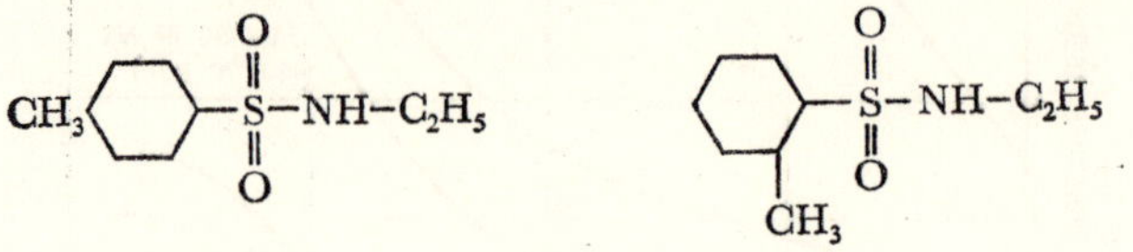

Fig. 15.18

example of a plasticiser used commercially is Santicizer 8, a blend o- and p-toluene ethyl sulphonamide (Fig. 15.18).

Self-lubricating grades are of particular value in some gear and bearing applications. One commercial nylon compound incorporates 0·20% molybdenum disulphide and 1% of graphite whilst many other commercial compounds contain only one of these two lubricants.

Glass-reinforced nylon compounds have become available in recent years. The properties of glass-filled nylon 66 are compared with the unfilled polymer in Table 15.6.

The glass-filled materials show a much greater resistance to creep, whilst coefficient of friction, hardness and abrasion resistance are of the same

Table 15.6 COMPARISON OF GLASS-FILLED AND UNFILLED NYLON 66

Property	Units	ASTM test	Glass filled	Unfilled
Specific gravity	—	D.792	1·38	1·14
Tensile stength	10^3 lb in^{-2}	D.638	23	11·5
E_B	%	D.638	3–5	80–100
Flexural modulus	10^5 lb in^{-2}	D.790	11	4·3
Izod impact	ft lb in^{-1}	(B.S.) 2782	1·3–1·8	1·3–1·8
% water absorption at saturation		(B.S.) 2782	5·6	8·9
Coefficient of linear expansion	10^{-5} in/in. degF	—	1·3	5·6

order as with unfilled polymers. Such compositions may be used to replace metals in applications which had been hitherto beyond the scope of thermoplastics.

The selection of pigments for nylons is restricted by the high processing temperatures involved in processing. There are however a number of materials, mainly inorganic, which withstand processing and normal conditions of service without degrading the polymer.

15.7 PROCESSING

In the processing of nylons consideration should be given to the following points:

1. The tendency of the material to absorb water.
2. The high melting point of the homopolymers.
3. The low melt viscosity of the homopolymers.
4. The tendency of the material to oxidise at high temperatures where oxygen is present.
5. The crystallinity of the solid polymer and hence the extensive shrinkage during cooling.

The above features are particularly marked with nylons 6, 66 and 610 and less marked with nylons 11 and 12. Providing they are dry the copolymers may be processed in much the same way as conventional thermoplastics.

In the injection moulding of nylon 66, for example, it is necessary that the granules be dry. The polymer is normally supplied in sealed containers but should be used within an hour of opening. If reworked polymer is being used, or the granules have become otherwise damp, the polymer should be dried in an oven at about 70–90°C. Too high a temperature will oxidise the surface of the granules and result in inferior mouldings.

Injection moulding cylinders should be free from dead spots and a temperature gradient along the cylinder is desirable. It is important that the plunger does not enter the melt since molten polymer may be forced through the annulus between plunger and cylinder. This may cause damage to the sliding surface or even worse a stream of molten material may be ejected rearwards and possibly injure an operator.

Because of the low melt viscosity of the polymer at processing temperatures it will 'drool' through normal injection nozzles even when the plunger is retracted. Several types of nozzle have been specially designed for use with nylon and all function by sealing the end of the nozzle, either by allowing a pip of polymer to harden, by the use of a spring-loaded valve or by the using of sliding side-closure nozzles. In designs in which solidified polymer is formed at the nozzle it is necessary to make provision for a cold-slug well in the mould, a feature frequently not possible with single-cavity tools. Where spring-loaded closing devices are used the spring should be kept as cool as possible if rapid thermal fatigue is to be avoided.

Because of the crystallisation that occurs on cooling from the melt the polymers show a higher moulding shrinkage than that generally observed with amorphous polymers. With average moulding conditions this is about 0·018 in/in with nylon 66 but by increasing the injection pressure and the injection time the shrinkage may be halved. This is because a high initial mould cavity pressure is developed and a large part of the crystallisation process will be complete before the cavity pressure has dropped to zero. The shrinkage will also be affected by the melt temperature, the mould temperature, the injection speed and the design of the mould as well as by the type of nylon used.

The nylons, nylon 66 in particular, may also exhibit a certain amount of after-shrinkage. Further dimensional changes may occur as a result of moulding stresses being relieved by the plasticising effect of absorbed water. It is consequently often useful to anneal mouldings in a non-oxidising oil for about 20 minutes at a temperature 20 degC higher than the maximum service temperature. Where this is not known a temperature of 170°C is suitable for nylon 66 with somewhat lower temperatures for the other nylons.

When dimensional accuracy is required in a specific application the effect of water absorption should also be taken into account. Manufacturers commonly supply data on their products showing how the dimensions change with the ambient humidity.

The particular features of the nylons should also be taken into account in extrusion. Dry granules must be used unless a devolatilising extruder is employed. Because of the sharp melting point it is found appropriate to use a screw with a very short compression zone. Polymers of the lowest melt viscosity are to be avoided since they are difficult to handle. Provision should be made to initiate cooling immediately the extrudate leaves the die.

15.8 APPLICATIONS

The nylons have found a steadily increasing application as plastics materials for speciality purposes where their toughness, rigidity, abrasion resistance, good hydrocarbon resistance and reasonable heat resistance are important. Because of their high cost they have not become general purpose materials such as polyethylene and polystyrene which are about a third of the price of the nylons.

The largest applications of the homopolymers (nylon 6, 66, 610, 11 and 12) have been in mechanical engineering. Well-known applications include gears, cams, bearings, bushes and valve seats. In addition to the advantageous properties cited above, nylon moving parts may be frequently operated without lubrication, are silent running and may often be moulded in one piece when previously a metal part required assembling of several parts, or alternatively, extensive machining with consequent waste of material.

In recent years the nylons have met increased competition from acetal resins (Chapter 16), the latter being superior in fatigue endurance, creep resistance and water resistance. Under average conditions of humidity the nylons are superior in impact toughness and abrasion resistance. When a nylon is considered appropriate it is necessary to consider the relative importance of mechanical properties, water resistance and ease of processing. For the best mechanical properties nylon 66 would be considered but this material is probably the most difficult to process and has a high water absorption value. Nylon 6 is easier to process but has slightly inferior mechanical properties and an even higher water absorption. Nylons 11 and 12 have the lowest water absorption, are easy to process but there is some loss in mechanical properties.

Sterilisable mouldings have found application in medicine and pharmacy. Because of their durability, nylon hair combs have found wide acceptance in spite of their higher cost.

Nylon film has been used increasingly for packaging applications for foodstuffs and pharmaceutical products. The value of nylon in this application is due to low odour transmission and to some extent in the ability to boil-in-the-bag, a feature useful to the housewife. Film of high brilliance and clarity, particularly from nylon 11, is available for point-of-sale displays.

Although the nylons are not generally considered as outstanding electrical insulators, their toughness and, to some extent, their temperature resistance, have led to applications in coil formers and terminal blocks. The advent of acetal resins and polycarbonates does however present a challenge to applications in this sphere.

Nylon monofilaments have found application in brush tufting, wigs, surgical sutures, sports equipment, braiding and in outdoor upholstery. Nylons 610 and 11 have found extensive application in these fields because of their flexibility but nylon 66 is also used for brush tufting less than 0·0035 in. in diameter. A nylon 66/610 copolymer is used in the manufacture of a monofilament for angling purposes (Luron 2, I.C.I.).

Extruded applications of nylon, other than film and monofilament, are less commonly encountered because of the low melt viscosity of the polymers. Uses include cable sheathing which requires resistance to abrasion and/or chemical attack, flexible tubing for conveying petrol and other liquids, piping for chemical plant, rods for subsequent machining, as the tension member of composite belts for high-duty mechanical drive and for bottles requiring resistance to hydrocarbons. Nylons, 11, 66/610 and 66/610/6 are frequently preferred because of their ease of processing but the high molecular weight 6, 66 and 610 polymers find occasional use.

Nylon 11 is also used in powder form in spraying and fluid bed dipping to produce chemical resistant coatings. Although more expensive than the polyolefin and p.v.c. powders, it is of interest because of its hardness, abrasion resistance and petrol resistance.

As previously mentioned, mouldings have been produced by the polymerisation casting of caprolactam. The ability to produce large objects in this way enables one to envisage new horizons for the use of plastics in engineering and other applications.

Since large tonnage production is desirable in order to minimise the cost of a polyamide and since the consumption of nylons as plastics materials remains rather small, it is important that any new material introduced should also have a large outlet as a fibre. There are a number of polyamides in addition to those already mentioned that could well be very useful plastics materials but which would be uneconomical for all but a few applications if they were dependent on a limited outlet in the sphere of plastics.   Both nylon 7 and nylon 9 are such examples but their availability as plastics is only likely to occur if they become established fibre-forming polymers.   This in turn will depend on the economics of the telomerisation process and the ability to find outlets for the telomers produced other than those required for making the polyamides.

15.9 POLYAMIDES OF ENHANCED SOLUBILITY

Polyamides such as nylon 6, nylon 66, nylon 610, nylon 11 and nylon 12 exhibit properties which are largely due to their high molecular order and the high degree of inter-chain attraction which is a result of their ability to undergo hydrogen bonding.

It is however possible to produce polymers of radically different properties by the following modifications to the molecular structure:

1. Replacement of some or all of the —CONH— hydrogens by alkyl or alkoxy-alkyl groups to reduce hydrogen bonding which results in softer, lower melting point and even rubbery polymers (*N*-substitution).
2. Use of acids or amines containing large bulky side groups which prevent close packing of the molecules.
3. Use of trifunctional acids or amines to give branched structures.
4. Copolymerisation to give irregular structures.
5. Reduction in molecular weight.

The techniques of *N*-alkylation may be effected by the use of *N*-alkylated or *N*:*N'*-dialkylated diamines, or by the use of an ω-*N*-alkylamino carboxylic acid of type $R_1NHRCOOH$.   The polymers thus have repeating units of the general form

$$\sim\!\!(CH_2)_n \; CON\!\!\sim$$
$$|$$
$$R$$

Such *N*-alkyl compounds are not known to be of any current application although fibres from a partially *N*-alkylated derivative of nylon 610 have been described.

Treatment of a nylon with formaldehyde leads to the formation of *N*-methylol groups but the polymers are unstable.   If however the nylon is dissolved in a solvent such as 90% formic acid and then treated with formaldehyde and an alcohol in the presence of an acidic catalyst such as phosphoric acid a process of alkoxymethylation occurs (Fig. 15.19).

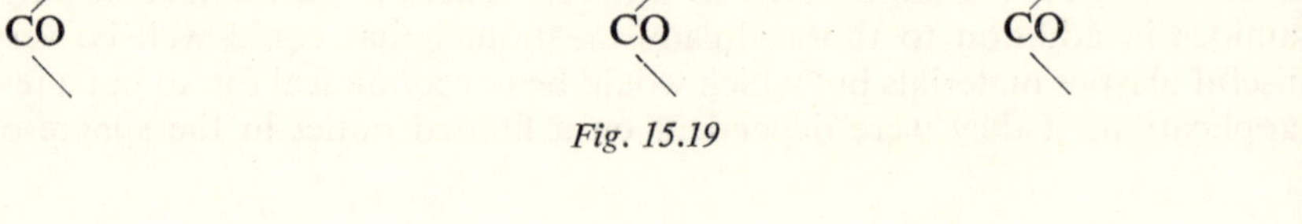

$$\diagdown NH + CH_2O + ROH \longrightarrow \diagdown N\!\!-\!\!CH_2OR \text{ or } \diagdown N(CH_2O)_2R$$

Fig. 15.19

Methyl methoxy nylons are commercially available in which about 33% of the —NH— groups have been substituted.

Such materials are soluble in the lower aliphatic alcohols, e.g. ethanol, and in phenols. They also absorb up to 21% of moisture when immersed in water. If this material is heated with 2% citric acid at elevated tempera-

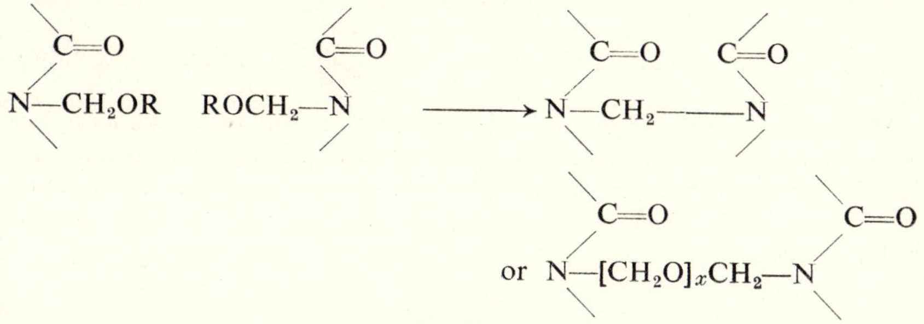

$$
\begin{array}{cccc}
\diagdown & \diagup & \diagdown & \diagup \\
C{=}O & C{=}O & C{=}O & C{=}O \\
\diagup & \diagdown & \diagup & \diagdown \\
N{-}CH_2OR & ROCH_2{-}N & \longrightarrow N{-}CH_2{-}{-}{-}N
\end{array}
$$

$$
\begin{array}{cc}
\diagdown & \diagup \\
C{=}O & C{=}O \\
\diagup & \diagdown
\end{array}
$$

$$
\text{or } N{-}[CH_2O]_xCH_2{-}N
$$

Fig. 15.20

tures, typically for 20 minutes at 120°C, cross linking will take place (Fig. 15.20).

These materials find a limited application in films and coatings which require good abrasion and flexing resistance. Some typical properties of cross-linked and uncross-linked polymers are given in Table 15.7.

In the early 1950's a new class of polyamides became available differing from the nylons in that they contained bulky side groups, had a somewhat irregular structure and were of low molecular weight (2,000–5,000). They are marketed under such trade names as 'Versamids' and 'Beckamides'.

A typical example of this class of polymer may be obtained by reacting ethylene diamine and 'dimer fatty acid', a material of inexact structure

Table 15.7 THE EFFECT OF CROSS LINKING ON THE PROPERTIES OF METHOXYMETHYL NYLON FILM

Property	*Uncross linked*	*Cross linked)* *(conditions as in text)*
Tensile strength (lb in⁻²)		
dry, 52% RH 22°C	3,980	6,500
wet, water-saturated 22°C	1,670	4,045
Elongation at break % dry 52% RH 22°C	385	355
Elastic recovery % from 100% elongation	75	73
Resistance to boiling water	gelatinises in less than 5 min	excellent
Colour	colourless	yellow

obtained by fractionating heat-polymerised unsaturated fatty oils and esters. An idealised structure for this acid is shown in Fig. 15.21. These materials are dark coloured, ranging from viscous liquids to brittle resins and with varying solubility.

They have found use as hardeners-cum-flexibilisers for epoxide resins

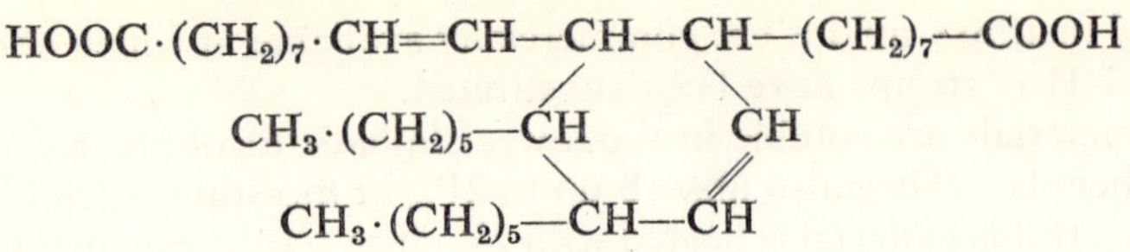

$$HOOC \cdot (CH_2)_7 \cdot CH = CH - CH - CH - (CH_2)_7 - COOH$$

$$CH_3 \cdot (CH_2)_5 - CH \qquad CH$$

$$CH_3 \cdot (CH_2)_5 - CH - CH$$

Fig. 15.21

(see Chapter 22) and are of interest in the production of thixotropic paints and adhesives.

As has been mentioned earlier a number of copolymers such as nylon 66/610/6 are available. Such a copolymer has an irregular structure and thus inter-chain bonding and crystallisation are limited. As a consequence this copolymer is soluble in alcohols and many other common polar solvents.

15.10 OTHER POLYAMIDES[7]

Although less than a dozen polyamide types together with a few miscellaneous copolymers have become available commercially, a very large number have been prepared and investigated. Of the many diamine–dibasic acid combinations those based on intermediates with less than 5 carbon atoms are generally unsuitable either because of the tendency to form ring structures or because the melting points are too high for melt spinning (important in fibre production). For example nylon 46 melts at 275°C. The many nylons based on amines and acids with 6–10 carbon atoms might also be of interest as fibres and plastics but are not yet attractive commercially because of the costs of synthesis. Similar remarks must also apply to nylons 8, 9, and 10.

For various reasons, such as the inability to fuse without decomposition, slow amidation and tendency to colour during polymerisation, aromatic polyamides have not become important. However Dynamit Nobel have recently introduced a terephthalic acid–diamine product known as Trogamid T. Interesting products have also been prepared from cyclic amines such as piperazine and bis(p-aminocyclohexyl) methanes. The German copolymer Igamid IC was based on the latter polymer. Polyamides have also been produced from intermediates with lateral side groups. The effect of such groups is similar to that of N-substitution in that there is a decrease in intermolecular cohesion and reduction in the ability of the molecules to pack in a crystal lattice. In some cases the polymers are still fibre-forming but they have much lower melting points. For example the polymer from 12-aminostearic acid is fibre forming but has a low melting point (109°C) and a low moisture absorbing capacity.

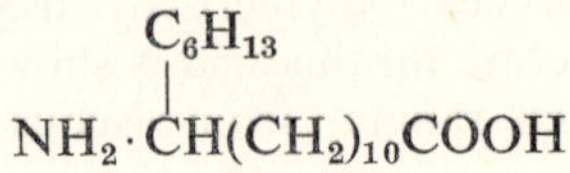

$$\overset{\displaystyle C_6H_{13}}{\underset{\displaystyle NH_2 \cdot CH(CH_2)_{10}COOH}{|}}$$

Other polyamides produced experimentally include polymers with active lateral groups (hydroxy, keto groups etc.), polymers with hetero-atoms (sulphur and oxygen) in the polyamide-forming intermediates, polymers with tertiary amino groups in the main chain and polymers with unsaturation in the main chain. There does not however appear to have been any serious attempt to develop unsaturated polyamide analogues to the polyester laminating resins.

15.11 POLYIMIDES[13,14]

Although polyimides have been known for many years they have remained as laboratory curiosities until the recent development of polymers based on pyromellitic dianhydride and diamines. These polymers are in general obtained by the reaction shown in Fig. 15.22.

The pyromellitic dianhydride is itself obtained by vapour phase oxidation of durene (1,2,4,5-tetramethyl benzene), using a supported vanadium

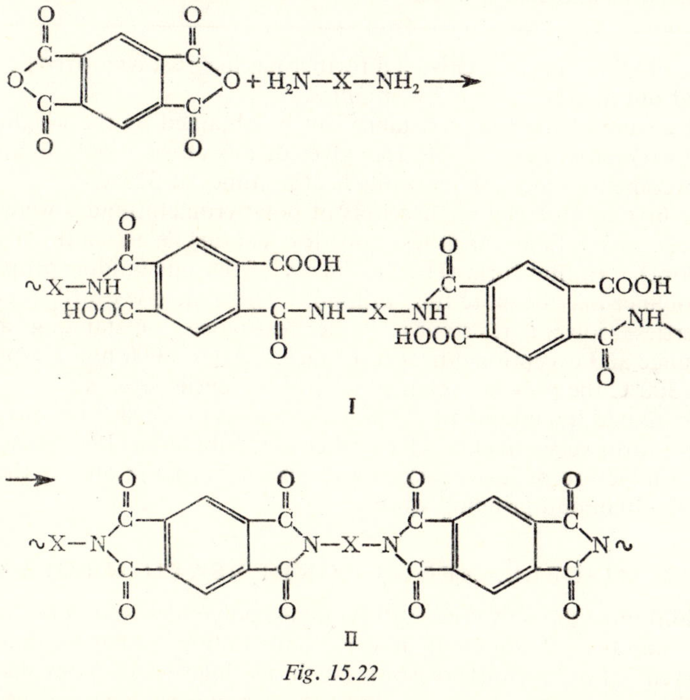

Fig. 15.22

oxide catalyst. A number of amines have been investigated and it has been found that certain aromatic amines give polymers with a high degree of oxidative and thermal stability. Such amines include *m*-phenylene diamine, benzidine and 4,4'-diaminodiphenylether, the last of these being employed in the manufacture of H-Film (Du Pont).

For convenience of application it is usual to utilise the two stage preparation shown above. Initially the soluble polymer (I) is formed which is then converted into the insoluble thermally stable polyimide (II).

Suitable solvents for the high molecular weight prepolymer (I) include dimethyl formamide and dimethylacetamide.

In order to prevent premature gelation the reaction mixture should be anhydrous, free from pyromellitic acid and reacted at temperatures not exceeding 50°C.

Films may be made by casting (I) and heating to produce the polyimide (II). Tough thin film may be obtained by heating for 1–2 hours at 150°C but thicker products tend to become brittle. A substantial improvement

Table 15.8 WEIGHT LOSS ON HEATING POLYPYROMELLITIMIDES AT 325°C[14]

Polymer based on	Film	Weight loss at 325°C			
		100 hr	*200 hr*	*300 hr*	*400 hr*
m-phenylene diamine	brittle	3·3	4·3	5·0	5·6
Benzidine	flexible	2·2	3·6	5·1	6·5
4,4'-diaminodiphenylether	flexible	3·3	4·0	5·2	6·6

can be obtained in some cases if a further baking of solvent-free polymer is carried out at 300°C for a few minutes.

A measure of the heat resistance can be obtained by the weight loss at various temperatures. Table 15.8 gives details of the weight loss of three polypyromellitimides after various heating times at 325°C.

The first commercial applications of polypyromellitimides were as wire enamels, insulating varnishes and for coating glass-cloth (Pyre.ML, Du Pont). In film form (H-Film) many of the outstanding properties of the polymer may be more fully utilised. These include excellent electrical properties, solvent resistance, flame resistance, outstanding abrasion resistance and exceptional heat resistance. After 1,000 hours exposure to air at 300°C the polymer retained 90% of its tensile strength.

The limited tractability of the polymer makes processing in conventional plastics form very difficult. Nevertheless the materials have been used in the manufacture of seals, gaskets and piston rings and also as the binder resin for diamond grinding wheels.

15.12 MISCELLANEOUS CONDENSATION POLYMERS

In addition to the polyamides, polyimides, polyethers and polysulphides a great number of different linear condensation polymers have been prepared. For various reasons (such as high cost, poor hydrolytic stability and tendency to cross link) these polymers have not become of technical significance.

Examples of such polymers with their general methods of preparation include:

1. Polyanhydrides

$$n \ HOOC \cdot (CH_2)_x COOH + n \ (CH_3CO)_2O$$
$$\longrightarrow \sim [\sim OC(CH_2)_x CO \cdot O \sim]_n^\sim + 2n \ CH_3COOH$$

2. Polyureas.   There are at least three routes including

$$n \text{ OCNRNCO} + n \text{ H}_2\text{NR}_1\text{NH}_2$$

$$\longrightarrow \sim\!\!\sim(\sim\!\!\sim\text{RNHCONHR}_1\text{NHCONH}\sim\!\!\sim)_{\overline{n}}\sim\!\!\sim$$

3. Polythioureas.   At least six routes have been investigated including

$$n \text{ R}_1\text{SCSR}_1 + n \text{ H}_2\text{NRNH}_2 \longrightarrow \sim\!\!\sim(\sim\!\!\sim\text{RNHCSNH}\sim\!\!\sim)_{\overline{n}}\sim\!\!\sim + 2n \text{ R}_1\text{SH}$$
$$\overset{\|}{\text{S}}$$

4. Polyamines

$$n \text{ Cl RCl} + n \text{ H}_2\text{NR}_1\text{NH}_2 \longrightarrow \sim\!\!\sim(\sim\!\!\sim\text{RNHR}_1\text{NH}\sim\!\!\sim)_{\overline{n}}\sim\!\!\sim + 2n \text{ HCl}$$

5. Polytriazoles

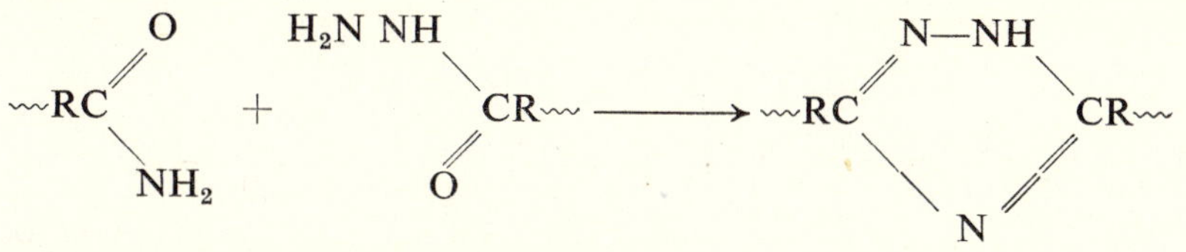

REFERENCES

1. *Plastics*, **17**, 64 (1952)
2. FREIDLNA, R. K., and KARAPETYAN, S. A., *Telomerization and New Synthetic Materials*, Pergamon, Oxford (1961)
3. KRALICEK, J., SEBENDA, J., ZADAK, Z., and WICHTERLE, O., *Chem. Prumsyl.*, **11**, 377–381 (1961)
4. AÉLION, R., *Ann. Chimq. (Paris)*, **3**, 5 (1948)
5. AÉLION, R., *Ind. Eng. Chem.*, **53**, 826 (1961)
6. COFFMAN, D. D., BERCHET, G. J., PETERSON, W. R., and SPANAGEL, E. W., *J. Polymer Sci.*, **2**, 306 (1947)
7. HILL, R. (Ed.) *Fibres from Synthetic Polymers*, Elsevier, Amsterdam/Houston/New York/London (1953)
8. GEIL, P., *Polymer Single Crystals*, Interscience, New York (1963)
9. MANDELKERN, L., *Crystallization of Polymers*, McGraw-Hill, New York (1964)
10. HOLMES, D. R., BUNN, C. W., and SMITH, D. J., *J. Polymer Sci.*, **17**, 159 (1955)
11. MÜLLER, A., and PFLÜGER, R., *Plastics*, **24**, 350 (1959)
12. RILEY, J. L., *Eng. Mater. Design*, **1**, 132 (1958)
13. JONES, J. I., OCHYUSKI, F. W., and RACKLEY, F. A., *Chem Ind (London)*, 1686 (1962)
14. BOWER, G. M., and FROST, L. W., *J. Polymer Sci.*, Part A, **1**, 3135 (1963)

BIBLIOGRAPHY

FLOYD, D. E., *Polyamide Resins*, Reinhold, New York (1958)
HILL, R. (Ed.) *Fibres from Synthetic Polymers*, Elsevier, Amsterdam/Houston/New York/London (1953)

Polyacetals and Related Materials

16.1 INTRODUCTION

From the time that formaldehyde was first isolated by Butlerov[1] in 1859 polymeric forms have been encountered by those handling the material. Nevertheless it is only since the late 1950's that polymers have been available with the requisite stability and toughness to make them useful plastics. In this short period these materials (referred to by the manufacturers as acetal resins or polyacetals) have achieved rapid acceptance as engineering materials competitive not only with the nylons but also with metals and ceramics.

The first commercially available acetal resin was marketed by Du Pont in 1959 under the trade name Delrin after the equivalent of ten million pounds had been spent in research on polymers of formaldehyde. The Du Pont monopoly was unusually short lived as Celcon, an acetal copolymer produced by the Celanese Corporation, became available in small quantities in 1960. This material became commercially available in 1962 and later in the same year Farbwerke Hoechst combined with Celanese to produce similar products in Germany (Hostaform). In 1963 Celanese also combined with the Dainippon Celluloid Company of Osaka, Japan and Imperial Chemical Industries to produce acetal copolymers in Japan and Britain under the trade names Duracon and Alkon respectively. Patents dealing with aldehyde polymers have been granted to firms in many countries, mainly involving polyformaldehyde but in some cases, particularly in Japan, polyacetaldehyde.

Of the many other polyethers, polythioethers and polysulphides prepared in the laboratory, some have become commercially available. Those of interest to the plastics industry are dealt with later in this chapter.

16.2 PREPARATION OF FORMALDEHYDE

Formaldehyde is an important chemical in the plastics industry, being a vital intermediate in the manufacture of phenolic and amino resins. It was also used by Reppe during the Second World War as an important starting point for the preparation of a wide range of organic chemicals. Consumption of formaldehyde in acetal resins is still a minor outlet for the material but exceptionally pure material is required for this purpose.

The most important route for the production of formaldehyde is from methanol, this normally being prepared by interaction of carbon monoxide and hydrogen.

$$CO + 2H_2 \xrightarrow[\substack{\text{Catalyst} \\ \sim 200 \text{ atm}}]{300-400°C} CH_3OH$$

The two gases involved can be obtained by the 'water-gas reaction' which involves passing water vapour over hot coke.

$$H_2O + C \longrightarrow H_2 + CO$$

Methanol is converted into formaldehyde by catalytic vapour phase oxidation over a metal oxide catalyst. In one variation of the process methanol is vaporised, mixed with air and then passed over the catalyst at 300–600°C. The formaldehyde produced is absorbed in water and then fed to a fractionating column. A 37% solution of formaldehyde in water is removed from the bottom of the column with some methanol as a stabiliser whilst excess methanol is taken from the top of the column and recycled.

Formaldehyde is also produced by the oxidation of light petroleum gases, a process which also yields methanol and acetaldehyde. This process is currently used in the Celanese Corporation plant for the production of Celcon.

Formaldehyde is a gas with a boiling point of −21°C. It is usually supplied as a stabilised aqueous solution (∼40% formaldehyde) known as formalin. When formalin is used as the source of the aldehyde, impurities present generally include water, methanol, formic acid, methylal, methyl formate and carbon dioxide. The first three of these impurities interfere with polymerisation reactions and need to be removed as much as possible. In commercial polymerisation the low polymers trioxane and paraformaldehyde are convenient sources of formaldehyde since they can be obtained in a greater state of purity.

16.3   ACETAL RESINS

16.3.1   Polymerisation of formaldehyde

Formaldehyde will polymerise in a number of ways as indicated in Fig. 16.1

The cylic trimer (trioxane) and tetramer are obtained by the action of a trace of sulphuric acid on hot formaldehyde vapour. Linear polymers with degrees of polymerisation of about 50 and a terminal hydroxyl group are obtained by evaporation of aqueous solutions of formaldehyde. In the presence of strong acid the average chain length may be doubled.

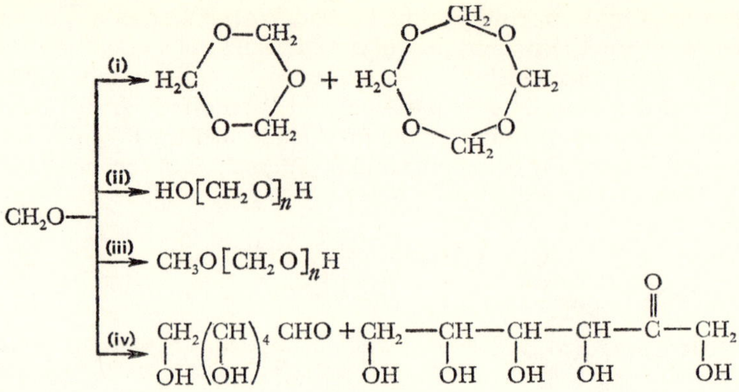

Fig. 16.1

Evaporation of methanol solution leads to products of type (iii) indicated above.

In the presence of lime water more complex reactions occur leading to the formation of aldoses and hexoses (iv). This particular reaction is of interest to the biochemist as it is now generally held that optically active plant carbohydrates are obtained from carbon dioxide and water via formaldehyde.

During the 1920's Staudinger and his collaborators[2] prepared linear polymers of formaldehyde in some classic researches which demonstrated for the first time the molecular structure of high polymers. When prepared by a solution polymerisation technique, brittle, pulverisable and thermally unstable products were obtained, but Staudinger also prepared polymers by allowing the material to polymerise in bulk at $-80°C$. These products though still thermally unstable possessed some degree of toughness.

In the early 1940's an intensive research programme on the polymerisation of formaldehyde was initiated by the Du Pont Company. As a consequence of this work polymers, both tough and adequately stable to processing conditions, were prepared and eventually marketed[3,4] (Delrin).

In order to manufacture such polymers, it is first necessary to produce a very pure form of formaldehyde. This is typically produced from an alkali-precipitated low molecular weight polyformaldehyde which has been carefully washed with distilled water and dried for several hours under vacuum at about 80°C. The dried polymer is then pyrolysed by heating at 150–160°C, and the resultant formaldehyde passed through a number of cold traps (typically four) at $-15°C$. Some prepolymerisation occurs in these traps and removes undesirable impurities from the monomer. The monomer is then introduced into the polymerisation vessel over a rapidly stirred and carefully dried inert medium such as heptane. A number of polymerisation initiators have been cited in the literature and include Lewis acids, amines, phosphines,[5] arsines and stibines. A typical initiator is triphenyl phosphine used to the extent of 20 p.p.m. based on the inert medium. A polymer stabiliser such as diphenylamine may also be present to a concentration of 100 p.p.m. Polymerisation is carried out until a 20% solids content is obtained. The polymer is then isolated by filtration, washed in turn with heptane and pure acetone and then dried in a vacuum oven at 80°C. Control of molecular weight may be made by adding traces of water which is an effective chain transfer agent. It is because of this particular property of water that it is necessary to work under conditions where the water content is carefully controlled.

Polymers produced by methods as described above have thermal stabilities many times greater than those obtained by the earlier bulk and solution methods of Staudinger. Staudinger had however shown that the diacetates of low molecular weight polyoxymethylenes (I) (polyformaldehydes) were more stable than the simple polyoxymethylene glycols (II).

$$CH_3COO \cdot [CH_2O]_{n<20} CH_2OOC \cdot CH_3 \qquad (I)$$

$$HO \cdot [CH_2O]_{n<20} CH_2OH \qquad (II)$$

On the other hand, Staudinger found that polyoxymethylenes with a degree of polymerisation of about 50 were less stable. Truly high molecular weight polyoxymethylenes (degree of polymerisation $\sim 1,000$) were not esterified by Staudinger; this was effected by the Du Pont research team and was found to improve the thermal stability of the polymer substantially.

The esterification reaction may be carried out with a number of different anhydrides but the literature[4,6] indicates that acetic anhydride is preferred. The reaction is catalysed by amines and the soluble salts of the alkali metals. The presence of free acid has an adverse effect on the esterification reaction, the presence of hydrogen ions causing depolymerisation by an unzipping mechanism. Reaction temperatures may be in the range of 130–200°C. Sodium acetate is a particularly effective catalyst. Esterification at 139°C, the boiling point of acetic anhydride, in the presence of 0·01% sodium acetate (based on the anhydride) is substantially complete within 5 minutes. In the absence of such a catalyst the percentage esterification is only of the order of 35% after 15 minutes.

The following extract is taken from an example in British Patent 770,717 to the Du Pont Company as an illustration of a typical method of esterification:

'Into a reaction vessel there is placed 500 g of a high molecular weight formaldehyde polymer, 4 litres of acetic anhydride and 1·6 g of anhydrous sodium acetate. The mixture is stirred and heated to 160°C. Nitrogen gas at 12 to 15 p.s.i. gauge pressure is maintained in the space above the reaction mixture during the heating period to prevent boiling. The polymer is completely dissolved in the reaction mixture at this temperature. The mixture is allowed to cool slowly with stirring and the polymer precipitates from the solution at about 133°C, the total time in solution being about 90 minutes. The acetylated polymer is removed by filtration and washed on the filter with 3 litres of acetone. It is then re-slurried in 3 litres of water using high speed agitation and the slurry is filtered again. The water washing is repeated two more times. It is then washed with 3 litres of acetone and 3 litres of acetone containing 2·0 g of beta-conidendrol. The product is then dried in a vacuum oven at 67°C.'

The beta-conidendrol is incorporated as an antioxidant and is frequently referred to in the patent literature, as is also di-β-napthhyl-p-phenylene diamine for this purpose. It is claimed that in the example given above the degradation rate at 222°C is only 0·09% per minute compared with typical values of 0·6–0·8% for unesterified polymer.

An alternative approach to the production of thermally stable poly-oxymethylenes has been made by chemists of the Celanese Corporation of America. The commercial products are marketed as Celcon, Alkon, Hostaform and Duracon. The principle of thermal stability in this case is the copolymerisation of formaldehyde with a second monomer[7] which is a cyclic ether of the general form shown in Fig. 16.2 (I).

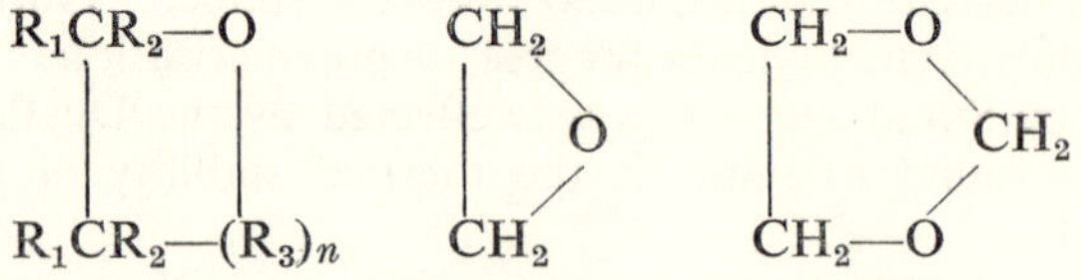

$$R_1CR_2\!-\!O \qquad CH_2 \qquad CH_2\!-\!O$$

(I)   (II)   (III)

Fig. 16.2

It is stated in the basic patent that ethylene oxide (II) and 1,3-dioxolane (III) are the preferred materials. By the occasional incorporation of molecules containing two successive methylene groups the tendency of the molecules to unzip is markedly reduced.

In one example[7] 25·0 g of cyclohexane were added to 25·0 g of trioxane (a cyclic trimer of formaldehyde) and cooled to −70°C; 0·03 ml of dioxolane were added together with 0·10 ml boron fluoride ethereate (stated in the basic patent to be the preferred catalyst). The tube was then sealed and immersed in a water bath at 66–68°C for four hours. After washing the product the polymer was dried and a 20% yield obtained. On heating this sample for 2 hours at 225°C there was a weight loss of 27·8%. Experiments were also carried out using 0·25 parts and 1·25 parts of dioxolane, but in these cases there was a higher weight loss and in addition, a lower melting point.

In another example, trioxane and dioxolane were blended in such a ratio as 'to provide one oxyethylene group for each 8·45 oxymethylene groups'. The boron fluoride ethereate comprised 0·089% by weight of the mixture which was then heated in a tube in a bath of boiling water for 2·16 hours. A polymer was produced in a yield of 42·5% by weight, had a melting point of 158–163°C and a degradation rate at 222°C of only 0·06% by weight per minute.

16.3.2 Structure and properties of acetal resins

It is difficult to resist a comparison between the structure and properties of acetal polymers and those of polyethylene.

Both polymers are linear, with a flexible chain backbone and are thus both thermoplastic. Both the structures shown (Fig. 16.3) are regular and since there is no question of tacticity arising both polymers are

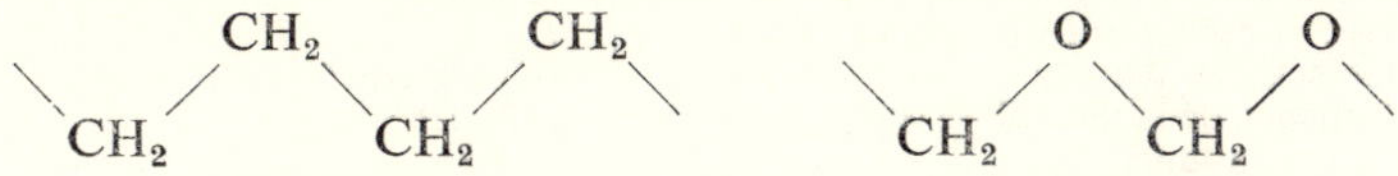

Polyethylene   Polyacetal

Fig. 16.3

capable of crystallisation. In the case of both materials polymerisation conditions may lead to structures which slightly impede crystallisation; with the polyethylenes this is due to a branching mechanism, whilst with the polyacetals this may be due to copolymerisation.

The acetal polymer molecules have a shorter backbone (—C—O—)—bond and they pack more closely together than those of polyethylene. The resultant polymer is thus harder and has a higher melting point. In spite of the close packing the glass transition temperature is as low as −73°C, owing largely to the high bond flexibility.

As is typical for crystalline polymers incapable of specific interactions with liquids, there are no solvents at room temperature but liquids which have a similar solubility parameter ($\delta = 11\cdot1$) will cause a measure of swelling, principally in the amorphous region.[8]

At room temperature there is only a small decrease in free energy on conversion of monomer to polymer. At higher temperatures the magnitude of the free energy change decreases and becomes zero at 127°C;[9] above this temperature the thermodynamics indicate that depolymerisation will take place. Thus it is absolutely vital to stabilise the polyacetal resin both internally and externally to form a polymer which is sufficiently stable for processing at the desired elevated temperatures.

The backbone bonds are polar but the structure is balanced and the polymer is quite a good dielectric. Reported data on resistivity indicate only moderate values presumably because of ionic fragments, impurities and additives.

Both the molecular[10] and fine structure[11] of the Du Pont polyoxymethylenes have been investigated and reported. Koch and Lindvig[10] have demonstrated that the repeating unit of the polymer is —$CH_2\cdot O$— and that the end groups are either acetate or methoxyl (derived from methanol which is present in trace in the formaldehyde during polymerisation). The molecular weights of these polymers are normally in the range of 20,000–110,000. Values for molecular weight determined by end group analysis and by osmotic methods show close agreement. This agreement, together with the fact that no structures which could be possible branch points in the molecule have been discovered, indicates that the polymers are substantially linear.

The acetal polymers exhibit a high crystallinity.[11] The percentage crystallinity will depend on the quench temperature and will range from about 77%, when quenched at 0°C, to about 80% when quenched at

Table 16.1   CRYSTALLINITY DATA FOR HOMOPOLYMER ACETAL RESINS[11]

Unit cell dimensions	$a = 4 \cdot 46$ Å $c = 17 \cdot 30$ Å
Molecular configuration	helical
No. of repeat units in identity period	9
Crystal cell density	$1 \cdot 506$ g cm^{-3}
Polymer densities (observed)	$1 \cdot 40 – 1 \cdot 45$
% crystallinity	75–85

160°C.   Annealing will increase the crystallinity, this being most marked at 150°C (Fig. 16.4).

The greater the percentage crystallinity the higher the yield point and tensile modulus.   It has also been shown that by raising the quench temperature the spherulite size is increased and that this greatly decreases the impact toughness.

6.3.3.   Properties of acetal resins[9,12,13]

The principal features of acetal resins leading to commercial application may be summarised as follows:

1. Stiffness.
2. Fatigue endurance.
3. Resistance to creep.
4. Low coefficient of friction (with equal dynamic and static coefficients).
5. Good appearance.

In many respects similar to the nylons, the acetal resins may be considered to be superior to the former in their fatigue endurance, creep resistance, stiffness and water resistance.   The nylons (except under dry conditions) are superior in impact toughness and abrasion resistance.

Some mechanical and thermal properties of acetal polymers are listed in Table 16.2.   The values quoted are those supplied by the manufacturers.

In addition to these more elementary properties, the creep and fatigue characteristics, together with service properties such as abrasion resistance, are of interest.

The data presented in Fig. 16.5 were obtained on a Sonntag-Universal machine which flexes a beam in tension and compression.   Whereas the acetal resin was subjected to stresses at 1,800 cycles per minute at 75°C and at 100% RH, the nylons were cycled at only 1,200 cycles per minute and had a moisture content of $2 \cdot 5\%$.   The polythene sample was also

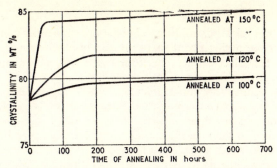

Fig. 16.4. *Effect of annealing on room temperature crystal-linity of acetal homopolymer resin. (Films moulded at 210°C, quenches to 50°C). (After Hammer et al[11])*

Table 16.2

Property	ASTM test	Acetal homopolymer	Acetal copolymer	Unit
Specific gravity	D.792	1·425	1·410	
Tensile strength (73°C)	D.638	10,000	8,500	lb in⁻²
Flexural modulus (73°C)	D.790	410,000	360,000	lb in⁻²
E_B (73°F)	D.638	15–75	23–35	%
Deflection temperature 264				
264 p.s.i. lb in⁻²	D.648	100	110	°C
66 p.s.i. lb in⁻²		170	158	°C
Vicat softening point	D.569	185	162	°C
Impact strength (73°C)	D.256	1·4–2·3	1·1	ft lb per in. notch
Crystalline melting point	—	175	163	°C
Water immersion				
(24 hr immersion)	D.570	0·4	0·22	%
(50% RH equilibrium)	—	0·2	0·16	%
(continuous immersion—equilibrium)		0·9	0·8	%
Coefficient of friction	—	0·1–0·3	0·2	—

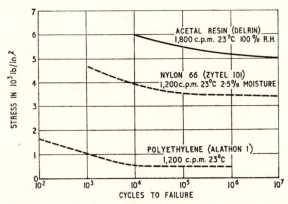

Fig. 16.5. *Fatigue resistance of acetal resin compared with nylon 66 and with polythene. Measured as the cycles to failure at a given applied stress. (Du Pont trade literature)*

flexed at 1,200 cycles per minute. Whilst the moisture content has not been found to be a significant factor it has been observed that the geometry of the test piece and, in particular, the presence of notches has a profound effect on the fatigue endurance limit.

The acetal resins show superior creep resistance to the nylons but are inferior in this respect to the polycarbonates. It is to be noted, however, that limitations in the load bearing properties of the polycarbonates restrict their use in engineering applications (see Chapter 17). Another property of importance in engineering is abrasion resistance—a property that is extremely difficult to assess. Results obtained from various tests indicate that the acetal polymers are superior to most plastics and die cast aluminium, but inferior to nylon 66 (see also Section 16.3.5 and Chapter 15).

The electrical insulation properties of the acetal resins may be described as good but not particularly outstanding. There are available alternative materials which are better insulators and are also less expensive. There are, however, applications where impact toughness and rigidity are required in addition to good electrical insulation characteristics, and in these instances acetal resins would be considered. Table 16.3 lists

Table 16.3

	ASTM test method	Value
Dielectric strength (short time)	D.149	(0·125 in. thick) 500 volts per 0·001 in. (0·010 in. thick) 1700 volts per 0·001 in.
Volume resistivity	D.257	6×10^{14} ohm cm (0·2% water) $4·6 \times 10^{13}$ ohm cm (0·9% water)
Surface resistivity	D.257	10^{16} ohm
Dielectric constant (73°F)	D.150	3·7 (10^2–10^4 c/s)
Power factor	D.150	0·004 (10^2–10^4 c/s)

some of the more important electrical characteristics of 'Delrin' acetal resin. Data for the trioxane based copolymer resin (e.g. 'Alkon') are virtually identical.

Acetal homopolymer resins show outstanding resistance to organic solvents, no effective solvent having yet been found for temperatures below 70°C. Above this temperature some phenolic materials such as the chlorophenols are effective. Stress cracking has not been encountered in organic solvents. Swelling occurs with solvents of similar solubility parameter to that of the polymer ($\delta = 11.1$).

The resistance of these polymers to inorganic reagents is not however so outstanding and they should not be used in strong acids, strong alkalies or oxidising agents. Staining resistance is generally good although hot coffee will cause staining. Acetal copolymer resins are somewhat more resistant to hot alkalies but resistance to acids is still comparatively poor. There do not appear to be any toxic or dermatitic hazards under normal conditions of use with either homopolymers or copolymers. Water does not cause any significant degrading hydrolysis of the polymer but may

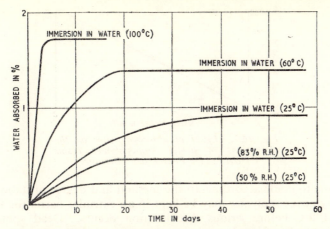

Fig. 16.6. *Effect of humidity, time and temperature on the water absorption of acetal homopolymer resin (Delrin). Du Pont Trade Literature)*

swell it or permeate through it. Fig. 16.6 shows the relation between humidity, time and temperature on the water absorption whilst Fig. 16.7 shows the effect of water absorption on dimensions for homopolymer resins.

Prolonged exposure to ultraviolet light will induce surface chalking and reduce the molecular weight leading to gradual embrittlement. As with the polyolefins it is found that the incorporation of a small amount of well dispersed carbon black increases resistance to ultraviolet degradation. Amongst miscellaneous properties it may be noted that the resins do not appear to be attacked by fungi, rodents and insects. The polymer burns slowly with a soot-free flame.

The homopolymer and the trioxane-based copolymers are generally similar in properties. The copolymer is claimed to have better thermal stability, better hydrolytic stability at elevated temperatures, easier mouldability and better alkali resistance. The homopolymer has slightly better mechanical properties, e.g. higher tensile strength, higher flexural modulus and greater surface hardness. As may be expected the homopolymer has a slightly higher crystalline melting point.

16.3.4 Processing

Acetal resins may be processed without difficulty on conventional injection-moulding, blow moulding and extrusion equipment. The main points to be considered are:

1. Overheating leads to the production of formaldehyde gas and if produced in sufficient quantities within the confines of an injection cylinder or extruder barrel the gas pressure may become sufficiently high that there is a risk or damage or injury. The time for which acetal resin may be heated at any given temperature will vary from

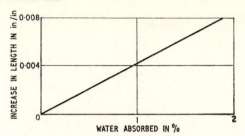

Fig. 16.7. *Effect of absorbed water on the dimensions of acetal homopolymer resins ('Delrin') (Du Pont Trade Literature)*

grade to grade according to the method and degree of stabilisation. A typical copolymer may be kept in an extruder barrel for 110 min at 190°C before serious discoloration occurs. Dead spots must be carefully avoided.

2. Although less hygroscopic than the nylons, acetal resins must be stored in a dry place.

3. With most homopolymers and copolymers the apparent viscosity is less dependent on temperature and shear stress (up to 10^6 dyn cm^{-2}) than the polyolefins thus simplifying die design. On the other hand the melt has a low elasticity and strength and this requires that extruded sections be supported and brought below the melting point as soon as possible consistent with obtaining a satisfactory crystalline texture.

   The lack of melt strength leads to particular problems with blow moulding because of the extensive drawing down of the parison under gravity. To overcome these problems copolymers have become available with slightly branched molecules which have a greater melt elasticity and tenacity. Such materials which also have more stress-dependent viscosities are not only of specific value in blow moulding but extrudates generally are easier to handle.

4. The high crystallinity which develops on cooling, results in a shrinkage of about 0·020 in/in. Because of the low glass transition temperature, crystallisation can take place quite rapidly at room temperatures and after-shrinkage is usually complete within 48 hours of moulding or extrusion. In processing operations injection moulds, blow moulding moulds and sizing dies should be kept at about 80–120°C in order to obtain the best results.

5. Also because of the low glass transition temperature it is not possible to make clear film, stable at room temperature, by quenching. Some improvement in clarity may be obtained by cold rolling as this tends to dispose the crystal structure into layers (see Chapter 6).

16.3.5   Additives

In spite of the techniques of using comonomers and of capping the chain ends, there is still room for improving the stability of the acetal resins against heat, oxygen and light. A review of the patent literature[9] mentions the use of phenolic bodies, amines, mercaptothiazoles, urea, thiourea, hydrazine derivates and sulphides amongst other stabilisers against heat and oxygen. Carbon black is the preferred ultraviolet stabiliser but alternative ultraviolet absorbers may have to be used where a black product is objectionable.

16.3.6 Applications of the Acetal Polymers and Copolymers

The acetal resins may best be considered as engineering materials. They therefore become competitive with a number of plastics materials, nylon in particular, and with metals.

Because of their light weight, ability to be moulded into intricate shapes in one piece, low coefficients of friction and absence of slip–stick behaviour, acetal resins find use as bearings.

An approximate rule of thumb guide as to whether acetal resin may be suitable for a particular bearing application may be obtained from PV data. The PV value is the product of pressure on the projected bearing area (in lb in^{-2}) and the velocity (surface speed in ft min^{-1}). For sleeve bearings the projected area is that of a plane running through the axis of a bearing of the same length as, and of width equal to the diameter of, the bearing. Therefore a bearing 2 in. long and 1 in. in diameter would have a projected area of 2 in^{-2}. It can easily be calculated that for a shaft rotating at 200 rev min^{-1} with a bearing load of 50 lb the PV value will be about 1,300. Approximate maximum PV values for a steel shaft on an acetal homopolymer bearing are:

	Max. PV
Completely unlubricated	1,600–2,000
Lubricated at installation	3,000–5,000
Repeated lubrication	5,000–10,000
Continuous lubrication	10,000–15,000

These figures are somewhat higher than those obtained with the nylons.

The lowest coefficients of friction and wear are obtained with acetal resin against steel. With other metals, in particular with aluminium, greater wear and higher friction occur. From the design point of view it is not generally desirable to use acetal to acetal in bearings because of the tendency to heat build-up, except with very light loads. Where the use of a non-metallic material is desirable it is found that better results are obtained using acetal and nylon in conjunction rather than either on its own.

Acetal resins find a number of applications in gears where they come into competition with the nylons. Acetal gears are superior to those made from nylon in fatigue resistance, dimensional stability and stiffness, whereas nylon gears in conditions of average humidity have greater resistance to impact fatigue and abrasion.

Amongst the many other applications for acetal resins should be mentioned links in conveyor belts, moulded sprockets and chains, blower wheels, cams, fan blades, check valves, pump impellers, blow-moulded aerosol containers and plumbing components such as valve stems and shower heads.

It may therefore be seen that acetal resins are primarily engineering materials being used to replace metals because of such desirable properties as low weight, corrosion resistance, resistance to fatigue, ease of fabrication and low coefficient of friction. Because of their comparatively high cost they cannot yet be considered as being general purpose thermoplastics alongside polyethylene, polypropylene, p.v.c. and polystyrene.

16.4 MISCELLANEOUS ALDEHYDE POLYMERS

A large number of polymers from aldehydes have been reported in the literature[14,15,16] but, apart from those polymers already described, are not of commercial importance.

With the exception of formaldehyde, the aldehydes may polymerise to give varying molecular configurations and depending on the stereoregulating influence of the catalysts, either amorphous rubbery or crystalline polymers may be obtained. It may however be mentioned that in such cases as isobutyraldehyde, *n*-heptaldehyde and chlorinated acetaldehydes the steric hindrance of side groups only allows polymerisation to proceed when the molecules are in certain configurations. In these cases a degree of stereoregularity may be imposed.

Care has to be taken in the polymerisation of aldehyde polymers in order to achieve reproducible results. It is also difficult to stabilise most of the products since thermodynamics frequently favour depolymerisation at temperatures a little above or at room temperature.

In the immediate future it is unlikely that any of these polymers will attain commercial significance. Bevington[15] has however suggested that certain polymers such as polychloral and dimethyl ketene–acetone copolymers may merit further attention.

16.5 POLYETHERS FROM GLYCOLS AND ALKYLENE OXIDES

If ethylene glycol is subjected to vigorous dehydrating conditions, simple molecules such as dioxan and acetaldehyde may be prepared (Fig. 16.8).

Under appropriate conditions it is possible to obtain linear polymers, the poly(ethylene oxides), from either ethylene glycol or oxide (Fig. 16.9).

Controlled polymerisation of ethylene oxide under alkaline conditions will produce a range of polymers marketed under the trade name Carbowax. These have molecular weights in the range 1,500–20,000 and are greases or waxes according to their degree of polymerisation. Lower molecular weight polymers have also been prepared, of which carbitol, $HO \cdot CH_2 \cdot CH_2 \cdot O \cdot CH_2 \cdot CH_2 \cdot OH$, may be considered as the limiting dimer.

In 1958 the Union Carbide Corporation introduced high molecular weight, highly crystalline ethylene oxide polymers under the trade name

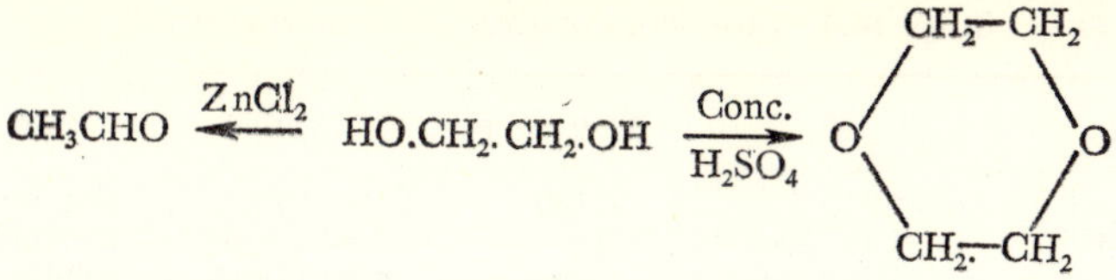

$$CH_3CHO \xleftarrow{ZnCl_2} HO.CH_2.CH_2.OH \xrightarrow[H_2SO_4]{Conc.}$$

Fig. 16.8

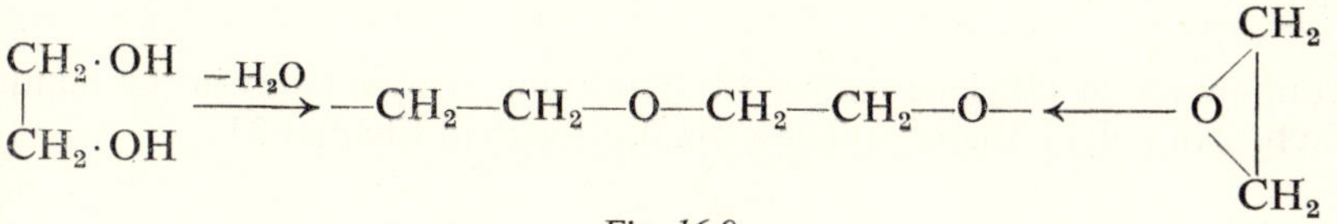

$$\begin{array}{l} CH_2 \cdot OH \\ | \qquad\qquad \xrightarrow{-H_2O} \quad -CH_2-CH_2-O-CH_2-CH_2-O- \longleftarrow \\ CH_2 \cdot OH \end{array}$$

Fig. 16.9

Polyox.[17] Although similar in appearance to polyethylene they are miscible with water in all proportions at room temperature.

Unlike the lower molecular weight poly(ethylene oxide)s these materials are tough and extensible, owing to their high molecular weight and their crystallinity. Typical mechanical properties of the polymer are given in Table 16.4.

The tensile strength will depend to a large extent on the rate of extension and the relative humidity. There is a severe drop in tensile values as the relative humidity exceeds 80%.

Unlike other water-soluble resins the poly(ethylene oxide)s may be injection moulded, extruded and calendered without difficulty. The

Table 16.4 SOME PROPERTIES OF A MEDIUM-HIGH MOLECULAR WEIGHT ETHYLENE OXIDE POLYMER[17]

Tensile strength	1,800–2,400 lb in^{-2}
Yield strength	1,000–1,500 lb in^{-2}
Ultimate elongation	700–1,200 %
Shore A hardness	99
Melting point	66°C
Brittle temperature	−50°C

viscosity is highly dependent on shear rate and to a lesser extent on temperature. Processing temperatures in the range 90–130°C may be used for polymers with an intrinsic viscosity of about 2·5. (The intrinsic viscosity is used as a measure of molecular weight.)

The polymers are of interest as water soluble packaging films for a wide variety of domestic and industrial materials. (Additional advantages of the poly(ethylene oxide)s are that they remain dry to the feel at high humidities and may be heat sealed.) The materials are also of use in a number of solution applications such as textile sizes and thickening agents. As a water soluble film they are competitive with poly(vinyl alcohol) whereas in their solution applications they meet competition from many longer established natural and synthetic water-soluble polymers.

A number of other polyethers derived from polyfunctional hydroxy

Table 16.4   PROPERTIES OF SOME OXETANE POLYMERS[18]

R	R_1	Melt pt °C	Nature	Solubility
CH_2Cl	CH_2Cl	180	crystalline	gen. insoluble
CH_2Cl	CH_2OAc	—	amorphous	insoluble in H_2O
CH_2OAc	CH_2OAc	—	amorphous	soluble in H_2O
CH_2OH	CH_2OH	>280	crystalline	insoluble
CH_3	CH_3	47	crystalline	insoluble

compounds or alkylene oxides are important intermediates in the manufacture of polyurethanes. These are dealt with in Chapter 23.

16.6 OXETANE POLYMERS

In the early 1950's novel polyethers were prepared in the laboratories of the Hercules Powder Company and of Imperial Chemical Industries Limited from oxacyclobutane derivatives. One such polyether, that from 3,3-bis (chloromethyl)-1-oxacyclobutane was commercially marketed by the first named company in 1959 under the trade name of Penton.

The polymers are of the general form

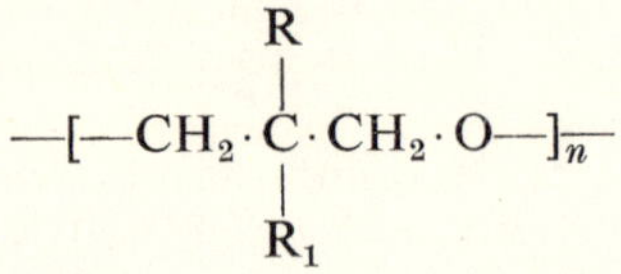

$$-[-CH_2 \cdot \underset{\underset{R_1}{|}}{\overset{\overset{R}{|}}{C}} \cdot CH_2 \cdot O-]_n-$$

and some general properties of specific polymers are given in Table 16.4.

The chloromethyl derivatives may be prepared from pentaerythritol via the trichloride or trichloride monoacetate (Fig. 16.10).

This monomer, 3,3-bis (chloromethyl)-1-oxacyclobutane, has the following characteristics[19]

Boiling point	83°C at 11 mmHg
Melting point	18·75°C
Density (25°C)	1·2951
Refractive index (20°C)	1·4858

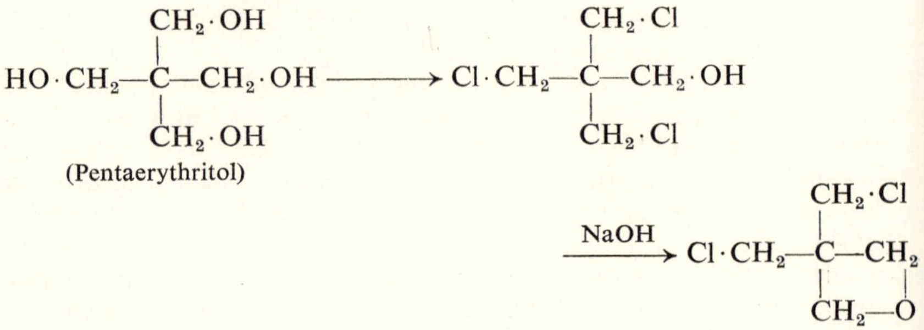

Fig. 16.10

The patent literature[20] indicates that polymerisation is carried out in the range −80 to +25°C using boron trifluoride or its ethereate as catalyst. Solvents mentioned include hexane, benzene, liquid sulphur dioxide, chloroform, methylene chloride and ethyl bromide. Where chlorinated solvents are employed the polymer is separated by addition of methanol,

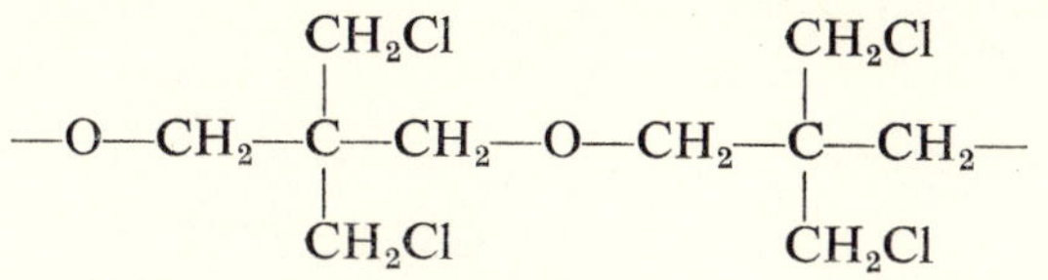

$$-O-CH_2-\overset{\displaystyle CH_2Cl}{\underset{\displaystyle CH_2Cl}{C}}-CH_2-O-CH_2-\overset{\displaystyle CH_2Cl}{\underset{\displaystyle CH_2Cl}{C}}-CH_2-$$

Fig. 16.11

filtered, washed with methanol and the product dried *in vacuo* at 60°C.

The commercial polymer is said to have a number average molecular weight of 250,000–350,000.[21] Because of its regular structure (Fig. 16.11) it is capable of crystallisation

Two crystalline forms have been observed.[22] One is formed by slow cooling from the melt and the other by slow heating of the amorphous polymer. The properties of the commercial products will therefore be

Table 16.5 PROPERTIES OF A COMMERCIAL OXETANE POLYMER

Property	ASTM test method	Value	Units
Tensile strength (73°C)	D.638	6,000	lb in^{-2}
Modulus of elasticity (73°F)	D.638	150,000	lb in^{-2}
E_B (73°F)	D.638	130	%
Flexural strength (73°C)	D.790	5,000	lb in^{-2}
Flexural modulus (73°F)	D.790	130,000	lb in^{-2}
Izod impact strength (73°F)	D.758	0·4	ft lb per in notch
Deflection temperature	D.648		
264 lb in^{-2}		99	°C
66 lb in^{-2}		140	°C
Thermal conductivity	D.696	3·13 × 10^{-4}	c.g.s. units
Volume resistivity	D.257	1·5 × 10^{16}	ohm cm
Power factor			
60–10^6 c/s (23°C)		∼0·011	
60 c/s (100°C)		0·045	
10^4–10^6 c/s (100°C)		∼0·003	
Dielectric constant (varying according to temp. and frequency)		2·8–3·5	
Dielectric strength		400	volts per 0·001 in.
Flammability	D.635		self extinguishing

to some extent dependent on their heat history. Glass transition temperatures observed range from 7 to 32°C and depend on the time scale of the method of measurement.[22]

Typical properties of a commercial chlorinated polyether ('Penton') are given in Table 16.5.[21]

Consideration of the figures given in Table 16.5 shows that the physical properties of the chlorinated polyethers are not particularly outstanding when compared with other plastics materials. On the other hand, apart

Table 16.6

	Dihalide	Trade names of polymer
(1)	Di-2-chloroethyl ether $Cl \cdot (CH_2)_2 \cdot O \cdot (CH_2)_2Cl$	Thiokol B Perduren G Novoplas
(2)	Di-2-chloroethyl formal $Cl \cdot (CH_2)_2O \cdot CH_2 \cdot O \cdot (CH_2)_2Cl$	Perduren H Thiokol ST
(3)	1,3-glycerol dichlorohydrin $Cl \cdot CH_2 \cdot CH \cdot CH_2 \cdot Cl$ $\qquad\qquad\qquad\qquad\quad \overset{\displaystyle |}{OH}$	Vulcaplas

from a somewhat low impact strength, these figures reveal no particular limitation.

The principal applications of these plastics arise from their very good chemical resistance being resistant to mineral acids, strong alkalis and most common solvents. They are however not recommended for use in conjunction with oxidising acids such as fuming nitric acid, fuming sulphuric acid or chlorosulphonic acid, with fluorine or with some chlorinated solvents particularly at elevated temperatures.

It is claimed that the maximum continuous operating temperature in most chemical environments is 120°C and even 140–150°C in some instances.[21]

The major chemical applications occur in the form of pipe and tank linings and injection moulded valve and pump parts. Coatings may be applied to metals by means of fluidised bed, water suspension and organic dispersion techniques.

16.7 POLYSULPHIDES

In 1928 J. C. Patrick attempted to produce ethylene glycol by reacting ethylene dichloride with sodium polysulphide. In fact a rubbery polymer was formed by the reaction.

$$n\,Cl \cdot CH_2 \cdot CH_2 \cdot Cl + n\,Na_2S_4 \rightarrow \sim\!\!\left(\!\sim CH_2\!-\!CH_2\!-\!\underset{\overset{\|}{S}}{S}\!-\!\underset{\overset{\|}{S}}{S}\!\sim\!\right)_{\!\!\overline{n}}\!\!\sim + 2n\,NaCl$$

This polymer became the first commercially successful synthetic rubber with the trade name Thiokol A.

Other commercial products were produced using different dihalides as indicated in Table 16.6.

These materials could all be vulcanised into rubbers with good oil resistance but with a high compression set.

Further variations in the properties of the polysulphides were also achieved by the following means:

1. The use of mixtures of dihalides. Thiokol FA which has less odour and a lower brittle point than Thiokol A is produced from a mixture of ethylene dichloride and di-2-chloroethyl formal.

2. By varying the value of the ranking x in the polysulphide Na_2S_x. The tetrasulphide is necessary to obtain a rubber with ethylene dichloride, but disulphides may be used with other dichlorides.
3. By incoporation of some trihalide to give a branched polymer such as Thiokol ST (about 2% of 1,2,3-trichloropropane is used in this instance). The resultant vulcanisates have lower cold flow and compression set than obtained with Thiokol A.
4. By reduction in the degree of polymerisation. To produce processable rubbers the original polymers are masticated with substances such as benzthiazoldisulphide and tetramethyl thiuram disulphide. The more severe degradation techniques to produce liquid polysulphides are mentioned below.

The general method of preparation of the polysulphides is to add the dihalide slowly to an aqueous solution of sodium polysulphide. Magnesium hydroxide is often employed to facilitate the reaction which takes 2–6 hours at 70°C.

Vulcanised rubbers may be prepared by heating the rubber with a mild oxidising agent such as zinc oxide. This probably causes chain linkage through terminal —SH groups.

$$\text{\textasciitilde\textasciitilde SH} + \text{HS\textasciitilde\textasciitilde} \xrightarrow[\text{ZnO}]{[O]} \text{\textasciitilde\textasciitilde S—S\textasciitilde\textasciitilde} + H_2O$$

In the case of linear polymers, e.g. Thiokol A, this reaction leads primarily to chain extension; with branched polymers actual cross linking occurs. Providing the molecular weight is sufficiently high and the molecules sufficiently flexible, cross linking is not essential for rubbery properties (this is observed with raw natural rubber), but such materials would have high cold flow, poor heat resistance and would swell extensively and often dissolve in solvents of a similar solubility parameter.

Other mild oxidising agents which abstract the terminal hydrogen atoms and thus facilitate disulphide formation may be used as vulcanising agents. They include benzoyl peroxide, *p*-nitroso benzene and *p*-quinone dioxime.

The applications of polysulphide rubbers are due to their excellent oil and water resistance and their impermeability to gases. Because of other factors including their unpleasant odour, particularly during processing, they are however much less used than the two major oil resistant synthetic rubbers, the polychloroprenes and the nitrile rubbers.

From the point of view of the plastics technologist the most useful products are the low molecular weight polysulphides[23,24] with molecular weights ranging from 300 to 4,000. These are produced by reductive cleavage of the disulphide linkage of high molecular weight polysulphides by means of a mixture of sodium hydrosulphide and sodium sulphite. The reaction is carried out in water dispersion and the relative amounts of the hydrosulphide and sulphite control the extent of cleavage.

$$\text{\textasciitilde\textasciitilde S—S\textasciitilde\textasciitilde} + \text{NaSH} + Na_2SO_3 \longrightarrow \text{\textasciitilde\textasciitilde SNa} + \text{HS\textasciitilde\textasciitilde} + Na_2S_2O_3$$

The sodium salt of the polysulphide is converted back to the free thiol on coagulation with acid.

These polymers are liquids which may usefully be cast or used for impregnation and caulking compounds. In addition they may be 'vulcanised' by a variety of agents, ostensibly by a chain lengthening process. It should however be noted that these polymers normally contain small quantities of trichloropropane in the original monomer mix so that the three-dimensional chain extension will lead to cross linking.

Similar reactions also occur with organic peroxides, dioximes, paint driers such as cobalt naphthenate and furfural. It is interesting to note that the cure time is dependent on the humidity of the atmosphere. With lead peroxide the rate doubles by increasing the relative humidity from 40 to 70%. The most important reaction is however that with epoxy resins (Chapter 22). These contain epoxide rings and chain linkage occurs by a 'polyaddition' mechanism.

$$\sim\!\!SH \quad + \quad \underset{\displaystyle CH_2-CH\sim}{\overset{\displaystyle O}{\diagup\diagdown}} \quad\longrightarrow\quad \sim\!\!S-CH_2-\overset{\displaystyle OH}{\underset{\displaystyle |}{CH}}\!\!\sim$$

Terminal Group   Terminal Group
of Polysulphide   of Epoxide

The normal amine hardeners used with epoxy resins are employed in conjunction with the polysulphides. The resulting products become more flexible the greater the percentage of polysulphide present and therefore a range of materials may be produced which vary from brittle resins to soft rubbers. Incorporation of relatively small amounts of polysulphide (20%) considerably increase the impact strength of the epoxy resins. The viscosity of the uncured epoxy resin is also reduced and so polysulphides can be used with advantage for laminating, adhesives and for castings. In the latter case the slightly inferior electrical properties resulting may be of some limitation.

16.8 POLYSULPHONES

In May 1965 Union Carbide announced a novel type of linear polymer, the polysulphone. It was stated that the material to be marketed consisted of phenylene units linked by isopropylidene, ether and sulphone groups.

The presence of ring structures in the main chain leads to the polymer having a deflection temperature under load (heat distortion temperature) as high as 174°C, whilst the polar groups present enhance the interchain attraction giving the polymer a high tensile strength and rigidity. Published data indicates that its properties are very similar to those of the polycarbonates (Chapter 17) and the phenoxies (Chapter 22), this being expected because of the similar structures.

The Union Carbide polysulphone is claimed to have excellent creep resistance, to be self-extinguishing and to be resistant to acids, alkalis and

Table 16.7 TYPICAL PROPERTIES OF POLYSULPHONES (AS MEASURED BY ASTM TESTS)

Specific gravity	1·24	Yield strength	10,000 lb in^{-2}
Deflection temperature			
under 264 lb in^{-2} load	174°C	Volume resistivity	10^{16} ohm cm
24 hr water absorption	0·22	Dielectric constant	3·14
		(60 c/s, 72°F)	

aliphatic hydrocarbons but soluble in aromatic and chlorinated hydro-carbons. Some typical properties of this transparent light amber coloured polymer are given in Table 16.7.

Anticipated applications of these materials include housings for engineering, electrical and domestic appliances where heat and/or creep resistance are important requirements.

REFERENCES
Acetal polymers

1. BUTLEROV, A., *Ann.*, **111**, 242 (1859)
2. STAUDINGER, H., *Die Hochmolekularen Organischen Verbindungen*, Julius Springer, Berlin (1932)
3. *U.S. Patent*, 2,768,994 (October 1956)
4. *British Patent*, 770,717 (July 1957)
5. *British Patent*, 742,135 (1956)
6. SCHWEITZER, C. E., MACDONALD, R. N., and PUNDERSON, J. O., *J. Appl. Polymer Sci.*, **1** (2), 158–163 (1959)
7. *U.S. Patent*, 3,027,352
8. ALSUP, R. G., PUNDERSON, J. O., and LEVERETT, G. F., *J. Appl. Polymer Sci.*, **1** (2), 185–191 (1959)
9. SITTIG, M., *Polyacetal Resins*, Gulf Publishing, Houston (1963)
10. KOCH, T. A., and LINDVIG, P. E., *J. Appl. Polymer Sci.*, **1** (2), 164–168 (1959)
11. HAMMER, C. F., KOCH, T. A., and WHITNEY, J. F., *J. Appl. Polymer Sci.*, **1** (2), 169–178 (1959)
12. LINTON, W. H., and GOODMAN, H. H., *J. Appl. Polymer Sci.*, **1** (2), 179–184 (1959)
13. LINTON, W. H., *Trans. Plastics Inst.*, **28** (75), 131 (June 1960)
14. FURUKAWA, J., and SAEGUSA, T., *Polymerisation of Aldehydes and Oxides*, Inter-science, New York (1963)

Miscellaneous aldehyde polymers and polyethers

15. BEVINGTON, J. C., *Brit. Plastics*, **35**, 75 (1962)
16. BEVINGTON, J. C., *Chem. Ind.*, 2025–29 (1961)
17. DAVIDSON, R. L., and SITTIG, M., *Water Soluble Resins*, Reinhold, New York (1962)

Oxetane polymers

18. FARTHING, A. C., *J. Appl. Chem.*, **8**, 186–8 (1958)
19. FARTHING, A. C., and REYNOLDS, R. J. W., *J. Polymer Sci.*, **12**, 503 (1954)
20. *British Patent* 723,777 (I.C.I.), *British Patent* 769,116 (Hercules): *U.S. Patent* 2,722,340; *U.S. Patent* 2,722,492; *U.S. Patent* 2,722,487; *U.S. Patent* 2,722,493; *U.S. Patent* 2,722,520 (Hercules); *British Patent* 758,450; *British Patent* 764,053; *British Patent* 764,284 (Hercules)
21. FLETCHER, F. T., *Trans. Plastics Inst.*, **30**, 127 (1962)
22. SANDIFORD, D. J. H., *J. Appl. Chem.*, **8**, 188–196 (1958)

Polysulphides

23. FETTES, E. M., and JORCZAK, J. S., *Ind. Eng. Chem.*, **42**, 2217 (1950)
24. JORCZAK, J. S., and FETTES, E. M., *Ind. Eng. Chem.*, **43**, 324 (1951)

BIBLIOGRAPHY

AKIN, R. B., *Acetal Resins*, Reinhold, New York (1962)
DAVIDSON, R. L., and SITTIG, M., *Water Soluble Resins*, Reinhold, New York (1962)
FURUKAWA, J., and SAEGUSA, T., *Polymerisation of Aldehydes and Oxides*, Interscience, New York (1963)
SITTIG, M., *Polyacetal Resins*, Gulf Publishing, Houston (1963)

17

Polycarbonates

17.1 INTRODUCTION

Reaction of polyhydroxy compounds with polybasic acids gives rise to condensation polymers containing ester (—COO—) groups. Because of the presence of these groups such polycondensates are known as polyesters and find use in such diverse applications as fibres, surface coatings, plasticisers, rubbers and laminating resins. These materials are discussed in detail in Chapter 21.

By reaction of polyhydroxy compounds with a carbonic acid derivative, a series of related polymers may be produced with carbonate (—O·CO·O—) linkages, the polymers being referred to as polycarbonates. Carbonic acid, $CO(OH)_2$, itself does not exist in the free state but by means of ester exchange (Fig. 17.1 (I)) and phosgenation techniques (II) it is possible to produce useful products.

Polycarbonates were first prepared by Einhorn[1] in 1898 by reacting the dihydroxy benzenes, hydroquinone and resorcinol separately with phosgene in solution in pyridine. The hydroquinone polycarbonate was an infusible and insoluble crystalline powder whereas the resorcinol polymer was an amorphous material melting at about 200°C. The third dihydroxy benzene, catechol yields a cyclic carbonate only, which is not surprising bearing in mind the proximity of the two hydroxy groups with each other. By the use of diphenyl carbonate, Bischoff and von Hedenström[2] prepared similar products by an ester exchange reaction in 1902.

In 1930 W. H. Carothers and F. J. Natta[3] prepared a number of aliphatic polycarbonates using ester interchange reactions. These materials had a low melting point, were easily hydrolysed and did not achieve commercial significance.

Carothers also produced a number of aliphatic linear polyesters but these did not fulfil his requirements for a fibre-forming polymer which were eventually met by the polyamide, nylon 66. As a consequence the polyesters were discarded by Carothers. However, in 1941 Whinfield and Dickson working at the Calico Printers Association in England announced

$$n \; \text{HOROH} + n \; R_1 O \cdot CO \cdot O \cdot R_1 \longrightarrow \text{---}(\text{---}R \cdot O \cdot \overset{\text{O}}{\underset{\|}{C}} \cdot O\text{---})_{\overline{n}} + 2n \; R_1 OH$$

(I)

$$n \; \text{HOROH} + n \; \text{COCl}_2 \longrightarrow \text{---}(\text{---}R \cdot O \cdot \overset{\text{O}}{\underset{\|}{C}} \cdot O\text{---})_{\overline{n}} + 2n \; \text{HCl}$$

(II)

Fig. 17.1

the discovery of a fibre from poly(ethylene terephthalate). Prompted by the success of such a polymer, Farbenfabriken Bayer initiated a programme in search of other useful polymers containing aromatic rings in the main chain. Carbonic acid derivatives were reacted with many dihydroxy compounds and one of these, bis-phenol A, produced a polymer of immediate promise.

Independently at the General Electric Company in America, work was being carried out in search of thermally and hydrolytically stable thermo-setting resins. As a by-product from this work the research team at General Electric also produced polycarbonates from bis-phenol A so that by 1958 production of bis-phenol A polycarbonates was being carried out in both Germany and the U.S.A.

Today, commercial polycarbonate resins are being marketed in Germany by Bayer (Makrolon), in the United States by General Electric (Lexan) and Mobay (Merlon) and in Japan by the Taijin Chemical Co. (Panlite). Additionally a number of patents have been taken out by various com-panies including Distillers, Gevaert and Eastman Kodak.

17.2 PRODUCTION OF INTERMEDIATES

As already indicated, the polycarbonates may be produced from a wide range of polyfunctional hydroxy compounds. In practice only the diphenyl compounds have proved of interest and the only polycarbonate of commercial significance is derived from bis-phenol A (4,4-dihydroxy-2,2-diphenyl-propane).

Bis-phenol A may be produced by the condensation of phenol with acetone under acidic conditions (Fig. 17.2).

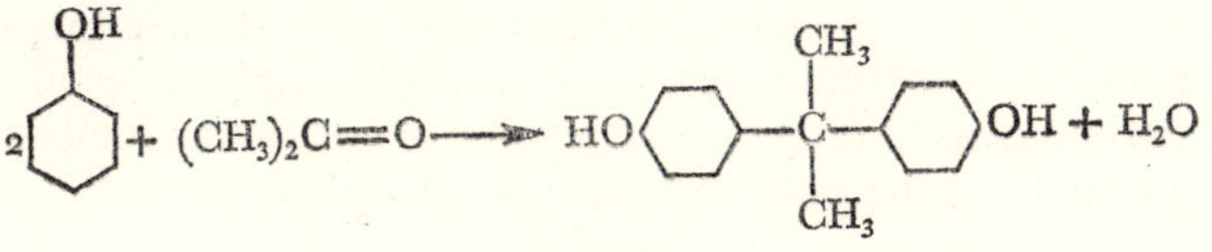

Fig. 17.2

Schnell[4] states that the initial product is isopropenylphenyl which then reacts with a further molecule to form the bis-phenol A.

At elevated temperatures the second stage of the reaction takes place in the reverse direction and so reactions are carried out below 70°C. In order to achieve a high yield, an excess of phenol is employed and the initial reaction product is a bis-phenol A–phenol adduct. The bis-phenol A may be separated from the adduct by crystallisation from appropriate solvents or by distilling off the phenol.

Improvements in the rate of the condensation reaction have been claimed with the use of co-catalysts such as an ionisable sulphur compound[5] and by pre-irradiation with actinic light.[6]

Unless great care is taken in control of phenol–acetone ratios, reaction conditions and in the use of catalysts, a number of undesirable by-products may be obtained such as the *o,p-* and *o-,o-* isomers of bis-phenol A and certain chroman type structures. Although tolerable when the bis-phenol A is used in epoxy resins, these have adverse effects on both physical properties and the colour of polycarbonate resins.

Residual traces of these impurities must thus be removed by some technique such as recrystallisation from chlorbenzene or aqueous alcohol. The melting point is a useful measure of purity and for polycarbonate resins the melting point should be in the range 154–157°C compared with values of 140–150°C for epoxy resin grade bis-phenol A.

Phosgene, employed in both of the main processes, is prepared commercially from carbon monoxide and chlorine.

Diphenyl carbonate, an alternative source of the carbonate group to phosgene may be obtained by reacting phenol with phosgene in aqueous caustic soda solution, the reaction being accelerated by tertiary amines. The diphenyl carbonate can be purified by redistillation.

17.3 POLYMER PREPARATION

There are four possible practical routes to linear polycarbonates:

1. Ester exchange of dihydroxy compounds with diesters of carbonic acid and monofunctional aromatic or aliphatic hydroxy compounds.
2. Ester exchange of bis-alkyl or bis-aryl carbonates of dihydroxy compounds with themselves or with other dihydroxy compound.
3. Reaction of dihydroxy compounds with phosgene in the presence of acid acceptors.
4. Reaction of the bis-chlorcarbonic acid ester of dihydroxy compounds with dihydroxy compounds in the presence of acid acceptors.

Of these four routes the first and third have been studied intensively, in particular in the preparation of the bis-phenol A polycarbonates. Since this polycarbonate is at present the only one of commercial importance the following remarks will apply to this only unless otherwise stated.

Ester exchange

The equation for the ester exchange reaction (1) is shown in Fig. 17.3.

In this method the reaction is carried out at 180–220°C at 20–30 mmHg pressure until 80–90% of the phenol of condensation has been removed.

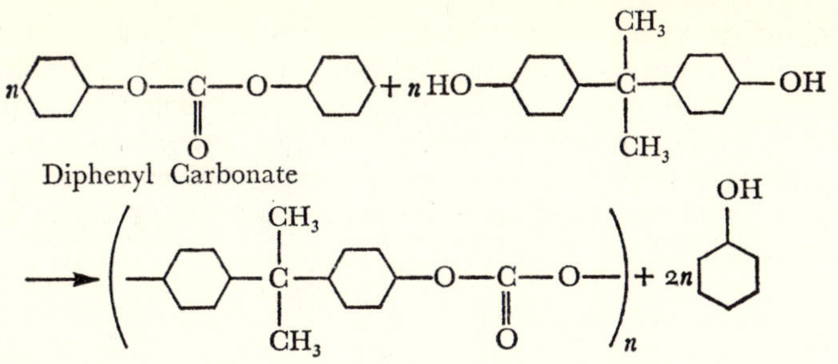

Fig. 17.3

The temperature is then gradually raised to 290–300°C and the pressure reduced to 1 mmHg or below. The melt viscosity increases considerably during this period and the reaction is stopped while the material can still be forced out of the kettle by an inert gas.

The high melt viscosity limits the molecular weights obtainable and although number-average molecular weights of 50,000 can be obtained it is difficult to attain values of above 30,000 without special equipment.

Because bis-phenol A is somewhat unstable at elevated temperature it is desirable to work with an excess of diphenyl carbonate so that the bis-phenol A is rapidly used up. The reaction may be conveniently carried out using twice or more than twice the theoretical quantity of diphenyl carbonate so that the initial reaction product is the bis-(phenyl carbonate) of bis-phenol A (Fig. 17.4 (*a*)).

Polymerisation then proceeds by splitting out of diphenyl carbonate to give the polycarbonate resin (Fig. 17.4 (*b*))

This variation has the obvious disadvantage that the less volatile diphenyl carbonate is more difficult to remove than phenol.

A number of basic materials such as hydroxides, hydrides and amides of alkaline and alkaline earth metals and metal oxides such as zinc oxide and

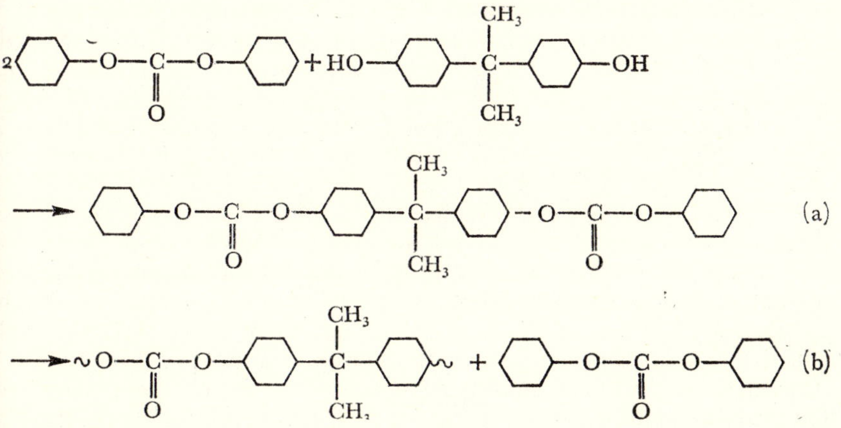

Fig. 17.4

antimony oxide are useful catalysts for the reaction.[7]   Acid ester exchange catalysts such as boric acid, *p*-toluene sulphonic acid and zinc chloride are less effective.   Catalyst systems are not essential when diaryl carbonates are used as carbonate sources and do in fact cause problems in their subsequent removal.   High molecular weight polycarbonates may be produced without undue difficulty by the phosgenation process.   The basic reaction is as shown in Fig. 17.5.

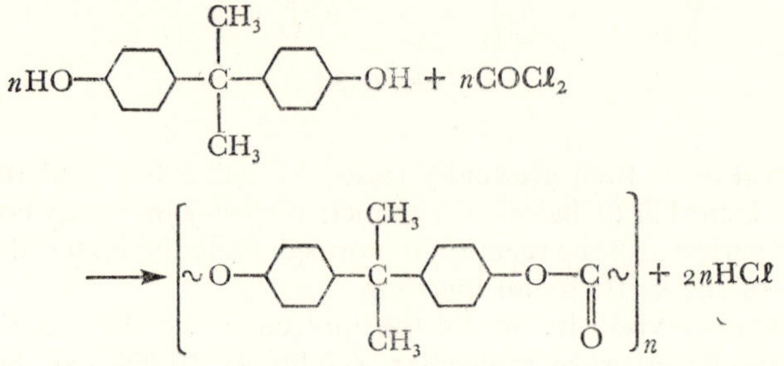

$$n\text{HO}-\!\!\!\bigcirc\!\!\!-\overset{\displaystyle \text{CH}_3}{\underset{\displaystyle \text{CH}_3}{\text{C}}}-\!\!\!\bigcirc\!\!\!-\text{OH} + n\text{COC}\ell_2$$

$$\longrightarrow \left[\sim\!\text{O}-\!\!\!\bigcirc\!\!\!-\overset{\displaystyle \text{CH}_3}{\underset{\displaystyle \text{CH}_3}{\text{C}}}-\!\!\!\bigcirc\!\!\!-\text{O}-\underset{\displaystyle \text{O}}{\text{C}}\sim \right]_n + 2n\text{HC}\ell$$

Fig. 17.5

For this reaction to proceed it is obviously necessary to remove the hydrochloric acid formed, preferably by means of a hydrohalide acceptor. The attractive possibility of dissolving the bis-phenol A in caustic soda solution and bubbling phosgene into it is not practical since the polymer is insoluble in the caustic soda and precipitates out at a low and variable molecular weight.

Greater success has been achieved with organic solvents which are also hydrohalide acceptors, pyridine being a specific example.

Typically in such a process the bis-phenol A is dissolved in about ten times its weight of pyridine and vigorously stirred at 25–35°C.   Phosgene is then bubbled into the solution and in a few minutes the pyridine hydro-chloride starts to precipitate.   As polymer is formed the viscosity of the solution increases and eventually becomes too great for stirring.   The polymer is then recovered by the addition of a solvent such as methyl alcohol which dissolves the pyridine hydrochloride but precipitates the polymer.

A variation of this process involves the formation of a preformed pyridine–phosgene complex.   Polymerisation will then be effected by adding a solution of bis-phenol A.

Because of the cost of pyridine the phosgenation process may be carried out with a mixture of pyridine and a non-hydrohalide accepting solvent for the polymer and the growing complexes.   Suitable solvents include methylene chloride, tetrachlorethane and chloroform.   Although unsubstituted aromatic hydrocarbons may dissolve the solvent they are not effective solvents for the acid chloride–pyridine complexes.

As an alternative to solution polymerisation one Bayer patent[8] describes the use of an interfacial type of reaction.   The monomer (bis-phenol A) is dissolved in aqueous caustic soda and then stirred with a solvent for the polymer, the solvent and the caustic soda remaining in two phases. Phosgene is then introduced in the system and reaction occurs at the interface or just inside the aqueous phase, the ionic ends of the growing molecule being soluble in the catalytic caustic soda solution. At the same time the rest of the molecule dissolves into the organic solvent. A catalytic quantity of a quaternary compound is necessary for the reaction to proceed. A high molecular weight polymer is produced by this reaction which may be recovered by washing the organic phase with water, neutralisation of the caustic soda and either precipitation of the polymer by a non-solvent or evaporation of the solvent after thorough washing.

Unless mono-functional reactants are added during the phosgenation reaction very high molecular weight polymers will be obtained of little commercial value. Phenol is a convenient monofunctional material for this purpose. Alternatively the use of a deficiency of one of the ingredients, e.g. the phosgene, is useful in controlling the molecular weight.

Both the ester exchange and the phosgenation reaction possess specific advantages and disadvantages. The ester exchange method, which is used commercially, avoids the use of solvent and hence costs of solvent and solvent recovery are avoided. The hazards of fire and toxicity associated with many organic materials are also non-existent. Finally, the esterification process yields a polymer which does not have to be densified before use in extruders and injection moulding machines. The disadvantages of the process are the long reaction times and the problems faced in attaining satisfactory equipment design and operation. The molten resin has a very high viscosity and this leads to severe problems of mixing and heat transfer. The reaction temperatures and the vacuum required are high and so heating systems must be adequate and the equipment leak-proof. In spite of such precautions it is difficult to obtain polymers with molecular weights above 50,000.

In contrast the phosgenation process gives high molecular weight polymers under moderate preparation conditions using fairly simple equipment. Balanced against this are the hazards associated with phosgene and the solvents, the need for solvent recovery and the need for purifying and densifying the polymer.

17.4 RELATION OF STRUCTURE AND PROPERTIES

A study of the molecular structure of bis-phenol A polycarbonates enables one to make fairly accurate predictions of the bulk properties of the polymer. The relevant factors to be considered are:

1. The molecule has a symmetrical structure and therefore questions of stereospecificity do not arise.
2. The carbonate groups are polar but separated by aromatic hydrocarbon groups.
3. The presence of benzene rings in the chain restricts flexibility of the molecule.
4. The repeating unit of the molecule is quite long.

Table 17.1   CRYSTAL STRUCTURE DATA OF BIS-PHENOL A
POLYCARBONATES
(after Prietschk)[9]

Crystal type	rhombic
Cell constants	$a = 11 \cdot 9$ $b = 10 \cdot 1$ c $21 \cdot 5$
No. of units in elementary cell	8
Crystal density	$1 \cdot 3$
Macroscopic density	$1 \cdot 2$

Because of its regularity it would be expected that the polymer would
be capable of crystallisation. In practice however the X-ray pattern
characteristic of crystalline polymer is absent in conventionally fabricated
samples. On the other hand films which have been prepared by slow
evaporation from solvent or by heating for several days at 180°C do exhibit both haziness and the characteristic X-ray diagram. The amount of
crystallisation and the size of the crystallite structures decrease with an

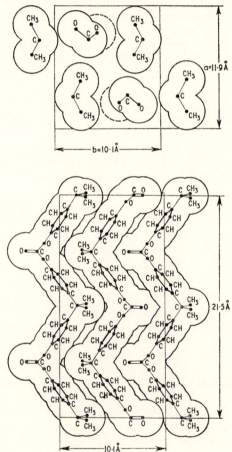

Fig. 17.6.   Unit cell diagrams for polycarbonate based on bis-phenol A. (After Prietschk[9])

increase in the molecular weight of the polymer. These effects are no doubt associated with both the stiffness of the molecule and its long repeat unit.

The crystalline structure of bis-phenol A polymers has been thoroughly studied by Prietschk[9] and some of the data he obtained on the crystal structure are summarised in Table 17.1.

Fig. 17.6 shows the disposition of the molecules in the elementary cell as determined by Prietschk. It will be seen that in the crystalline zone the

Table 17.2

Property	Units	ASTM standard	Value
Specific gravity	—	D.792	1·20
Tensile yield strength			
−25°F	10^3 lb in^{-2}	D.638	12–13
73°F	10^3 lb in^{-2}	D.638	8–9
212°F	10^3 lb in^{-2}	D.638	4·5–5·5
Tensile modulus	10^3 lb in^{-2}	D.638	320
Elongation	%	D.638	60–100
Poissons ratio	—	—	0·38
Flexural strength	10^3 lb in^{-2}	D.790	11–13
Flexural modulus	10^3 lb in^{-2}	D.790	375
Compressive strength	10^3 lb in^{-2}	D.695	11
Compressive modulus	10^3 lb in^{-2}	D.695	240
Izod impact strength			
$\frac{1}{8}$ in × $\frac{1}{2}$ in bar, notched, 77°F	ft lb per in. notch	D.256	12–16

molecules pack in such a way that the methyl groups attached to the pivotal carbon atom extend toward the back of the carbonate group of the neighbouring chain.

The limited degree of crystallinity in these resins appears to be the structural feature that makes the material so tough. Carefully crystallised samples of high crystallinity are generally quite brittle. It would seem that the interchain attraction, particularly in the crystalline zones, gives a high strength while some molecular flexibility in the amorphous zones give a degree of ductility which assists in the absorption of shock.

The rigid nature of the molecules leads to the high heat distortion and glass transition temperatures. That this rigidity, and not the polar interchain attraction of the ester group or of the benzene rings, is the predominating influence can be indicated by comparison of the poly-carbonate resin with poly(ethylene terephthalate). This latter polymer has similar chemical groups but has a much lower glass transition temperature

Both the chemical solubility and the electrical properties are consistent with that of a lightly polar polymer, whilst chemical reactivity is consistent with that of a polymer containing hydrolysable carbonate ester linkages partially protected by aromatic hydrocarbon groupings.

The influence of these factors on specific properties are amplified in the following sections.

Mechanical properties are affected by molecular weight but above number average (osmotic) molecular weights of 30,000 the influence is quite small. Commercial polymers have molecular weights of about 30,000 with extrusion grades having slightly higher molecular weights than injection moulding grades. Polymers with molecular weights substantially above this value have unduly high melt viscosities and there is little advantage in using them for normal purposes.

17.5 GENERAL PROPERTIES

Although somewhat more expensive than the general purpose thermoplastics, polycarbonates have established themselves in a number of applications. The desirable features of the polymer may be listed as follows:

1. Rigidity up to 140°C.
2. Toughness up to 140°C.
3. Transparency.
4. Very good electrical insulation characteristics.
5. Virtually self-extinguishing.
6. Physiological inertness.

The principal disadvantages may be listed as

1. Comparatively high cost.
2. Special care required in processing.
3. Pale yellow colour.
4. Limited resistance to chemicals and ultraviolet light.
5. Notch sensitivity and susceptibility to crazing under strain.

Such a tabulation of advantages and limitations is an oversimplification and may in itself be misleading. It is therefore necessary to study some of these properties in somewhat more detail.

Typical mechanical properties for bis-phenol A polycarbonates are listed in Table 17.2.

Of these properties the most interesting is the figure given for impact strength. Such high impact strength figures are in part due to the ductility of the resin.

Great care must be taken, as always, in the interpretation of impact test results. It is important to be informed on the influence of temperature, speed of testing and shape factor on the tough-brittle transitions and not to rely on results of a single test. A number of examples of the misleading tendency of quoting single results may be given. In the first instance while $\frac{1}{2}$ in. $\times$ $\frac{1}{8}$ in. bars consistently give Izod values of about 16, the values for $\frac{1}{2}$ in. $\times$ $\frac{1}{2}$ in. bars are of the order of 2·5 ft lb per inch notch. There appears to be a critical thickness for a given polycarbonate below which high values ($\sim$ 16 ft lb per in. notch) are obtained but above which much lower figures are to be noted. The impact strengths of bis-phenol A polycarbonates are also temperature sensitive. A sharp discontinuity occurs at about -10°C to -15°C for above this temperature $\frac{1}{2}$ in. $\times$ $\frac{1}{8}$ in. bars give numerical values of about 16 whilst below it values of 2 to $2\frac{1}{2}$ are to be obtained. Heat aging will cause similar drops in strength.

It should however be realised that this lower value (2·5 ft lb per in. notch) is still high compared with many other plastics.   Such values should not be considered as consistent with brittle behaviour.   Comparatively brittle mouldings can however be obtained if specimens are badly moulded.

An illustration of the toughness of the resin is given by the fact that when 5 kg weights were dropped a height of 3 metres on to polycarbonate bowls,

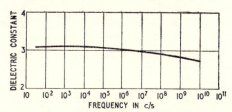

Fig. 17.7.   *Effect of frequency on dielectric constant of bis-phenol A polycarbonate*

the bowls although dented did not fracture.   It is also claimed that an $\frac{1}{8}$ in. thick moulded disc will stop a 0·22 calibre bullet, causing denting but not cracking.

The resistance of polycarbonate resins to 'creep' or deformation under load is markedly superior to acetal and polyamide thermoplastics.   A sample loaded at a rate of one ton per square inch for a thousand hours at 100°C only deformed 0·013 in/in.   Because of the good impact strength and creep resistance it was felt at one time that the polycarbonates would become important engineering materials.   Such hopes have been frustrated by the observations that where resins are subjected to tensile strains of 0·75% or more cracking or crazing of the specimen will occur. This figure applies to static loading in air.   When there are frozen-in stresses due to moulding, or at elevated temperatures, or in many chemical environments and under dynamic conditions crazing may occur at much lower strain levels.   Aging of the specimen may also lead to similar effects.   As a result moulded and extruded parts should only be subjected to very light loadings, a typical maximum value for static loading in air being 2,000 lb in^{-2}.

The electrical insulation characteristics of bis-phenol A polycarbonates are in line with those to be expected of a lightly polar polymer (see Chapter 6).

Because of a small dipole polarisation effect the dielectric constant is somewhat higher than that for p.t.f.e. and the polyolefins but lower than those of polar polymers such as the phenolic resins.   The dielectric constant is almost unaffected by temperature over the normal range of operations and little affected by frequency changes up to 10^6 c/s.   Above this frequency however the dielectric constant starts to fall as is common with polar materials (see Fig. 17.7).

In common with other dielectrics the power factor is dependant on the presence of polar groups.   At low frequencies and in the normal

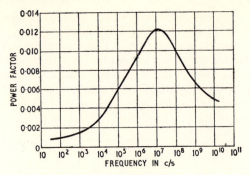

Fig. 17.8. Effect of frequency on the power factor of bis-phenol A polycarbonate

working temperature range (20–100°C) the power factor is almost surprisingly low for a polar polymer ($\sim$ 0·0009). As the frequency increases the power loss increases and the power factor reaches a maximum value of 0·012 at 10^7 c/s (see Fig. 17.8). The polycarbonates have a high volume resistivity and because of the low water absorption these values obtained are little affected by humidity.

A summary of the electrical properties of bis-phenol A polycarbonates is given in Table 17.3.

Although the general electrical properties of the polycarbonates are less impressive than those observed with polyethylene they are more than adequate for many purposes. These properties coupled with the heat and flame resistance, transparency and toughness have led to the extensive use of these resins in electrical applications.

Polycarbonate carefully prepared from bis-phenol A of very high purity is water white in colour. Commercial mouldings are however quite yellow because of the lower purity of commercial bis-phenol A and because

Table 17.3

Property	Units	ASTM test	Value
Power factor at 73°F (23°C)			
60 c/s	—	D.150	0·0009
1 kc/s	—		0·0011
10 kc/s	—		0·0021
100 kc/s	—		0·0049
1 Mc/s	—		0·010
Dielectric constant			
60 c/s	—	D.150	3·17
1 kc/s	—		3·02
10 kc/s	—		3·00
100 kc/s	—		2·99
1 Mc/s	—		2·96
Volume resistivity (23°C)	ohm cm		$2·1 \times 10^{16}$
Dielectric strength short time, $\frac{1}{8}$ in. sample	volts per 0·001 in.	D.149	400

of the darkening which occurs owing to chemical changes during processing and service. The polymer has a refractive index of 1·586 at 25°C. As may be expected of a polar polymer the dielectric constant (3·17 at 60 c/s) is greater than the square of the refractive index (2·51) but does tend towards this value at very high frequencies (see Chapter 6).

Typical figures for the basic thermal properties of polycarbonates are summarised in Table 17.4.

Peilstöcker[10] has studied in some detail the dependence of the properties of bis-phenol A polycarbonate on temperature. He found that if the

Table 17.4

Property	Standard	Units	Value
Deflection temp. under load			
method a	ASTM D.648	°C	135–140
method b		°C	140–146
Martens heat distortion point	DIN 53458	°C	115–127
Vicat heat distortion point	VDE 0302	°C	164–166
Specific heat	—		0·28
Thermal conductivity	—	c.g.s. units	$4·6 \times 10^{-4}$
Coeff. of thermal expansion (linear)	ASTM D.696	in in^{-1} degC^{-1}	7×10^{-5}
Glass transition temperature	—	°C	~ 145°C
Crystal melting point (by optical methods)	—	°C	220–230°C

resin is heated to just below the glass transition temperature some stiffening of the sample takes place owing to some ordering of the molecules. The degree of molecular ordering did not however affect the form of the X-ray diagram. The annealing effect takes place quite rapidly and is complete within 80 minutes at 135°C. This effect may be partially reversed by heating at about the transition temperature, viz 140–160°C, and completely reversed by raising the temperature of the sample to its optical melting point. The rubbery range extends from the glass transition temperature to the optical melting point. Samples maintained at this temperature will slowly crystallise. The maximum rate of crystallisation occurs at about 190°C, spherulitic structures being formed at this temperature within 8 days.

The chemical resistance of polyester materials is well recognised to be limited because of the comparative ease of hydrolysis of the ester groups. Whereas this ease of hydrolysis was also observed in aliphatic poly-carbonates produced by Carothers and Natta in 1930 the bis-phenol A polycarbonates are somewhat more resistant. This may be ascribed to the protective influence of the hydrophobic benzene rings on each side of the carbonate group. The resin thus shows a degree of resistance to dilute (25%) mineral acids and dilute alkaline solutions other than caustic soda and caustic potash. Where the resin comes into contact with organophilic hydrolysing agents such as ammonia and the amines the benzene rings give little protection and reaction is quite rapid.

The absence of both secondary and tertiary C—H bonds lead to a high measure of oxidative stability. Oxidation does take place when thin films are heated in air to temperatures above 300°C and causes cross linking but this is of little practical significance. The absence of double bonds gives a very good but not absolute resistance to ozone.

Although moulded polycarbonate parts are substantially amorphous, crystallisation will develop in environments which enable the molecules to move into an ordered pattern. Thus a liquid that is capable of dissolving amorphous polymer may provide a solution from which polymer may precipitate out in a crystalline form because of the favourable free energy conditions.

For solvation to take place it is first of all necessary for the solvent to have a solubility parameter within about 0·7 of the solubility parameter of the polycarbonate (9·5–9·7). A number of solvents (see Chapter 5) meet the requirement but some are nevertheless poor solvents. The reason for this is that although they may tend to dissolve the amorphous polymer they do not interact with the polycarbonate molecule which for thermodynamic reasons will prefer to crystallise out. If, however, some specific interaction between the resin and the solvent can be achieved then the two species will not separate and solution will be maintained. This can be effected by using a solvent which has a proton donating ability (e.g. symtetrachlorethane $\delta = 9\cdot4$ or methylene chloride, $\delta = 9\cdot7$) as a weak bond can be formed with the proton-accepting carbonate group thus preventing crystallisation. Other good solvents are cis-1,2-dichlorethylene, chloroform and 1,1,2-trichloroethane. Thiophene, dioxane and tetrahydrofuran are rated as fair solvents.

A number of materials exist which neither attack the polymer molecule chemically nor dissolve it but which cannot be used because they cause cracking of fabricated parts. It is likely that the reason for this is that such

Table 17.5

Environment	Water absorption %			Equilibrium swelling (in/in.)
	25 hr	*50 hr*	*150 hr*	
50% RH 23°C	0·05	0·1	0·15	0·0004
Water immersion 23°C	0·2	0·27	0·35	0·0008
Boiling water immersion	0·58	0·58	0·58	0·0013

media have sufficient solvent action to soften the surface of the part to such a degree that the frozen-in stresses tend to be released but with consequent cracking of the surface.

The very low water absorption of bis-phenol A polycarbonates contributes to a high order of dimensional stability. Table 17.5 shows how the water absorption of $\frac{1}{8}$ in. thick samples changes with time and environmental conditions and the consequent influence on dimensions.

The permeability characteristics of the bis-phenol A polycarbonates are shown in Table 17.6.

Stannett and Myers[11] have reported that crystallisation may reduce the nitrogen permeability by 50%. Schnell[7] has quoted the moisture vapour permeability of the polycarbonate from 4,4-dihydroxy diphenyl-1,1-

cyclohexane as being somewhat below half that of the bis-phenol A polymer (1–7, cf. 3·8 units).

When fabricated polycarbonate parts are exposed to ultraviolet light, either in laboratory equipment or by outdoor exposure, a progressive

Table 17.6 PERMEABILITY

Water vapour	$3·8 \times 10^{-8}$ g cm hr^{-1} cm^{-2} mmHg^{-1}
Nitrogen	$0·012 \times 10^{-8}$ cm^3 (S.T.P.) mm sec^{-1} cm^{-2}cmHg^{-1}
Carbon dioxide	$0·32 \times 10^{-8}$ cm^3 (S.T.P.) mm sec^{-1} cm^{-2} cmHg^{-1}

dulling is observed on the exposed surface. The dullness is due to microscopic cracks on the surface of the resin. If the surface resin is analysed it is observed that it has a significantly lower molecular weight than the parent polymer.

Such degradation of the surface causes little effect on either flexural strength or flexural modulus of elasticity but the influence on the impact properties is more profound. In such instances the minute cracks form centres for crack initiation and samples struck on the face of samples opposite to the exposed surface show brittle behaviour. For example a moulded disc which will withstand an impact of 12 ft lb without fracture before weathering will still withstand this impact if struck on the exposed side but may only resist impacts of 0·75 ft lb when struck on the unexposed face.

Because polycarbonates are good light absorbers, ultraviolet degradation does not occur beyond a depth of 0·030–0·050 in. Whilst this is often not serious with moulded and extruded parts, film may become extremely brittle. Improvements in the resistance of cast film may be made by addition of an ultraviolet absorber but common absorbers cannot be used in moulding compositions because they do not withstand the high processing temperatures.

Heat aging effects are somewhat complex. Heating at 125°C will cause reduction in elongation at break to 5–15% and in Izod impact strength from 16 down to 1–2 ft lb per in. notch and a slight increase in tensile strength in less than four days. Further aging has little effect on these properties but will cause progressive darkening. Heat aging in the presence of water will lead to more severe adverse effects.

17.6 PROCESSING CHARACTERISTICS

Satisfactory production of polycarbonate parts may only be achieved if consideration is given to certain characteristics of the polymer.

In the first place although the moisture pick-up of the resin is small it is sufficient to cause problems in processing. In the extruder or injection moulding machine it will volatilise into steam and frothy products will emerge from die and nozzle. It is therefore necessary to keep all materials scrupulously dry. Commercial materials are supplied in tins that have been vacuum sealed at elevated temperatures. These tins should only be opened after heating for several hours in an oven at 110°C and the granules should be used immediately. The use of heated hoppers is advocated.

The melt viscosity of the resin is very high and processing equipment should be rugged. The use of in-line screw plasticisers is to be particularly recommended. The effect of increasing temperature on viscosity is less marked with polycarbonates than with other polymers (see Fig. 17.9[12]). The apparent melt viscosity is also less dependent on the rate of shear than usual with thermoplastics (Fig. 17.10).

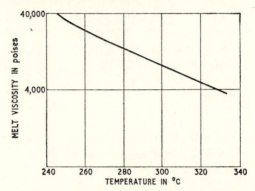

Fig. 17.9. *Influence of temperature on the melt viscosity of a typical bis-phenol A polycarbonate (shear stress = ~1 × 10[6] dyn/cm[-2]). (After Christopher and Fox[12])*

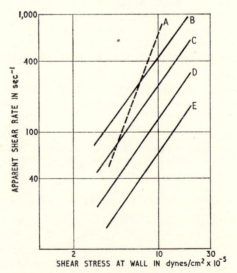

Fig. 17.10. *Shear stress–shear rate relationships for a polystyrene at 440°F (A) and polycarbonate resin at 650°F (B), 600°F (C), 550°F (D) and 500°F (E). (After Fiedler et al[13])*

Processing temperatures are high and fall between the melting point ($\sim 230°C$) and $300°C$, at which temperature degradation occurs quite rapidly.

Polycarbonate melts adhere strongly to metals and if allowed to cool in an injection cylinder or extrusion barrel may, on shrinkage, pull pieces of metal away from the wall. It is therefore necessary to purge all equipment free of the resin with a polymer such as polyethylene, after processing.

There is little crystallisation on cooling and after-crystallisation has not been observed. Mould shrinkage is consequently of the order of 0·006 in/in. and is the same both along and across the flow.

The rigidity of the molecule means that molecules may not have time to relax before the temperature drops below the glass transition point. Frozen-in strain may be gauged by noting how well the sample will withstand immersion in carbon tetrachloride. In general, moulding strain will be reduced by using high melt temperatures, preplasticising machines, high injection rates, hot moulds ($\sim 100°C$) and, where used, inserts should be hot. Annealing at $135°C$ for 24 hours will be of some value.

Provided due care is taken with respect to predrying and to crazing tendencies, polycarbonates may also be thermoformed, used for fluidised bed coating and be machined and cemented. Like metals, but unlike most thermoplastics, polycarbonates may be cold formed by punching and cold rolling. Cold rolling can in fact improve the impact resistance of the resin.

17.7 APPLICATIONS OF BIS-PHENOL A POLYCARBONATES

As mentioned earlier the most important advantages of polycarbonate resins are rigidity, toughness, transparency, self-extinguishing characteristics, good electric insulation characteristics and heat resistance. The principal limitations are cost, colour, difficulties in processing, limited chemical and light resistance and notch sensitivity.

If the advantageous properties are considered separately it will be realised that there are other plastics materials which are tough, that are rigid, or transparent, or self-extinguishing and so on. The applications of polycarbonates are thus restricted to applications where more than one of the desirable properties are required in combination and where there is no cheaper alternative.

Several examples may be found in the electrical and electronics industries which consume about half of the total polycarbonate produced. Covers for relays and other electronic equipment need to have good insulation characteristics coupled with transparency, flame-resistance and durability, a combination unique to polycarbonates. Several hundred types of coil formers have been produced from the resin. In this case the ability to wind the wire tightly without deformation of the former, the heat stability and oxidation resistance coupled with the good electrical characteristics have proved invaluable. Polycarbonate mouldings have also been used for terminals, housings, battery parts, contact strips, fuse covers and a host of other miscellaneous electrical applications. Polycarbonate films find use in the manufacture of capacitors. The use of bis-phenol A polycarbonate film for cable wrapping has been somewhat restricted because the tensile strains set up in winding the film may be sufficient to cause crazing and cracking. Somewhat related to electrical applications is the use in Germany of television implosion guards moulded from polycarbonate instead of the acrylic types favoured elsewhere.

The toughness and transparency of polycarbonates has also led to a number of other industrial applications. In Great Britain one of the first established uses was for compressed air lubricator bowls. In the first five years of commercial production it is estimated that over 100,000 breeding cages for rats were produced. Transparent milking pail lids have also been moulded.

Polycarbonates have also found applications in domestic mouldings. Cups, saucers and tumblers are adequately tough and are not stained by the usual domestic beverages and fruit juices. They are thus competitive with melamine–formaldehyde mouldings, the latter having superior resistance to scratching. Tough transparent babies' bottles may be blow moulded at very high rates because of the high setting-up temperatures. The polymer has also been used for food mixer housings and for transparent saucepan lids.

Polycarbonate film was first developed for photographic applications by the Agfa division of Farbenfabriken Bayer in 1956 and is today also in commercial production in the United States. The film has a high dimensional stability, may be readily cemented and because of its high strength may be made available in thinner films than is normal practice. This material is especially suitable for quality colour separations and fine engravings.

17.8 MISCELLANEOUS CARBONIC ESTER POLYMERS

Unless the hydroxyl groups have such proximity that cyclisation takes place, polycarbonates will normally be produced whenever phosgene or a carbonate ester is reacted with a polyhydroxy compound. This means that a very large range of polycarbonate resins are possible and in fact many hundreds have been prepared.

Aliphatic polycarbonates have few characteristics which make them potentially valuable materials but study of various aromatic polycarbonates is instructive even if not of immediate commercial significance. Although bis-phenol A polycarbonates still show the best all-round properties other carbonic ester polymers have been prepared which are outstandingly good in one or two specific properties. For example some materials have better heat resistance, some have better resistance to hydrolysis, some have greater solvent resistance whilst others are less permeable to gases.

It is particularly interesting to consider the influence of the substituents R and R_1, in diphenylol alkanes of the type shown in Fig. 17.11. Such variations will influence properties because they will affect the flexibility of the molecule about the central C-atom, the spatial symmetry of the molecule and also the interchain attraction, the three principal factors determining the physical nature of a high polymer.

Thus where R and R_1 are hydrogen the molecule is symmetrical, the absence of bulky side groups leads to high intermolecular attraction and the flexibility of the molecule enables crystallisation to take place without

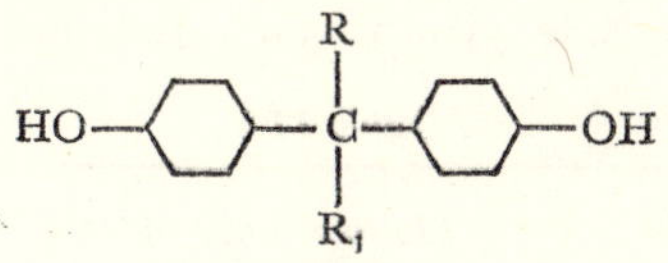

Fig. 17.11

difficulty. The resultant material is highly crystalline with a melting point of above 300°C and is insoluble in known solvents.

Where R is hydrogen and R_1 a methyl group the molecule is less symmetrical, less flexible and the intermolecular attraction would be slightly less. The melting point of this polymer is below 200°C. In the case where R and R_1 are both methyl groups the molecule is more symmetrical but the flexibility of the molecule about its central carbon atom is reduced. Because of these two factors this polymer, the commercial bis-phenol A polycarbonate, has glass transition temperatures and melting points slightly above that of the aforementioned material.

The higher aliphatic homologues in this series show lower melting points the reduction depending on symmetry and on the length of the side group. The symmetrical methyl, ethyl and propyl disubstituted materials have similar glass transition temperatures presumably because the molecules have similar degrees of flexibility.

Introduction of aromatic or cycloaliphatic groups at R and/or R_1 gives further restriction to chain flexibility and the resulting polymers have

Table 17.7 MELTING RANGE AND GLASS TRANSITION TEMPERATURE OF POLY-CARBONATES FROM 4,4′DIHYDROXYDIPHENYLMETHANE DERIVATIVES

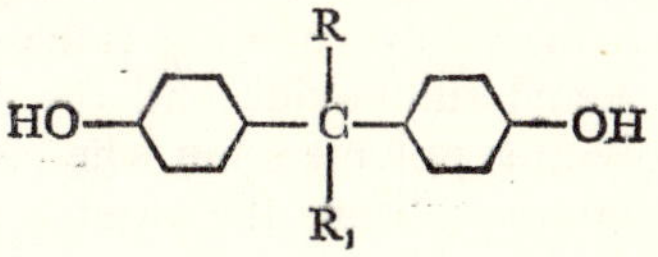

R	R_1	Melting range (°C)	Glass temperature (°C)
—H	—H	300	—
—H	—CH$_3$	185–195	130
—CH$_3$	—CH$_3$	220–230	149
—CH$_3$	—C$_2$H$_5$	205–222	134
—C$_2$H$_5$	—C$_2$H$_5$	175–195	149
—CH$_3$	—CH$_2$—CH$_2$—CH$_3$	200–220	137
—CH$_2$—CH$_2$—CH$_3$	—CH$_2$—CH$_2$—CH$_3$	190–200	148
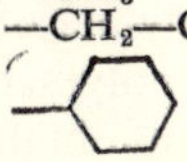 (ring)	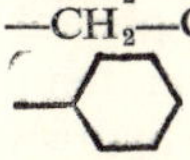 (ring)	250–275	200

transition temperatures markedly higher than that of the bis-phenol A polycarbonate.

The melting ranges and glass transition temperatures of a number of polycarbonates from 4,4′dihydroxydiphenylmethane derivatives are given in Table 17.7.

Polycarbonates have also been prepared from diphenylol compounds where the benzene rings are separated by more than one carbon atom.   In the absence of bulky side groups such polymer molecules are more

Table 17.8

Diphenylol compound (i.e. linkage between rings	Melting range of polymer (°C)	Solubility of polymer in in ordinary solvents
—CH$_2$—	over 300	insoluble
—CH$_2$—CH$_2$—	290–300	insoluble
—CH$_2$—(CH$_2$)$_8$—CH$_2$—	135–150	insoluble

flexible and crystallise very rapidly.   As is to be expected the more the separating carbon atoms the lower the melting range.   This effect is shown in data supplied by Schnell[4] (Table 17.8).

Polymers have been prepared from nuclear substituted 4,4-dihydroxy diphenyl alkanes of which the halogenated materials have been of particular interest.   The symmetrical tetrachlorbisphenol A yields a polymer with a glass transition temperature of 180°C and melting range of 250–260°C but soluble in a variety of solvents.

Crystallisable polymers have also been prepared from diphenylol compounds containing sulphur or oxygen atoms or both between the aromatic rings.   Of these the polycarbonates from 4,4-dihydroxy diphenyl ether and from 4,4-dihydroxydiphenyl sulphide crystallise sufficiently to form opaque products.   Both materials are insoluble in the usual solvents. The diphenyl sulphide polymer also has excellent resistance to hydrolysing agents and very low water absorption.   Schnell quotes a water absorption of only 0·09% for a sample at 90% relative humidity and 250°C.   Both the sulphide and ether polymers have melting ranges of about 220–240°C. The 4,4-dihydroxy diphenyl sulphoxide and the 4,4-dihydroxy diphenyl sulphoxide yield hydrolysable polymers but whereas the polymer from the former is soluble in common solvents the latter is insoluble.

Further variations in the polycarbonate system may be achieved by copolymerisation.   The reduced regularity of copolymers compared with the parent homopolymers would normally lead to amorphous materials. Since however the common diphenylol alkanes are identical in length they can be interchanged with each other in the unit cell providing the side groups do not differ greatly in their bulkiness.

Christopher and Fox[12] have given examples of the way in which polycarbonate resins may be tailor-made to suit specific requirements.   Whereas the bis-phenol from *o*-cresol and acetone (bis-phenol C) yields a polymer of high hydrolytic stability and low transition temperature, the polymer from phenol and cyclohexanone has average hydrolytic stability but a high heat distortion temperature. By using a condensate of *o*-cresol and cyclohexanone a polymer may be obtained with both hydrolytic stability and a high heat distortion temperature.

Finally mention may be made of the phenoxy resins. These do not contain the carbonate group but are otherwise similar in structure, and to some extent in properties, to the bis-phenol A polycarbonate. They are dealt with in detail in Chapter 22.

REFERENCES

1. EINHORN, A., *Ann.*, **300,** 135 (1898)
2. BISCHOFF, C. A., and VON HEDENSTRÖM, H. A., *Ber.*, **35,** 3431 (1902)
3. CAROTHERS, W. H., and NATTA, F. J., *J. Am. Chem. Soc.*, **52,** 314 (1930)
4. SCHNELL, H., *Trans. Plastics Inst.*, **28,** 143 (1960)
5. *U.S. Patent* 2,468,982
6. *U.S. Patent* 2,936,272
7. SCHNELL, H., *Angew. Chem.*, **68,** 633 (1956)
8. *German Patent* 959,497
9. PRIETSCHK, A., *Kolloid-Z.*, **156,** (1), 8, Dr. Dietrich Steinkopff Verlag, Darmstadt (1958)
10. PEILSTÖCKER, G., *Kunstoffe Plastics*, **51,** 509 (September 1961)
11. STANNETT, V. T., and MEYERS, A. W., Unpublished, quoted in reference 12
12. CHRISTOPHER, W. F., and FOX, D. W., *Polycarbonates*, Reinhold, New York (1962)
13. FIEDLER, E. F., CHRISTOPHER, W. F., and CALKINS, T. R., *Mod. Plastics*, **36,** 115 (1959)

BIBLIOGRAPHY

CHRISTOPHER, W. F., and FOX, D. W., *Polycarbonates*, Reinhold, New York (1962)
SCHNELL, H., *Chemistry and Physics of Polycarbonates*, Interscience, New York (1964)

18

Cellulose Plastics

18.1 NATURE AND OCCURRENCE OF CELLULOSE

Cellulose is the most abundant of naturally occurring organic compounds for, as the chief constituent of the cell walls of higher plants, it comprises at least one-third of the vegetable matter of the world. The cellulose content of such vegetable matter varies from plant to plant. For example oven-dried cotton contains about 90% cellulose, while an average wood has about 50%. The balance is composed of lignin, polysaccharides other than cellulose and minor amounts of resins, proteins and mineral matter. In spite of its wide distribution in nature, cellulose for chemical purposes is derived commercially from only two sources, cotton linters and wood pulp.

Cotton linters are the short fibres removed from cotton seeds after the long fibres for use in textiles have been taken off by the process of ginning. Digestion under pressure at temperatures in the range 130–180°C with a 2–5% aqueous solution of sodium hydroxide will remove the bulk of the impurities and after a bleaching operation to remove coloured bodies the residual cotton contains about 99% alpha-cellulose, the term given to pure cellulose of high molecular weight. The viscosity average molecular weight of native cellulose is in excess of 500,000 but the purification stage is accompanied by some degradation so that the resultant material usually has a molecular weight in the range 100,000–500,000 (600–3,000 repeating glucose units).

Alternatively cellulose is produced from wood via wood pulp. A number of processes are used in which the overall effect is the removal of the bulk of the non-cellulosic matter. The most widely used are the sulphite process, which uses a solution of calcium bisulphite and sulphur dioxide, the soda process using sodium hydroxide and the sulphate process using a solution of sodium hydroxide and sodium sulphide. (The term sulphate process is used since sodium sulphate is the source of the sulphide.) For chemical purposes the sulphite process is most commonly used. As normally prepared these pulps contain about 88–90% alpha cellulose but this may be increased by alkaline purification and bleaching.

Analysis of pure cellulose indicates an empirical formula $C_6H_{10}O_5$ corresponding to a glucose anhydride. There is ample evidence to indicate that in fact cellulose is a high molecular weight polyanhydroglucose. In particular it may be mentioned that controlled hydrolysis of cellulose yields cellobiose, cellotriaose and cellotetraose which contain respectively two, three and four anhydroglucose units. Complete hydrolysis will give yields of glucose as high as 95–96%.

The fact that, whereas glucose is a strongly reducing sugar, cellulose is almost non-reducing indicates that the linkage between the anhydroglucose units occurs at the reducing carbon atom. As cellobiose, known to consist of two glucose units joined by β-linkage, rather than maltose

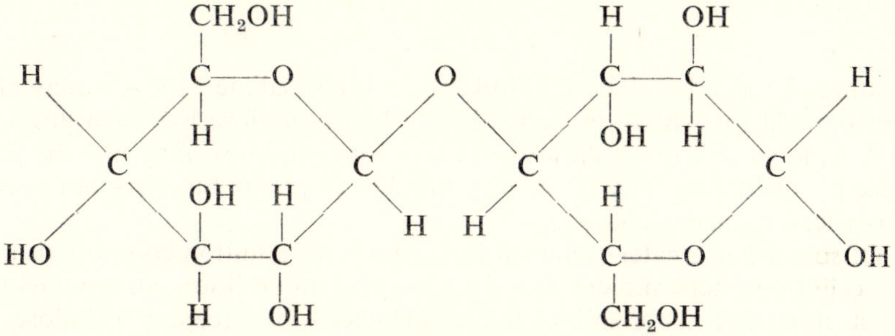

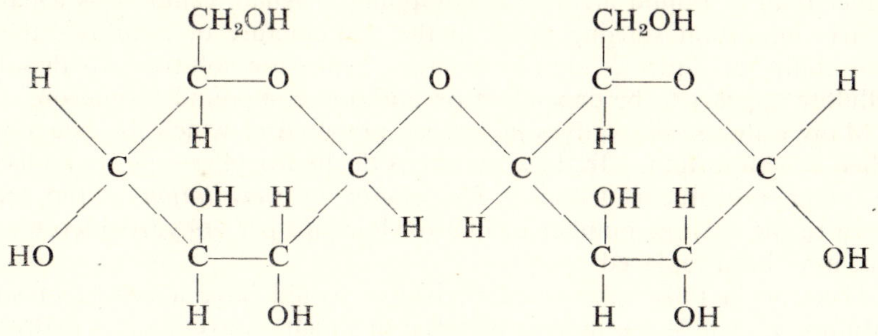

Fig. 18.1

with the α-linkage, is one of the stepwise degradation products the evidence is that cellulose molecules are made up of many anhydroglucose units joined together by beta-glucosidic linkages (Fig. 18.1).

Study of the structure of cellulose (Fig. 18.2) leads one to expect that the molecules would be essentially extended and linear and capable of existing in the crystalline state. This is confirmed by X-ray data which indicate that the cell repeating unit (10·25 Å) corresponds to the cellobiose repeating unit of the molecule.

Although it might be anticipated that, because of the abundance of hydroxyl groups, cellulose would be water soluble this is not the case. This is because the regular spacing of hydroxyl groups, particularly in the crystalline zones, facilitates extensive hydrogen bonding.   Thus although cellulose is somewhat hygroscopic intermolecular bonds are too great for solution to occur.

Cellulose may be degraded by a number of environments.   For example acid-catalysed hydrolytic degradation will eventually lead to glucose by

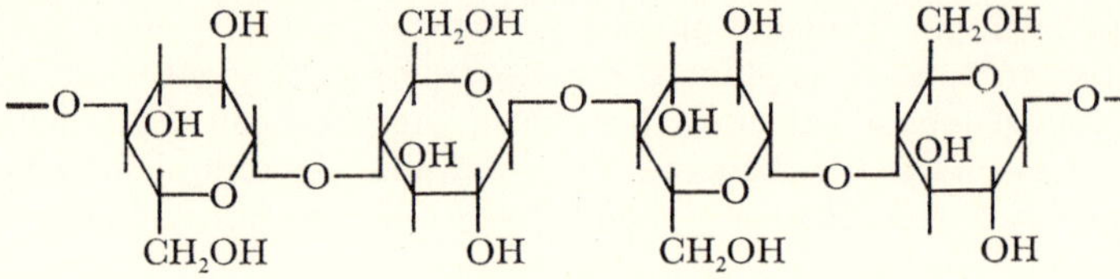

Fig. 18.2

rupture of the 1,4-β-glucosidic linkages.   Intermediate products may also be obtained for which the general term hydrocellulose has been given.

A wide variety of oxidation products, oxycelluloses, may also be produced.   Oxidation may occur at a number of points but does not necessarily lead to chain scission.

Of somewhat greater technical interest are the addition compounds and the cellulose esters and ethers.   Of the apparent addition compounds the most important is alkali cellulose produced by steeping cellulose in caustic soda and considered to be of general form $(C_6H_{10}O_5)x$ $(NaOH)_y$ rather than a sodium alcoholate compound.   Alkali cellulose is a particularly important starting point in the manufacture of cellulose ethers. The ability of aqueous cuprammonium hydroxide solutions to dissolve cellulose appears to be dependent on addition compound formation.

Many cellulose derivatives have been prepared of which the esters and ethers are important.   In these materials the hydroxyl groups are replaced by other substituent groups.   The *degree of substitution* is the term given to the average number of hydroxyl groups per anhydroglucose unit that have been replaced.

Therefore a fully substituted derivative would have a degree of substitution 3·0 whilst a cellulosic material in which on average 1·8 hydroxyl groups per glucose unit had been replaced would have a degree of substitution 1·8.   Commercial derivatives usually have a degree of substitution of less than 3·0 the actual value chosen being determined by the end-use.

The likelihood of any given hydroxyl group reacting will be determined largely by its position in the molecule and the position of the molecule in the fibrous structure.   The reaction rate is largely determined by the rate of diffusion of the reagent and this is much greater in amorphous regions than in the crystalline areas.   It is desirable in the preparation of derivatives that uniform substitution should occur, or at least that the hydroxyl groups in one molecule should have the same chance of reaction as those in another molecule.   If this is not the case molecules on the surface of a cellulose fibre may well be fully substituted while molecules disposed in the centre of the fibre will be completely unreacted.

When reaction is carried out homogeneously in solution this state of affairs more or less exists and it is possible to achieve a statistically random degree of substitution. (It is to be noted that the primary hydroxyl groups will be more reactive than the secondary hydroxyl groups.)

The nitration of cellulose is unusual in that uniform reaction takes place even though the fibrous structure is retained. This is explained by the fact that nitration is an equilibrium reaction unaffected by fibre structure, the extent of nitration being determined by the strength of the nitrating acid.

Because of the insolubility of cellulose it is not possible to carry out uniform esterification with the lower organic acids (acetic acid, propionic acid etc.) and in those cases where incompletely substituted derivatives are required a two-stage reaction is employed. This involves first total esterification in a medium in which the ester dissolves, followed by the uniform removal of some of the substituent groups (this now being possible in solution) by hydrolysis.

18.2 CELLULOSE ESTERS

The cellulose esters are useful polymers for the manufacture of plastics. Until about 1950 they did in fact form the most important group of thermoplastics materials. The historical importance and significance of these materials have been discussed more fully in the first chapter of this book.

The most important of the esters is cellulose acetate. This material has been extensively used in the manufacture of films, moulding and extrusion compounds, fibres and lacquers. As with all the other cellulose polymers it has however become of small importance to the plastics industry compared with the polyolefins, p.v.c. and polystyrene. In spite of their higher cost cellulose acetate–butyrate and cellulose propionate appear to have retained their smaller market because of their excellent appearance and toughness.

The doyen of the ester polymers is cellulose nitrate. Camphor modified cellulose nitrate has been known for over 100 years and still retains its use in a few specialised applications.

18.2.1 Cellulose nitrate

Preparation

The reaction between cellulose and nitric acid is one of esterification. It is possible to achieve varying degrees of esterification according to the number of hydroxyl groups that have been replaced by the nitrate group. Complete substitution at all three hydroxyl groups on the repeating anhydroglucose units will give the explosive cellulose trinitrate and will contain 14·14% nitrogen. This material is not made commercially but esters with lower degrees of nitrate are of importance as indicated in Table 18.1.

It will be observed from Table 18.1 that industrial cellulose nitrates or 'nitrocelluloses' (as they are often erroneously called) have a degree of

Table 18.1

Degree of nitration	Nitrogen content (%)	Typical usage
Cellulose mononitrate	6·76% (theor.)	—
Cellulose dinitrate	11·11 (theor.)	
	10·7–11·1	plastics, lacquers
	11·2–12·3	films, lacquers
	12·4–13·5	cordite
Cellulose trinitrate	14·14 (theor.)	

substitution somewhere between 1·9 and 2·7 and that materials with lower degrees of substitution are used for plastics applications.

The nitration process involves the steeping of cotton linters into a mixture of nitric and sulphuric acids and subsequent removal, stabilisation, bleaching and washing of the product. Subsequent conversion into plastics materials involves displacing residual water by alcohol, mixing the alcohol-wet nitrate with camphor and other ingredients, seasoning the rolled hides, pressing and finally cutting to shape.

Before nitration the moisture content of the purified linters is reduced to well below 5% since the presence of water will modify the progress of the reaction and tends to produce undesirable products. The drying operation is carried out by breaking open the cotton linters and passing along a hot air drier.

The nitration bath normally contains sulphuric acid as a condensing agent and a typical bath for producing a cellulose nitrate with a nitrogen content of 11% would be

Nitric acid	25%
Sulphuric acid	55%
Water	20%

In a typical process 1,200 lb of the mixed acids are run into the reaction vessel and 30 lb of the dried cotton linters are added. The mixture is agitated by a pair of contra-rotating stirrers and nitration is allowed to proceed at about 35–40°C for 20 minutes. It is interesting to note that the cellulosic material retains its fibrous form throughout the nitration process.

On the completion of nitration the batch is dropped from the reaction vessel into a centrifuge and the acid mixture spun-off and recovered. The nitrated linters, which still contain appreciable quantities of acid are then plunged into a drowning tank, where the nitric acid is diluted with a large volume of water. The resultant ester is then pumped, as a slurry, into storage vats which may hold the products of several nitrations.

The product at this stage is unstable. It has been shown that some of the sulphuric acid reacts with the cellulose hydroxyl group to form sulphates. These tend to split off reforming sulphuric acid to initiate an autocatalytic decomposition which can, and has in the past, led to disastrous explosions. The remedy lies in removing the sulphate groups in a stabilisation process by boiling the cellulose ester with water containing a controlled trace of acid for several hours. Side effects of this process are a reduction in molecular weight and in the nitrogen content.

The stabilised nitrate may then be bleached with sodium hypochlorite, centrifuged to remove much of the water in which the polymer has been slurried and dehydrated by displacement with alcohol while under pressure in a press. It is interesting to note that in these processes approximately 35,000 gallons of water are used for every ton of cellulose nitrate produced. Control of purity of the water is important, in particular the iron content should be as low as 0·03 parts per million since iron can adversely affect both the colour and heat stability of the polymer.

Manufacture of celluloid sheets

Although originally a trade name the term celluloid has come into general use to describe camphor-plasticised cellulose nitrate compositions.

The rather unexpected plasticising effect of camphor was first appreciated by Hyatt over a hundred years ago and, in spite of all that has been learned about polymers since then, no superior plasticiser has yet been discovered.

Camphor was originally obtained from the camphor tree *laurus camphora* in which it appeared in the optically active dextro-rotary form. Since about 1920 the racemic (d,1) mixture derived from oil of turpentine has been more generally used. By fractional distillation of oil of turpentine the product pinene is obtained. By treating this with hydrochloric acid, pinene hydrochloride (also known as bornyl chloride) may

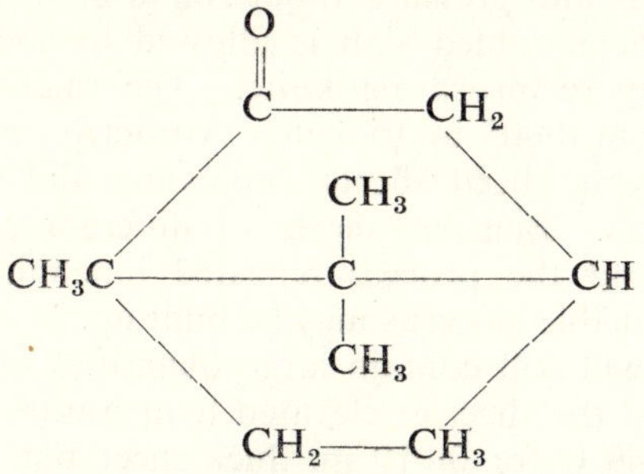

Fig. 18.3

be produced. This is then boiled with acetic acid to hydrolyse the material to the racemic borneol which on oxidation yields camphor. Camphor is a white crystalline solid (Melt. Pt 175°C) with the structure shown in Fig. 18.3.

It has a low solubility parameter ($\sim 7\cdot5$) which differs considerably from that of the cellulose dinitrate ($\sim 10\cdot7$). This indicates that compatibility is not simply due to similarities of cohesive forces but also to some form of interaction probably involving the carbonyl group.

Mixing of the ingredients is carried out in steam heated dough mixers fitted with solvent extraction hoods. A typical charge would consist of

240 lb of cellulose nitrate and 80 lb of camphor. The residual alcohol in the nitrate develops a powerful solvent action and the cellulose nitrate loses its fibrous form and the whole mix becomes a gelatinous mass. Typical mixing temperatures are of the order of 40°C and mixing times are approximately 1 hour.

Pigments or dyes may be added at this stage and where clear water-white sheet is required a small amount of a soluble violet dye is added to offset the faintly yellow colour of the natural mix. Stabilisers such as zinc oxide, zinc acetate or urea may be added to prevent the composition from developing acidity.

The celluloid dough is then filtered by forcing through a pad of calico and brass gauze backed by a heavy brass plate at a press of about 1·5 tons per square inch. Any undesirable foreign matter is thus separated from the dough.

The filtered dough is then returned to a mixer and the alcohol content reduced to 25% by kneading under vacuum. Further reduction in the alcohol content is brought about by rolling the compound on a hooded two-roll mill. The milled product is then consolidated on a two-bowl calender and sheeted off in hides about $\frac{1}{2}$ in. thick. At this stage the solvent content is between 12 and 16%.

A number of hides are then laid up in a box mould with a grooved base. The mould is then loaded into a press and heated with hot water to consolidate the mass. Great care must be taken to avoid overheating since this can cause disastrous explosions. It is not unknown for the head of a hydraulic press to have been blown through the roof of the press shop as a result of overheating the celluloid. Press temperatures are typically about 750°C and pressure about 500 lb in^{-2}.

After pressing, the moulded slab is allowed to cool and then sliced using a horizontally reciprocating knife. The thickness of the sliced sheet may range from 0·005 in. to 1 in. Attractive mottled effects may be produced by plying sliced sheets, repressing and then subsequently reslicing on the bias. Bias-cut sheets of different patterns may then again be pressed and the process repeated indefinitely. In this way complex but reproducible patterns may be built up.

The sliced sheet will still contain large quantities of alcohol and it is necessary to 'season' the sheet at elevated temperatures. This may only take three days at 49°C for 0·010 in. thick sheet but will last about 56 days for 1 in. thick blocks. The removal of alcohol, as might be expected, is accompanied by considerable shrinkage. Fully seasoned sheet has a volatile content of 2% the bulk of which is water but there is some residual alcohol. The sheets may be fully polished by heating in a press between glazed plates under pressure for a few minutes. Because the material is thermoplastic it is necessary to cool it before removal from the press.

It is also possible to extrude alcohol-containing celluloid compositions through either ram or screw extruders under carefully controlled conditions. The process is now believed to be universally obsolete.

Sheet and block may be machined with little difficulty providing care is taken to avoid overheating and to collect the inflammable swarf.

Structure and properties of celluloid

Nitration of cellulose followed by plasticisation of the product with camphor has the effect of reducing the orderly close packing of the cellulose molecules. Hence whereas cellulose is insoluble in solvents, except in certain cases where there is chemical reaction, celluloid is soluble in solvents such as acetone and amyl acetate. In addition the camphor present may be dissolved out by chloroform and similar solvents which do not dissolve the cellulose nitrate.

The solvation by plasticiser also gives celluloid thermoplastic properties owing to the reduction in interchain forces. On the other hand since the cellulose molecule is somewhat rigid the product itself is stiff and does not show rubbery properties at room temperature, cf. plasticised p.v.c.

As may be expected from such a polar material it is not a particularly good electrical insulator particularly at high frequencies. The high dielectric constant is particularly noteworthy.

The chemical resistance of celluloid is not particularly good. It is affected by acids and alkalis, discolours on exposure to sunlight and tends to harden on aging. More seriously it is extremely inflammable, this being by far the greatest limitation of the material.

Typical physical properties of celluloid are compared with other cellulose plastics in Table 18.2.

The high inflammability and relatively poor chemical properties of celluloid severely restrict its use in industrial applications. Consequently, the material is used because of the following desirable characteristics.

1. Water-white transparency of basic composition but capable of forming highly attractive multi-coloured sheeting.
2. Rigidity.
3. Reasonable toughness.
4. Capable of after-shrinkage around inserts.

Applications

The annual production of celluloid is small compared with the total world production of plastics. In Great Britain the material is marketed as Xylonite (BX Plastics, Ltd.) of which about 1,000 tons per annum are produced.

The one-time important applications in photographic film, in bicycle parts (pump covers and mudguards) and in toys manufactured by a blowing process from flat sheet are no longer of importance.

Today the principal outlets are knife handles, table-tennis balls and spectacle frames. The continued use in knife handles is due to the pleasant appearance and the ability of the material to after-shrink around the extension of the blade. Table-tennis balls continue to be made from celluloid since it has been difficult to match the 'bounce' and handle of the celluloid ball, the type originally used, with balls fabricated from newer polymers. Even here celluloid is now meeting the challenge of certain ethenoid polymers. Spectacle frames are still of interest because of the

Table 18.2 TYPICAL PHYSICAL PROPERTIES OF CELLULOSIC PLASTICS

(It is necessary to quote a range of figures in most instances since the value of a particular property is very dependent on formulation)

	Cellulose nitrate	Cellulose acetate	Cellulose acetate butyrate	Cellulose acetate propionate	Cellulose propionate	Ethyl cellulose
Specific gravity	1·35-1·40	1·27-1·32	1·15-1·22	1·19-1·23	1·18-1·24	1·12-1·15
Refractive index (25°C)	1·5	1·47-1·5	1·47-1·48	1·46-1·49	1·46-1·49	1·47
Tensile strength (10^3 lb in^{-2})	5-10	3·5-11	2·5-7·5	3·5-7·3	2-6	6-9
Elongation at break (%)	10-40	5-55	8-80	30-100	45-65	10-40
Izod impact ft lb/in. notch	2·0-8·0	0·5-5·0	0·4-5·0	0·9-5·0	1·2-11·0	1·0-6·5
Rockwell M hardness 70°F (ASTM D.229-39)	25-50	-30 to +75	0-80	60-120 (R-scale)	-15 to +106	-25 to +80
Flow temperature (°C) (ASTM D.569-40T)	145-152	115-165	115-165	150-180	145-180	100-150
Heat distortion temperature (°C)	—	50-100	56-94	45-110	51-70	50-66
Volume resistivity (ohm cm) 50% RH, 25°C	$\sim 10^{11}$	$\sim 10^{12}$	$\sim 10^{12}$	$\sim 10^{12}-10^{14}$	$\sim 10^{12}-10^{15}$	10^{15}
Dielectric constant 60 c/s	6·7-7·3	3·5-7·5	3·7-4·5	3·7-4·0	3·0-3·5 (10^6 c/s)	2·7
Power factor 60 c/s	0·06-0·15	0·01-0·07	0·008-0·012	0·01-0·04	0·017-0·02 (10^6 c/s)	0·007
% Water absorption 2 in $\times \frac{1}{8}$ in. disc 24 hr immersion ASTM D.570-40T	0·6-2·0	1·0-3·0	0·9-2·4	1·5-2·8	1·6-2·0	0·5-1·5

attractive colour. There are however restrictions to their use for this application in certain countries and cellulose acetate is often preferred.

18.2.2 Cellulose acetate

Preparation

The earliest preparation of cellulose acetate is credited to Schützenberger in 1865. The method used was to heat the cotton with acetic anhydride in sealed tubes at 130–140°C. The severe reaction conditions led to a white amorphous polymer but the product would have been severely degraded and the process difficult to control. Subsequent studies made by Liebermann, Francimont, Miles, the Bayer Company and by other workers led to techniques for controlled acetylation under less severe conditions.

The methods available today may be considered under two headings, homogeneous acetylation in which the acetylated cellulose dissolves into a solvent as it is formed, and the heterogeneous technique in which the fibre structure is retained.

As mentioned in Section 18.1 the probability of acetylation of any one cellulosic group is strongly dependent on its position in the fibre. Since this cannot be dissolved before acetylation it will be realised that some molecules will be completely acetylated whilst others may be untouched. It is thus necessary first to acetylate completely the cellulose and the resultant triacetate material, which is soluble in certain solvents, may then be back-hydrolysed in solution. Under these conditions the probabilities of hydrolysis of any acetyl groups in one molecule will be similar to the reaction probabilities of these groups in another molecule and products with a reasonably even degree of substitution less than three may be obtained.

The preparation of the acetate by homogeneous acetylation may be considered in three stages:

1. Pretreatment of the cellulose.
2. Acetylation.
3. Hydrolysis.

The aim of pretreatment is to open up the cellulosic matter in order to achieve more even substitution and to accelerate the main acetylation reaction. A large number of pretreatments have been described in the patent literature but in practice exposure to glacial acetic acid is that most commonly employed.

The acetylation is usually carried out in bronze stirred mixers. The acetylating mixture normally contains three components, an acetylating agent, a catalyst and a diluent.

Although acetyl chloride and ketene ($CH_2{=}C{=}O$) have been described in the literature acetic anhydride has been the commonly employed acetylating agent.

The reaction between one of the hydroxyl groups of the cellulose molecule (X·OH) and the anhydride is:

$$X \cdot OH + CH_3CO \cdot O \cdot CO \cdot CH_3 \longrightarrow X \cdot O \cdot CO \cdot CH_3 + CH_3 \cdot COOH$$

Similarly a number of catalysts have been suggested but concentrated sulphuric acid, first suggested by Francimont in 1879, is almost universally employed today.

The diluent, which is usually a solvent for the acetate, facilitates the reaction particularly in respect of temperature control. Acetic acid is generally employed either alone or in conjunction with other materials for this purpose. It may be added initially but is also formed during the acetylation of the cellulose. A mild acetylating agent in itself it is not merely a diluent but influences the course of the reaction. The low boiling solvent methylene chloride (Boil. Pt. 40°C) is now commonly used in conjunction with acetic acid. An advantage in using methylene chloride is that excessive exothermic heat may be removed as latent heat of evaporation as the methylene chloride boils. Bubbles formed during boiling or 'simmering' may also assist in mixing of the reaction blend.

In the so-called Dormagen process developed by I. G. Farben the cotton was first preheated with 30–40% of its own weight with glacial acetic acid for 1–2 hours. The pretreated material was then fed to the acetylisers which consisted of horizontal bronze cylinders. For every 100 parts of pretreated cellulose there was added the following acetylating mixture, previously cooled to 15–20°C.

300 parts	acetic anhydride
400 parts	methylene chloride
1 part	sulphuric acid

Cooling water was passed through jackets surrounding the reactor in order to prevent the temperature from exceeding 50°C.

Esterification is complete in 5–6 hours and the product at this stage is known as *primary cellulose acetate*.

Current acetylation techniques may be based on continuous production similar to that employed in the Dormagen process or batchwise in dough-type mixers.

Whatever acetylating technique is used this is then followed by the ripening operation. The 'ripening' is carried out without isolating the triacetate. Dilute acetic acid or water is added to the acetylising mixture, about 20–25% based on the weight of the cellulose. Hydrolysis is allowed to proceed for a number of days, typically for about 72 hours. The progress of hydrolysis is followed by checking the solubility of the acetate in alcohol–benzene and acetone solutions. When the required degree of substitution has been reached the cellulose acetate is precipitated by judicious addition of water to the stirred mixture. It is then washed thoroughly and dried in an electric or vacuum oven. Care must be taken in the operation of the precipitation stage since otherwise lumps are likely to be formed.   For certain applications, such as for photographic film, further purification operations may be carried out.

The product at this stage is referred to as *secondary cellulose acetate.* Different degrees of acetylation are required for different end products and these are indicated in Table 18.3.

Compounding of cellulose acetate

The cellulose molecule is rigid and forms strong hydrogen bonds with adjacent molecules.   It is thus insoluble and decomposes before softening on heating.   Partial replacement of hydroxyl groups by acetyl groups has a number of effects

1. It reduces inter-chain hydrogen bonding.
2. It increases inter-chain separation.
3. It makes the polymer less polar—the polarity depending on the degree of acetylation.

Because of these influences cellulose acetate can be dissolved in a variety of media, although a liquid suitable as a solvent for cellulose acetate with a degree of substitution of two would be unlikely to be a solvent for acetates with degrees of substitution of either one or three.

Although acetylation thus renders the cellulosic structure soluble cellulose acetate will still decompose below its softening point.   It is thus

Table 18.3   INFLUENCE OF DEGREE OF SUBSTITUTION ON THE PROPERTIES AND USES OF CELLULOSE ACETATE

Degree of substitution	Acetyl content (%—COCH₃)	Acetic acid yield (%)	Solubility	Uses
2·2–2·3	36·5–38·0	52·1–54·3	soluble in acetone	injection moulding
2·3–2·4	38·0–39·5	54·3–56·4		film
2·4–2·6	39·5–41·5	56·4–59·3		lacquers
2·8–3·0	42·5–44	60·7–62·8	insoluble in acetone, soluble in chloroform	triacetate film and fibre

necessary to compound cellulose acetate with plasticisers in order to obtain plastics materials of suitable flow properties.   Other ingredients are also added at the same time.

Although the prime function of plasticisers in cellulose acetate is to bring the processing temperature of the compound below the polymer decomposition temperature it has additional values.   An increase in the plasticiser content will reduce the melt viscosity at a given temperature and simplify processing.   The physical properties of the finished product will be modified, increasing toughness but reducing the heat distortion temperature, the latter not being an important property in most cellulose acetate applications.

Although many plasticisers have been suggested for cellulose acetate very few have been used in practice.   The most important of these are dimethyl phthalate ($\delta = 10 \cdot 5$), triacetin ($\delta = 9 \cdot 95$) and triphenyl phosphate ($\delta = 9 \cdot 8$) each of which have a solubility parameter within one unit of that of cellulose diacetate ($\sim 10 \cdot 8$).

Of these *dimethyl phthalate* (d.m.p.) is used in most compositions. It is cheap, has a high compatibility with secondary cellulose acetate and is efficient in increasing flexibility, toughness and the ease of flow at a given temperature. Its principal disadvantages are its high volatility and the fact that it increases the flammability of the compound. Similar in compatibility but rather less volatile is *diethyl phthalate*. This material has less of an influence on flexibility and flow properties than the methyl ester.

Triphenyl phosphate is a crystalline solid which has less compatibility with the polymer. This may be expected from solubility parameter data. It is often used in conjunction with dimethyl phthalate and has the added virtues of imparting flame resistance and improved water resistance. It is more permanent than d.m.p. *Triacetin* is less important now than at one time since, although it is compatible, it is also highly volatile and lowers the water resistance of the compound. Today it is essential to prepare low cost compounds to allow cellulose acetate to compete with the synthetic polymers, and plasticisers such as *ethyl phthalyl ethyl glycollate*, which are superior in some respects, are now rarely used.

Small amounts of stabiliser (1–5%) are normally added to improve weather resistance. These materials are the usual ultraviolet light absorbers such as phenyl salicylate and various benzoates. Triphenyl phosphate also has a beneficial influence.

Other ingredients may include cheapening extenders such as castor oil, colouring agents, lubricants and, rarely, fillers.

Compounding may be carried out by either a wet or a dry process. In the wet process, now obsolescent, the ingredients are mixed as a viscous solution in acetone in a dough mixer. The resulting dough is then rolled on a hot two-roll mill to evaporate the bulk of the solvent. It is then necessary to 'season' the resulting hides until the solvent content is reduced to a tolerably low level.

Dry processes which obviate solvent difficulties are now preferred[1,2] and are similar to those employed with the major thermoplastics. They include the use of two-roll mills, internal mixers, extruders and extrusion compounders. The use of dry blend techniques similar to that used more recently with p.v.c. have also been used.[2]

Properties of cellulose acetate plastics

Cellulose acetate plastics have no really outstanding properties. Their continued use for mouldings and extrusions depends on their toughness and good appearance at a reasonable cost, although somewhat above the prices ruling for the major ethenoid plastics p.v.c., polythene and polystyrene. In common with most other plastic materials they are capable of unlimited colour variations, including water-white transparency. Processing is quite straightforward provided the granules are dry.

Compared with the major ethenoids cellulose acetate plastics have a high water absorption, poor electrical insulation characteristics, limited aging resistance, limited heat resistance and are attacked or dissolved by a wide variety of reagents.

A wide range of cellulose acetate compounds are commercially available. The properties of these compounds depend on three major factors:

1. The chain length of the cellulose molecule.
2. The degree of acetylation.
3. The type and amount of plasticiser(s).

During production of cellulose acetate from cellulose a certain amount of chain degradation takes place. As a result the degree of polymerisation of commercial acetate esters is usually within the range 175–360. It is convenient to assess the chain length by solution viscosity methods. Products differing in viscosity will be produced by varying the source of the original cellulose and by modifying reaction conditions. Since marked batch-to-batch variations in the viscosity of the finished product occur in practice, products of specified viscosity are obtained by blending.

The greater the molecular weight the higher is the flow temperature and the heat distortion temperature. Variations in molecular weight, in the normal range, however have less effect than do variations in the degree of acetylation and in the plasticiser used.

Increasing the degree of acetylation from that corresponding to a diacetate will obviously reduce the hydroxyl content and this will increase the water resistance. The polymer also becomes less polar and the solvent properties correspondingly alter. It is also observed that an increase in the degree of acetylation reduces the hardness, impact strength water absorption and increases the 'flow temperature'. The influence of the degree of acetylation on these properties is shown clearly in Fig. 18.4 (*a*)–(*d*).

Cellulose acetate plastics are generally produced using polymers from a fairly narrow range of molecular weights and degrees of acetylation. In practice the greatest variation in properties is achieved by modifying the type and amount of plasticiser. Table 18.4[3] shows the influence of varying the amount of plasticiser on several important properties of cellulose acetate.

The so-called flow temperature cannot be considered to be either the processing temperature or the maximum service temperature. It is

Table 18.4 INFLUENCE OF AMOUNT OF PLASTICISER (DIMETHYL PHTHALATE) ON SOME PHYSICAL PROPERTIES OF CELLULOSE ACETATE COMPOSITIONS

Parts d.m.p. per 100 pts polymer	37·8	30·0	22·6
Flow temperature (°F)	266	285	307
Elongation (%)	8·0	6·5	5·0
Tensile strength (lb in^{-2})	6,630	8,100	9,930
Flexural strength (lb in^{-2})	11,140	14,200	17,000
Rockwell hardness (*M*)	79·5	90·5	101·1
Water absorption (%)	2·13	2·47	3·24
Leaching (48 hr) %	0·58	0·52	0·45
Loss in weight (1 week 150°C) %	2·40	0·86	0·53

obtained using the highly arbitrary Rossi–Peakes flow test (B.S.1524) and is the temperature at which the compound is forced down a capillary of fixed dimension by a fixed load at a specified rate.   It is thus of use only for comparison and for quality control purposes.   Since the rates of shear and temperatures used in processing are vastly different from those used in this test extreme caution should be taken when assessing the results of flow temperature tests.

Typical values for the principal properties of cellulose acetate compounds are tabulated in Table 18.2 in comparison with other cellulosic plastics.   Since cellulose acetate is seldom used today in applications where detailed knowledge of physical properties are required these are given without further comment.

Applications

As already indicated cellulose acetate is used because of its reasonable toughness, transparency and wide colour range.   It is not suitable when good electric insulation properties, heat resistance, weathering resistance, chemical resistance and dimensional stability are important.

The main outlets are for films and sheeting.   Because of its clarity the film is used extensively for photographic purposes and for packaging. Sheeting is used for a variety of purposes.   Thin sheet is useful for high quality display boxes whilst thicker sheet is used for spectacle frames.

The use of cellulose acetate for moulding and extrusion is now becoming small in Great Britain owing largely to the competition of the styrene polymers and polyolefins.   The major outlets at the present time are in the fancy goods trade as toothbrushes, combs, hair slides etc.   Processing provides no major problem provided care is taken to avoid overheating and the granules are dry.   The temperatures and pressures used vary, from 160 to 250°C and 7 to 15 ton in^{-2} respectively, according to grade.   The best injection mouldings are obtained using a warm mould.

Secondary cellulose acetate has also been used for fibres and lacquers whilst cellulose triacetate fibre has been extensively marketed in Great Britain under the trade name Tricel.

18.2.3   Other cellulose esters

Homologues of acetic acid have been employed to make other cellulose esters and of these cellulose propionate, cellulose acetate–propionate and cellulose acetate–butyrate are produced on a commercial scale.   These materials have larger side chains than cellulose acetate and with equal degrees of esterification, molecular weights and incorporated plasticiser, they are slightly softer, of lower density, have slightly lower heat distortion temperatures and flow somewhat more easily.   The somewhat greater hydrocarbon nature of the polymer results in slightly lower water-absorption values (see Table 18.2).

It should however be realised that some grades of cellulose acetate may be softer, be easier to process and have lower softening points than some

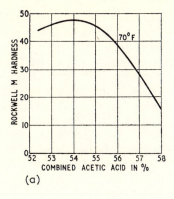

(a)

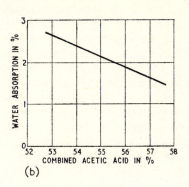

(b)

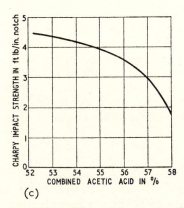

(c)

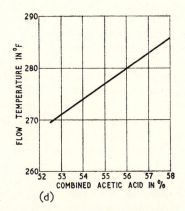

(d)

*Fig. 18.4. Effect of degree of acelytation on (a) hardness, (b) water absorp-
tion, (c) impact strength and (d) flow temperature. (31% plasticiser content)
(Hercules Powder Co. literature)*

grades of cellulose acetate–butyrate, cellulose acetate–propionate and
cellulose propionate since the properties of all four materials may be
considerably modified by chain length, degree of substitution and in particular by the type and amount of plasticiser.

Cellulose acetate-butyrate (c.a.b.) has been manufactured for a number
of years in the United States (Tenite Butyrate–Kodak) and in Germany
(Cellidor B-Bayer). It is not manufactured in Great Britain but imports
of this material are well over 1,000 tons per annum.

In a typical process for manufacture on a commercial scale bleached
wood pulp or cotton linters are pretreated for 12 hours with 40–50%
sulphuric acid and then, after drying, with acetic acid. Esterification of
the treated cellulose is then carried out using a mixture of butyric acid
and acetic anhydride, with a trace of sulphuric acid as catalyst. Commercial products vary extensively in the acetate–butyrate ratios employed.

The lower water absorption, better flow properties and lower density
of c.a.b. compared with cellulose acetate are not in themselves clear

justification for their continued use.   There are other completely synthetic thermoplastics which have an even greater superiority at a lower price and do not emit the slight odour of butyric acid as does c.a.b.   Its principal virtues which enable it to compete with other materials are its toughness, excellent appearance and comparative ease of mouldability (providing the granules are dry).   The material also lends itself to use in fluid bed dip-coating techniques giving a coating with a hard glossy finish which can only be matched with more expensive alternatives.

A number of injection mouldings have been prepared from c.a.b. with about 19% combined acetic acid and 44% combined butyric acid.   Their principal end products have been for tabulator keys, automobile parts, toys and tool handles.   In the United States c.a.b. has been used for telephone housings.   Extruded c.a.b. piping has been extensively used in America for conveying water, oil and natural gas, while c.a.b. sheet has been able to offer some competition to acrylic sheet for outdoor display signs.

In Great Britain consumption has been maintained over the past few years.   It would however appear to be severely challenged by the a.b.s. polymers and to a lesser extent the propylene and acrylic polymer.

In the mid 1950's *cellulose propionate* became commercially available (Forticel–Celanese).   This material is very similar in both cost and properties to c.a.b.   Like c.a.b. it may take on an excellent finish, providing a suitable mould is used, is less hygroscopic then cellulose acetate, and is easily moulded.

As with the other esters a number of grades are available differing in the degree of esterification and in type and amount of plasticiser.   Thus the differences in properties between the grades are generally greater than any differences between 'medium' grades of cellulose propionate and c.a.b.   Whereas a soft grade of the propionate may have a tensile strength of 2,000 lb in^{-2} and a heat distortion temperature of 51°C a hard grade may have tensile strengths as high as 6,000 lb in^{-2} and a heat distortion temperature of 70°C.

It has been used for similar purposes as c.a.b. but appears to be less used in Great Britain than the older-established mixed ester.

Cellulose acetate propionate (Tenite Propionate–Kodak) is similar to cellulose propionate.

Many other cellulose esters have been prepared in the laboratory and some have reached pilot plant status.   Of these the only one believed to be of current importance is *cellulose caprate* (decoate).   According to the literature, degraded wood pulp is activated by treating with chloroacetic acid and the product is esterified by treating with capric anhydride, capric acid and perchloric acid.   The material is said to be useful as optical cement.[1]

18.3 CELLULOSE ETHERS

By use of a modification of the well-known Williamson synthesis it is possible to prepare a number of cellulose ethers.   Of these materials ethyl cellulose has found a small limited application as a moulding material and somewhat greater use for surface coatings. The now obsolete benzyl cellulose was used prior to the Second World War as a moulding material whilst methyl cellulose, hydroxyethyl cellulose and sodium carboxymethyl cellulose are useful water-soluble polymers.

With each of these materials the first step is the manufacture of alkali cellulose (soda cellulose). This is made by treating cellulose (either bleached wood pulp or cotton linters) with concentrated aqueous sodium hydroxide in a nickel vessel at elevated temperature. After reaction excess alkali is pressed out, and the resultant 'cake' is then broken-up and vacuum dried until the moisture content is in the range 10–25%. The moisture and combined alkali contents must be carefully controlled as variations in them will lead to variations in the properties of the resultant ethers.

18.3.1 Ethyl cellulose

Ethyl cellulose is prepared by agitating the alkali cellulose with ethyl chloride in the presence of alkali at about 60°C for several hours. Towards the end of the reaction the temperature is raised to about 130–140°C. The total reaction time is approximately twelve hours. The reaction is carried out under pressure.

If the etherification were taken to completion the product would be the compound shown in Fig. 18.5.

It is essential that there be sufficient alkali present, either combined with the cellulose, or free, to neutralise the acid formed by both the main

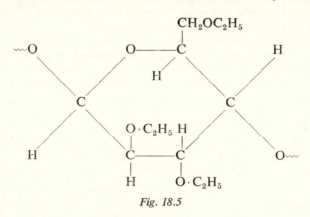

Fig. 18.5

reaction and in a side reaction which involves the hydrolysis of ethyl chloride.

Ethyl ether and ethyl alcohol which are formed as by-products are removed by distillation and the ethyl cellulose is precipitated by hot water. The polymer is then carefully washed to remove sodium hydroxide and sodium chloride and dried.

The properties of the ethyl cellulose will depend on:
1. The molecular weight.
2. The degree of substitution.
3. Molecular uniformity.

The molecular weight may be regulated by controlled degradation of the alkali cellulose in the presence of air. This can be done either before or during etherification. The molecular weight of commercial grades is usually expressed indirectly as the viscosity of a 5% solution in an 80:20 toluene–ethanol mixture.

The completely etherified material with a degree of substitution of 3 has an ethoxyl content of 54·88%. This material has little strength and flexibility, is not thermoplastic, has limited compatibility and solubility and is of no commercial value. A range of commercial products are however available with a degree of substitution between 2·15 and 2·60 corresponding to a range of ethoxyl contents from 43 to 50%.

The ethoxyl content is controlled by the ratio of reactants and to a lesser degree by the reaction temperature.

Whereas mechanical properties are largely determined by chain length, the softening point, hardness, water absorption and solubility are rather more determined by the degree of substitution (see Fig. 18.6).

Typical physical properties of ethyl cellulose are compared with those of the cellulose esters in Table 18.2.

The solubility of ethyl cellulose depends on the degree of substitution. At low degrees of substitution (0·8–1·3) the replacement of some of the hydroxyl groups by ethoxyl groups reduces the hydrogen bonding across the cellulosic chains to such an extent that the material is soluble in water. Further replacement of hydroxyl groups by the less polar and more hydrocarbon ethoxyl groups increases the water resistance. Fully etherified ethyl cellulose is soluble only in non-polar solvents.

The relationship between degree of substitution and solubility characteristics is predictable from theory and is summarised in Table 18.5.

Ethyl cellulose is subject to oxidative degradation when exposed to sunlight and elevated temperatures. It is therefore necessary to stabilise

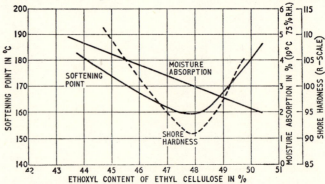

Fig. 18.6. *Influence of the ethyoxyl content of ethyl cellulose on softening point moisture absorption and hardness.* (*Hercules Power Co. literature*)

the material against degrading influences during processing or service. In practice three types of stabiliser are incorporated, an antioxidant such as the phenolic compound 2,2′-methylene bis (4-methyl-6-tert-butyl phenol), an acid acceptor such as an epoxy resin for use where plasticisers may give rise to acidic degradation products and an ultraviolet absorber such as 2,4-dihydroxy benzophenone for outdoor use. Plasticisers such as tritolyl phosphate and diamyl phenol have a beneficial stabilising effect.

Ethyl cellulose has never become well known in Europe and apart from one or two specific applications has not been able to capture any significant proportion of the market held by the cellulose esters. Although it

Table 18.5 SOLUBILITY OF ETHYL CELLULOSE

Average number of ethoxyl groups per glucose unit	Solubility
~ 0·5	soluble in 4–8% sodium hydroxide
0·8–1·3	soluble in water
1·4–1·8	swelling in polar–non-polar solvent mixtures
1·8–2·2	increasing solubility in above mixtures
2·2–2·4	increasing solubility in alcohol and less polar solvents
2·4–2·5	widest range of solubilities
2·5–2·8	soluble only in non-polar solvents

has the greatest water resistance and the best electrical insulating properties amongst the cellulosics this is of little significance since when these properties are important there are many superior non-cellulosic alternatives. The principal use for ethyl cellulose injection mouldings are in those applications where good impact strength at low temperatures are required such as refrigerator bases and flip lids and ice-crusher parts.

Ethyl cellulose is often employed in the form of a 'hot melt' for strippable coatings. Such strippable coatings first became prominent during the Second World War for packaging military equipment. Since then they have been extensively used for protecting metal parts against corrosion and marring during shipment and storage. A typical composition consists of 25% ethyl cellulose, 60% mineral oil, 10% resins and the rest stabilisers and waxes. Coating is performed by dipping the cleaned metal part into the molten compound. The metal part is withdrawn and an adhering layer of the composition is allowed to harden by cooling. Hot melts have also been used for casting and paper coating.

The ether is also used in paint, varnish and lacquer formulations. A recent development is the use of ethyl cellulose gel lacquers. These are permanent coatings applied in a similar way to the strippable coatings. They have been used in the United States for coating tool handles, door knobs and bowling pins.

18.3.2 Miscellaneous ethers

Only one other cellulose ether has been marketed for moulding and extrusion applications, *benzyl cellulose*. This material provides a very rare example of a polymer which although available in the past is no longer commercially marketed.   The material had a low softening point and was unstable to both heat and light and has thus been unable to compete with the many alternative materials now available.

A number of water soluble cellulose ethers are marketed.[4]   *Methyl cellulose* is prepared by a method similar to that for ethyl cellulose.   A degree of substitution of $1 \cdot 6$–$1 \cdot 8$ is usual since the resultant ether is soluble in cold water but not in hot.   It is used as a thickening agent and emulsifier in cosmetics, as a paper size, in pharmaceuticals, ceramics and in leather tanning operations.

Hydroxyethyl cellulose, produced by reacting alkali cellulose with ethylene oxide, is employed for similar purposes.

Reaction of alkali cellulose with the sodium salt of chloracetic acid yields *sodium carboxymethyl cellulose*, (s.c.m.c.).   Commercial grades usually have a degree of substitution between $0 \cdot 50$ and $0 \cdot 85$.   The material which appears to be physiologically inert is very widely used.   Its principal application is as a soil-suspending agent in synthetic detergents.   It is also the basis of a well-known proprietary wallpaper adhesive.   Miscellaneous uses include fabric sizing and as a surface active agent and viscosity modifier in emulsions and suspensions.   Purified grades of s.c.m.c. are employed in ice cream to provide a smooth texture and in a number of pharmaceutical and cosmetic products.

Schematic equations for the production of the fully substituted varieties of the above three ethers are given below (R represents the cellulose skeleton).

$$R(ONa)_{3n} + CH_3Cl \longrightarrow R(OCH)_{3n}$$
Methyl Cellulose

$$R(ONa)_{3n} + CH_2\!\!-\!\!CH_2 \text{ (O)} \longrightarrow R(OCH_2CH_2OH)_{3n}$$
Hydroxyethyl Cellulose

$$R(ONa)_{3n} + ClCH_2 \cdot COONa \longrightarrow R(OCH_2COONa)_{3n} + NaCl$$
Sodium Carboxymethyl Cellulose

18.4 REGENERATED CELLULOSE

Because of high interchain bonding cellulose is insoluble in solvents and is incapable of flow on heating, the degradation temperature being reached before the material starts to flow.   It is thus somewhat intractable in its native form.   Cellulose however may be chemically treated so that the modified products may be dissolved and the solution may then either be cast into film or spun into fibre.   By treatment of the film or fibre the cellulose derivative may be converted back (regenerated) into cellulose although the processing involves reduction in molecular weight.

In the case of fibres three techniques have been employed:

1. Dissolution of the cellulose in cuprammonium solution followed by acid coagulation of extruded fibre ('cuprammonium rayon'—no longer of commercial importance). In this case the acid converts the cuprammonium complex back into cellulose.
2. Formation of cellulose acetate, spinning into fibre and subsequent hydrolysis into cellulose.
3. Reaction of alkali cellulose with carbon disulphide to produce a cellulose xanthate which forms a lyophilic sol with caustic soda. This may be extruded into a coagulating bath containing sulphate ions which hydrolyses the xanthate back to cellulose. This process is known as the viscose process and is that used in the manufacture of rayon.

By modification of the viscose process a regenerated cellulose foil may be produced which is known under the familiar trade name Cellophane.

The first step in the manufacture of the foil involves the production of alkali cellulose. This is then shredded and allowed to age in order that oxidation will degrade the polymer to the desired extent. The alkali cellulose is then treated with carbon disulphide in xanthating churns at 20–28°C for about three hours.

The xanthated cellulose contains about one xanthate group per two glucose units. The reaction may be indicated schematically as

$$R \cdot ONa + CS_2 \longrightarrow R \cdot O \cdot \underset{\underset{S}{\|}}{C} \cdot SNa$$

The resultant yellow sodium cellulose xanthate is dispersed in an aqueous caustic soda solution where some hydrolysis occurs. This process is referred to as 'ripening' and the solution as 'viscose'. When the hydrolysis has proceeded sufficiently the solution is transferred to a hopper from which it emerges through a small slit on to a roller immersed in a tank of 10–15% sulphuric acid and 10–20% sodium sulphate at 35–40°C. The viscose is coagulated and by completion of the hydrolysis the cellulose is regenerated. The foil is subsequently washed, bleached, plasticised with ethylene glycol or glycerol and then dried.

The product at this stage is 'plain' foil and has a high moisture vapour transmission rate. Foil which is more moisture proof may be obtained by coating with pyroxylin (cellulose nitrate solution) containing dibutyl phthalate as plasticiser or with vinylidene chloride–acrylonitrile copolymers. A range of foils are available differing largely in their moisture impermeability and in heat sealing characteristics.

Regenerated cellulose foil has been extensively and successfully used as a wrapping material particularly in the food and tobacco industries. As with other cellulose materials it is now having to face the challenge of the completely synthetic polymers. Although the foil has been able to compete in the past the advent of polypropylene film in the early 1960's has produced a serious competitor which may well lead to a marked reduction in the use of the cellulosic materials.

18.5 VULCANISED FIBRE

This material has been known for many years, being used originally in the making of electric lamp filaments. In principle vulcanised fibre is produced by the action of zinc chloride on absorbent paper. The zinc chloride causes the cellulosic fibres to swell and be covered with a gelatinous layer. Separate layers of paper may be plied together and the zinc chloride subsequently removed to leave a regenerated cellulose laminate.

The removal of zinc chloride involves an extremely lengthy procedure. The plied sheets are passed through a series of progressively more dilute zinc chloride solutions and finally pure water in order to leach out the gelatinising agent. This may take several months. The sheets are then dried and consolidated under light pressure.

The sheets may be formed to some extent by first softening in hot water or steam and then pressing in moulds at pressures of 200–500 lb in^{-2}. Machining, using high speed tools, may be carried out on conventional metal working machinery.

A number of grades are available according to the desired end use. The principal applications of vulcanised fibre are in electrical insulation, luggage, protective guards and various types of materials handling equipment. The major limitations are dimensional instability caused by changes in humidity, lack of flexibility and the long processing times necessary to extract the zinc chloride.

REFERENCES

1. PAIST, W. D., *Cellulosics*, Reinhold, New York (1958)
2. STANNETT, V., *Cellulose Acetate Plastics*, Temple Press Ltd., London (1950)
3. FORDYCE, C. R., and MEYER, L. W. A., *Ind. Eng. Chem.*, 33, 597 (1940)
4. DAVIDSON, R. L., and SITTIG, M., *Water Soluble Resins*, Reinhold, New York (1962)

BIBLIOGRAPHY

DAVIDSON R. L., and SITTIG, M., *Water Soluble Resins*, Reinhold, New York (1962)
MILES, F. D., *Cellulose Nitrate*, Oliver and Boyd, London (1955)
OTT, G., SPURLIN, H. M., and GRAFFLIN, M. W., *Cellulose and its Derivatives*, 2nd Edn, Interscience, New York (1954) (3 vols)
PAIST, W. D., *Cellulosics*, Reinhold, New York (1958)
STANNETT, V., *Cellulose Acetate Plastics*, Temple Press Ltd., London (1950)
YARSLEY, V. E., FLAVELL, W., ADAMSON, P. S., and PERKINS, N. G., *Cellulosic Plastics*, Iliffe, London (also published as a Plastics Institute Monograph) (1964)

19

Phenolic Resins

19.1 INTRODUCTION

The phenolic resins may be considered to be the first polymeric products produced commercially from simple compounds of low molecular weight, i.e. they were the first truly synthetic resins to be exploited. Their early development has been dealt with briefly in Chapter 1 and more fully elsewhere.[1]

Today these materials are widely used as moulding powders laminating resins, casting resins, as binders and impregnants, in surface coatings and adhesives and in a miscellany of other applications. Although their growth rate has been less spectacular than those of the major ethenoid thermoplastics in recent years, there is a continued expansion in consumption in most fields of application, although in a few specific instances, such as in telephones and domestic plugs and switches they have been partially replaced by newer materials. There have also been comparatively few new end-uses of the phenolics and thus increased consumption is largely due to the greater use of established products in which phenolics are used.

19.2 RAW MATERIALS

The phenolics are resinous materials produced by condensation of a phenol, or mixture of phenols, with an aldehyde. Phenol itself and the cresols are the most widely used phenols whilst formaldehyde and, to a much less extent furfural, are almost exclusively used as the aldehyde.

19.2.1 Phenol

At one time the requirement for phenol (melting point 41°C), could be met by distillation of coal tar and subsequent treatment of the middle oil with caustic soda to extract the phenols. Such tar acid distillation products, sometimes containing up to 20% o-cresol, are still used in resin manufacture but the bulk of phenol available today is obtained synthetically from benzene or other chemicals by such processes as the sulphonation process, the Raschig process and the cumene process.   Synthetic phenol is a purer product and thus has the advantage of giving rise to less variability in the condensation reactions.

In the sulphonation process vaporised benzene is forced through a mist of sulphuric acid at 100–120°C and the benzene sulphonic acid formed

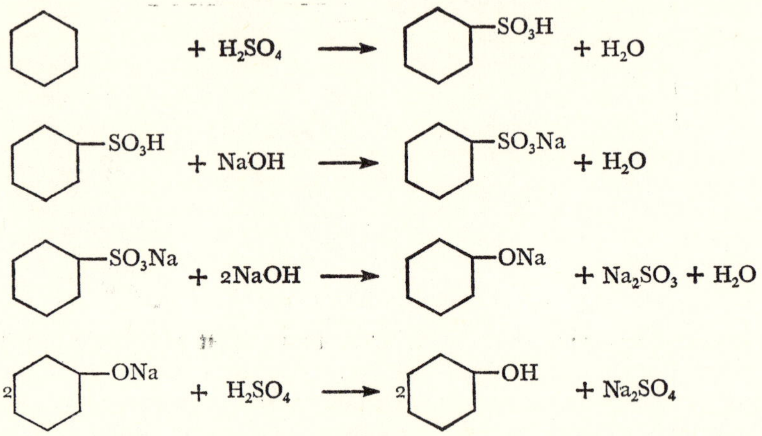

Fig. 19.1

neutralised with soda ash to produce benzene sodium sulphonate.   This is fused with a 25–30% excess of caustic soda at 300–400°C.   The sodium phenate obtained is treated with sulphuric acid and the phenol produced is distilled with steam (Fig. 19.1).

Today the sulphonation route is somewhat uneconomic and is being replaced by newer routes.   Processes involving chlorination, such as the Raschig process, are used on a large scale commercially.   A vapour phase reaction between benzene and hydrochloric acid is carried out in the presence of catalysts such as an aluminium hydroxide–copper salt complex. Monochlorobenzene is formed and this is hydrolysed to phenol with water in the presence of catalysts at about 450°C at the same time regenerating the hydrochloric acid.   The phenol formed is extracted with benzene, separated from the latter by fractional distillation and purified by vacuum distillation.   In recent years developments in this process have reduced the amount of by-product dichlorobenzene formed and also considerably increased the output rates.

A third process, now the principal synthetic process in use in Europe, is the cumene process.

In this process liquid propylene, containing some propane, is mixed with benzene and passed through a reaction tower containing phosphoric acid on kieselguhr as catalyst.   The reaction is exothermic and the propane present acts as a quench medium.   A small quantity of water is injected into the reactor to maintain catalyst activity.   The effluent from the reactor is then passed through distillation columns.   The propane is partly recycled, the unreacted benzene returned to feed and the cumene taken off (Fig. 19.2). The cumene is then oxidised in the presence of alkali at about 130°C (Fig. 19.3). The hydroperoxide formed is decomposed in a stirred vessel by addition of dilute sulphuric acid. The mixture is passed to a separator and the resulting organic layer fractionated (Fig. 19.4). Some benzophenone is also produced in a side reaction.

The economics of this process are to some extent dependent on the value of the acetone which is formed with the phenol. The process is however generally considered to be competitive with the modified Raschig process in which there is no by-product of reaction. In all of the above processes benzene is an essential starting ingredient. At one time this was obtained exclusively by distillation of coal tar but today it is being produced to an ever-increasing extent from petroleum.

A route to phenol has recently been developed starting from cyclohexane which is first oxidised to a mixture of cyclohexanol and cyclohexanone.

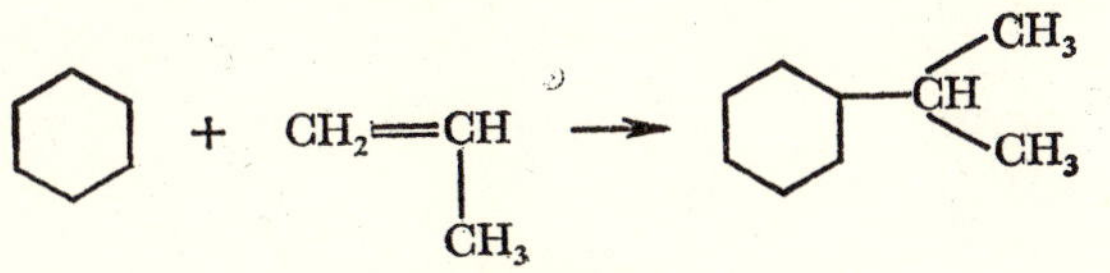

Fig. 19.2

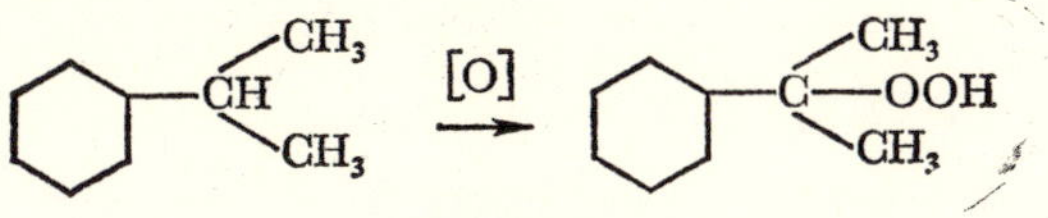

Fig. 19.3

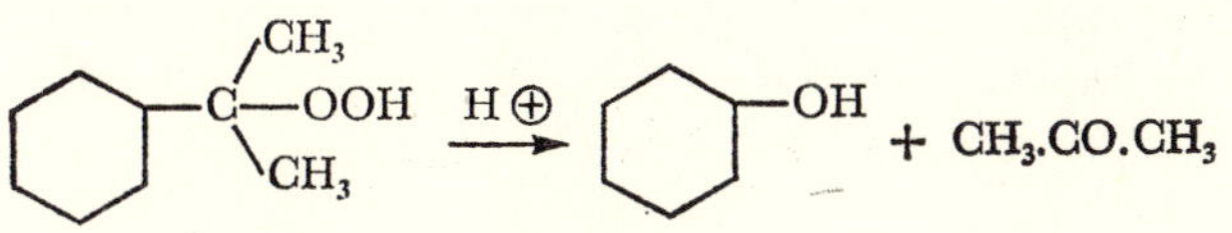

Fig. 19.4

In one process the oxidation is carried out in the liquid phase using cobalt naphthenate as catalyst. The cyclohexanone present may be converted to cyclohexanol, in this case the desired intermediate, by catalytic hydrogenation. The cyclohexanol is converted to phenol by a catalytic process using selenium or with palladium on charcoal. The hydrogen produced in this process may be used in the conversion of cyclohexanone to cyclohexanol. It also may be used in the conversion of benzene to cyclohexane in processes where benzene is used as the precursor of the cyclohexane.

Other routes for the preparation of phenol are under development and include the Dow process based on toluene. In this process a mixture of toluene, air and catalyst are reacted at moderate temperature and pressure to give benzoic acid. This is then purified and decarboxylated, in the presence of air, to phenol (Fig. 19.5).

Pure phenol crystallises in long colourless needles which melt at 41°C.

It causes severe burns on the skin and care should be taken in handling the material.   Phenol is supplied commercially either in the solid (crystalline) state or as a 'solution' in water (water content 8–20%).   Where supplied as a solid it is usually handled by heating the phenol, and the molten material is pumped into the resin kettles or into a preblending tank.   If the 'solution' is used care must be taken to avoid the phenol crystallising out.

19.2.2   Other phenols

A number of other phenols obtained from coal tar distillates are used in the manufacture of phenolic resins.   Of these the cresols are the most important (Fig. 19.6).

The cresols occur in cresylic acid, a mixture of the three cresols together with some xylenols and neutral oils, obtained from coal tar distillates. Only the *m*-cresol has the three reactive positions necessary to give cross-linked resins and so this is normally the desired material.   The *o*-isomer is

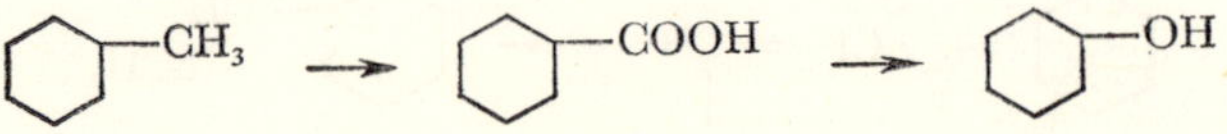

Fig. 19.5

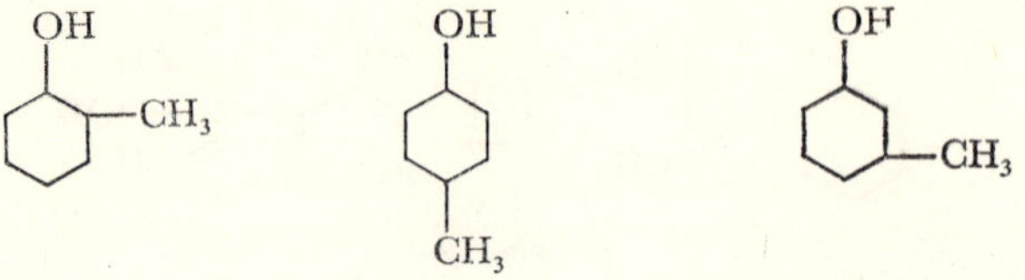

o -Cresol   p -Cresol   m -Cresol
Boiling Point 191·0°C   Boiling Point 201·9°C   Boiling Point 202·2°C

Fig. 19.6

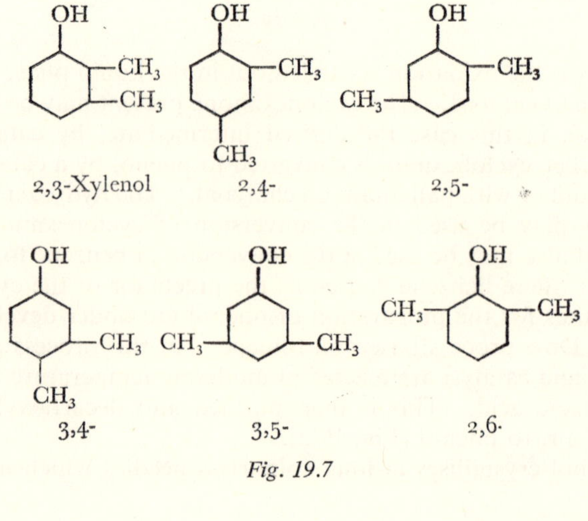

2,3-Xylenol   2,4-   2,5-

3,4-   3,5-   2,6-

Fig. 19.7

easily removed by distillation but separation of the close-boiling *m*- and *p*-isomers is difficult and so mixtures of these two isomers are used in practice.

Xylenols, also obtained from coal tar, are sometimes used in oil-soluble resins. Of the six isomers (Fig. 19.7) only 3,5-xylenol has the three reactive positions necessary for cross linking and thus mixtures with a high proportion of this isomer are generally used.

Other higher boiling phenolic bodies obtainable from coal tar distillates are sometimes used in the manufacture of oil-soluble resins. Mention may also be made of cashew nut shell liquid which contains phenolic bodies and which is used in certain specialised applications.

A few synthetic substituted phenols are also used in the manufacture of oil-soluble resins. They include *p*-tert-butyl phenol, *p*-tert-amyl phenol, *p*-tert-octyl phenol, *p*-phenyl phenol and diphenylol propane (bis-phenol A).

19.2.3 Aldehydes

Formaldehyde is by far the most commonly employed aldehyde in the manufacture of phenolic resins. Its preparation has been described in Chapter 16. It is normally used as an aqueous solution, known as formalin, containing about 37% by weight of formaldehyde. From $\frac{1}{2}$–10% of methanol may be present to stabilise the solution and retard the formation of polymers. When the formalin is used soon after manufacture only low methanol contents are employed since the formalin has a higher reactivity. Where a greater storage life is required the formalin employed has a higher methanol content, but the resulting increased stability is at the expense of reduced reactivity.

Furfural (see Chapter 24) is occasionally used to produce resins with good flow properties for use in moulding powders.

19.3 CHEMICAL ASPECTS

Although phenolic resins have been known and widely utilised for over 50 years their detailed chemical structure remains to be established. It is now known that the resins are very complex and that the various structures present will depend on the ratio of phenol to formaldehyde employed, the pH of the reaction mixture and the temperature of the reaction. Phenolic resin chemistry has been discussed in detail elsewhere[2,3,4,5,6,7] and will only be discussed briefly here.

Reaction of phenol with formaldehyde involves a condensation reaction which leads, under appropriate conditions, to a cross-linked polymer structure. For commercial application it is necessary first to produce a tractable fusible low molecular weight polymer which may, when desired, be transformed into the cross-linked polymer. For example, in the manufacture of a phenolic (phenol–formaldehyde, P–F) moulding a low molecular weight resin is made by condensation of phenol and formaldehyde. This resin is then compounded with other ingredients, the mixture ground to a powder and the product heated under pressure in a mould. On heating, the resin melts and under pressure flows in the mould. At the same time further chemical reaction occurs leading to cross linking. It is obviously desirable to process under such conditions that the required amount to flow has occurred before the resin hardens.

The initial phenol–formaldehyde reaction products may be of two types, *novolaks* and *resols*.

19.3.1 Novolaks

The novalaks are prepared by reacting phenol with formaldehyde in a molar ratio of approximately $1:0\cdot8$ under acidic conditions. Under these conditions there is a slow reaction of the two reactants to form the o- and p-methylol phenols (Fig. 19.8).

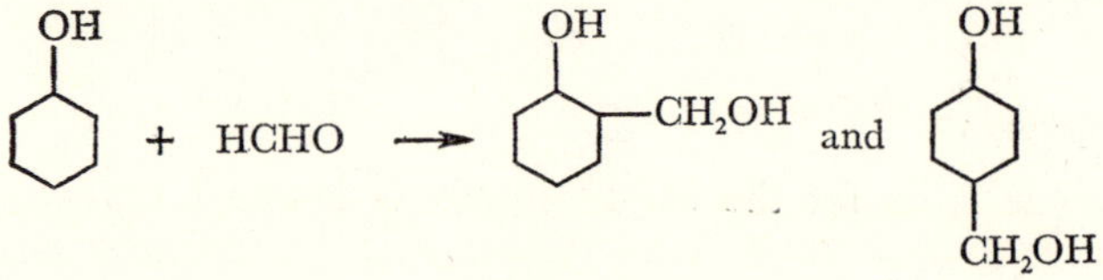

Fig. 19.8

These then condense rapidly to form products of the dihydroxydiphenyl methane (d.p.m.) type (e.g. Fig. 19.9).

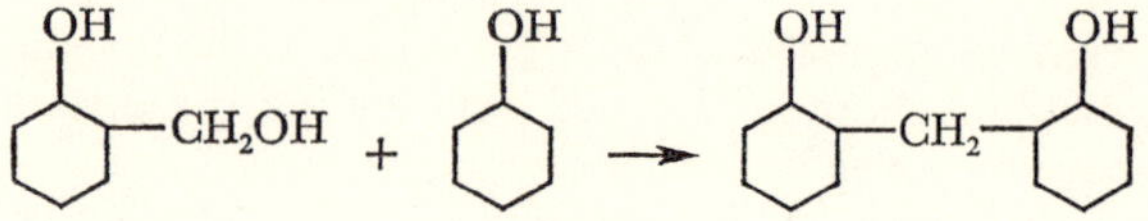

Fig. 19.9

There are three possible isomers and the proportions in which they are formed will depend on the pH of the reaction medium. Under the acid conditions normally employed in novolak manufacture the 2,4- and 4,4-'-d.p.m. compounds are the main products (Fig. 19.10).

These materials will then slowly react with further formaldehyde to form their own methylol derivatives which in turn rapidly react with further

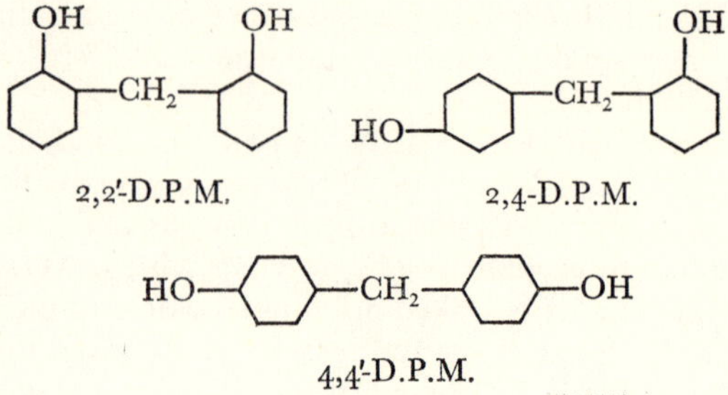

2,2'-D.P.M.   2,4-D.P.M.

4,4'-D.P.M.

Fig. 19.10

phenol to produce higher polynuclear phenols. Because of the excess of phenol there is a limit to the molecular weight of the product produced, but on average there are 5–6 benzene rings per molecule. A typical example of the many possible structures is shown in Fig. 19.11.

The novolak resins themselves contain no reactive methylol groups and do not form cross-linked structures on heating. If however they are

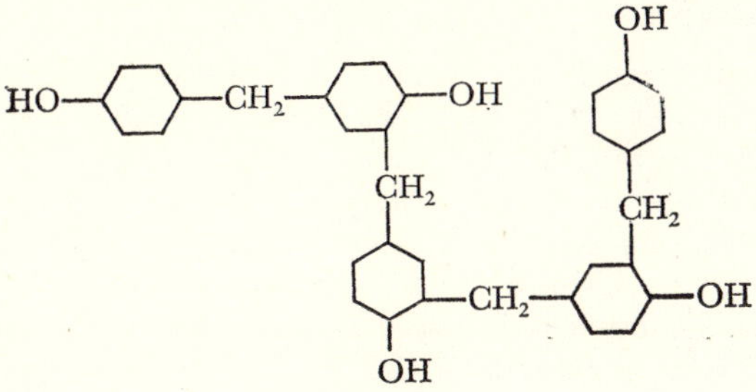

Fig. 19.11

mixed with compounds capable of forming methylene bridges, e.g. hexa-methylene tetramine or paraformaldehyde, they cross link on heating to form infusible, 'thermoset' structures.

In general it is considered essential that the bulk of the phenol used initially should not be substituted, i.e. should be reactive, at the *o*- and *p*-positions and is thus trifunctional with respect to the reaction with formaldehyde.

19.3.2 Resols

A resol is produced by reacting a phenol with an excess of aldehyde under basic conditions.

In this case the formation of phenol–alcohols is rapid but their subsequent condensation is slow. Thus there is a tendency for polyalcohols, as well as monoalcohols, to be formed. The resulting polynuclear

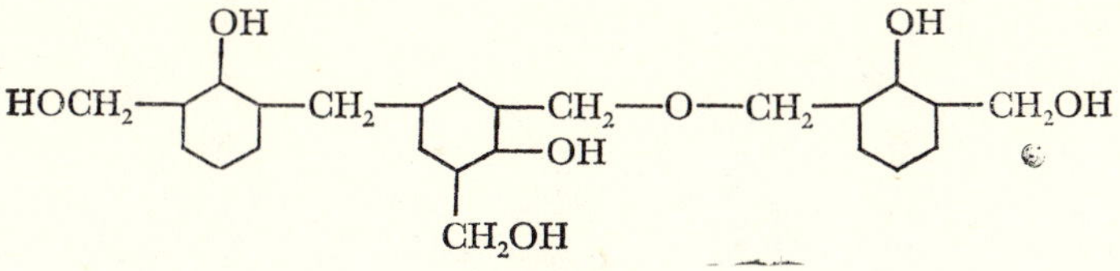

Fig. 19.12

polyalcohols are of low molecular weight. Liquid resols have an average of less than two benzene rings per molecule, while a solid resol may have only three to four. A typical resol would have the structure shown in Fig. 19.12.

Heating of these resins will result in cross linking via the uncondensed methylol groups or by more complex mechanisms. The resols are sometimes referred to as one-stage resins since cross-linked products may be made from the initial reaction mixture solely by adjusting the pH. On the other hand the novolaks are sometimes referred to as two stage resins as here it is necessary to add some agent which will enable additional methylene bridges to be formed.

19.3.3 Hardening

The novolaks and resols are soluble and fusible low molecular weight products. They were referred to by Baekeland as A-stage resins. On hardening, these resins pass through a rubbery stage in which they are swollen, but not dissolved, by a variety of solvents. This is referred to as the B-stage. Further reaction leads to rigid, insoluble, infusible, hard products known as C-stage resins. When prepared from resols the B-stage resin is sometimes known as a *resitol* and the C-stage product a *resit*. The terms A-, B- and C-stage resins are also sometimes used to describe anologous states in other thermosetting resins.

The mechanism of the hardening processes has been investigated by Zinke in Austria, von Euler in Sweden and Hultzsch in Germany using blocked methylol phenols so that only small isolable products would be obtained.

In general their work indicates that at temperatures below 160°C cross linking occurs by phenol methylol–phenol methylol and phenol methylol–phenol condensations, viz Fig. 19.13.

As these condensation reactions can occur at the two *ortho* and the *para* positions in phenol, *m*-cresol and 3,5-xylenol, cross-linked structures will

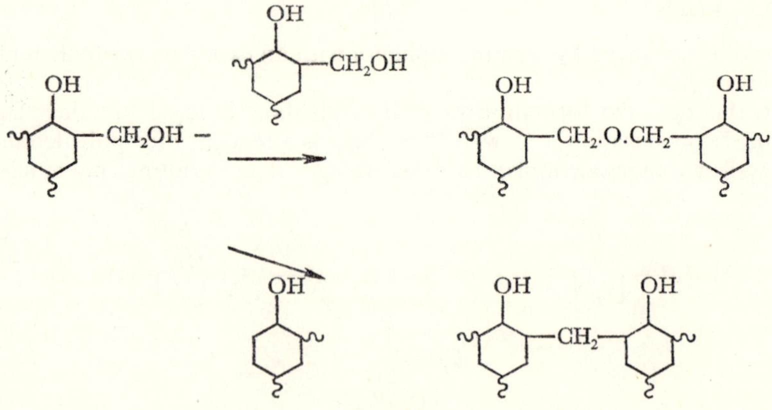

Fig. 19.13

be formed. It has been pointed out by Megson[5] that because of steric hindrance the amount of cross linking that can take place is much less than would involve the three reactive groups of all the phenolic molecules. It is now generally considered that the amount of cross linking that actually takes place is less than was at one time believed to be the case.

Above 160°C it is believed that additional cross-linking reactions take place involving the formation and reaction of quinone methides by condensation of the ether linkages with the phenolic hydroxyl groups (Fig. 19.14).

These quinone methide structures are capable of polymerisation and of other chemical reactions.

It is likely that the quinone–methide and related structures formed at these temperatures account for the dark colour of phenolic compression

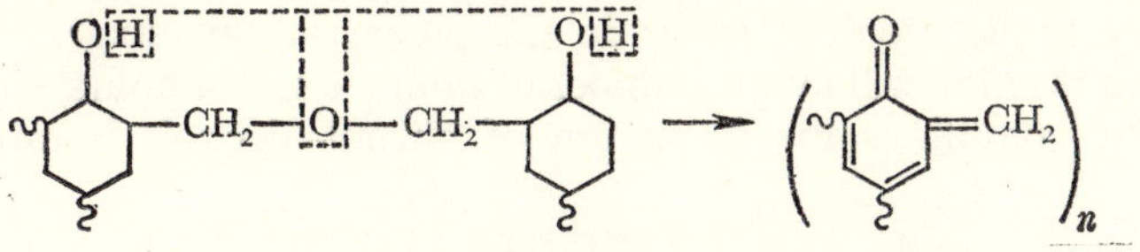

Fig. 19.14

mouldings.   It is to be noted that cast phenol–formaldehyde resins, which are hardened at much lower temperatures, are water-white in colour.   If however these castings are heated to about 180°C they darken considerably.

In addition to the above possible mechanisms the possibility of reaction at *m*-positions should not be excluded.   For example it has been shown by Koebner that *o*- and *p*-cresols, ostensibly difunctional, can, under certain conditions, react with formaldehyde to give insoluble and infusible resins. Furthermore, Megson has shown that 2,4,6-trimethyl phenol, in which the two *ortho*- and the one *para*- positions are blocked, can condense with formaldehyde under strongly acidic conditions.   It is of interest to note that Redfarn produced an infusible resin from 3,4,5-trimethyl phenol under alkaline conditions.   Here the two *m*- and the *p*-positions were blocked and this experimental observation provides supplementary evidence that additional functionalities are developed during reaction, for example in the formation of quinone methides.

The importance of the nature of the catalyst on the hardening reaction must also be stressed.   Strong acids will sufficiently catalyse a resol to cure thin films at room temperature, but as the pH rises there will be a reduction in activity which passes through a minimum at about pH 7.   Under alkaline conditions the rate of reaction is related to the type of catalyst and to its concentration.   The effect of pH value on the gelling time of a casting resin (phenol–formaldehyde ratio $\sim 1:2\cdot25$) is shown in Fig. 19.15.

19.4 RESIN MANUFACTURE

Both novolaks and resols are prepared in similar equipment, shown diagrammatically in Fig. 19.16.   The resin kettle may be constructed from copper, nickel or stainless steel where novolaks are being manufactured.   Stainless steel may also be used for resols but where colour formation is unimportant the cheaper mild steel may be used.

In the manufacture of novolaks, 1 mole of phenol is reacted with about

0·8 mole of formaldehyde (added as 37% w/w formalin) in the presence of some acid as catalyst.   A typical charge ratio would be:

Phenol	100 parts by weight
Formalin (37% w/w)	70 parts by weight
Oxalic acid	1·5 parts by weight

The reaction mixture is heated and allowed to reflux, under atmospheric pressure at about 100°C.   At this stage valve A is open and valve B closed.   Because the reaction is strongly exothermic initially it may be necessary to use cooling water in the jacket at this stage.   The condensation reaction will take a number of hours, e.g. 2–4 hours, since under the acidic conditions the formation of phenol–alcohols is rather slow.

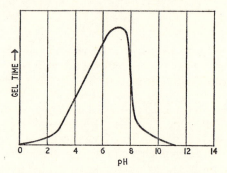

Fig. 19.15.   *Effect of pH on the gel time of a P–F cast resin.   (After Apley[8])*

When the resin separates from the aqueous phase and the resin reaches the requisite degree of condensation, as indicated by refractive index measurements, the valves are changed over (i.e. valve A is closed and valve B opened) and water present is distilled off.

In the case of novolak resins the distillation is normally carried out without the application of vacuum.   Thus, as the reaction proceeds and the water is driven off, there is a rise in the temperature of the resin which may reach as high as 160°C at the end of the reaction.   At these temperatures the fluid is less viscous and more easily stirred.   In cases where it is important to remove the volatiles present, a vacuum may be employed after the reaction has been completed, but for fast curing systems some of the volatile matter (mainly low molecular weight phenolic bodies) may be retained.

The end point may be checked by noting the extent of flow of a heated pellet down a given slope or by melting point measurements.   Other control tests include alcohol solubility, free phenol content and gelation time with 10% hexa.

In the manufacture of resols a molar excess of formaldehyde (1·5–2·0:1) is reacted with the phenol in alkaline conditions.   In these conditions the formation of the phenol alcohols is quite rapid and the condensation to a resol may take less than an hour. A typical charge for a laboratory scale preparation would be :

Phenol	94 g (1 mole)
Formalin (40%)	112 cm³ (1·5 moles formaldehyde)
0·88 ammonia	4 cm³

The mixture is refluxed until the reaction has proceeded sufficiently. It may then be neutralised and the water formed distilled off, usually under reduced pressure to prevent heat-hardening of the resin. Because of the

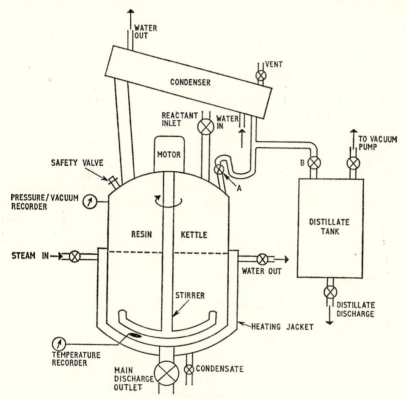

Fig. 19.16. *Diagrammatic representation of resin kettle and associated equipment used for the preparation of phenolic resins. (After Whitehouse and Pritchett[2])*

presence of methylol groups the resol has a greater water-tolerance than the novolak.

The reaction may be followed by such tests as melting point, acetone or alcohol solubility, free phenol content or loss in weight on stoving at 135°C.

Two classes of resol are generally distinguished, water-soluble resins prepared using caustic soda as catalyst, and spirit-soluble resins which are catalysed by addition of ammonia. The water soluble resins are usually only partially dehydrated during manufacture to give an aqueous resin solution with a solids content of about 70%. The solution viscosity can critically affect the success in a given application. Water-soluble resols are used mainly for mechanical grade paper and cloth laminates and in decorative laminates.

In contrast to the caustic soda-catalysed resols the spirit-soluble resins have good electrical insulation properties. In order to obtain superior insulation characteristics a cresol-based resol is generally used. In a typical reaction the refluxing time is about 30 minutes followed by dehydration under vacuum for periods up to 4 hours.

19.5 MOULDING POWDERS

Novolaks are most commonly used in the manufacture of moulding powders although resols may be used for special purposes such as in minimum odour grades and for improved alkali resistance. The resins are generally based on phenol since they give products with the greatest mechanical strength and speed of cure, but cresols may be used in acid-resisting compounds and phenol–cresol mixtures in cheaper compositions. Xylenols are occasionally used for improved alkali resistance.

The resols may be hardened by heating and/or by addition of catalysts. Hardening of the novolaks may be brought about by addition of hexamethylene tetramine (hexa, hexamine). Because of the exothermic reaction on hardening (cure) and the accompanying shrinkage, it is necessary to incorporate inert materials (fillers) to reduce the resin content. Fillers are thus generally necessary to produce useful mouldings and are not incorporated simply to reduce cost. Fillers may give additional benefits such as improving the shock resistance.

Other ingredients may be added to prevent sticking to moulds (lubricants), to promote the curing reaction (accelerators), to improve the flow properties (plasticisers) and to colour the product (pigments).

19.5.1   Compounding ingredients

It is thus seen that a phenol–formaldehyde moulding powder will contain the following ingredients:

> Resin
> Hardener (with Novolaks)
> Accelerator
> Filler
> Lubricant
> Pigment
> Plasticiser (not always used)

In addition to the selection of phenol used and the choice between novolak and resol there are a number of further variations possible in the resin used. For example in the manufacture of a novolak resin slight adjustment of phenol–formaldehyde ratio will affect the size of novolak molecule produced. Higher molecular weight novolaks give a stiff flow moulding powder but the resin being of lower reactivity, the powders have a longer cure time. A second variable is the residual volatile content. The greater the residual volatiles (phenolic bodies) the faster the cure. Thus a fast-curing, stiff-flow resin may be obtained by using a phenol–formaldehyde ratio leading to larger molecules and leaving some of the low molecular weight constituents in the reaction mixture. Yet another modification may be achieved by changing the catalyst used. Thus whereas in the normal processes using oxalic acid catalysts, the initial products are *p-p-* and *o-p-* diphenyl methanes, under other conditions it is possible to achieve products which have reacted more commonly in the *ortho-* position. Such resins thus have the *p*-position free and, since this is very reactive to hexa, a fast curing resin is obtained.

Hexa is used almost universally as the hardener. It is made by passing a slight excess of ammonia through a lightly stabilised aqueous solution of

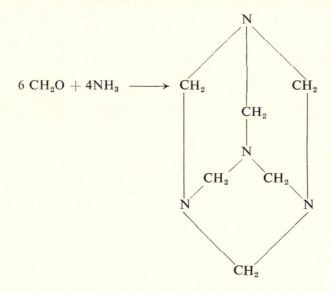

$$6 \ CH_2O + 4NH_3 \longrightarrow$$

Fig. 19.17

formaldehyde, concentrating the liquor formed and crystallising out the hexa (Fig. 19.17).

Between 10 and 15 parts of hexa are used in typical moulding compositions. The mechanism by which it cross links novolak resins is not fully understood but it appears capable of supplying the requisite methylene bridges required for cross linking. It also functions as a promoter for the hardening reaction.

Basic materials such as lime or magnesium oxide increase the hardening rate of novolak–hexa compositions and are sometimes referred to as accelerators. They also function as neutralising agents for free phenols and other acidic bodies which cause sticking to, and staining of, moulds and compounding equipment. Such basic substances also act as hardeners for resol-based compositions.

Woodflour, a fine sawdust preferably obtained from soft woods such as pine, spruce and poplar is the most commonly used filler. Somewhat fibrous in nature it is not only an effective diluent for the resin to reduce exotherm and shrinkage, but it is also cheap and improves the impact strength of the mouldings. There is a good adhesion between phenol–formaldehyde resin and the woodflour and it is possible that some chemical bonding may occur.

Another commonly employed low cost organic filler is coconut shell flour. This can be incorporated into the moulding composition in large quantities and this results in cheaper mixes than when woodflour is used. The mouldings also have a good finish. However, coconut shell flour-filled mouldings have poor mechanical properties and hence the filler is generally used in conjunction with wood flour.

For better impact strength cotton flock, chopped fabric or even twisted cord and strings may be incorporated. The cotton flock-filled compounds have the greatest mouldability but the lowest shock resistance whilst the twisted cords and strings have the opposite effect. Nylon fibres and fabrics are sometimes used to confer strength and flexibility and glass fibres may be used for strength and rigidity.

Asbestos may be used for improved heat and chemical resistance and silica, mica and china clay for low water absorption grades. Iron-free mica powder is particularly useful where the best possible electrical

Table 19.1

	G.P. grade	Electrical grade	Medium shock-resisting grade	High shock-resisting grade
Novalak resin	100	100	100	100
Hexa	12·5	14	12·5	17
Magnesium oxide	3	2	2	2
Magnesium stearate	2	2	2	3·3
Nigrosine dye	4	3	3	3
Wood flour	100	—	—	—
Mica	—	120	—	—
Cotton flock	—	—	110	—
Textile shreds	—	—	—	150
Asbestos	—	40	—	—

insulation characteristics are required but because of the poor adhesion of resin to the mica it is usually used in conjunction with a fibrous material such as asbestos. Organic fillers are commonly used in a weight ratio of 1:1 with the resin and mineral fillers in the ratio 1·5:1.

Stearic acid and metal stearates such as calcium stearate are generally used as lubricants at a rate of about 1–3% on the total compound. Waxes such as carnauba and ceresin or oils such as castor oil may also be used for this purpose.

In order that the rate of cure of phenolic moulding compositions is sufficiently rapid to be economically attractive, curing is carried out at a temperature which leads to the formation of quinone–methides and their derivatives which impart a dark colour to the resin. Thus the range of pigments available is limited to blacks, browns and relatively dark blues, greens, reds and oranges.

In some moulding compositions other special purpose ingredients may be incorporated. For example naphthalene, furfural and dibutyl phthalate are occasionally used as plasticisers or more strictly as flow promoters. They are particularly useful where powders with a low moulding shrinkage are required. In such formulations a highly condensed resin is used so that there will be less reaction, and hence less shrinkage, during cure. The plasticiser is incorporated to the extent of about 1% to give these somewhat intractable materials adequate flow properties.

Some typical formulations are given in Table 19.1.

19.5.2 Compounding of phenol–formaldehyde moulding compositions

Although there are many variants in the process used for manufacturing moulding powders, they may conveniently be classified into dry processes and wet processes.

In a typical dry process, finely ground resin is mixed with the other ingredients for about 15 minutes in a powder blender. This blend is then fed on to a heated 2-roll mill. The resin melts and the powdery mix is fluxed into a leathery hide which bands round the front roll. The temperatures chosen are such that the front roll is sufficiently hot to make the resin tacky and the rear roll somewhat hotter so that the resin will melt and be less tacky. Typical temperatures are 70–100°C for the front roll and 100–120°C for the back. As some further reaction takes place on the mill, resulting in a change of melting characteristics, the roll temperatures should be carefully selected for the resin used. In some processes two mills may be used in series with different roll temperatures to allow greater flexibility in operation. To achieve consistency in the end-product a fixed mixing schedule must be closely followed. Milling times vary from 10 minutes down to a straight pass through the mill.

The hide from the mill is then cooled, pulverised with a hammer-mill and the resulting granules are sieved. In a typical general purpose composition the granules should pass a 14×26 sieve. For powders to be used in automatic moulding plant fine particles are undesirable and so particles passing a 100×41 sieve (in a typical process) are removed. In addition to being more suitable for automatic moulding machines these powders are also more dust-free and thus more pleasant to use. For ease of pelleting, however, a proportion of 'fines' is valuable.

For the manufacture of medium-shock resisting grades the preblend of resin, filler and other ingredients does not readily form a hide on the mill rolls. In this case the composition is preblended in an internal mixer before passing on to the mills.

Extrusion compounders such as the Buss Ko-Kneader have been used for mixing phenolic resins. It is claimed that they produce in some respects a better product and are more economical to use than mill-mixers.

High-shock grades cannot be processed on mills or other intensive mixers without destroying the essential fibrous structure of the filler.   In these cases a wet process is used in which the resin is dissolved in a suitable solvent, such as industrial methylated spirits, and blended with the filler and other ingredients in a dough mixer.   The resulting wet mix is then laid out on trays and dried in an oven.

19.5.3   Processing characteristics

As it is a thermosetting material, the bulk of phenol–formaldehyde moulding compositions are processed on compression and transfer moulding plant with a very small amount being extruded.

Moulding compositions are available in a number of forms, largely determined by the nature of the fillers used.   Thus mineral-filled and wood-flour filled grades are generally powders whilst fibre-filled grades may be of a soft–lumpy texture.   Fabric-filled grades are sold in the form of shredded impregnated 'rag'.   The powder grades are available in differing granulations.   Very fine grades are preferred where there is a limited flow in moulds and where a high gloss finish is required.   Fine powders are however dusty and a compromise may be sought.   For mouldings in which extensive flow will occur, comparatively coarse (and thus dust free) powders can be used and a reasonable finish still obtained.

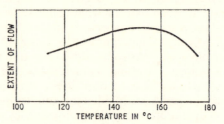

Fig. 19.18.   Dependence of the extent of flow on temperature for a general purpose phenolic resin.   Curves of this type may be obtained from measurements made on widely different pieces of equipment, e.g. the Resci–Peakes flow tester and the flow disc.   Thus no scale has been given for the vertical axis

For the best pelleting properties it would appear that some 'fines' are desirable for good packing whilst 'fines' are generally undesirable in powders employed in automatic compression moulding.

Since the resins cure with evolution of volatiles, compression moulding is carried out using moulding pressures of 1–2 ton in^{-2} at 155–170°C.   As with other thermosetting materials an increase in temperature has two effects.   Firstly, it reduces the viscosity of the molten resin and, secondly, it increases the rate of cure.   As a result of these two effects it is found that in a graph of extent of flow plotted against temperature there is a temperature of maximum flow.   (Fig. 19.18).

There is no entirely satisfactory way of measuring flow. In the B.S. 2782 flow cup test an amount of moulding powder is added to the mould to provide between 2 and 2·5 g of flash. The press is closed at a fixed initial rate and at a fixed temperature and pressure. The time between the onset of recorded pressure and the cessation of flash (i.e. the time at which the mould has closed) is noted. This time is thus the time required to move a given mass of material a fixed distance and is thus a measure of viscosity. It is not a measure of the time available for flow. This property, or rather the more important 'length of flow' or extent of flow, must be measured by some other device such as the flow disc or by the Rossi–Peakes flow test, neither of which are entirely satisfactory. Cup flow times are normally of the order of 10–25 seconds if measured by the B.S. specification. Moulding powders are frequently classified as being of 'stiff flow' if the cup flow time exceeds 20 seconds, 'medium flow' for times of 13–19 seconds and 'soft flow' or 'free flow' if under 12 seconds.

The bulk factor (i.e. ratio of the density of the moulding to the apparent powder density) of powder is usually about 2–3 but the high-shock grades may have bulk factors of 10–14 when loose, and still as high as 4–6 when packed in the mould. Powder grades are quite easy to pellet, but this is difficult with the fabric filled grades.

Phenol–formaldehyde moulding compositions may be preheated by high-frequency methods without difficulty. Preheating, by this or other techniques, will reduce cure time, shrinkage and required moulding pressures. Furthermore, preheating will enhance the ease of flow with consequent reduction in mould wear and danger of damage to inserts.

Moulding shrinkage of general purpose grades is in the order of 0·005–0·008 in/in. Highly loaded mineral-filled grades have a lower shrinkage whilst certain grades based on modified resins, e.g. acid resistant and minimum odour grades, may have somewhat higher shrinkage values.

Cure times will depend on the type of moulding powder used, the moulding temperature, the degree of preheating employed and, most important, on the end-use envisaged for the moulding. The time required to give the best electrical insulation properties may not coincide with the time required, say, for greatest hardness. However, one useful comparative test is the minimum time required to mould a blister-free flow cup under the B.S.771 test conditions. For general purpose material this is normally about 60 seconds but may be over twice this time with special purpose grades.

19.5.4 Properties of phenolic mouldings

Since the polymer in phenolic mouldings is cross linked and highly interlocked, phenolic mouldings are hard, heat resistant insoluble materials.

The chemical resistance of the mouldings depends on the type of filler and resin used. Simple phenol–formaldehyde materials are readily attacked by aqueous sodium hydroxide solution but cresol and xylenol based resins are more resistant. Provided the filler used is also resistant, phenolic mouldings are resistant to acids except 50% sulphuric acid, formic acid and oxidising acids. The resins are stable up to 200°C.

The mechanical properties are strongly dependent on the type of filler used and typical figures are given in Table 19.2.

Being polar the electrical insulation properties are not outstanding but are adequate for many purposes. At 100°C a typical wood flour–phenolic moulding has a dielectric constant of 18 and a power factor of 0·7 at 800 c/s.

One disadvantage of phenolics compared with the aminoplastics and the alkyd resins is their poor tracking resistance under conditions of high humidity. This means that phenolics have a tendency to form a conductive path through carbonisation along a surface between two metal electrodes at differing potential. Whether tracking will occur depends on the separation of the electrodes, the humidity of the atmosphere, the potential difference and on the presence and nature of surface contaminants. For many applications the poor tracking resistance is not a serious problem and the wide use of phenolic laminates and mouldings for electrical insulation applications is evidence of this.

19.5.5   Applications

Since the advent of Bakelite over 50 years ago phenol–formaldehyde moulding compositions have been used for a great variety of purposes. Perhaps the most well-known applications are in domestic plugs and switches. It should however be pointed out that since the Second World War, in Britain at least, urea–formaldehyde plastics have partially replaced phenol–formaldehyde for these purposes because of their better anti-tracking properties and wider colour range. There are nevertheless, many applications where the phenolics have proved quite adequate and continue to be used as insulators. In general it may be said that the phenolics have better heat and moisture resistance than the urea–formaldehyde mouldings (see Chapter 20). Phenol–formaldehyde mouldings have also found many other applications in the electrical industry, in some instances where high electrical insulation properties are not so important. These include instrument cases, knobs, handles and telephones. In some of these applications they have now been replaced by urea–formaldehydes, melamine–formaldehydes, alkyds or the newer thermoplastics because of the need for bright colours or in some cases in an attempt to produce tougher products. In the car industry phenol–formaldehyde mouldings are used in fuse-box covers, distributor heads and in other applications where good electrical insulation together with good heat resistance are required.

Phenol–formaldehyde mouldings continue to be used in many industrial applications where heat resistance, low cost and adequate shock resistance (varying of course with the type of powder used) are important features. Bottle caps and closures also continue to be made from phenolics in large quantities. For some applications minimum-odour grades based on resols are used. The development of automatic presses together with the

Table 19.2 PROPERTIES OF PHENOLIC MOULDINGS

Property	Units	General purpose	Medium shock	High shock	Electrical low loss	Acid resistant	Minimum odour	Heat resistant	B.S.2782 test method
B.S.771 classification	—	GX	MS	HS	L	—	—	HR	
Specific gravity	—	1·35	1·37	1·40	1·85	1·42	1·38	1·94	509A
Shrinkage	in/in.	0·006	0·005	0·002	0·002	0·009	0·007	0·002	106A
Impact strength	ft lb	0·16	0·29	0·8–1·4	0·14	0·14	0·17	0·10	305A
Cross-breaking strength	lb in^{-2}	11,500	11,000	12,000	11,000	8,750	11,250	8,750	304A
Tensile strength	lb in^{-2}	8,000	7,000	6,500	8,500	6,250	7,250	5,000	301A
Blister temperature	°C	175	170	175	190	185	190	195	—
Power factor 800 c/s	—	0·1–0·4	0·1–0·35	0·1–0·5	0·03–0·05	0·03–0·14	0·15–0·3	0·1–0·3	207A
Power factor 10^6 c/s	—	0·03–0·05	—	—	0·01–0·02	—	—	—	—
Dielectric constant 800 c/s	—	6·0–10·0	5·5–7·5	6·0–10·0	4·0–6·0	5·0–6·0	7·0–9·0	8·0–16·0	207A
Dielectric constant 10^6 c/s	—	4·5–5·5	—	—	4·3–5·4	—	—	—	—
Electric strength (20°C)	V/0·001 in.	150–300	200–275	150–250	275–350	225–300	175–225	250–350	—
Electric strength (90°C)	V/0·001 in.	100–250	75–175	50–150	250–350	200–275	75–150	200–300	201A
Water absorption 24 hr 23°C	mg	45–65	30–50	50–100	2–6	15–25	45–70	3–8	502F
Volume resistivity	ohm cm	10^{10}–10^{12}	10^{10}–10^{12}	$10^{9 \cdot 5}$–$10^{11 \cdot 5}$	$10^{11 \cdot 5}$–10^{14}	$10^{11 \cdot 5}$–10^{13}	10^{10}–$10^{11 \cdot 5}$	10^{11}–10^{12}	202A

advent of fast-curing grades has stimulated the use of phenol–formaldehydes for many small applications in spite of the competition from the major thermoplastics.

Today the phenol–formaldehyde moulding compositions do not have the eminent position they held until about 1950.   In some important applications they have been replaced by other materials, thermosetting and thermoplastic, whilst they have in the past decade found use in relatively few new outlets.   However, the general increase in standards of living has increased the sales of many products which use phenolics and consequently the overall use of phenol–formaldehyde moulding powders has been well maintained.

19.6 PHENOLIC LAMINATES

There are now commercially available a large range of laminated plastics materials.   Resins used include the phenolics, the aminoplastics, polyesters, epoxies, silicones and the furane resins, whilst reinforcements may be of paper, cotton fibre, other organic fibres, asbestos or glass fibre.   Of these the phenolics were the first to achieve commercial significance and they are still of considerable importance.

One-stage resins (resols) in which there are sufficient methylol groups to enable cross linking to occur without the need for formaldehyde donors are invariably used.   Resins based on phenol, or phenol-cresol mixtures are used in fabric laminates where the greatest mechanical strength is required, whereas cresylic acid (*m*-cresol content 50–55%) is generally used for electrical grade laminating resins because of the better electrical properties which result.   Caustic soda is commonly used as the catalyst for mechanical laminates but is not used in electrical laminates because it affects the electrical insulation properties adversely and ammonia is the usual catalyst in this instance.

For laminating, the ammonia-catalysed resins are usually dissolved in industrial methylated spirits (I.M.S.) or less commonly, isopropyl alcohol. Resins which have a high methylol content (i.e. made by using a high ratio of formaldehyde to phenol) and in which caustic soda is used as the catalyst are water-soluble and the aqueous solutions are useful where a high degree of impregnation is desirable.   They are commonly used in mechanical and decorative laminates.

The reinforcement may be a paper or a fabric.   Many different papers are used being selected according to the end-use of the laminate.   For example the Kraft papers are strong and produce laminates of high mechanical strength, the relatively non-porous sulphite wood pulp papers are used for electrical tubes whilst cotton paper and α-cellulose paper which are highly absorbent and of good colour are used in decorative laminates.   A wide range of fabrics are now used in conjunction with phenolic resins.   They include cotton, linen, rayon, glass fabrics and asbestos mat and cloth.

Although certain solventless processes have been used the resin is usually applied to the reinforcement by passing the latter through a varnish (40–50% solids content) of the resin in solvent. To ensure consistency of impregnation it is important to control the solids content, the viscosity and the specific gravity of the resin. At the same time the thickness, absorbency and density of the reinforcement, or base material, should also be kept within narrow limits. Fig. 19.19 shows a typical

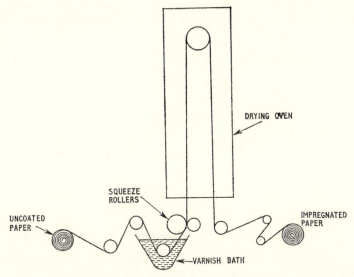

Fig. 19.19. *Impregnation plant fitted with vertical drying oven. (After Brown[9])*

arrangement for applying the resin to the reinforcement. The reinforcement is led into a tank of varnish and the resulting wet base is led through pressure rollers to squeeze out the excess varnish. The coated base material is then passed through either a vertical or horizontal drying oven. In a typical arrangement the temperature at the inlet end of the oven is at about 50–90°C and at the outlet about 145°C. The evaporating solvent is recovered and the resin taken to the required degree of polymerisation before emerging from the oven. The oven temperature must thus be dependent on the curing characteristics of the resin, the length of the oven and the coating rate employed. The impregnated paper is commonly checked for resin content and degree of cure. Control of degree of cure is important as the resin must have precise flow properties. If the viscosity is too high it will not flow sufficiently to consolidate the resin; conversely if it is too low the resin will spew out and leave a dry and inferior laminate. The degree of cure is perhaps most conveniently assessed by the practical test of preparing a small laminate in the laboratory by pressing at some controlled temperature and pressure. The weight of resin which spews out of the laminate is thus inversely related to the degree of cure, whilst more directly it will give an assessment of the laminating behaviour of the paper.

Flat laminates are prepared by plying up pieces of impregnated paper and pressing in a multi-daylight press between metal plates under pressure of 1,000–2,000 lb in⁻² and a temperature of 150–160°C.   After curing, which may take about 30 minutes for $\frac{1}{4}$ in. thick sheet, the platens are partially cooled before removal of the laminates in order to reduce blistering and warping.   Where the impregnated paper has a high volatile content it may also be necessary to heat the press after loading in order to control the rate of volatilisation and thus reduce blistering.

By the use of carefully tailored pieces of impregnated reinforcement, it is possible to produce laminated mouldings.   Such mouldings are tough and

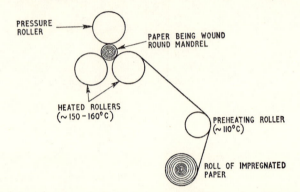

Fig. 19.20.   *Three roller tube winding machine.   (After Brown[9])*

have a high mechanical strength but take considerably longer to cure than corresponding products prepared from moulding powders.

Tubes and bushings are prepared by winding coated or impregnated paper around a mandrel and in pressure contact with heated rollers.   A typical three-roller tube winding machine is shown in Fig. 19.20.   A number of other simple shapes may be prepared by laminating under low pressure using hand-clamped tools or rubber bags.

19.6.1   The properties of phenolic laminates

The properties of a phenolic laminate will obviously depend on a great many factors.   Of these the following are perhaps the most important:

1. The type of resin used, including the nature of the catalyst, the concentration of methylol groups and the average molecular weights.
2. The properties of the varnish such as the nature of the solvent and the viscosity and resin content of the varnish.
3. The type of reinforcement.   In the case of fabric reinforcement, factors such as cloth weight and crimp will have a large effect on mechanical properties.
4. Moulding conditions, i.e. moulding pressure, temperature and time.

Figs. 19.21 and 19.22 show how two variables, moulding pressure and resin content affect the mechanical properties of a laminate.

From the above comments it will be seen that it is rather difficult to

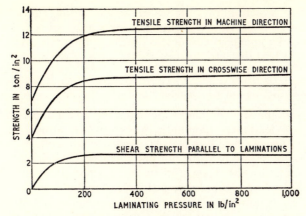

Fig. 19.21. *Effect of moulding pressure on the tensile and shear strength of Kraft paper laminates. (After Pepper and Barwell[10])*

quote typical figures for laminates based on phenolic resins. Thus the figures given in Table 19.3 can only be considered as a general guide.

In the manufacture of a laminate for electrical insulation, paper, which is the best dielectric, is normally selected as the base reinforcement. An 'electrical' grade of paper is in fact a better dielectric than the resin and thus in conditions of low humidity the resin content of the laminate can be quite low, particularly if the surfaces of the laminate are protected with an insulating varnish. For humid conditions a high resin content is used since this will lead to laminates with low water absorption, an essential property for a good insulator. Tubular laminates normally show superior insulation properties to the flat sheets since they are cured

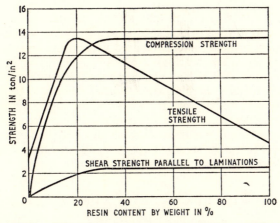

Fig. 19.22. *Effect of resin content on strength properties of a high-density Kraft paper laminate. (After Pepper and Barwell[10])*

layer-by-layer which allows water of condensation to escape during manufacture. As previously mentioned cresylic-based resins are usually used for such laminates in conjunction with ammonia as catalyst in order to achieve the best dielectric properties.

19.6.2 Applications of phenolic laminates

Phenolic resin–paper laminates are extensively used for high-voltage insulation applications. Laminates from other reinforcements are less suitable for this purpose but may be used for low voltage applications. Phenolic laminates are of value not only because of their good insulation properties but also because of their good strength, high rigidity and machinability. Sheet, tubular and moulded laminates are all employed.

Cotton fabric laminates are used in the manufacture of gear wheels which are quiet running and which withstand shock loading. Since the

Table 19.3 AVERAGE PHYSICAL PROPERTIES OF PHENOLIC RESIN BASED LAMINATES

Property	Unit	Paper for h.t. insulation	Paper for l.t. insulation	Fine fabric	Asbestos felt	Glass fabric
Tensile strength	10^3 lb in^{-2}	10	14	16	7–15	12–35
Cross breaking strength	10^3 lb in^{-2}	20	27	19	16	25
Impact strength (Izod edgewise)	ft lb	0·2–0·35	0·35–0·45	1·5	1·0	10·0
Specific gravity	—	1·37	1·40	1·36	1·6–1·8	1·4–1·7
Power factor 800 c/s	—	0·018	0·020	0·25	—	0·01–0·04
Power factor 10^6 c/s	—	0·032	0·042	0·10	0·11	0·01–0·02
Dielectric constant 10^6 c/s	—	4·6	5·2	6·5	6·1	4·5–5·5
Electric strength normal to laminate at 90°C (0·062 in. thick)	V/0·001 in.	450	380	40	100	—

laminates have a lower strength than steel, gear wheels made from them should be used at lower working stresses and designed with a greater face width for load transmission as compared with a similar gear made from steel. Water-lubricated bearings from phenolic–cotton or phenolic–asbestos laminates are used as bearings for steel rolling mills called to sustain bearing loads as high as 3,000 lb in^{-2}.

Although phenolic resins are too dark for use in the surface layers of decorative laminates these resins are employed in impregnating the core paper. (In these cases a melamine–formaldehyde resin is used for impregnating the top decorative layer.) Phenolic laminates have also been used in aircraft construction and in chemical plant.

19.7 MISCELLANEOUS APPLICATIONS

Although the two most well-known applications of phenolic resins are in mouldings and laminates they are also used in a very large number of other applications. At one time *cast phenolic resins*[8] were also an important class of plastics materials. These are made by reacting 1 mole of phenol with about 2·25 moles of formaldehyde using an alkaline catalyst. The reaction is carried out for about 3 hours at 70°C to produce a resinous material which remains soluble in the water present. To prevent the reaction going to completion in the reaction vessel the alkali present is neutralised by lactic acid or phthalic acid. The large quantity of water present (from the formalin solution and a small amount of water of condensation) is then removed by vacuum distillation during which time the polymerisation reaction continues rather more slowly under the acidic conditions (pH $\sim$ 4–5). Plasticisers and dyestuffs are usually incorporated before vacuum distillation. When the resin water content has reached the minimum possible consistent with pourability, the resin is cast into moulds. The resin is then hardened by heating for 3–10 days at 70–85°C. During cure, water is liberated and this is trapped in small droplets within the resin. The lower the amount of water present at the casting stage and the lower the phenol–formaldehyde ratio the smaller the droplets. Where reagents such as glycerol are incorporated into the resin, the droplets become smaller than the wavelength of visible light and the product becomes transparent.

The selection of the phenol–formaldehyde ratio of about 1:2·25 is a compromise in balancing mechanical properties (for which phenol–formaldehyde ratios of about 1:2·5 are most suitable) and curing rates (for which the optimum phenol–formaldehyde ratio is about 1:1·75). The use of a pH of 4–5 during distillation and hardening leads to a hardening rate sufficiently slow for the distillation of the water to be carried out without gelling in the reaction vessel, but not such that final hardening takes an infinitely slow time. Even so cure times are very long. It is not desirable to exceed cure temperatures of 75–80°C as this would lead to vaporisation of occluded water with subsequent void formation and blistering. Cast phenolic resins have been used for umbrella handles, knobs, propelling pencil bodies and for other purposes where attractive appearance is of importance. The advent of the cheaper ethenoid thermoplastics since the Second World War has however largely ousted the cast resins from such outlets and they are thus today of only minor importance.

There has been some interest in *phenolic resin foams* during recent years but these have yet to achieve large scale usage. These foams are self-extinguishing but currently more expensive to produce than the well-established expanded polystyrene. For good control of properties mechanical mixing devices, similar to those employed with polyurethanes, are used. A resol is first blended with the foaming agent such as sodium bicarbonate. The mixture is then stirred rapidly with an acid hardener such as *p*-toluene sulphonic acid and the product metered into a trough. An exothermic resin hardening reaction takes place generating sufficient heat to decompose the blowing agent and foam the product. Other ingredients may be included to control pore structure and density. By varying the formulations a range of foams may be produced with densities varying from about 1 to 20 lb ft^{-3}. The foaming times vary from only a few seconds for the very low density foams to 5–10 minutes for the higher density products.

A structural grade of foam has a thermal conductivity of 0·28 b.t.u. in ft^{-2} hr^{-1} degF^{-1} as measured by B.S. 874:1956, a figure which is of the same order as established thermal insulating materials. The foams may be classed as self-extinguishing but in a sustained flame they char and the surface is slowly eroded by the hot gases.[11] The foams may be made either as large blocks which can be subsequently sliced or alternatively foaming may be carried out *in situ*, for example between cavity walls of a building.

Somewhat intermediate in nature between the moulded powders and the laminates are the *fibre–resin preform mouldings*. These are produced by making a resin-containing fibre preform, usually of the same shape as the finished moulding, and then subjecting the preform to a consolidating

Table 19.4　SOME PHYSICAL PROPERTIES OF FIBRE–RESIN PREFORM SHEET[12]

Tensile strength	14,000–18,000 lb in^{-2}
Impact strength	1·5–3·0 ft lb on $\frac{1}{2}$ in. notched Izod
Shear strength	9,000–15,000 lb in^{-2}
Bending strength	15,000–19,000 lb in^{-2}
Water absorption (24 hr immersion)	1–10%

pressure in a compression mould. There are two principal variants of the process, the impregnation process and the beater process. In the impregnation process the selected fibre, normally of the cellulosic type and typically based on wood pulp, is fed with water to a beating or pulping machine of the type commonly used in paper making. After beating, the resulting slurry is fed to a stock tank where it is diluted with water from which it may be fed to the felting vat as required. A perforated metal felting tool is then lowered into the vat and by the application of a vacuum the pulp is deposited on to the screen. The amount deposited will depend on the degree of beating, the pulp consistency, the felting time and the degree of suction applied. The resultant felt is then presqueezed and transferred to a drying oven. The dried felt is then impregnated with a solution of resin, the solvent is evaporated and the preform may be moulded at pressures of about 500–800 lb in^{-2} at the usual moulding temperatures employed with phenolic resins. In the alternative beater process the resin, which can be in the form of an emulsion, dispersion, or solution, is added to the fibre at the beater stage. A preform is then made from the slurry, dried and moulded. The beater process is generally the more versatile and normally leads to products of greater impact strength.

Pulp–resin preform mouldings have the merits of light weight, low cost and good strength (see Table 19.4). They do however have a high water absorption and because of the limitations of the felting process are generally restricted to products of constant cross section.

Preform mouldings are particularly useful in carrying containers and protective covers. Examples of their use include television receiver backs, moulded suitcases and typewriter cases. Although the finish obtained during moulding is frequently adequate in industrial applications some improvement is necessary where a good appearance is desired. Methods used include painting or vacuum forming a thermoplastics sheet material over the outside of a moulded preform which has been coated with a suitable adhesive.

Phenolic resins are useful surface coating materials. Resols are useful for stoving lacquers for coating chemical plant, textile equipment, razor blades, brassware and food cans. Phenolic resins are used with poly(vinyl formal) as a flexible, tough and solvent resistant wire enamel. Oil-soluble resins based on synthetic phenols form the basis of some gloss paints.

A variety of adhesives based on phenolic resins are available. These include metal cements made by combining a resol with a vinyl polymer such as poly(vinyl formal). Resols are also used for plywood glues which may be cured using alkaline catalysts at 135°C. If resorcinol and/or *para*formaldehyde are included in the formulation, slightly lower curing temperatures may be used. These glues have good resistance to aging, moisture and bacteria. Resorcinol–formaldehyde novolak resins themselves are useful wood glues. Highly-filled novolak–hexa compositions form the basis of useful lamp capping cements.

Impregnated wood products are used in pattern making for foundry use and as stretching blocks and other similar tooling applications in the aircraft industry. These may be made by bonding wood veneers with a phenolic glue at sufficiently high pressures so that there is some impregnation of the adhesive into the wood. An alternative approach which gives a denser product is to immerse wood veneers in a resol solution under pressure. The impregnated veneers are then dried and plied together in a press. In this process the resin content of the impregnated wood may reach as high as 40%.

During the past decade phenolic resins have become of increased significance in rubber compounding. For example the resin based on cashew nut shell liquid, which contains phenolic bodies such as anacardic

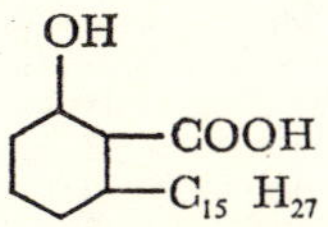

OH

—COOH
—$C_{15} H_{27}$

Anacardic Acid

Fig. 19.23

acid (Fig. 19.23) may, when blended with hexamine, be incorporated into nitrile rubber (butadiene–acrylonitrile rubber).

The resins act as a plasticiser during processing but cross link while the rubber is vulcanising to give a harder product with improved oxidation resistance, oil resistance and tensile strength. The addition of sufficient resin will lead to an ebonite-like product.

Similar products from natural rubber may be made by fluxing a cresol novolak containing hexamine with the rubber at 150°C.

In many applications the phenolic resins function primarily as binders, this being to some extent their purpose in moulding compositions, laminates and fibre–resin preform mouldings.   Further examples of this function are found in brake linings, grinding wheels and flexible abrasives, resin pulp board and wallboard, sand core bonding and shell moulds for metal castings.   The resins are also used in binding glass fibre thermal insulation and in the manufacture of waste-wood board.

Other uses include impregnation of wood to improve dimensional stability and reduce water absorption, sealing of porous metal castings by impregnation, and coil impregnation, to give a rigid structure both heat and water resistant.

REFERENCES

1. KAUFMAN, M., *The First Century of Plastics*, The Plastics Institute, London (1963)
2. WHITEHOUSE, A. A. K., and PRITCHETT, E. G. K., *Phenolic Resins*, Plastics Institute Monograph No. C3, 2nd Edn, London (1955)
3. ROBITSCHEK, P., and LEWIN, A., *Phenolic Resins*, Iliffe, London (1950)
4. MARTIN, R. W., *The Chemistry of Phenolic Resins*, John Wiley, New York (1956)
5. MEGSON, N. J. L., *Phenolic Resin Chemistry*, Butterworths, London (1958)
6. CARSWELL, T. S., *Phenoplasts*, Interscience, New York (1947)
7. HULTZSCH, K., *Chemie der Phenolharze*, Springer-Verlag, Berlin (1950)
8. APLEY, M., *Trans PI*, **20**, 7 (April 1952)
9. BROWN, W. J., *Laminated Plastics*, Plastics Institute Monograph No. E4, London (1961)
10. PEPPER, K. W., and BARWELL, F. T., *J. Soc. Chem. Ind.*, **63**, 150 (1944)
11. MITCHELL, R. G. B., and SMITH, D., *Plastics (London)*, **24**, 44, 85 (1959)
12. LEWIN, A., *Trans PI*, **28**, 224 (1960)

BIBLIOGRAPHY

APLEY, M., *Cast Resins*, Plastics Institute Monograph No. 4A, London (1946)
BROWN, W. J., *Fabric Reinforced Plastics*, Cleaver-Hume Press, London (1947)
BROWN, W. J., *Laminated Plastics*, Plastics Institute Monograph No. E4, London (1961)
CARSWELL, T. S., *Phenoplasts*, Interscience, New York (1947)
DUFFIN, D. J., *Laminated Plastics*, Reinhold, New York (1958)
GOULD, D. F., *Phenolic Resins*, Reinhold, New York (1959)
HULTZSCH, K., *Chemie der Phenolharze*, Springer-Verlag, Berlin (1950)
KAUFMAN, M., *The First Century of Plastics*, The Plastics Institute, London (1963)
LEARMONTH, G. S., *Laminated Plastics*, Leonard Hill, London (1951)
MARTIN, R. W., *The Chemistry of Phenolic Resins*, John Wiley, New York (1956)
MEGSON, N. J. L., *Phenolic Resin Chemistry*, Butterworths, London (1958)
ROBITSCHEK, P., and LEWIN, A., *Phenolic Resins*, Iliffe, London (1950)
SORREL, S. E., *Paper Base Laminates*, Cleaver-Hume Press, London (1950)
WHITEHOUSE, A. A. K., and PRITCHETT, E. G. K., *Phenolic Resins*, Plastics Institute Monograph No. C3, 2nd Edn, London (1955)

20

Aminoplastics

20.1 INTRODUCTION

The term aminoplastics has been coined to cover a range of resinous polymers produced by interaction of amines or amides with aldehydes. Of the various polymers of this type that have been produced there are two of current commercial importance, the urea–formaldehyde and the melamine–formaldehyde resins. There has in the past also been some commercial interest in aniline–formaldehyde resins and in systems containing thiourea but today these are of little or no importance. Recently melamine–phenol–formaldehyde resins have been introduced for use in moulding powders and benzoguanamine-based resins for surface coating applications.

Interest in aminoplastics dates from the publication of a patent by John in 1918[1] which disclosed resinous materials prepared by heating urea with commercial formalin and which suggested the use of the resultant viscous solutions as adhesives and as impregnants for fabrics. In the following years patents were taken out by Pollak and Ripper,[2] and Goldschmidt and Neuss,[3] the former pair directing their efforts towards the manufacture of an 'organic glass', the latter group towards moulding compositions. In 1926, as a result of work by E. C. Rossiter, moulding powders based on urea–thiourea–formaldehyde were marketed under the trade name Beetle by The British Cyanides Co. Ltd. (later known as British Industrial Plastics Ltd.). Similar products were subsequently produced in other countries.

During the next fifteen years the urea resins were also developed for use as adhesives, textile finishing agents and in the production of surface coatings and wet-strength paper. Since the war the development of chipboard has resulted in a large new outlet for urea-based resins which have also found other uses, such as in firelighters.

In 1935 Henkel[4] patented the production of resins based on melamine. Today these resins are important in the manufacture of decorative laminates and in tableware.

20.2 UREA–FORMALDEHYDE RESINS

Of the various amino-resins that have been prepared, the urea–formaldehyde (U–F) resins are by far the most important commercially.   Like the phenolic resins, they are, in the finished product, cross-linked (thermoset) insoluble, infusible materials.   For application, a low molecular weight product or resin is first produced and this is then cross linked only at the end of the fabrication process.

In a general comparison with phenolic resins, the U–F materials are cheaper, light in colour, are lacking in odour, have better resistance to electrical tracking but have an inferior heat resistance and a higher water absorption.

20.2.1   Raw materials

Urea is a white crystalline compound with a melting point of 132·6°C and is highly soluble in water.   It is the cheaper of the two resin intermediates.

Urea is prepared commercially by the reaction of liquid carbon dioxide and ammonia in silver lined autoclaves, at temperatures in the range 135–195°C and pressures of 70–230 atm.   The reaction proceeds by way of ammonium carbamate

$$2NH_3 + CO_2 \longrightarrow NH_2 \cdot CO_2 \cdot NH_4 \longrightarrow CO(NH_2)_2 + H_2O$$

A 40–60% conversion per pass is achieved and unreacted feedstocks are returned to the compressors

Urea may also be obtained from calcium cyanamide via cyanamide

$$CaCN_2 + H_2SO_4 \longrightarrow NH_2 \cdot CN + CaSO_4$$

$$NH_2 \cdot CN + H_2O \longrightarrow CO(NH_2)_2$$

The method for producing formaldehyde was described in Chapter 16. In aminoplastics manufacture it is used in the form of formalin (36–37% w/w CH_2O).   As in the case of phenolic resin production, formalin with both high and low methanol content is used according to the needs of the manufacturer.   The low methanol content formalin is more reactive but is also less stable and must be used soon after use. For this reason some resin manufacturers prefer to use formalin with a high (7–10%) methanol content.

20.2.2   Theories of resinification

The mechanism of reaction of urea and formaldehyde to give infusible insoluble products is far from understood.   In practice reaction is initially carried out using neutral or mildly alkaline conditions leading to the formation of mono- and di-methylol ureas (Fig. 20.1).

Where the ratio of formaldehyde to urea is sufficiently high to produce a reasonable proportion of dimethylol urea, infusible products may be obtained by subjecting the initial reaction product to acidic conditions.

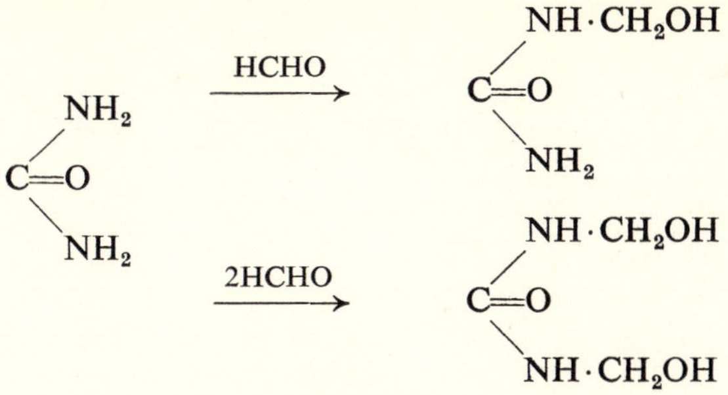

$$
\begin{array}{c}
\text{NH}_2 \\
| \\
\text{C}{=}\text{O} \\
\backslash \\
\text{NH}_2
\end{array}
\quad
\begin{array}{c}
\xrightarrow{\text{HCHO}} \\
\\
\xrightarrow{\text{2HCHO}}
\end{array}
\quad
\begin{array}{c}
\text{NH}\cdot\text{CH}_2\text{OH} \\
/ \\
\text{C}{=}\text{O} \\
\backslash \\
\text{NH}_2 \\
\\
\text{NH}\cdot\text{CH}_2\text{OH} \\
/ \\
\text{C}{=}\text{O} \\
\backslash \\
\text{NH}\cdot\text{CH}_2\text{OH}
\end{array}
$$

Fig. 20.1

The nature of further reactions is not known but possible reactions include methylol–methylol condensation (Fig. 20.2) (I), methylol–imino reactions (II) and formaldehyde–imino condensations (III). Thus dimethylol urea, with two methylol and two imino groups, has a potential functionality of four.

More complex alternative theories have been proposed by Brookes,[5] Kadowaki,[6] Marvel,[7] Redfarn[8] and Thurston.[9] Such theories often

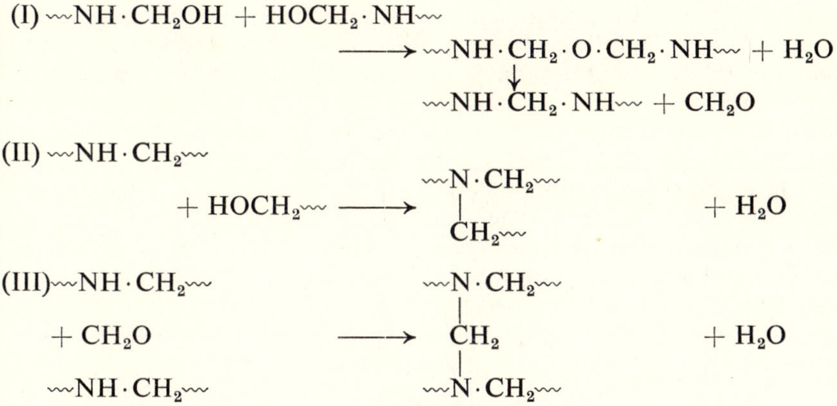

(I) $\sim$NH$\cdot$CH$_2$OH $+$ HOCH$_2\cdot$NH$\sim$

$$\longrightarrow \sim\text{NH}\cdot\text{CH}_2\cdot\text{O}\cdot\text{CH}_2\cdot\text{NH}\sim + \text{H}_2\text{O}$$
$$\downarrow$$
$$\sim\text{NH}\cdot\text{CH}_2\cdot\text{NH}\sim + \text{CH}_2\text{O}$$

(II) $\sim$NH$\cdot$CH$_2\sim$

$$
+ \text{HOCH}_2\sim \longrightarrow
\begin{array}{c}
\sim\text{N}\cdot\text{CH}_2\sim \\
| \\
\text{CH}_2\sim
\end{array}
\quad + \text{H}_2\text{O}
$$

(III) $\sim$NH$\cdot$CH$_2\sim$

$+$ CH$_2$O

$\sim$NH$\cdot$CH$_2\sim$

$$
\longrightarrow
\begin{array}{c}
\sim\text{N}\cdot\text{CH}_2\sim \\
| \\
\text{CH}_2 \\
| \\
\sim\text{N}\cdot\text{CH}_2\sim
\end{array}
\quad + \text{H}_2\text{O}
$$

Fig. 20.2

stipulate the production of more specific complex intermediates such as methylene ureas or trimeric ring structures which have never yet been isolated in resins. Since these theories have, as yet, in no way contributed to the practical technology of urea–formaldehyde resins they will not be dealt with in detail here. The interested reader is referred to the original papers and to critical reviews elsewhere.[10–13]

20.2.3 U–F moulding materials

Thermosetting compositions based on urea-formaldehyde are widely employed because of their low cost, wide colour range, rigidity and good electrical properties.

Manufacture

A moulding powder based on urea–formaldehyde will contain a number of ingredients. Those most commonly employed include the following:

1. Resin.
2. Filler.
3. Pigment.
4. Hardener.
5. Stabiliser.
6. Plasticiser.
7. Lubricant.

The first stage of resin preparation is to dissolve urea into the 36% w/w formalin which has been adjusted to a pH of 8 with caustic soda. Since formaldehyde interferes with the normal functioning of universal indicators a pH meter is used when making pH adjustments. The blending may be carried out without heating in a glass-lined or stainless-steel reactor for about 90 minutes. In an alternative process the blending is carried out at about 40°C for 30 minutes. In some cases the pH, which may drop during reaction, is adjusted by addition of small quantities of hexamine. The urea–formaldehyde ratios normally employed are in the range 1:1:3 to 1:1:5. The reaction product, which is probably little more than a solution of urea in formalin is used immediately in the manufacture of powders.

Only a limited range of fillers is used commercially with U–F resins. Bleached wood pulp is employed for the widest range of bright colours and in slightly translucent mouldings. Wood flour, which is significantly cheaper, is also widely used but gives an opaque moulding and, because of its inherently brown colour, leads to some limitation in the colour range. For mouldings of enhanced translucency, chopped regenerated cellulose (Cellophane) film which is free from voids and has a refractive index (1·565) close to that of the resin (1·55–1·56), may be used. Fabric fillers and mineral fillers are not commonly employed with U–F resins.

A wide variety of pigments is now used in U–F moulding compositions. Their principal requirements are that they should not affect the stability or moulding characteristics of the powder, that they should be stable to processing conditions, be unaffected by conditions of service including insolubility in any solvents with which the mouldings might come into contact and should not interfere with the electrical properties.

In order to obtain a sufficient rate of cure at moulding temperatures it is usual to add about 0·2–2·0% of a 'hardener'. This functions by decomposing at moulding temperatures to give an acidic body that will accelerate the cure rate. A very large number of such latent acid catalysts have been described in the literature of which some of the more prominent are ammonium sulphamate, ammonium phenoxyacetate, ethylene sulphite and trimethyl phosphate.

Urea–formaldehyde powders have a limited shelf-life but some improvement is made by incorporating a stabiliser such as hexamine into the moulding powder. In some formulations the cure rate and the related time for flow are controlled by keeping the latent acid catalyst fixed and adjusting the stabiliser.

Plasticisers are used in special grades of moulding powders. Their main virtue is that they enable more highly condensed resins to be used and thus reduce curing shrinkage whilst maintaining good flow properties. Glyceryl-α-tolyl ether (monocresyl glycidyl ether) is often used for this purpose. Plasticisers may also be used in small quantities to improve the flow of other grades.

Metal stearates such as zinc, magnesium or aluminium stearates are commonly used as lubricants at about 2% concentration. Other materials that have been used successfully include oxidised paraffin wax and sulphonated castor oil.

In typical manufacturing processes the freshly prepared urea–formaldehyde initial reaction product is mixed with the filler (usually with a resin–filler ratio of about 2:1) and other ingredients except pigment in a trough mixer. This process, which takes about 2 hours at 60°C, enables thorough impregnation of the wet base with the resin solution and also advances the resinification reaction. After a check has been made that it is slightly alkaline the resulting wet base is then fed to a drier which may be either of the turbine or rotary type. The turbine drier consists of a number of slowly rotating circular trays stacked one above the other in a large oven. Each of the trays has a number of radial slits. The powder is fed to the top tray where it rests for one revolution when, by means of scraper blades, it is pushed through the slits on to the second tray where the process is repeated. In a typical process the residence time of the mixture in the drier is about 2 hours at 100°C. In an alternative process the wet base is fed into a rotary drier in which it remains for about $\frac{3}{4}$–1 hour whilst being subjected to counterblast air at 120–130°C. This process reduces the water content from about 40% to about 6% and also advances the condensation.

On emerging from the drier the base is hammer-milled and then fed to a ball-mill for 6–9 hours. The pigments are added at the ball-mill stage. During this process samples are taken and checked for colour and processing characteristics. It is frequently necessary to make slight adjustments to the formulation by adding further pigment or other ingredient at this stage. The ball-milling process ensures a good dispersion of pigment and gives a fine powder that will produce mouldings of excellent finish. On the other hand the powder has a high bulk factor and problems of air and gas trapping will occur during moulding. These problems are overcome by densifying the product.

One method of densification is to heat the powder as it passes along a belt and to drop the heated powder into the nip of a two-roll mill. In this process the material passes directly through the rolls to form a strip which is then hammer-milled to give powder which is in the form of tiny flat flakes. In another process the fine powder is slowly stirred in a large pot and water, or a water–methanol blend, or steam, run into the mixture. The particles partly cohere in the damp conditions and on subsequent drying give densified granules.   A third process is to charge the powder into an internal mixer which is at a temperature of about 100°C.   The particles cohere and after about two minutes the batch is fed to a hammer-mill to give a coarse granule.   More recent processes involve the use of continuous compounders, such as the Buss Ko-Kneader.

As an alternative to the wet process described above, moulding compositions may be made by mixing a powdered resin or a dimethylol derivative with other ingredients on a two-roll mill or in an internal mixer.   The condensation reaction proceeds during this process and when deemed sufficiently advanced, the composition is sheeted off and disintegrated to the desired particle size.

Control tests on the moulding powder include measurement of water content, flow, powder density and rate of cure.

From the above discussion it will be recognised that in addition to differences in colour, commercial urea–formaldehyde moulding powders may differ in the following respects:

1. The nature of the filler used.
2. The ease of flow (dependent on the degree of heating during the drying stages, and in some cases on the heating operations associated with densifying).
3. The speed of cure, partly related to the ease of flow but also associated with the amounts of hardener and stabiliser.
4. The type of grind.
5. The presence or absence of plasticiser.

It is these differences which determine the range of grades at present commercially available.

Processing

Urea–formaldehyde moulding powders may be moulded without difficulty on conventional compression and transfer moulding equipment.   The powders however have a limited storage life.   They should thus be stored in a cool place and, where possible, used within a few months of manufacture.

Moulding temperatures in the range 125–160°C are employed.   The low temperatures are used with thick sections and the high temperatures for thin sections.   Mouldings may easily be over-cured by moulding for too long and/or at too high a temperature and this is made manifest by blistering, bleaching and a distinct fishy smell.   Compression moulding pressures recommended range from 1 to 4 ton in^{-2}, the higher pressures being usually employed for deep draw articles.   The cure time will depend on the thickness of the moulding and on the mould temperature.   Using a typical powder, an $\frac{1}{8}$ in. thick moulding will require about 55 seconds cure time at 145°C.   Much shorter times ($\sim$10–20 seconds) are now employed industrially for such articles as bottle caps (which have a section thickness somewhat less than $\frac{1}{8}$ in.) and which are moulded at the higher end of the moulding temperature range.   The amount of cure carried out should depend on the properties required of the moulding and on the economics of the process. It has been shown[14] that for the best balance of mechanical and electrical properties the degree of cure required coincides with that giving the best water resistance. Thus in practice a moulding is deemed properly cured if, sawn through its thickest section, it is unaffected in appearance, or cannot be scratched with a fingernail, after 10 minutes immersion in boiling water.

Preheating techniques are commonly employed since these lead to shorter cures, easier flow and generally better products. The high power factor of the material enables high frequency preheaters to be used successfully. It is also frequently advantageous to pellet the powders as in the case of phenolics.

Urea–formaldehyde moulding powders may be transfer moulded. Pressures of 4–10 ton in^{-2}, calculated on the area of the transfer pot, are generally recommended.

Properties and applications

When first introduced the value of U–F moulding powders lay in their availability in a wide range of colours, at that time a novelty amongst thermosetting moulding composition. The wide colour range possible continues to be a reason for the widespread use of the material but other useful features have also become manifest.

The major desirable features of U–F mouldings are:

1. Low cost. The cheaper grades are sometimes lower in weight cost than the general purpose phenolics. (It is to be noted that U/F's have a somewhat higher density.)
2. Wide colour range.
3. Do not impart taste and odour to foodstuffs and beverages with which they come in contact.
4. Good electrical insulation properties with particularly good resistance to tracking.
5. Resistance to continuous heat up to a temperature of 70°C.

Some typical values of physical properties of mouldings from urea–formaldehyde compositions are given in Table 20.1.

The bulk of U–F moulding compositions are used in two applications, bottle caps and electrical fittings. Their use in bottle caps is due to their low cost, wide colour range and because they do not impart taste and odour. Both α-cellulose and wood flour-filled compositions are used for this purpose. The good electrical insulation properties at low frequencies and the particularly good tracking resistance together with their low cost are the reasons for their dominance over phenolics and other materials in electrical fittings. Wood flour-filled compositions are normally used for the darker colour plugs and fixtures and α-cellulose filled materials for the ligher coloured mouldings.

Urea–formaldehyde is the preferred material for such applications as coloured toilet seats, hair-drier housings and vacuum flask cups and jugs.

Because of their non-thermoplastic nature and resistance to detergents and solvents, many buttons are made from U–F moulding powders. Imitation horn effects may be achieved by blending normal grades with grades of high translucency.

Miscellaneous uses include meat trays, toys, knobs, switches and lampshades. U–F lampshades are generally strictly utilitarian in design and of limited aesthetic appeal. It is important in this application to ensure adequate ventilation of the air space above the lamp in order to prevent overheating and subsequent cracking of the shade. For similar reasons fittings for ceiling light bowls, as often used in bathrooms and kitchens, may fail through lack of adequate ventilation.

20.2.4   Adhesives and related uses

Important uses for U–F resins are as wood adhesives for furniture glues, plywood and chipboard manufacture.

To prepare a suitable resin, formalin is first neutralised to a pH of 7·5 and urea then dissolved into it. (U–F ratio $\sim 1:2$). Sodium formate may be added as a buffer to regulate the pH. The mixture is boiled under reflux, typically for about 15 minutes, to give dimethylol urea and other low molecular weight products. The resin is then acidified to pH 4, conveniently with formic acid, and reacted for a further 5–20 minutes. The resulting resin is then stabilised by neutralising to a pH 7·5 with alkali to give a water-soluble resin with an approximately 50% solids content. When the resin is to be used in aqueous solution, as is normally the case, it it then partially dehydrated to give a 70% solids content by vacuum distillation. For some uses, for example for application in tropical countries, the resin is spray dried to ensure greater stability.

The finished product is checked for viscosity, solids content, pH value (which must be in the range of 7·3–7·5) and for its reactivity with a standard hardener.

Resins are commonly available with U–F ratios ranging from 1:1·4 to 1:2·2. Since formaldehyde is more expensive than urea the high F–U ratio resins are more expensive. They do however have greater clarity, the best water resistance, marginally superior mechanical properties, longer shelf life (up to two years) and greatest reactivity. The degree of condensation is quite important since, if insufficient, the resin would be absorbed into the wood and would thus be unavailable to act as an adhesive.

The resins are hardened by acidic conditions. Phosphoric acid, or more commonly ammonium chloride, an acid donor, is employed. The ammonium chloride functions by reaction with formaldehyde to give hydrochloric acid. Hexamine is also formed during this reaction.

$$4NH_4Cl + 6CH_2O \longrightarrow N_4(CH_2)_6 + 6H_2O + 4HCl$$

About 1·5 parts ammonium chloride per 100 parts of the resin solution are generally used. The hardener is added as an aqueous solution.

Table 20.1 PROPERTIES OF MOULDINGS PREPARED FROM UREA–FORMALDEHYDE AND MELAMINE–FORMALDEHYDE MOULDING COMPOSITIONS
(Testing according to B.S.2782)

Property	Units	Urea-formaldehyde				Melamine-formaldehyde				
		α-cellulose filled	Wood flour filled	Plasticised	Translucent	Cellulose filled	Glass filled	Mineral filled	Melamine-phenolic	G.P. phenolic
Specific gravity	–	1·5-1·6	1·5-1·6	1·5-1·6	~1·5	1·5-1·55	~2·0	~1·8	1·5-1·6	1·35
Tensile strength	10^3 lb in^{-2}	7·5-11·5	7·5-11·5	7-9·5	7-10	8-12	6-10	4-6	6-8	8
Impact strength	ft lb	0·20-0·35	0·16-0·35	0·16-0·24	0·14-0·2	0·15-0·24	0·16-0·23	0·12-0·22	0·12-0·15	0·16
Cross-breaking strength	10^3 lb in^{-2}	11-17	11-16·5	13·5-15·5	13-17	13-21	9-14	6-11	10-12	11·5
Dielectric strength (90°C)	V/0·001 in.	120-200	60-180	100-200	70-130	160-240	150-250	200-250	30-150	75-175
Volume resistivity	ohm cm	10^{11}-10^{13}	10^{11}-10^{13}	10^{12}-10^{13}	–	10^7-10^8	10^{13}-10^{14}	10^{13}-10^{14}	10^9-10^{10}	10^{10}-10^{11}
Water absorption										
24 hr at 20°C	mg	50-130	40-170	50-90	50-100	10-50	10-20	7-14	–	–
30 min at 100°C	mg	180-460	250-600	300-450	300-600	40-110	20-35	15-40	–	–

At one time urea–formaldehyde was used extensively in the manufacture of plywood but the product is today less important than heretofore. For this purpose a resin (typically U–F ratio 1:1·8)–hardener mixture is coated on to wood veneers which are plied together and pressed at 95–110°C under pressure of 200–800 lb in^{-2}. U–F resin-bonded plywood is suitable for indoor application but is generally unsuitable for outdoor work where phenol–formaldehyde, resorcinol–formaldehyde or melamine modified resins are more suitable.

The resins continue to be used in large quantities in general wood assembly work. In most cases the resin–hardener mixture is applied to the surfaces to be joined and then clamped under pressure while hardening occurs. It is also possible to coat the resin on to one surface and the hardener on the other surface, allowing them to come into contact *in situ* and thus eliminating pot-life problems. Gap-filling resins may be produced by incorporating fillers to reduce shrinkage and consequent cracking and crazing.

One of the largest applications of U–F resins at the present time is in the manufacture of chipboard. Wood chips are mixed with about 10% of a resin–hardener solution and the mixture pressed in a multi-daylight press for about 8 minutes at 150°C. Since the odour of formaldehyde is disagreeable it is important that little of the pungent chemical be released into the press shop during the opening of the presses. For this reason the resin should have a low free formaldehyde content. Since a low degree of condensation is desirable to ensure good dispersion a rather low F–U ratio is necessary in order to achieve a low free formaldehyde content.

Wood chipboard is free from grain and is thus essentially isotropic in its behaviour. The mechanical properties are approximately the same as the average of the properties of the original wood measured along and across the grain. The water resistance of chipboard is poor but being isotropic it does not warp as long as it is able to swell freely in all directions.

20.2.5 Foams and firelighters

Foams may be made from urea–formaldehyde resins using simple techniques. In one process the resin is mixed with a foaming detergent, whipped up with air in a mixing device and blended with an acid, such as phosphoric acid, as it leaves the mixer. The foams may be formed *in situ* in building cavity walls but, because of the large amounts of water present, it is necessary that the foam be formed between porous surfaces. Typical products have a closed cell content of about 80% and have little mechanical strength, as they are very friable. They have a very low thermal conductivity with a K value of 0·15–0·20 b.t.u. in. ft^{-2} hr^{-1} degF^{-1}, thus comparing very favourably with other insulating materials. Foams ranging from 0·5 to 3·0 lb ft^{-3} may be produced; those with a density of about 0·75 lb ft^{-3} having the lowest conductivity. Such foams are very cheap and are now being made *in situ* in building applications in Britain. Foams have also been used as an aid to floral decoration (Floropak) and in ground form as an artificial snow in cinema and television productions.

A rather strange but nevertheless large scale application of U–F resins is in the manufacture of firelighters, made by a modification of the foam process. The resin solution is blended with a small amount of detergent and then whisked with paraffin. A hardener is added and the resin allowed to set. In effect the product is a U–F foam saturated with paraffin.

Another unorthodox application is to form a U–F foam at the bottom of certain oil-well drill holes. The foam acts as a filter allowing the oil out but holding back sand and other mineral products.

20.2.6 Other applications

Modification of urea–formaldehyde resins with other reagents gives rise to a number of useful materials. For example co-condensation of urea–formaldehyde and a mono-alcohol in the presence of small quantities of an acidic catalyst will involve simultaneous etherification and resinification. *n*-Propanol, *n*-butanol and iso-butanol are commonly used for

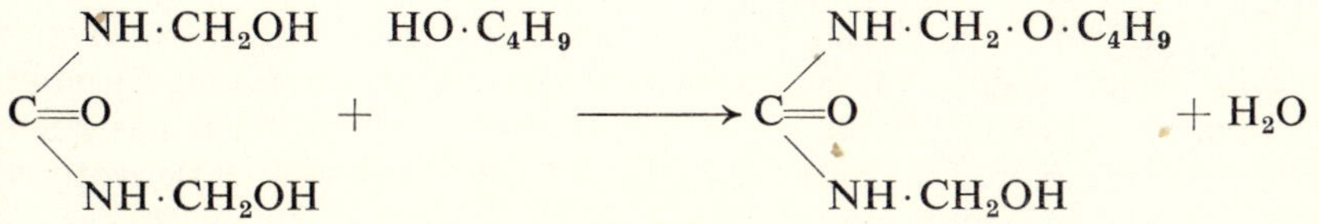

$$\underset{\substack{| \\ C=O \\ | \\ NH \cdot CH_2OH}}{NH \cdot CH_2OH} \quad + \quad HO \cdot C_4H_9 \quad \longrightarrow \quad \underset{\substack{| \\ C=O \\ | \\ NH \cdot CH_2OH}}{NH \cdot CH_2 \cdot O \cdot C_4H_9} \quad + H_2O$$

Fig. 20.3

this purpose. As an example *n*-butanol will react with the methylol urea as shown in Fig. 20.3.

By varying reaction conditions and reactant proportions differing products may be obtained. Many of the alkoxy groups are retained during cure and the resins have a degree of thermoplasticity. Soluble in organic solvents and used in conjunction with plasticising alkyd resins, these materials form useful stoving lacquers. Air-drying lacquers suitable as wood finishes, may be obtained by addition of acid hardeners.

Whereas the butylated resins have enhanced solubility in organic solvents, enhanced solubility in water (which is rather limited in resins of high molecular weight) is required for some purposes and this may be achieved in a number of ways. For example, in acid condensation of urea and formaldehyde in the presence of sodium bisulphite the following reaction takes place

$$\text{\textasciitilde\textasciitilde NH} \cdot CH_2OH + NaHSO_3 \longrightarrow \text{\textasciitilde\textasciitilde NH} \cdot CH_2 \cdot SO_3Na + H_2O$$

Ionisation occurs in aqueous solution to give a resin of negative charge, an example of a number of 'anionic resins'

$$\text{\textasciitilde\textasciitilde NH} \cdot CH_2 \cdot SO_3Na \longrightarrow NH \cdot CH_2SO_3^{\ominus} + Na^{\oplus}$$

Modification of urea resins with certain organic bases, e.g. triethylene tetramine, will give resins with basic groups which form ionisable salts in the presence of acids

$$\sim\!\!CNRR' + HCl \longrightarrow \sim\!\!CNHRR'^{\oplus} + Cl^{\ominus}$$

These resins are referred to as 'cationic resins'. Paper with improved wet strength may be obtained by adding an ionic resin at the beater stage of a paper-making operation. For the best results a high molecular weight resin is required.

Urea resins find extensive use in textile finishing. For example, cellulose fabrics may be padded into aqueous solutions of methylol ureas or their methyl ethers. Excess material is removed and the resins are hardened *in situ* by passing the fabric through ovens at 130–160°C. Although there is negligible difference in the appearance of the fabric, a considerable measure of crease resistance is acquired. Such resin treatment does however lead to two immediate problems. Firstly the cellulose fabric has lower tear and tensile strengths. This problem is partially overcome by mercerisation (steeping in sodium hydroxide solution) before resin treatment. The second problem occurs where the fabric is subjected to repeated bleaching action since the resin reacts with hypochlorite bleach to give chloramines which break down on ironing forming hydrochloric acid which tenderises the fabric. This problem has been progressively reduced in recent years by the use of cyclic urea derivatives which do not form chloramines.

20.3 MELAMINE–FORMALDEHYDE RESINS[10,11,15]

Melamine (1,3,5-triamino–2,4,6-triazine) was first prepared by Liebig in 1835. For a hundred years the material remained no more than a laboratory curiosity until Henkel[4] patented the production of resins by condensation with formaldehyde. Today large quantities of melamine-formaldehyde resins are used in the manufacture of moulding composition, laminates, adhesives, surface coatings and other applications. Although in many respects superior in properties to the urea-based resins they are also significantly more expensive.

20.3.1 Melamine

A number of methods of producing melamine have been described in the literature.[11] These include:

1. Heating dicyandiamide, either with ammonia or on its own under pressure.
2. Fusion of dicyandiamide with a guanidine salt.
3. From urea

$$6CO(NH_2)_2 \xrightarrow[\substack{400-600 \\ atm}]{300-600°C} 6NH_2 + 3CO_2 + \underset{\text{Melamine}}{C_3H_6N_6}$$

4. By catalytic interaction of ammonia and carbon monoxide at elevated temperatures and pressure.
5. Electrolysis of dilute solutions of hydrogen cyanide in ammonium bromide to give cyanogen bromide. This is then dissolved in a solvent such as tetrahydrofuran and reacted with gaseous ammonia to produce cyanamide. The cyanamide is then heated in an autoclave at about 190–200°C in the presence of ammonia and the melamine recovered by filtration.

$$HCN + NH_4Br \longrightarrow CN \cdot Br + H_2 + NH_3$$

$$CNBr + 2NH_3 \xrightarrow{\text{Solvent}} CN \cdot NH_2 + NH_4Br$$

$$3CN \cdot NH_2 \xrightarrow{NH_3} \underset{\text{Melamine}}{C_3H_6N_6}$$

Of these methods the first named is understood to be the most important commercially. Dicyandiamide ('dicy') is prepared by heating cyanamide solution at 70–80°C. The cyanamide itself is prepared from calcium cyanamide (Fig. 20.4).

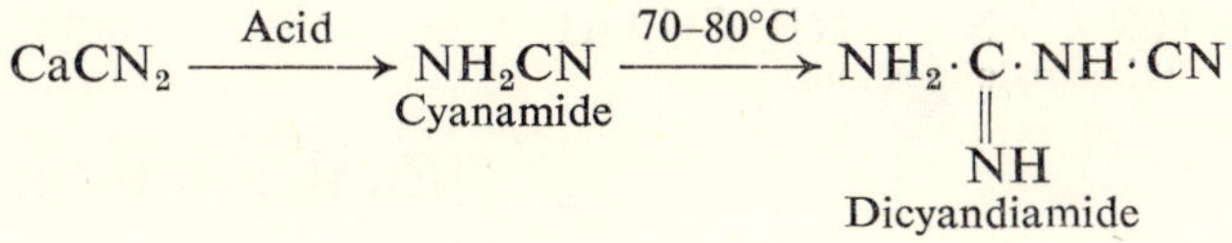

$$CaCN_2 \xrightarrow{\text{Acid}} \underset{\text{Cyanamide}}{NH_2CN} \xrightarrow{70-80°C} \underset{\underset{NH}{\overset{\|}{}}}{NH_2 \cdot C \cdot NH \cdot CN}$$

Dicyandiamide

Fig. 20.4

If 'dicy' is heated just above its melting point of 209°C there is a vigorous exothermic reaction which results in the evolution of ammonia and the formation of some melamine together with a number of complex water-insoluble de-ammoniation products. In order to achieve a high yield

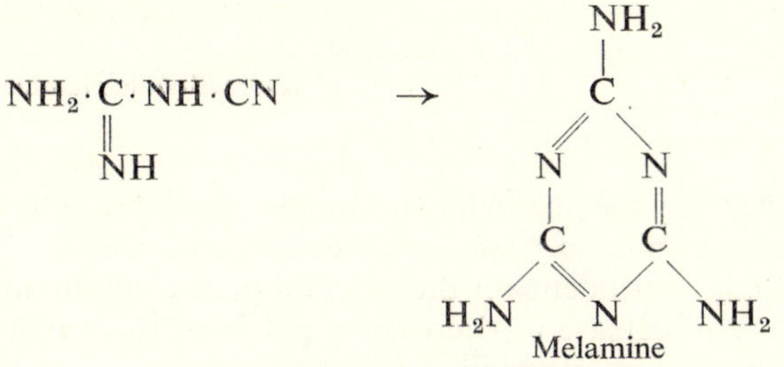

Melamine

Fig. 20.5

of melamine in commercial manufacture the reaction is carried out in the presence of ammonia at about 300°C under pressure (Fig. 20.5).

It will be noted that dicyandiamide is the dimer of cyanamide and melamine is the trimer.

Melamine, a non-hygroscopic, white crystalline solid, melts with decomposition above 347°C and sublimes at temperatures below the melting point. It is only slightly soluble in water, 100 ml of water dissolve 0·38 g at 20°C and 3·7g at 90°C. It is weakly basic and forms well-defined salts with acids.

20.3.2   Resinification

Reaction of melamine with neutralised formaldehyde at about 60°C leads to the production of a mixture of water-soluble methylol melamines. These methylol derivatives can possess up to six methylol groups per molecule and include trimethylolmelamine and hexamethylol melamine. The methylol content of the mixture will depend on the melamine–formaldehyde ratio and on the reaction conditions.

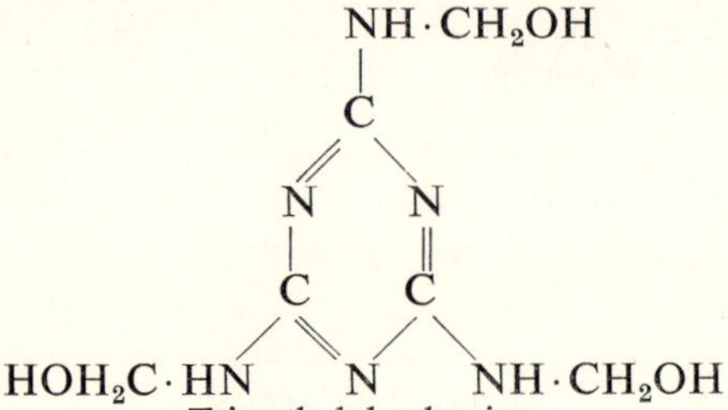

$$NH \cdot CH_2OH$$

Trimethylolmelamine

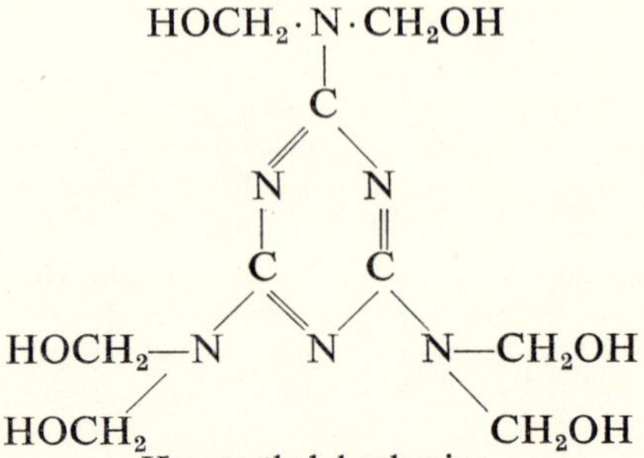

Hexamethylolmelamine

Fig. 20.6

On further heating the methylol melamines condense and a point is reached where hydrophobic resin separates out on cooling. The resinification is strongly dependent on the pH and is at a minimum at about pH 10·0–10·5. An increase or decrease of pH from this value will result in a considerable increase in resinification rates.

There is some evidence that the principal resinification reaction involves methylol–methylol condensations.[16]

$$\sim\!\!\sim\!\!NH \cdot CH_2OH + HO \cdot CH_2NH\!\!\sim\!\!\sim$$

$$\longrightarrow \sim\!\!\sim\!\!NH \cdot CH_2 \cdot O \cdot CH_2 \cdot NH\!\!\sim\!\!\sim + H_2O$$

Methylene links may also be formed by the following reactions

$$\sim\!\!NH\cdot CH_2OH + H_2N\!\!\sim \longrightarrow \sim\!\!NH\cdot CH_2\cdot NH\!\!\sim + H_2O$$

$$\sim\!\!NH\cdot CH_2\cdot O\cdot CH_2\cdot NH\!\!\sim \longrightarrow \sim\!\!NH\cdot CH_2\cdot NH\!\!\sim + CH_2O$$

In commercial practice the resin is condensed to a point close to the hydrophobe point and then either applied to the substrate, or converted into a moulding powder, before proceeding with the final cure.

In a typical process a jacketed still fitted with a stirrer and reflux condenser is charged with 240 parts 37% w/w (40% w/v) formalin and the pH adjusted to 8·0–8·5 using sodium carbonate solution with the aid of a pH meter. 126 parts of melamine (to give a melamine–formaldehyde ratio of 1/3) are charged into the still and the temperature raised to 50–85°C. The melamine goes into solution and forms methylol derivatives. For treatment of fabrics, paper and leather this product may be diluted and cooled for immediate use. It may also be spray dried to give a more stable product. Cooling the solution would yield crystalline trimethylol melamine which may be air dried but which is less soluble in water than the spray dried product.

For laminating and other purposes the initial product is further heated to about 85°C with continuous stirring. After about 30 minutes, and at regular intervals thereafter, samples of the resin are taken and added to ice-cold water. Diminished water tolerance is indicated when the resin solution becomes cloudy on entering the water. Reaction is then continued until the stage is reached when addition of 3 cm³ of water will cause 1 cm³ of resin to become turbid.

Reactions may be carried out at lower pH values and higher temperatures in order to achieve faster reactions. For some applications where a high degree of fibre impregnation is required hydrophilic resin may also be produced. Such hydrophilic resins have limited stability in aqueous solution and must either be used within a few hours of manufacture or spray dried.

The more hydrophobic resins have only a slightly greater stability in solution with a shelf life of only a few days. Some improvement may be achieved by diluting the resin content down to about 50% solids content with industrial methylated spirit. The diluted resin should then be adjusted to a pH of 9·0–9·5 to improve the stability. The addition of about 0·1% borax (anhydrous) calculated on the weight of the solids content is useful in obtaining this pH and maintaining it for several months.[17] It is conveniently added as an aqueous solution. The stabilised resins should be stored at 20–35°C. Too low a storage temperature will cause precipitation, too high a temperature, gelation. Precipitation may also occur if the resin is insufficiently condensed and gelation with over condensation.

20.3.3 Moulding powders

Melamine–formaldehyde moulding powders are generally prepared by methods similar to those used with urea–formaldehyde material. In a typical process an aqueous syrup, containing resin with a melamine–formaldehyde molar ratio of 1/2, is compounded with fillers, pigments, lubricants, stabilisers and in some cases accelerators in a dough-type mixer.   The product is then dried and ball-milled by processes similar to those described in Section 20.2.3.1.   In one process described in the literature magnesium carbonate is employed to act as a pH stabiliser during storage.   For the more common decorative moulding powders α-cellulose is used as a filler.   Some bleached wood flour is sometimes added to reduce shrinkage cracks near inserts.   Because of the high refractive index of the cured resin (~ 1.65) it is not possible to obtain highly translucent mouldings using regenerated cellulose fillers.

Industrial grade materials employ fillers such as asbestos, silica and glass fibre.   These are often incorporated by dry-blending methods similar to those used with wood flour-filled phenolic compositions.

Mouldings from melamine–formaldehyde powders are superior to the urea–formaldehyde plastics in a number of respects.

These include:

1. Lower water absorption, especially with mineral-filled resins.
2. Better resistance to staining by aqueous solutions such as fruit juices and beverages.   Further improvement in this respect is still desirable and somewhat better results are claimed using benzoguanamine with the melamine.
3. Electrical properties, which are initially similar to the urea–formaldehyde resins, are maintained better in damp conditions and at elevated temperatures.
4. Better heat resistance.
5. Greater hardness.

Compared with the phenolic resins they have a better colour range, track resistance and scratch resistance.   They have a similar order of heat resistance, although their dimensional stability when exposed to hot dry conditions is not so good.   Melamine–formaldehyde moulding materials are more expensive than general purpose urea–formaldehyde and phenol–formaldehyde resins.

For high duty electrical applications the mineral-filled melamine based compositions have superior electrical insulation and heat resistance to the cellulose-filled grades.   The use of glass fibre leads to mouldings of higher mechanical strength, improved dimensional stability and higher heat resistance than with other fillers.   Mineral filled melamine-based powders are used when phenolics and urea–formaldehyde compositions are unsuitable.   They are thus to some extent competitive with the melamine–phenolics, the alkyd moulding powders and, to some small extent, epoxy moulding materials.

An interesting use of melamine resins in compression moulding involves decorative foils.   A suitably printed or decorated grade of paper is impregnated with resin and dried.   A compression moulding is then prepared using a melamine–formaldehyde, or some other moulding powder.   Shortly before the cure is complete the mould is opened, the foil placed in position and the resin in the foil cured in that position so that the foil actually bonds on to the moulding.

Melamine-based compositions are easily moulded in conventional compression and transfer-moulding equipment. Moulding temperatures are usually in the range 145–165°C and moulding pressures 2–4 ton in^{-2}. In transfer moulding, pressures of 5–10 ton in^{-2} are used. An $\frac{1}{8}$ in. thick moulding required about $2\frac{1}{2}$ minutes cure at 150°C but shorter times are possible with preheated powder.

The curing time employed depends on the properties required of the finished product. For example cold water absorption increases and electrical breakdown decreases as the curing time increases. The effect of cure time on the properties of mouldings has been investigated by Morgan and Vale[18] (see Fig. 20.7). They have suggested that for an optimum combination of electrical and mechanical properties with minimum boiling-water absorption and low after-shrinkage the dye test provides a useful guide to cure. In this test[19,20] the mouldings are immersed for 10 minutes in a boiling 0·01% aqueous solution of Rhodamine B. The moulding is deemed adequately cured if the mouldings remain unchanged in colour except at flash lines or at other points where the resin skin has been removed.

The principal application of melamine–formaldehyde moulding compositions is for the manufacture of tableware, largely because of their wide colour range, surface hardness and stain resistance. The stain resistance does however leave something to be desired and one aim of current research is to discover alternative materials superior in this respect. Cellulose-filled compositions also find a small outlet for trays clock cases and radio cabinets and other purposes. The mineral-filled powders are used in electrical applications.

20.3.4 Laminates containing melamine–formaldehyde resin

The high hardness, good scratch resistance, freedom from colour and heat resistance of melamine–formaldehyde resins suggest possible use in laminating applications. The use of laminates prepared using only melamine resins as the bonding agent is however limited, because of the comparatively high cost (cf. phenolic resins) of the resin, to some electrical applications. On the other hand a very large quantity of decorative laminates are produced in which the surface layers are impregnated with melamine resins and the base layers with phenolic resins. These products are well known under such trade names as Formica and Warerite.

Resins for this purpose generally use melamine–formaldehyde ratios of 1/2·2 to 1/3. Where electrical grade laminates are required the condensing catalyst employed is triethanolamine instead of sodium carbonate.

Decorative laminates have a core or base of Kraft paper impregnated with a phenolic resin. A printed pattern layer impregnated with a melamine–formaldehyde or urea–thiourea–formaldehyde resin is then laid on the core and on top of this a melamine resin-impregnated

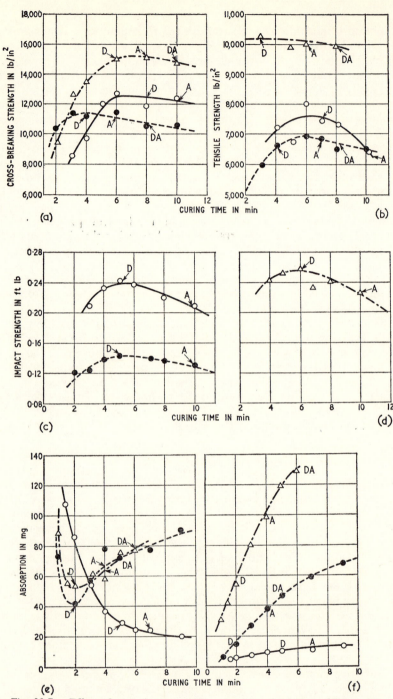

Fig. 20.7. *Effect of cure time on some properties of M–F mouldings. Cure temperature: cellulose filled 295–308°F; mineral filled 300–320°F; ○ glass-filled material, ● asbestos-filled material, △ cellulose-filled material; (a) Cross-breaking strength; (b) tensile strength; (c) impact strength; (d) impact strength; (e) water absorption (mg), boiling water 30 min; (f) cold water*

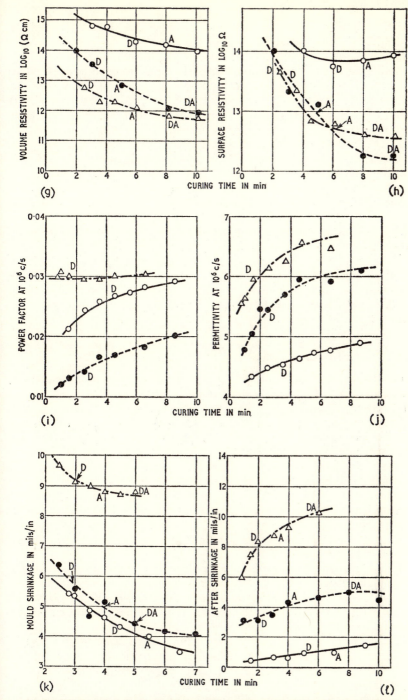

24 hr, 23°C; (g) volume resistivity; (h) surface resistivity; (i) power factor; (j) permittivity; (k) mould shrinkage; (l) after-shrinkage. The letters D, A and DA indicate the time of optimum cure indicated by the dye test D (see text) by boiling in 10% H_2SO_4 (A) and boiling in a mixture of 0·9% H_2SO_4 and 0·025% Kiton Red (DA). (After Morgan and Vale[18])

protective translucent outer sheet. The assembly is then cured at 125–150°C in multi-daylight presses in the usual way.

The electrical grade laminates are made by impregnating a desized glass cloth with a triethanolamine modified resin (as mentioned above). The dried cloth is frequently precured for about 1 hour at 100°C before the final pressing operation. A typical cure for a fifteen-ply laminate would be 10–15 minutes at 140°C under a pressure of 250–1,000 lb in^{-2}. Cloth based on alkali glass yields laminates with poor electrical insulation properties. Much better results are obtained using electrical-grade glass which has been flame-cleaned. The use of certain amino-silane treatments is claimed to give even better physical and electrical insulation properties.

Glass-reinforced melamine–formaldehyde laminates are valuable because of their good heat resistance (they can be used at temperatures up to 200°C) coupled with good electrical insulation properties including resistance to tracking.

Decorative laminates have achieved remarkable success because of their heat resistance, scratch resistance and solvent resistance. Their availability in a wide range of colours has led to their well-known applications in table tops and as a wall-cladding in public buildings and public transport vehicles.

20.3.5 Miscellaneous applications

In addition to their use in moulding powders and laminates, melamine–formaldehyde resins are widely used in many forms.

Hot setting adhesives, prepared in the same way as laminating resins, give colourless glue lines and are resistant to boiling water. Their use alone has been limited because of high cost but useful products may be made by using them in conjunction with a urea-based resin or with cheapening extenders such as starch or flour.

Melamine–formaldehyde condensates are also useful in textile finishing. For example they are useful agents for permanent glazing, rot proofing, wool shrinkage control and, in conjunction with phosphorus compounds, flame proofing.

Compositions containing water-repellent constituents such as stearamide may also improve water repellency.

Modified melamine resins are also employed commercially. Alkylated resins analogous to the alkylated urea–formaldehyde resins provide superior coatings but are more expensive than the urea-based products.

Paper with enhanced wet-strength may be obtained by incorporating melamine resin acid colloid into the pulp. Melamine resin acid colloid is obtained by dissolving a lightly condensed melamine resin or trimethylolmelamine, which are both normally basic in nature, in dilute hydrochloric acid. Further condensation occurs in solution and eventually a colloidal solution is formed in which the particles have a positive charge. Careful control over the constitution of the colloidal solution must be exercised in order to obtain products of maximum stability.

20.4 MELAMINE–PHENOLIC RESINS

Moulding powders based on melamine–phenol–formaldehyde resins were introduced by Bakelite Ltd. in the early 1960's. Some of the principal physical properties of mouldings from these materials are given in Table 20.1

The principal characteristic of these materials is the wide range of colours possible including many intense bright colours. The melamine phenolics may be considered to be intermediate between the phenolic moulding materials and those from melamine–formaldehyde. As a result they have better moulding latitude and mouldings have better dry heat dimensional stability than the melamine–formaldehyde materials. Their tracking resistance is not as good as melamine–formaldehyde materials but often adequate to pass tracking tests. The main applications of these materials are as handles for saucepans, frying pans, steam irons and coffee pots where there is a requirement for a coloured heat resistant material. It is considered unlikely that the melamine–phenolics will absorb much of the market held by melamine resins, irrespective of price, since this market is largely dependent on either the non-odorous nature or the good tracking resistance of the material used. Neither of these two requirements are fulfilled by the melamine–phenolics. Future developments thus seem to lie in the creation of new markets for a coloured, heat resistant material intermediate in price between the phenolic and melamine materials.

20.5 ANILINE–FORMALDEHYDE RESINS[21]

Although occasionally in demand because of their good electrical insulation properties, aniline–formaldehyde resins are today only rarely encountered. They may be employed in two ways, either as an unfilled moulding material or in the manufacture of laminates.

To produce a moulding composition, aniline is first treated with hydrochloric acid to produce water-soluble aniline hydrochloride. The aniline hydrochloride solution is then run into a large wooden vat and formaldehyde solution is run in at a slow but uniform rate, the whole mix being subject to continuous agitation. Reaction occurs immediately to give a deep orange–red product. The resin is still a water-soluble material and so it is fed into a 10% caustic soda solution to react with the hydrochloride thus releasing the resin as a creamy yellow slurry. The slurry is washed with a counter-current of fresh water, dried and ball-milled.

Because of the lack of solubility in the usual solvents, aniline–formaldehyde laminates are made by a 'pre-mix' method. In this process the aniline hydrochloride-formaldehyde product is run into a bath of paper pulp rather than of caustic soda. Soda is then added to precipitate the resin on to the paper fibres. The pulp is then passed through a paper-making machine to give a paper with a 50% resin content.

Aniline–formaldehyde resin has very poor flow properties and may only be moulded with difficulty and mouldings are confined to simple shapes. The resin is essentially thermoplastic and does not cross link with the evolution of volatiles during pressing. Long pressing times, about 90 minutes for a $\frac{1}{2}$ in. thick sheet, are required to achieve a suitable product.

Laminated sheets may be made by plying up the impregnated paper and pressing at 3,000 lb in^{-2} moulding pressure and 160–170°C, for 150 minutes followed by 75 minutes cooling in a typical process. A few shaped mouldings may also be made from impregnated paper, by moulding at higher moulding pressures. In one commercial example a hexagonal circuit breaker lifting rod was moulded at 7,000 lb in^{-2}.

As with the other aminoplastics, the chemistry of resin formation is incompletely understood. It is however believed that under acid conditions at aniline–formaldehyde ratios of about 1:1·2, which are similar

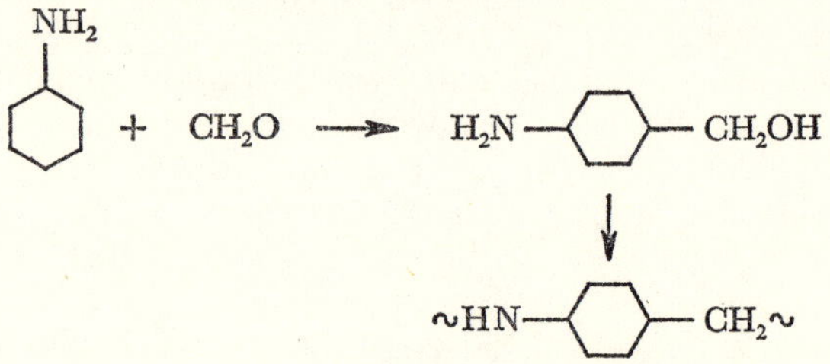

Fig. 20.8

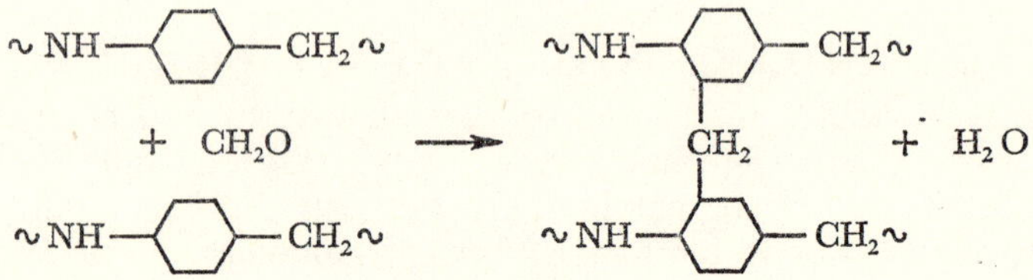

Fig. 20.9

to those used in practice, that reaction proceeds via *p*-aminobenzyl alcohol with subsequent condensation between amino and hydroxyl groups (Fig. 20.8).

It is further believed that the excess formaldehyde then reacts at the *ortho*- position to give a lightly cross-linked polymer with very limited thermoplasticity (Fig. 20.9).

Such condensation reactions occur on mixing the two components. The resultant comparative intractability of the material is one of the main reasons for its industrial eclipse.

Some typical properties of aniline–formaldehyde mouldings are given in Table 20.2.

20.6 RESINS CONTAINING THIOUREA

Thiourea may be produced either by fusion of ammonium thiocyanate or by the interaction of hydrogen sulphide and cyanamide.

$$NH_4SCN \rightleftharpoons CS(NH_2)_2$$

$$NH_2CN + H_2S \longrightarrow CS(NH_2)_2$$

The first process is an equilibrium reaction which yields only a 25% conversion of thiourea after about 4 hours at 140–145°C. Prolonged or excessive heating will cause decomposition of the thiourea whilst pressure changes and catalysts have no effect on the equilibrium. Pure thiourea is a crystalline compound melting at 181–182°C and is soluble in water.

Thiourea will react with neutralised formalin at 20–30°C to form methylol derivatives which are slowly deposited from solution. Heating of

Table 20.2   TYPICAL PROPERTIES OF ANILINE–FORMALDEHYDE MOULDINGS

Specific gravity	1·2
Rockwell hardness	M 100, 125
Water absorption (24 hr)	0·08% (ASTM D.570)
Tensile strength	10,500 lb in^{-2}
Impact strength	0·33 ft lb/in. notch (Izod)
Dielectric constant	3·56–3·72 (100 c/s–100 Mc/s)
Power factor 100 c/s	0·00226
1 Mc/s	0·00624
100 Mc/s	0·00318
Upper service temperature	~90°C
Track resistance	between phenolics and U–Fs
Resistant to alkalis, most organic solvents	
Attacked by acids	

methylol thiourea aqueous solutions at about 60°C will cause the formation of resins, the reaction being accelerated by acidic conditions. As the resin average molecular weight increases with further reaction the resin becomes hydrophobic and separates from the aqueous phase on cooling. Further reaction leads to separation at reaction temperatures in contrast to urea–formaldehyde resins which can form homogeneous transparent gels in aqueous dispersion.

Polymer formation is apparently due to methylol–methylol and methylol–amino reaction (Fig. 20.10).

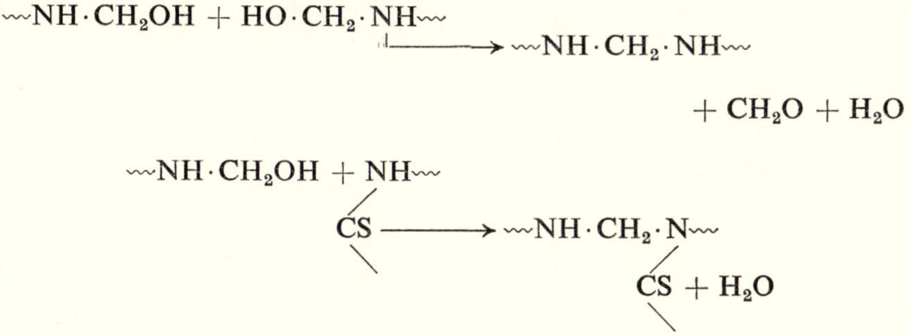

$$\sim NH \cdot CH_2OH + HO \cdot CH_2 \cdot NH \sim$$

$$\longrightarrow \sim NH \cdot CH_2 \cdot NH \sim$$

$$+ CH_2O + H_2O$$

$$\sim NH \cdot CH_2OH + NH \sim$$
$$CS \longrightarrow \sim NH \cdot CH_2 \cdot N \sim$$
$$CS + H_2O$$

Fig. 20.10

In comparison with urea–based resins, thiourea resins are slower curing and the products are somewhat more brittle. They are more water-repellent than U–F resins.

At one time thiourea–urea–formaldehyde resins were of importance for moulding powders and laminating resins because of their improved water resistance. They have now been almost completely superseded by melamine–formaldehyde resins with their superior water resistance. It is however understood that a small amount of thiourea-containing resin is still used in the manufacture of decorative laminates.

REFERENCES

1. *British Patent* 151,016
2. *British Patent* 171,094; *British Patent* 181,014; *British Patent* 193,420; *British Patent* 201,906; *British Patent* 206,512; *British Patent* 213,567; *British Patent* 238,904; *British Patent* 240,840; *British Patent* 248,729
3. *British Patent* 187,605; *British Patent* 202,651; *British Patent* 208,761
4. *British Patent* 455,008
5. BROOKES, A., *Plastics Monograph* No. 2, Institute of the Plastics Industry (now The Plastics Institute), London (1946)
6. KADOWAKI, H., *Bull. Chem. Soc. Japan*, **11**, 248 (1936)
7. MARVEL, C. S., *et al. J. Am. Chem. Soc.*, **68**, 1681 (1946)
8. REDFARN, C. A., *Brit. Plastics*, **14**, 6 (1942)
9. THURSTON, J. T., Unpublished paper given at the Gibson Island Conference on Polymeric Materials (1941)
10. VALE, C. P., *Aminoplastics*, Cleaver-Hume Press, London (1950)
11. VALE, C. P., and TAYLOR, W. G. K., *Aminoplastics*, Iliffe, London (1964)
12. RITCHIE, P. D., *A Chemistry of Plastics and High Polymers*, Cleaver-Hume, London (1949)
13. MOORE, W. R., *An Introduction to Polymer Chemistry*, University of London Press, London (1963)
14. HOFTON, J., *Brit. Plastics*, **14**, 350 (1942)
15. MILLS, F. J., Paper in *Plastics Progress* 1953 (Ed. P. MORGAN), Iliffe, London (1953)
16. GAMS, A., WIDMER, G., and FISCH, W., *Brit. Plastics*, **14**, 508 (1943)
17. *British Patent* 738,033
18. MORGAN, D. E., and VALE, C. P., Paper in S.C.I. Monograph No. 5 *The Physical Properties of Polymers*, Society of the Chemical Industry, London (1959)
19. VALE, C. P., *Trans. Plastics Inst.*, **20**, 29 (1952)
20. B.S.S. 1322
21. *Plastics* (London), **15**, 34 (1950)

BIBLIOGRAPHY

BLAIS, J. F., *Amino Resins*, Reinhold, New York (1959)
VALE, C. P., *Aminoplastics*, Cleaver-Hume Press, London (1950)
VALE, C. P., and TAYLOR, W. G. K., *Aminoplastics*, Iliffe, London (1964) (also available as the third edition of a Plastics Institute Monograph—earlier editions by A. BROOKES—1946 and BROOKES and VALE—1954)

Polyester Resins

21.1 INTRODUCTION

Polyester resins are encountered in many forms. They are important as laminating resins, moulding compositions, fibres, films, surface coating resins, rubbers and plasticisers. The common factor in these widely different materials is that they all contain a number of ester linkages in the main chain. (There are also a number of polymers such as poly(vinyl acetate) which contain a number of ester groups in side chains but these are not generally considered within the term polyester resins.)

These polymers may be produced by a variety of techniques of which the following are technically important:

1. Self-condensation of ω-hydroxy acids, commercially the least important route

$$HORCOOH + HORCOOH \text{ etc. } \longrightarrow \sim\!\!\!\sim ORCOORCOO\!\!\sim\!\!\!\sim$$

2. Condensation of polyhydroxy compounds with polybasic acids, e.g. a glycol with a dicarboxylic acid

$$HOROH + HOOCR_1COOH + HOROH$$
$$\longrightarrow \sim\!\!\!\sim OROOCR_1COORO\!\!\sim\!\!\!\sim$$

3. Ester exchange

$$R_1OOCRCOOR_1 + HOR_2OH$$
$$\longrightarrow \sim\!\!\!\sim OOCRCOOR_2OO\!\!\sim\!\!\!\sim + R_1OH$$

Credit for the preparation of the first polyester resin is given variously to Berzelius[1] in 1847 and to Gay-Lussac and Pelouze in 1833.[2] Their first use came about in the early years of this century for surface coatings where they are well known as *alkyd* resins, the word alkyd being derived somewhat freely from alcohol and acid. Of particular importance in coatings are the *glyptals*, glycerol–phthalic anhydride condensates. Although these materials were also used at one time for moulding materials they are now obsolete and quite different from present day alkyd moulding powders.

Linear polyesters were studied by Carothers during his classical researches into the development of the nylons but it was left to Whinfield and Dickson to discover poly(ethylene terephthalate) now of great importance in the manufacture of fibres (e.g. Terylene, I.C.I.) and films (e.g. Melinex, Mylar). The fibres were first announced in 1941.

At about the same time, an allyl resin known as CR39 was introduced in the United States as a low pressure laminating resin. This was followed about 1946 with the introduction of polyester laminating resins which are today of great importance in the manufacture of glass reinforced plastics. Alkyd moulding powders were introduced in 1948 and have since found specialised applications as electrical insulators.

21.2 POLYESTER LAMINATING RESINS

The polyester laminating resins are viscous, generally pale yellow coloured materials of a low degree of polymerisation ($\sim$8–10). They are produced by condensing a glycol with both an unsaturated and a saturated dicarboxylic acid. The unsaturated acid provides a site for subsequent cross linking whilst provision of a saturated acid reduces the number of sites for cross linking and hence reduces the cross-link density and brittleness of the end-product. In practice the polyester resin is mixed with a reactive diluent such as styrene. This eases working, often reduces the cost and may enhance the reactivity of the polyester. Before applying the resin to the reinforcement a curing system is blended into the resin. This may be so varied that curing times may range from a few minutes to several hours whilst the cure may be arranged to proceed either at ambient or elevated temperatures. In the case of cold curing systems it is obviously necessary to apply the resin to the reinforcement as soon as possible after the catalyst system has been added and before gelation and cure occur. The usual reinforcement is glass fibre, as a preform, cloth, mat or rovings but sisal or more conventional fabrics may be used.

Since cross linking occurs via an addition mechanism across the double bonds in the polyesters and the reactive diluent there are no volatiles given off during cure (cf. phenolic and amino-resins) and it is thus possible to cure without pressure (see Fig. 21.1). Since room temperature cures are also possible the resins are most useful in the manufacture of large structures such as boats and car bodies.

Small quantities of higher molecular weight resin in powder form are also manufactured. They are used in solution or emulsion form as binders for glass fibre preforms and also for the manufacture of pre-impregnated cloths.

21.2.1 Selection of raw materials

1,2-Propylene glycol, is probably the most important glycol used in the manufacture of the laminating resins. It gives resins which are less

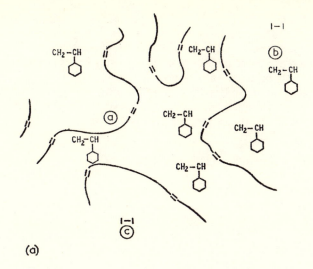

(a)

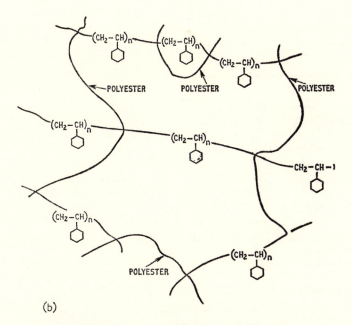

(b)

Fig. 21.1. *The nature of cured polyester laminating resins.*
(1) Structures present in polyester resin ready for laminating:
(a) low molecular weight unsaturated resin molecules
(b) reactive diuent (styrene) molecules
(c) initiator (catalyst) molecules
(2) Structures present in cured polyester resin.   Cross linking via
an addition copolymerisation reaction.   The value of n ∼ 2–3 on
average in general purpose resins

crystalline and more compatible with styrene than those obtained using ethylene glycol.   Propylene glycol is produced from propylene via propylene oxide.   The use of glycols higher in the homologous series gives products which are more flexible and have greater water resistance.   They do not appear to be used on a large sale commercially.

Products such as diethylene glycol and triethylene glycol, obtained by side reactions in the preparation of ethylene glycol are sometimes used but they give products with greater water absorption and inferior electrical properties (Fig. 21.2).

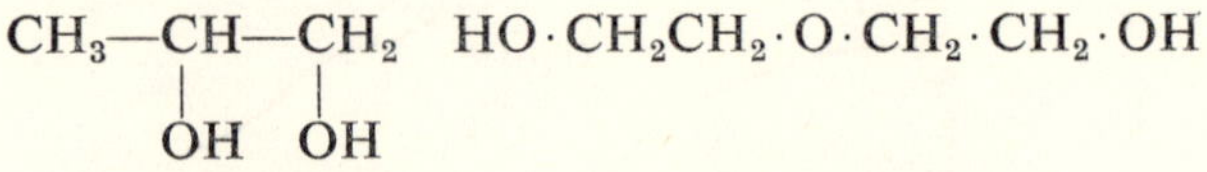

$$CH_3—CH—CH_2 \quad HO \cdot CH_2CH_2 \cdot O \cdot CH_2 \cdot CH_2 \cdot OH$$
$$\quad\quad\quad | \quad\quad |$$
$$\quad\quad OH \quad OH$$

1,2-Propylene Glycol   Diethylene Glycol

Fig. 21.2

Most conventional general purpose resins employ either maleic acid (usually as the anhydride) or its trans-isomer fumaric acid (which does not form an anhydride) as the unsaturated acid (Fig. 21.3).

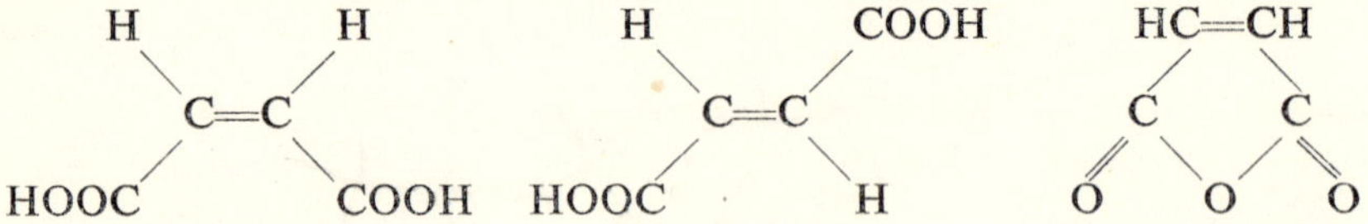

Maleic Acid   Fumaric Acid   Maleic Anhydride

Fig. 21.3

Maleic anhydride is commonly prepared by passing a mixture of benzene vapour and air over a catalyst (e.g. a vanadium derivative) at elevated temperatures (e.g. 450°C).   It is a crystalline solid melting at 52·6°C (the acid melts at 130°C).

Fumaric acid may be prepared by heating maleic acid, with or without catalysts.   It is also obtained as a by-product in the manufacture of phthalic anhydride from naphthalene.   The acid is a solid melting at 284°C.   Fumaric acid is sometimes preferred to maleic anhydride as it is less corrosive, tends to give lighter coloured products and the resins have slightly greater heat resistance.

Saturated acids

The prime function of the saturated acid is to space out the double bonds and thus reduce the intensity of cross linking.   Phthalic anhydride is most commonly used for this purpose because it provides an inflexible link and maintains the rigidity in the cured resin.   It has been used in increasing proportions during the past decade since its low price enables cheaper resins to be made.   The most detrimental effect of this is to reduce the heat resistance of the laminates but this is frequently unimportant.   It is produced by catalytic oxidation of naphthalene, usually over vanadium pentoxide and is a crystalline solid melting at 131°C.

Isophthalic acid (Melt. Pt 347°C), made by oxidation of *m*-xylene, has recently been introduced for resins. The resins have higher heat distortion temperatures and flexural moduli, better craze resistance and better

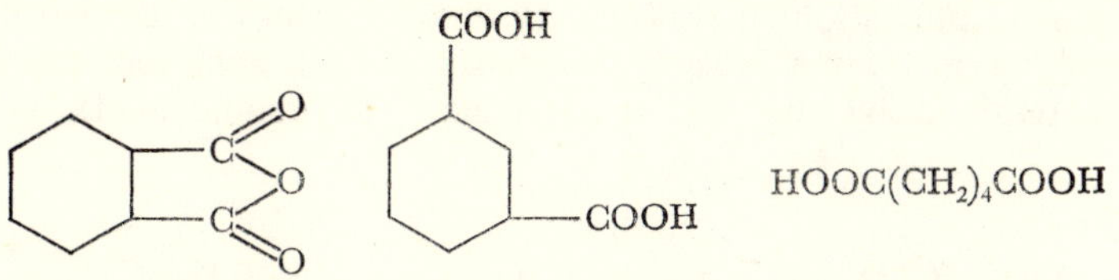

| Phthalic Anhydride | Isophthalic Acid | Adipic Acid |

HOOC(CH₂)₄COOH → $HOOC(CH_2)_4COOH$

Fig. 21.4

water and alkali resistance. They are also useful in the preparation of resilient gel coats.

Where a flexible resin is required adipic and sebacic acids are used. Whereas the phthalic acids give a rigid link these materials give highly flexible linkages and hence flexibility in the cured resin. Flexible resins are of value in gel coats.

Diluents

Because of its low price, compatability, low viscosity and ease of use styrene is the preferred reactive diluent in general purpose resins. Methyl methacrylate is sometimes used, usually in conjunction with styrene in resins for translucent sheeting. Vinyl toluene and diallyl phthalate are also occasionally employed. The use of many other monomers is described in the literature.

Special materials

A number of special purpose resins are available which employ somewhat unusual acids and diluents. A resin of improved heat resistance is obtained by using 'nadic' anhydride, the Diels–Alder reaction product

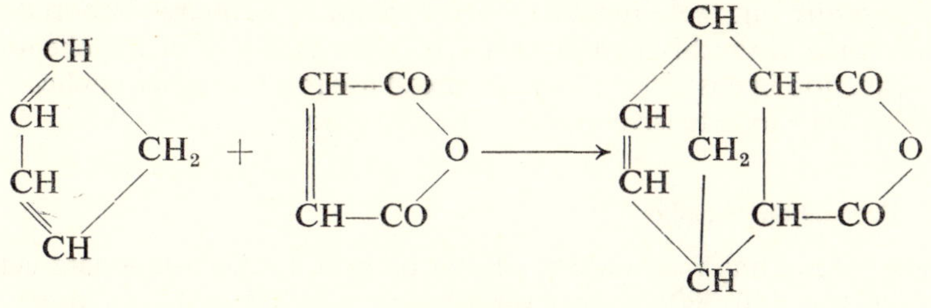

Fig. 21.5

of cyclopentadiene and maleic anhydride (Fig. 21.5).

A substantial improvement in heat resistance may also be obtained by replacing the styrene with triallyl cyanurate (Fig. 21.6).

This monomer is prepared by reacting cyanuric chloride with excess allyl alcohol in the presence of sodium hydroxide at 15–20°C. Laminates based on polyester resins containing triallyl cyanarate are claimed to be able to withstand a temperature of 250°C for short periods.

For many applications it is necessary that the resin has reasonable self-extinguishing properties. Such properties can be achieved and trans-parency retained by the use of HET-acid (chlorendic acid). This is

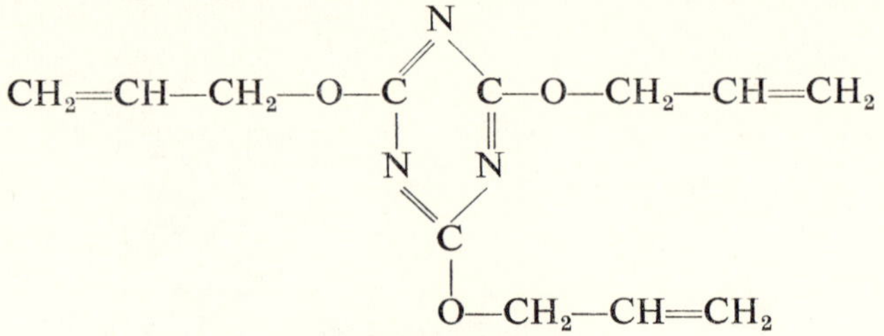

$$CH_2{=}CH{-}CH_2{-}O{-}C \overset{\displaystyle N}{\underset{\displaystyle N}{\Big\langle}} C{-}O{-}CH_2{-}CH{=}CH_2$$

Triallyl Cyanurate

Fig. 21.6

obtained by reacting hexachlorocyclopentadiene with maleic anhydride and converting the resulting anhydride adduct into the acid by exposure to moist air (Fig. 21.7).

The self extinguishing properties of the resin are due to the high chlorine content of the acid (54·8%). The double bond of the acid is unreactive and it is necessary to use it in conjunction with an unsaturated acid such as fumaric acid to provide for cross linking.

An alternative approach is first to produce a polyester resin containing an excess of maleic acid residues (maleate groups) and then to react this with the hexachlorocyclopentadiene to form the adduct *in situ* (Fig. 21.8).

Laminates prepared from highly chlorinated resins of this type tend to discolour on prolonged exposure to light and this retarded the early development of these resins. Stabilisers have however been developed and current resins are substantially superior to the early resins of this type.

Many other acids, glycols and reactive monomers have been described in the literature but these remain of either minor or academic importance. In a number of cases this is simply because of the high cost of the chemical and a reduction in cost due to its widespread use in some other application could well lead to extensive use in polyester resins.

21.2.2   Production of resins

Polyester laminating resins are produced by heating the component acids and glycols at 150–200°C for several hours, e.g. 8 hours. In order to obtain a good colour and to prevent premature gelation the reaction is carried out under an inert blanket of carbon dioxide or nitrogen. The reaction mixture is agitated to facilitate reaction and to prevent local overheating. A typical charge for a general purpose resin would be:

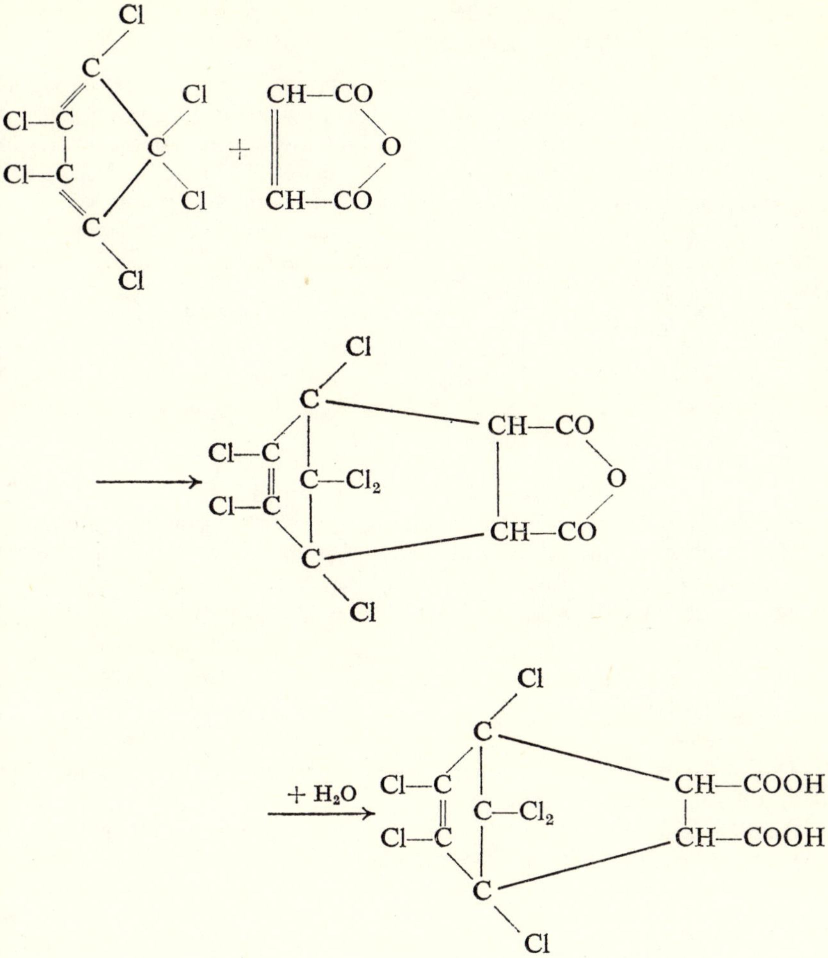

Fig. 21.7

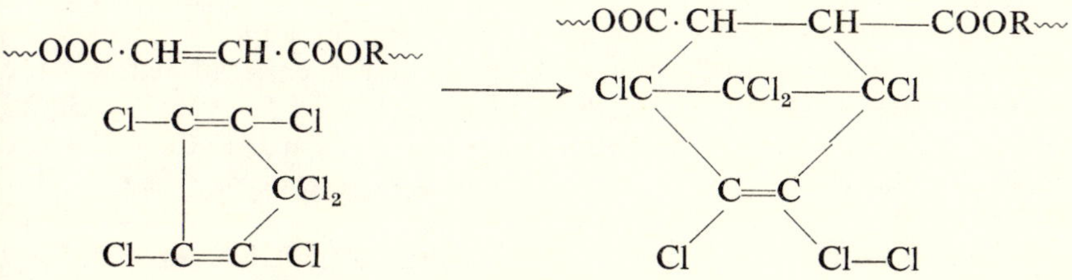

Fig. 21.8

Propylene glycol	159 parts
Maleic anhydride	114 parts
Phthalic anhydride	86 parts

The molar ratios of these three ingredients in the order above is $1\cdot2:0\cdot67:0\cdot33$. The slight excess of glycol is primarily to allow for evaporation losses. Xylene is often used to facilitate the removal of water of condensation by means of azeotropic distillation. The reaction is followed by measuring the acid number of small samples periodically removed from the reactor. (The acid number is the number of milligrams of potassium hydroxide equivalent to the acidity present in one gram of resin.) Where there are equimolecular proportions of glycol and acid the number average molecular weight is given by 56,000/acid number. Since there is some deviation from equimolecular equivalence in practice care should be taken in using this relationship. Reaction is usually stopped when the acid number is between 25 and 50, the heaters are switched off and any xylene present is allowed to boil off into a receiver.

When the resin temperature drops below the boiling point of the reactive diluent (usually styrene) the resin is pumped into a blending tank containing suitably inhibited diluent. It is common practice to employ a mixture of inhibitors in order to obtain a balance of properties in respect of colour, storage stability and gelation rate of catalysed resin. A typical system based on the above polyester formulation would be:

Styrene	148 parts
Benzyl trimethyl ammonium chloride	$0\cdot38$ parts
Hydroquinone	$0\cdot05$ parts
Quinone	$0\cdot005$ parts

The blend is allowed to cool further and the resin is transferred into drums for shipping and storage.

Quality control tests on the resins most commonly employed are for specific gravity, viscosity, colour and clarity.

21.2.3 Curing systems

The cross-linking reaction is carried out after the resin has been applied to the glass fibre. In practice the curing is carried out either at elevated temperatures of about 100°C where press mouldings are being produced, or at room temperature in the case of large hand lay-up structures.

Benzoyl peroxide is most commonly used for elevated temperature curing. The peroxide is generally supplied as a paste ($\sim50\%$) in a liquid such as dimethyl phthalate to reduce explosion hazards and to facilitate mixing. The curing cycle in pressure moulding processes is normally less than five minutes.

In the presence of certain amines such as dimethyl aniline, benzoyl peroxide will bring about the room temperature cure of general purpose polyester resins. This system is no longer however in common use.

Today either methyl ethyl ketone peroxide or cylohexanone peroxide is generally used for room temperature curing in conjunction with cobalt naphthenate. The peroxides (strictly speaking polymerisation initiators) are referred to as 'catalysts' and the cobalt compound as an 'accelerator'. Other curing systems have been devised but are seldom used.

Commercial *methyl ethyl ketone peroxide* (m.e.k.p.) is a mixture of compounds and is a liquid usually supplied blended into dimethyl phthalate the mixture containing about 60% peroxide. Its activity varies according to the composition of the mixture. It is useful in that it can easily be metered into the resin from a burette but great care must be taken in order to obtain adequate dispersion into the resin. It is also difficult to detect small quantities of this corrosive material which may have been spilt on the skin and elsewhere.

Cyclohexanone peroxide, a white powder, another mixture of peroxidic materials, has a similar reactivity to m.e.k.p. Usually supplied as a 50% paste in dimethyl or dibutyl phthalate it has to be weighed out but is easier to follow dispersion and to observe spillage. The quantity of peroxide used is generally 0·5–3% of the polyester.

Cobalt naphthenate is generally supplied in solution in styrene, the solution commonly having a cobalt concentration of 0·5–1·0%. The cobalt solution is normally used in quantities of 0·5–4·0% based on the polyester. The accelerator solution is rather unstable as the styrene will tend to polymerise and thus although the accelerator may be metered from burettes, the latter will block-up unless frequently cleaned. Cobalt naphthenate solution in white spirit and dimethyl phthalate have proved unsatisfactory. In the first case dispersion is difficult and laminates remain highly coloured whilst with the latter inferior end-products are obtained and the solution is unstable. Stable solutions of cobalt octoate in dimethyl phthalate are possible and these are often preferred because they impart less colour to the laminate.

An interest has developed recently in the use of vanadium naphthenates as accelerators. In 1956 the author[3] found that if m.e.k.p. was added to a polyester resin containing vanadium naphthenate the resin set almost immediately, that is, while the peroxide was still being stirred in. Whereas this effect was quite reproducible with the sample of naphthenate used, subsequent workers have not always obtained the same result. It would, thus appear that the curing characteristics are very dependent on the particular grade of resin and of vanadium naphthenate used. It was also observed by the author that the gelation rate did not always increase with increased temperature or accelerator concentration and in some instances there was a retardation. Recent workers[4] have found that whilst the behaviour of the naphthenate varies according to such factors as the resin and catalyst used, certain vanadium systems are of value where a high productivity in hand lay-up techniques is desired.

The peroxides and accelerator should not be brought into contact with each other as they form an explosive mixture. When the resin is to be used, first the accelerator and then the peroxide are carefully dispersed into the resin which may also contain inert fillers and thixotropic agents.

According to the concentration of catalyst and accelerator used the resin will gel in any time from 5 minutes to several hours. Gelation will be followed by a rise in temperature which may reach 200°C. Where the resin is applied to the glass mat before gelation the high surface–volume ratio facilitates removal of heat and little temperature rise is noted. Gelation and the exothermic reaction are followed by hardening and the resin becomes rigid. Maximum mechanical strength is not however attained for about a week or more. Hardening is accompanied by substantial volumetric shrinkage ($\sim 8\%$) and for this reason polyester resins are only infrequently used for casting purposes.

Unsaturated polyesters are invariably susceptible to air inhibition and surfaces may remain undercured, soft and in some cases tacky if freely exposed to air during the curing period. The degree of surface undercure varies to some extent with the resin formulation and the hardening system employed. Where the resin is to be used in hand lay-up techniques or for surface coatings air inhibition may cause problems. A common way of avoiding difficulties is to blend a small amount of paraffin wax (or other incompatible material) in with the resin. This blooms out on to the surface forming a protective layer over the resin during cure.

21.2.4 Structure and properties

The cured resins, being cross linked, are rigid and do not flow on heating. The styrene, phthalic anhydride, maleic anhydride and propylene glycol residues are predominantly hydrocarbon but are interspersed with a number of ester groups. These latter groups provide a site for hydrolytic degradation particularly in alkaline environments. The polar nature of the ester group leads to the resin having a higher power factor and dielectric constant than the hydrocarbon polymers and this limits their use as high frequency electrical insulators.

Many mechanical properties are dependent on the intensity of cross links and in the rigidity of the molecules between cross links. It has already been shown that cross-link intensity may be controlled by varying the ratio of unsaturated to saturated acids whereas rigidity is to a large extent determined by the structure of the saturated acid employed.

21.2.5 Polyester–glass fibre laminates

Glass fibres are the preferred form of reinforcement for polyester resins since they provide the strongest laminates. Fabrics from other fibres may however be used and can in some instances provide adequate reinforcement at lower cost. Glass fibres are available in a number of forms of which the following are the most important:

1. Glass cloth. A range of cloths are available and the finest of these are used in order to obtain the best mechanical properties. They are, however, expensive in use and they are only used in certain specialised applications such as in the aircraft industry and for decorative purposes.
2. Chopped strand mat. This consists of chopped strands (bundles of glass filaments) about 2 in. long bonded together by a resinous

binder. This type of mat is used extensively in glass reinforced polyester structures.

3. Needle mat. This is similar to chopped strand mat except that the mat is held together by a loose stitching rather than a binder.
4. Preforms. Preformed shapes may be made by depositing glass fibres on to a preform mould. The fibres are then held together by spraying them with a binder.

Other types of glass structures used include rovings, yarns, tapes, rovings fabrics and surfacing mats.

The glass used may be of two types, electrical ('E') glass, a low alkali borosilicate glass, or alkali ('A') glass with an alkali content of 10–15%. The former gives laminates with the best weathering and electrical properties whereas the latter is cheaper. In order that good adhesion should be achieved between resin and glass it is necessary to remove any size (in the case of woven cloths) and then to apply a finish to the fibres. The function of a finish is to provide a bond between the inorganic glass and the organic resin. Today the most important of these finishes are based on silane compounds, e.g. Garan treatment. In a typical system vinyl trichlorsilane is hydrolysed in the presence of glass fibre and this condenses with hydroxyl groups on the surface of the glass (Fig. 21.9).

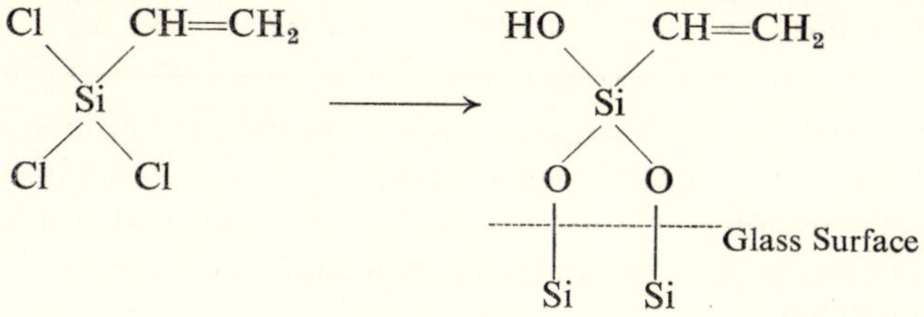

Fig. 21.9

A number of different binder materials are in use for chopped strand mat and include starch, polyvinyl acetate and polyesters. The binder used depends on the end use of the laminate and the method of fabrication.

Methods of producing laminates have been dealt with in detail in other publications[5,6,7,8,9] and so details will not be given here.

The major process today is the hand lay-up technique in which resin is stippled and rolled into the glass mat (or cloth) by hand. Moulds are easy to fabricate and large structures may be made at little cost.

For mass production purposes matched metal moulding techniques are employed. Here the preform or mat is placed in a heated mould and the resin poured on. The press is closed and light pressure ($\sim$50 lb in^{-2}) applied. Curing schedules are usually about 3 minutes at 120°C. It is possible to produce laminates using less resin with pressure moulding than with hand lay-up techniques and this results in higher mechanical properties.

A number of techniques intermediate between these two extreme processes also exist involving vacuum bags, vacuum impregnation, rubber plungers and other devices.

Inert fillers are sometimes mixed with the resin in an effort to reduce cost. However, many fillers increase the viscosity to such an extent that with hand lay-up methods much more of the resin–filler mix is required to impregnate the mat. Since greater difficulty in working may also prolong processing time and there is invariably a marked drop in mechanical properties care must be taken before making a decision whether or not to employ fillers.

Some typical properties of polyester–glass laminates are given in Table 21.1.

From the figures given in Table 21.1 it will be seen that laminates can have very high tensile strengths. On the other hand some laminates

Table 21.1

Property	Hand lay-up mat laminate	Press formed mat laminate	Fine square woven cloth laminate	Rod from rovings
Specific gravity	1·4–1·5	1·5–1·8	~2·0	2·19
Tensile strength (10^3 lb in^{-2})	8–17	18–25	30–45	150
Flexural strength (10^3 lb in^{-2})	10–20	20–27	40–55	155
Flexural modulus (10^6 lb in^{-2})	~0·5	~0·6	1–2	6·6
Power factor (10^6 c/s)	0·02–0·08	0·02–0·08	0·02–0·05	—
Dielectric constant (10^6 c/s)	3·2–4·5	3·2–4·5	3·6–4·2	—
Water absorption (%)	0·2–0·8	0·2–0·8	0·2–0·8	—

made by hand lay-up processes may have mechanical properties not very different from thermoplastics such as the polyacetals and unplasticised p.v.c.

The most desirable features of polyester–glass laminates are:

1. They can be used to construct large mouldings without complicated equipment.
2. Good strength and rigidity although much less dense than most metals.
3. They can be used to make large, tough, low density, translucent panels.
4. They can be used to make the materials fire retardant where desired.
5. Superior heat resistance to most rigid thermoplastics, particularly those that are available in sheet form.

Because of their favourable price polyesters are preferred to epoxide and furane resins for general purpose laminates and account for at least 95% of the low pressure laminates produced. The epoxide resins find specialised uses for chemical, electrical and heat-resistant applications and for optimum mechanical properties. The furane resins have a limited use in chemical plant. The use of high pressure laminates from phenolic, aminoplastic and silicone resins is discussed elsewhere in this book.

The largest single outlet for polyester–glass laminates is in sheeting for roofing and building insulation and accounts for about one third of the resin produced. For the greatest transparency it is important that the refractive indices of glass, cured resins and binder be identical. For this reason the glass fibre and resin suppliers provide raw materials which are specially made to approximate to these requirements. This outlet is now being severely challenged by rigid p.v.c. sheeting which is much cheaper than fire-retardant polyester laminates.

The second major outlet is in land transport where the ability to form large structures has been used in the building of sports car bodies, panelling for lorries, particularly translucent roofing panels, and in public transport vehicles. In such applications the number of mouldings required is quite small. The polyester–glass structures are less suitable for large quantity production since in these circumstances the equipment requirements rise steeply and it eventually becomes more economical to use the more traditional stamped metal shapings.

Polyester resins have been widely accepted in the manufacture of boat hulls up to 100 ft in length. Such hulls are competitive in price with those built from traditional materials and are easier to maintain and repair.

Aircraft radomes, ducting, spinners and other parts are often prepared from polyester resins in conjunction with glass cloth or mat. The principal virtue here is the high strength–weight ratio possible particularly when glass cloth is used. Land, sea and air transport applications account for almost half the polyester resin produced.

Other applications include such diverse items as chemical plant, stacking chairs, swimming pools, trays and sports equipment.

21.2.6 Allyl resins[10]

A number of useful resins have been prepared from allyl compounds, i.e. derivatives of allyl alcohol $CH_2{=}CH \cdot CH_2OH$. One of these, diethylene glycol bisallyl carbonate, was one of the first polyester-type materials to be developed for laminating and casting. It was introduced about 1941

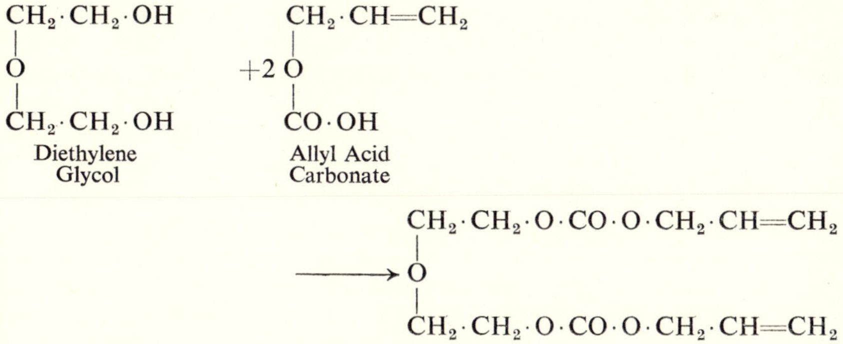

$$CH_2 \cdot CH_2 \cdot OH$$
$$|$$
$$O$$
$$|$$
$$CH_2 \cdot CH_2 \cdot OH$$

Diethylene Glycol

$$+2 \quad CH_2 \cdot CH{=}CH_2$$
$$|$$
$$O$$
$$|$$
$$CO \cdot OH$$

Allyl Acid Carbonate

$$CH_2 \cdot CH_2 \cdot O \cdot CO \cdot O \cdot CH_2 \cdot CH{=}CH_2$$
$$|$$
$$\longrightarrow O$$
$$|$$
$$CH_2 \cdot CH_2 \cdot O \cdot CO \cdot O \cdot CH_2 \cdot CH{=}CH_2$$

Fig. 21.10

by the Pittsburgh Plate Glass Company as Allymer CR39 and was produced by the reaction shown in Fig. 21.10. It could be cured with benzyl peroxide at 80°C.

Diallyl phthalate (see also Section 21.3) has also been used as a laminating resin but because of its higher price it has been largely replaced by the glycol-saturated acid-unsaturated acid polyesters.

Other allyl compounds described in the literature include diallyl carbonate, diallyl isophthalate and diallyl benzene phosphonate.

21.3 POLYESTER MOULDING COMPOSITIONS

Although phenolic and amino moulding powders remain by far the most important of the thermosetting moulding compositions a number of new materials have been introduced[11] over the last twenty years based on polyester, epoxide and silicone resins.

Four classes of polyester compound may be recognised:

1. Dough moulding compounds (d.m.c.).
2. Alkyd moulding compositions, sometimes referred to as 'polyester alkyds'.
3. Diallyl phthalate compounds.
4. Diallyl isophthalate compounds.

The dough moulding compounds were originally developed in an attempt to combine the mechanical properties of polyester–glass laminates with the speed of cure of conventional moulding powders.   In spite of their somewhat high cost they have now established themselves in a number of applications where a mechanically strong electrical insulant is required.

Dough moulding compositions are prepared by blending resin, powdered mineral filler, reinforcing fibre, pigment and lubricant in a dough mixer, usually of the Z-blade type.   The resins are similar to conventional laminating resins, a fairly rigid type being preferred so that cured mouldings may be extracted from the mould at 160°C without undue distortion. Organic peroxides such as benzoyl peroxide and tertiary butyl perbenzoate are commonly used as 'catalysts'.   The choice of 'catalyst' will influence cure conditions and will also be a factor in whether or not surface cracks appear on the mouldings.   Mineral fillers such as calcium carbonate are employed not only to reduce costs but to reduce shrinkage and to aid the flow since an incorrect viscosity may lead to such faults as fibre bunching and resin-starved areas.   Although glass fibre (E type) is most commonly employed as the reinforcing fibre sisal is used in cheaper compositions.   Stearic acid or a metal stearate are the usual lubricants.

Table 21.2

	Low cost general purpose	High grade mechanical	High grade electrical
Polyester resin	100	100	100
E glass $\frac{1}{4}$ in. length	20	—	90
E glass $\frac{1}{2}$ in. length	—	85	—
Calcium carbonate	240	150	—
Fine silica	—	—	100
Sisal	40	—	—
Benzoyl peroxide	1	1	1
Calcium stearate	2	2	2
Pigment	2	2	2

Formulations for three typical d.m.c. grades are given in Table **21.2**.

The non-fibrous components are first mixed together and the fibrous materials are then added. The properties of the components are critically dependent on the mixing procedures since these will affect dispersion and fibre degradation.

In common with all polyester moulding compositions the dough moulding compounds cure without evolution of volatiles and thus pressures as low as 200 lb in^{-2}, but normally about 1,000 lb in^{-2}, may be used. The material, of putty-like consistency is first preformed into a ball shape and loaded into the mould of a fast-acting press in such a way that there should be a minimum of weld lines and undesirable fibre alignment. Temperatures in the range 110–170°C may be employed and at the higher temperatures cure times of less than one minute are possible.

The 'polyester' alkyd moulding compositions are also based on a resin similar to those used for laminating. They are prepared by blending the resin with cellulose pulp, mineral filler, lubricants, pigments and peroxide curing agents on hot rolls until thoroughly mixed and of the desired flow properties. The resultant hide is removed, cooled, crushed and ground.

On heating with a peroxide, diallyl phthalate will polymerise and eventually cross link because of the presence of two double bonds (Fig. 21.11).

This monomer has been used as the basis of a laminating resin and as a reactive diluent in polyester laminating resin, but at the present time its

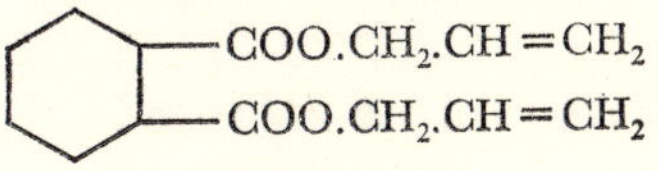

$$\text{—COO.CH}_2\text{.CH} = \text{CH}_2$$
$$\text{—COO.CH}_2\text{.CH} = \text{CH}_2$$

Fig. 21.11

principal value is in moulding compositions. It is possible to heat the monomer under carefully controlled conditions to give a soluble and stable partial polymer in the form of a white powder. The powder may then be blended with fillers, peroxide catalysts and other ingredients in the same manner as the polyester alkyds to form a moulding powder. Similar materials may be obtained from diallyl isophthalate.

The diallyl phthalate (d.a.p.) resins compare favourably with the phenolic resins in their electrical insulation characteristics under conditions of dry and wet heat. The diallyl isophalate (d.a.i.p.) compositions are more expensive but have better heat resistance and are claimed to be capable of withstanding temperatures of as high as 220°C for long periods. Both the d.a.p. and the d.a.i.p. materials are superior to the phenolics in their tracking resistance and in their availability in a wide range of colours. They do, however, tend to show a higher shrinkage on cure and in cases where this may be important, e.g. thin walls round inserts, it may be necessary to employ epoxide moulding compositions (see Chapter 22).

The 'polyester alkyd' resins are lower in cost than the d.a.p. resins but are weaker mechanically, have a lower resistance to cracking round inserts and do not maintain their electrical properties so well under severe humid conditions.   Fast-curing grades are available which will cure in as little as 20 seconds.

Some pertinent properties of the various polyester compounds are compared with those of a G.P. phenolic composition in Table 21.3.

The alkyd moulding compositions are used almost entirely in electrical applications where the cheaper phenolic and amino-resins are unsuitable.

21.4 FIBRE AND FILM-FORMING POLYESTERS

Although Carothers investigated many fibre-forming polyesters in his researches which led to the development of the nylons, it was left to J. R. Whinfield and J. T. Dickson working at the Calico Printers Association in England to discover the one polyester of importance as a synthetic fibre, poly(ethylene terephthalate).

Poly(ethylene terephthalate) fibres marketed as Dacron (Du Pont) and Terylene (I.C.I.) are widely used in both domestic and industrial applications where their high strength, crease resistance and low water absorption are of value.

Terephthalic acid is commonly prepared from *p*-xylene by an oxidation process.   Although the *p*-xylene may be obtained from coal tar the success

Table 21.3

	P/F G.P.	D.M.C. (G.P.)	Polyester alkyd	D.A.P. alkyd	D.A.I.P. alkyd	Units
Moulding temperature*	150–170	140–160	140–165	150–165	150–165	°C
Cure time (cup flow test)*	60–70	25–40	20–30	60–90	60–90	sec
Shrinkage	0·007	0·004	0·009	0·009	0·006	in/in.
Impact strength	0·12–0·2	2·0–4·0	0·13–0·18	0·12–0·18	0·09–0·13	ft lb
Specific gravity	1·35	2·0–2·1	1·7–1·8	1·64	1·8	
Power factor (800 c/s)	0·1–0·4	0·01–0·05	0·01–0·05	0·03–0·05	—	
Power factor (10^6 c/s)	0·03–0·05	0·01–0·03	0·02–0·04	0·02–0·04	0·04–0·06	
Dielectric constant (800 c/s)	6–10	5·5–6·5	4·5–5·5	4·0–5·5	—	
Dielectric constant (10^6 c/s)	4·5–5·5	5·0–6·0	4·5–5·0	3·5–5·0	4·0–6·0	
Volume resistivity	10^{10}–10^{12}	>10^{14}	>10^{14}	>10^{14}	>10^{14}	ohm cm
Electric strength (90°C)	100–250	200–300	240–350	300–400	300–400	V/0·001
Water absorption	45–65	15–30	40–70	5–15	10–20	in. mg

Except where marked by an asterisk these results were obtained by tests methods as laid down in B.S.771.

of poly(ethylene terephthalate) is due in no small measure to the development of processes which produce *p*-xylene from petroleum fractions by the process of reforming, and this is today the principal route to the acid.

Since *p*-xylene is difficult to obtain in pure form because of the closeness of its boiling point to the other isomeric xylenes (*o*-144°C, *m*-138·8°C, *p*-138·5°C) it is also difficult to obtain terephthalic acid, which sublimes at 300°C, in a pure form.   It is therefore convenient to produce the polymer by an ester-interchange reaction between the more easily purified dimethyl terephthalate and ethylene glycol. The low melting point of the dimethyl ester (142°C) and its greater reactivity provide further reasons for its use (Fig. 21.12)

The reaction is carried out in two stages, first ester-interchange in the presence of catalysts, such as antimony trioxide with cobaltous acetate,

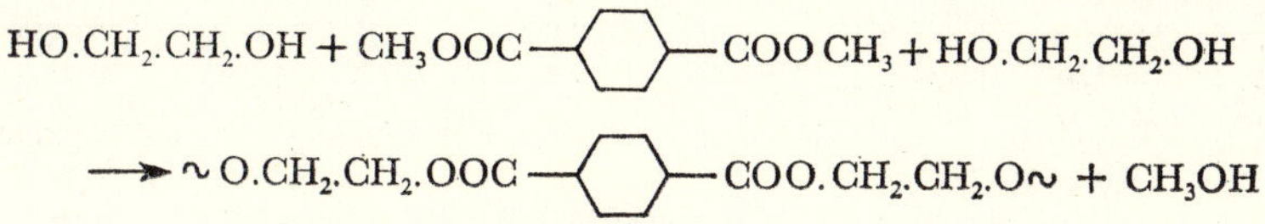

$$HO.CH_2.CH_2.OH + CH_3OOC-\left<\bigcirc\right>-COO\,CH_3 + HO.CH_2.CH_2.OH$$

$$\longrightarrow \sim O.CH_2.CH_2.OOC-\left<\bigcirc\right>-COO.CH_2.CH_2.O\sim + CH_3OH$$

Fig. 21.12

to produce a low molecular weight polymer, followed by heating at higher temperatures under reduced pressures in order to achieve further condensation.

Because of its structural regularity, poly(ethylene terephthalate) crystallises readily above its transition temperature. Whilst the intermolecular attraction is less than with nylon 66, the presence of a benzene ring in the main chain leads to a somewhat stiffer chain with the result that the polyester has both a higher glass transition temperature (80°C) and crystalline melting point (265°C). The drawing of fibres and films must be carried out between these temperatures in order to produce useful products.

The solubility parameter of poly(ethylene terephthalate) is about 10·7 but because it is a highly crystalline material only proton donors that are capable of interaction with the ester groups are effective. A mixture of phenol and tetrachloroethane is often used when measuring molecular weights, which are about 20,000 in the case of commercial polymers.

Although a polar polymer, its electrical insulating properties at room temperature are good even at high frequencies owing to the fact that since

Table 21.4 CRYSTALLINITY DATA FOR POLY(ETHYLENE TEREPHTHALATE)

Molecular configuration	almost completely extended
Cell dimensions	$a = 4·56$ Å
	$b = 5·94$ Å
	$c = 10·75$ Å (chain axis)
Density of cell	1·47 g cm^{-3}
Amorphous density	1·33 g cm^{-3}
Density of oriented polymer	1·38–1·39 g cm^{-3}
Density of crystalline polymer (unoriented)	$\sim 1·45$
Crystalline melting point	265°C
Maximum rate of crystallisation—at	170°C

room temperature is well below the transition temperature dipole orientation is severely restricted. Some data on the crystallinity of poly(ethylene terephthalate) are presented in Table 21.4.

Poly(ethylene terephthalate) film is produced by quenching extruded film to the amorphous state and then reheating and stretching the sheet

Table 21.5   TYPICAL PROPERTIES OF POLYESTER FILMS
(ASTM Test Methods)

Property	Units	Poly(ethylene terephthalate)	Kodel
Specific gravity	—	1·39	1·226
Tensile strength	10^3 lb in^{-2}	17–25	10–17
Elongation at break	%	50–130	45
Dielectric strength	V/0·001 in.	4,000	>4,000
Dielectric constant (60 c/s)		3·16	3·1
(20°C) (10^6 c/s)		2·98	2·9
Power factor (60 c/s)		0·002	0·005
(20°C) (10^6 c/s)		0·014	0·014
Volume resistivity	ohms cm	10^{19}	—
Water absorption (24 hr immersion)	%	0·55	0·30

approximately threefold in each direction at 80–100°C. In a two-stage process machine direction stretching induces 10–14% crystallinity and this is raised to 20–25% by transverse orientation. In order to stabilise the biaxially oriented film it is annealed under restraint at 180–210°C, this increasing the crystallinity to 40–42% and reducing the tendency to shrink on heating.

Some typical properties of commercial poly(ethylene terephthalate) film, e.g. Melinex (I.C.I.), Mylar (Du Pont), are given in Table 21.5.

The principal uses of poly(ethylene terephthalate) film are electrical, particularly in capacitors, as slot liners for motors and for recording tape. Its high strength and dimensional stability have led to a number of drawing office applications. The film is also a useful packaging material whilst metallised products have a number of uses as a decorative material.

With one exception no other high molecular weight linear polyesters have achieved any sort of commercial significance. The one exception is the condensation polymer of dimethyl terephthalate and 1,4-cyclo-hexylene glycol (also known as 1,4-cyclohexane-dimethanol) (Fig. 21.13).

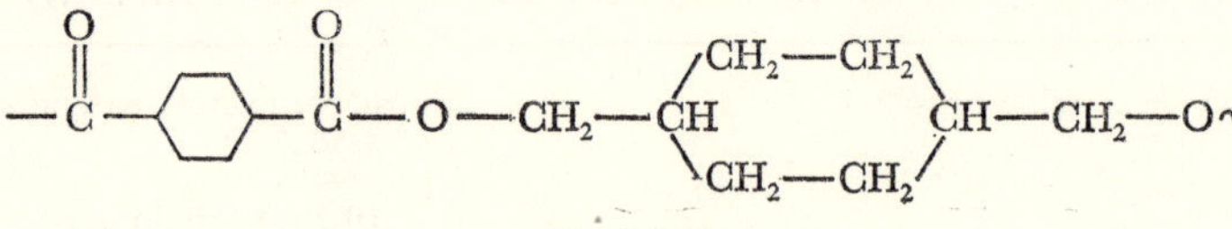

Fig. 21.13

This polymer has a slightly stiffer chain and hence a slightly higher melting point and heat distortion temperatures than poly(ethylene terephthalate). Films are available (Kodel–Kodak) which have been biaxially stretched about 200% from polymer with molecular weights of about 25,000. They are similar electrically to poly(ethylene terephthalate), weaker mechanically but have superior resistance to water and in weathering stability.[12] Some properties are given in Table 21.5.

Fibres are also available from poly(1,4-cyclohexylenedimethylene terephthalate) and are marketed as Kodar (Kodak) and Vestan (Hüls).

21.5 SURFACE COATINGS, PLASTICISERS AND RUBBERS

It has been estimated that in the United States somewhat over half of surface coatings are of the polyester (alkyd) type.

These resins are produced by reacting a polyhydric alcohol, usually ethylene glycol with a polybasic acid, usually phthalic acid and the fatty acids of various oils such as linseed oil, soya bean oil and tung oil. These oils are triglycerides of the type shown in Fig. 21.14. R_1, R_2 and R_3 usually contain unsaturated groupings. The alkyd resins would thus have structural units such as is shown in Fig. 21.15.

In modern manufacturing methods the oil is sometimes reacted directly with the glycol to form a monoglyceride and this is then reacted with

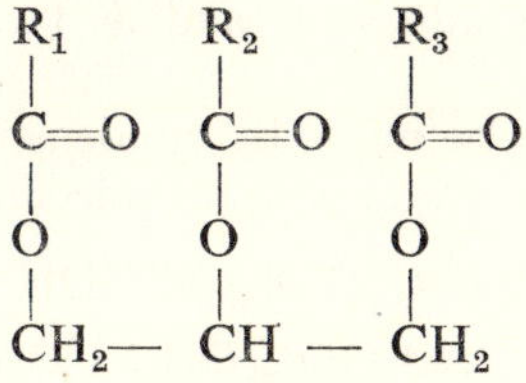

Fig. 21.14

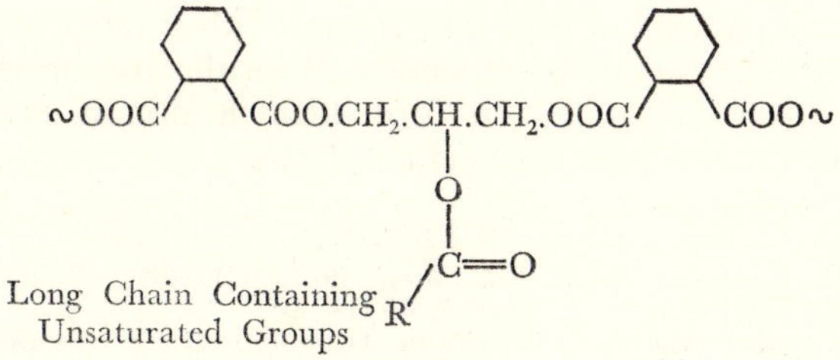

Fig. 21.15

the acid to form the alkyd resin. When the resulting surface coating is applied to the substrate the molecules are substantially linear. However in the presence of certain 'driers' such as lead soaps there is oxidative cross linking via the unsaturated group in the side chain and the resin hardens.

The alkyd resins are of value because of their comparatively low cost, durability, flexibility, gloss retention and reasonable heat resistance. Alkyd resin modified with rosin, phenolic resin, epoxy resins and monomers such as styrene is of current commercial importance.[13,14]

Low molecular weight liquid polyester resins are useful as plasticisers particularly for p.v.c. where they are less volatile and have greater resistance to extraction by water than monomeric plasticisers. Examples of such plasticisers are poly(propylene adipate) and poly(propylene sebacate). In some cases monobasic acids such as lauric acid are used to control the molecular weight.

Rubbery polyesters have been produced but are now of minor importance. Rubbery polyester–amides were introduced by I.C.I. under the

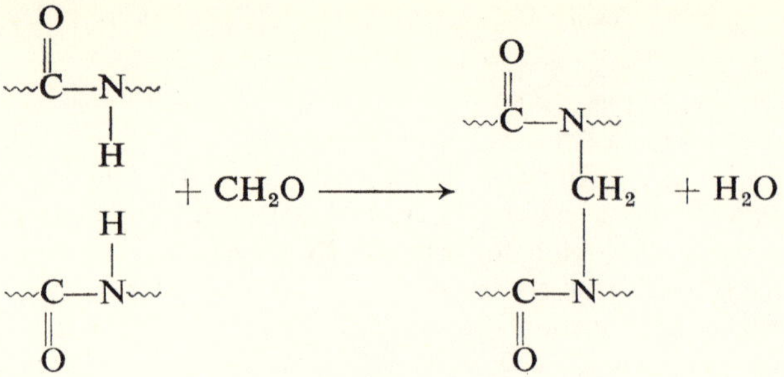

$$\text{~~C—N~~} \quad \text{(with } O \text{ double bond on C, } H \text{ on N)}$$

Fig. 21.16

trade name Vulcaprene as a leathercloth material but currently they are used primarily as leather adhesives and as flexible coatings for rubber goods.

A typical polymer may be made by condensing ethylene glycol, adipic acid and ethanolamine to a wax with a molecular weight of about 5,000.

$$\text{HO}\cdot(\text{CH}_2)_2\cdot\text{OH} + \text{HOOC}\cdot(\text{CH}_2)_4\cdot\text{COOH} + \text{NH}_2\cdot(\text{CH}_2)_2\cdot\text{OH}$$
$$\longrightarrow \text{—O(CH}_2)_2\text{OOC(CH}_2)_4\text{CONH(CH}_2)_2\text{O~~}$$

The chain length of the polymer is then increased by reacting the wax via the hydroxyl, amino or acid end groups with a diisocyanate such as hexamethylene diisocyanate (see Chapter 23 for the appropriate reaction).

The rubbers are vulcanised by formaldehyde donors reacting across —NH— groups in the main chain (Fig. 21.16).

REFERENCES

1. BERZELIUS, J., *Rap. Ann. Progr. Sci. Physq.*, **26** (1847)
2. LUSSAC, J. G., and PELOUZE, J., *Ann.*, **7**, 40 (1833)
3. BRYDSON, J. A., and WELCH, C. W., *Plastics* (London), **21**, 282 (1956)
4. SCOTT, K. A., and GALE, G. M., *Rubber and Plastics Research Association of Great Britain, Research Report* 132 (September 1964)
5. MORGAN, P., *Glass Reinforced Plastics*, 3rd Edn, Iliffe, London (1961)
6. SONNEBORN, R. H., *Fibreglass Reinforced Plastics*, Reinhold, New York (1954)
7. DE DANI, A., *Glass Fibre Reinforced Plastics*, Newnes, London (1960)
8. LAWRENCE, J. R., *Polyester Resins*, Reinhold, New York (1960)
9. HAGEN, H., *Glasfaserverstärkte Kunstoffe*, Springer-Verlag (1961)
10. RAECH, H., *Allylic Resins and Monomers*, Reinhold, New York (1965)
11. GREATREX, J. L., and HAYNES, I. E., *Brit. Plastics*, **35**, 340 (1962)
12. WATSON, M. T., *Soc. Plastics Engrs*, 1083 (October 1961)
13. MARTENS, C. R., *Alkyd Resins*, Reinhold, New York (1961)
14. PATTON, T. C., *Alkyd Resin Technology*, Interscience, New York (1962)

BIBLIOGRAPHY

BJORKSTEN, J., *Polyesters and their Applications*, Reinhold, New York (1956)
DE DANI, A., *Glass Fibre Reinforced Plastics*, Newnes, London (1960)
HAGEN, H., *Glasfaserverstärkte Kunstoffe*, Springer-Verlag, Berlin (1961)
HILL, R., *Fibres from Synthetic Polymers*, Elsevier, Amsterdam (1953)
LAWRENCE, J. R., *Polyester Resins*, Reinhold, New York (1960)
MARTENS, C. R., *Alkyd Resins*, Reinhold, New York (1961
MORGAN, P., *Glass Reinforced Plastics*, 3rd Edn, Iliffe, London (1961)
PATTON, T. C., *Alkyd Resin Technology*, Interscience, New York (1962)
PETUKHOV, B. V., *The Technology of Polyester Fibres*, Pergamon, Oxford (1963)
RAECH, H., *Allylic Resins and Monomers*, Reinhold, New York (1965)
SONNEBORN, R. H., *Fibreglas Reinforced Plastics*, Reinhold, New York (1956)

22

Epoxide Resins

22.1 INTRODUCTION

In spite of their comparatively high cost, marketing at two to three times the price of a general purpose polyester resin, the epoxide resins have achieved significant importance as industrial materials. Used primarily as surface coatings and for the encapsulation of electronic components they have also found important applications as adhesives, tooling compounds and laminating materials.

The commercial interest in epoxide (epoxy) resins was first made apparent by the publication of German Patent 676,117 by I.G. Farben[1] in 1939 which described liquid polyepoxides. In 1943 P. Castan[2] filed U.S. Patent 2,324,483 covering the curing of the resins with dibasic acids. This important process was subsequently exploited by the Ciba Company. A later patent of Castan[3] covered the hardening of epoxide resins with alkaline catalysts used in the range 0·1–5%. This however became of somewhat restricted value as the important amine hardeners are usually used in quantities higher than 5%.

In the early stage of their development the epoxy resins were used almost entirely for surface coatings and developments in this field are to a large extent due to the work of S. O. Greenlee and described in a number of patents. These included work on the modification of epoxy resins with glycerol[4], the esterification of the higher molecular weight materials with drying oil acids[5] and reactions with phenolic[6] and amino resins.[7]

Before the Second World War the cost of the intermediates for these resins (in most cases epichlorhydrin and bis-phenol A) would have prevented the polymers from becoming of commercial importance. Subsequent improvements in the methods of producing these intermediates and improved techniques of polymerisation have however led to wide commercial acceptance.

The properties of the cross-linked resins depend very greatly on the curing system used but in general it may be said that the outstanding features of the resins are their toughness, low shrinkage on cure, high adhesion to many substrates, good alkali resistance and the versatility of curing systems available.

The resins are marketed by many companies including Shell Chemical (Epikote Resin), Ciba (A.R.L.) (Araldite) and Bakelite–Xylonite Ltd., to name three British companies involved in their commercial exploitation.

The resins are variously described as ethoxyline, epoxy and epoxide resins. The term 'epoxy' is favoured in the United States and 'epoxide' in Great Britain.

22.2 PREPARATION OF RESINS FROM BIS-PHENOL A

The first, and still the most important, commercial epoxy resins are reaction products of bis-phenol A and epichlorhydrin. Other types of epoxide resins were introduced in the late 1950's and early 1960's prepared by epoxidising unsaturated structures. These materials will be dealt with in Section 22.4. The bis-phenol A is prepared by reaction of acetone and phenol (Fig. 22.1).

Since both phenol and acetone are readily available and the bis-phenol A

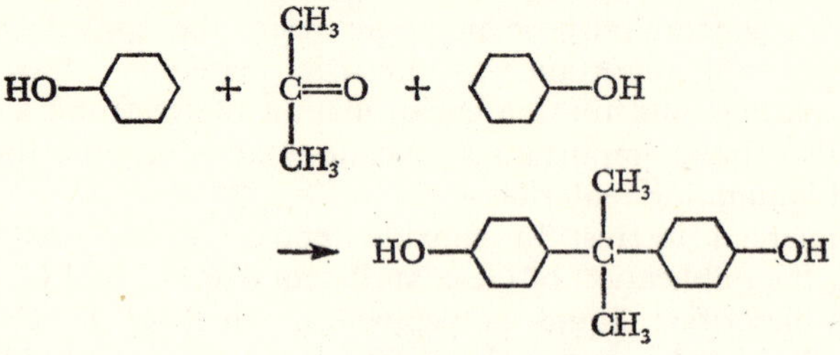

Fig. 22.1

Fig. 22.2

is easy to manufacture this intermediate is comparatively inexpensive. This is one of the reasons why it has been the preferred dihydric phenol employed in epoxy resin manufacture. Since most epoxide resins are of low molecular weight and because colour is not particularly critical the degree of purity of the bis-phenol A does not have to be so great as when used in the polycarbonate resins. Bis-phenol A with a melting point of

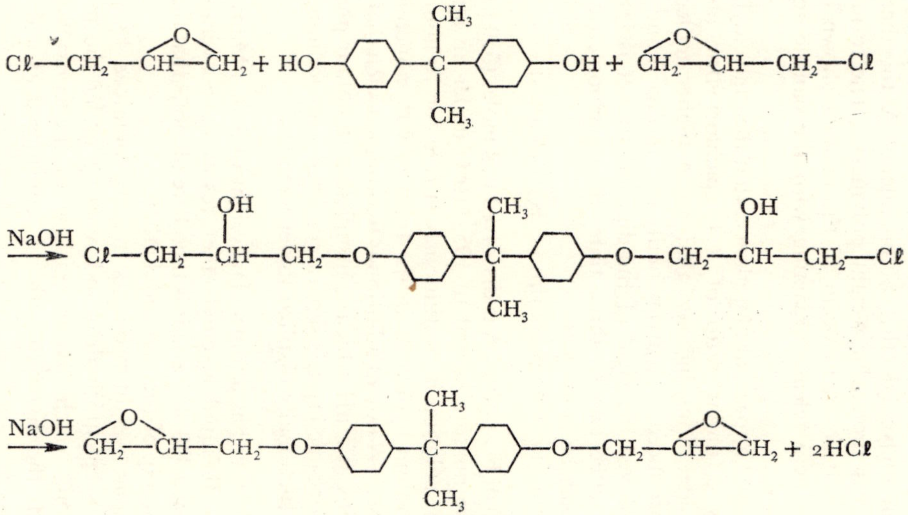

Cl—CH₂—CH—O + HO ... OH ... C(CH₃)₂ ... OH + CH₂—CH—CH₂—Cl

NaOH → Cl—CH₂—CH—CH₂—O ... C(CH₃)₂ ... O—CH₂—CH—CH₂—Cl, OH

NaOH → CH₂—CH—CH₂—O ... C(CH₃)₂ ... O—CH₂—CH—CH₂ + 2HCl

Fig. 22.3

153°C is considered adequate for most applications whilst less pure materials may often be employed.

Epichlorhydrin, the more expensive compound, is derived from propylene by the sequence of reactions shown in Fig. 22.2.

It will be noticed that the initial steps correspond with those used in the manufacture of glycerol. The material is available commercially at 98% purity and is a colourless mobile liquid.

Many of the commercial liquid resins consist essentially of the low molecular weight diglycidyl ether of bis-phenol A together with small quantities of higher molecular weight polymers. The formation of the diglycidyl ether is believed to occur in the manner shown in Fig. 22.3, the hydrochloric acid released reacting with the caustic soda to form sodium chloride.

Although it would appear, at first glance, that the diglycidyl ether would be prepared by a molar ratio of 2:1 epichlorhydrin–bis-phenol A, probability considerations indicate that some higher molecular weight species will be produced. Experimentally it is in fact found that when a 2:1 ratio is employed the yield of the diglycidyl ether is less than 10%. Therefore in practice two to three times the stoichiometric quantity of epichlorhydrin may be employed.

A typical laboratory scale preparation[8] is as follows:

'1 mole (228 g) of bis-phenol A is dissolved in 4 moles (370 g) of epichlorhydrin and the mixture heated to 105–110°C under an atmosphere of nitrogen. The solution is continuously stirred for 16 hours while 80 g 2 moles) of sodium hydroxide in the form of 30% aqueous solution is added dropwise. A rate of addition is maintained such that the reaction mixture remains at a pH which is insufficient to colour phenophthalin. The resulting organic layer is separated, dried with sodium suphate and may then be fractionally distilled under vacuum.'

The diglycidyl ether has a molecular weight of 340. Many of the well-known commercial liquid glycidyl ether resins have average molecular weights in the range 340–400 and it is therefore obvious that these materials are composed largely of the diglycidyl ether.

Higher molecular weight products may be obtained by reducing the amount of excess epichlorhydrin and reacting under the more strongly alkaline conditions which favour reaction of the epoxide groups with bis-phenol A. If the diglydidyl ether is considered as a diepoxide and represented as

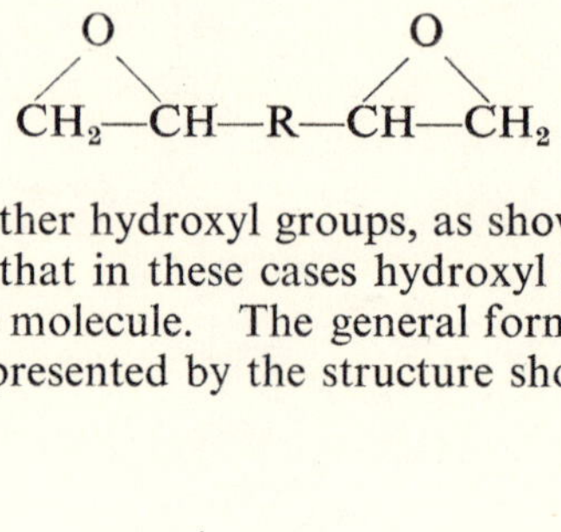

$$\underset{\displaystyle CH_2-CH-R-CH-CH_2}{\overset{\displaystyle O \qquad\qquad O}{\diagup\diagdown \qquad\qquad \diagup\diagdown}}$$

this will react with further hydroxyl groups, as shown in Fig. 22.4

It will be observed that in these cases hydroxyl groups will be formed along the chain of the molecule. The general formulae for glycidyl ether resins may thus be represented by the structure shown in Fig. 22.5.

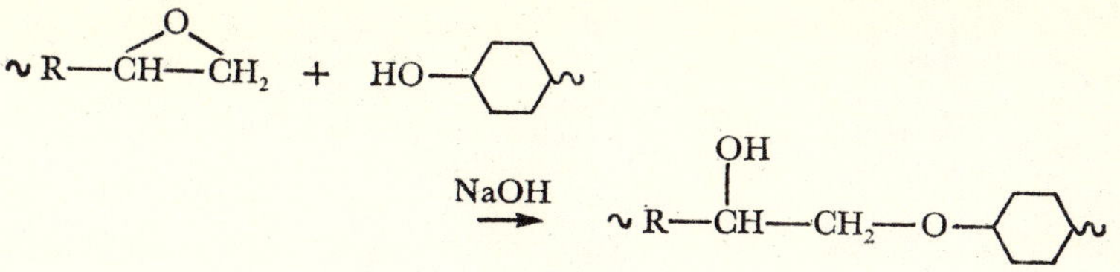

$$\sim R\text{—}\overset{O}{\overbrace{CH\text{—}CH_2}} \; + \; HO\text{—}\langle \rangle \sim$$

$$\xrightarrow[]{NaOH} \qquad \sim R\text{—}\overset{OH}{\overset{|}{CH}}\text{—}CH_2\text{—}O\text{—}\langle \rangle \sim$$

Fig. 22.4

When $n = 0$, the product is the diglycidyl ether, and the molecular weight is 340. When $n = 10$ molecular weight is about 3,000. Since commercial resins seldom have average molecular weights exceeding 4,000 it will be realised that in the uncured stage the epoxy resins are polymers with a low degree of polymerisation.

Table 22.1 shows the effect of varying the reactant ratios on the molecular weight of the epoxide resins.[9]

It is important that care should be taken to remove residual caustic soda and other contaminants when preparing the higher molecular weight

Table 22.1 EFFECT OF REACTANTS RATIOS ON MOLECULAR WEIGHTS

Mol. ratio epichlorhydrin bis-phenol A	Mol. ratio NaOH/ epichlorhydrin	Softening point (°C)	Molecular weight	Epoxide equivalent	Epoxy groups per molecule
2·0	1·1	43	451	314	1·39
1·4	1·3	84	791	592	1·34
1·33	1·3	90	802	730	1·10
1·25	1·3	100	1,133	862	1·32
1·2	1·3	112	1,420	1,176	1·21

resins and in order to avoid the difficulty of washing highly viscous materials these resins may be prepared by a two-stage process.

This involves first the preparation of lower molecular weight polymers with a degree of polymerisation of about 3. These are then reacted with bis-phenol A in the presence of a suitable polymerisation catalyst such that the reaction takes place without the evolution of by-products.[10]

The epoxide resins of the glycidyl ether type are usually characterised by six parameters:

1. Resin viscosity (of liquid resin).
2. Epoxide equivalent.
3. Hydroxyl equivalent.
4. Average molecular weight (and molecular weight distribution).
5. Melting point (of solid resin).
6. Heat distortion temperature (deflection temperature under load) of cured resin.

Resin viscosity is an important property to consider in handling the resins. It depends on the molecular weight, molecular weight distribution, the chemical constitution of the resin and the presence of any modifiers or diluents. Since even the diglycidyl ethers are highly viscous materials

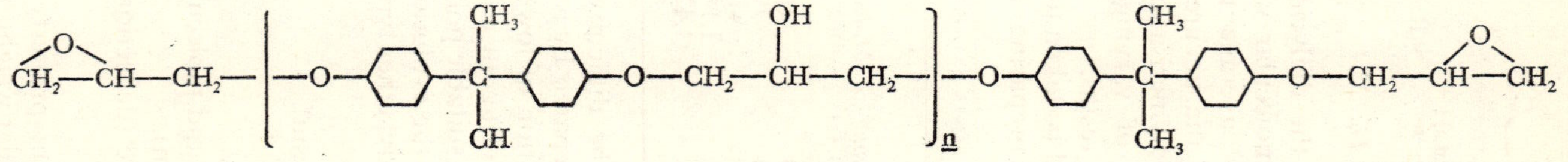

Fig. 22.5

with viscosities of about 40–100 poise at room temperature it will be appreciated that the handling of such viscous resins can present serious problems.

The epoxide equivalent is a measure of the amount of epoxy groups. This is the weight of resin (in grammes) containing 1 gramme chemical equivalent epoxy. For a pure diglycidyl ether with two epoxy groups per molecule the epoxide equivalent will be half the molecular weight (i.e. epoxide equivalent = 170). The epoxy equivalent is determined by reacting a known quantity of resin with hydrochloric acid and measuring the unconsumed acid by back titration. The reaction involved is

$$\text{\textasciitilde CH}\overset{O}{\underset{}{\diagup\diagdown}}\text{CH}_2 + HCl \longrightarrow \text{\textasciitilde CH}\overset{OH}{\underset{|}{}}\text{CH}_2\text{—Cl}$$

It is possible to correlate epoxy equivalent for a given class of resin with infrared absorption data.

The hydroxyl equivalent is the weight of resin containing one equivalent weight of hydroxyl groups. It may be determined by many techniques but normally by reacting the resin with acetyl chloride.

The molecular weight and molecular weight distribution may be determined by conventional techniques. As the resins are of comparatively low molecular weight it is possible to measure this by ebullioscopic and by end-group analysis techniques.

It is useful to measure the melting point of the solid resins. This can be done either by the ring and ball technique or by Durrans mercury method. In the latter method a known weight of resin is melted in a test tube of fixed dimensions. The resin is then cooled and solidifies. A known weight of clean mercury is then poured on to the top of the resin and the whole assembly heated, at a fixed rate, until the resin melts and the mercury runs through the resin. The temperature at which this occurs is taken as the melting point.

The A.S.T.M. heat distortion temperature (deflection temperature under load) test may be used to characterise a resin. Resins must, however, be compared using identical hardeners and curing conditions.

Typical data for some commercial glycidyl ether resins are given in Table 22.2.

Recently solid resins have been prepared having a very closely controlled molecular weight distribution.[11] These resins melt sharply to give low

Table 22.2

Resin	Average Mol. Wt.	Epoxide equivalent	Viscosity cP at 25°C	Melting point °C (Durrans)
A	350–400	175–210	4–10,000	—
B	450	225–290	—	—
C	700	300–375	—	40–50
D	950	450–525	—	64–76
E	1,400	870–1,025	—	95–105
F	2,900	1,650–2,050	—	125–132
G	3,800	2,400–4,000	—	145–155

viscosity liquids. It is possible to use larger amounts of filler with the resins with a consequent reduction in cost and coefficient of expansion so that such resins are useful in casting operations.

22.3 CURING OF GLYCIDYL ETHER RESINS

The cross linking of epoxy resins may be carried out either through the epoxy groups or the hydroxy groups. Two types of curing agent may also be distinguished, catalytic systems and polyfunctional cross-linking agents that link the epoxide resin molecules together. Some systems used may involve both the catalytic and cross-linking systems.

Whilst the curing mechanisms may be quite complex and the cured resins too intractable for conventional analysis some indication of the mechanisms involved has been achieved using model systems.

It has been shown in the course of this work[12] that the reactivity of the epoxy ring is enhanced by the presence of the ether linkage separated from it by a methylene link.

$$\underset{\displaystyle CH_2-CH-CH_2-O\text{\tiny\char`\~}}{\overset{\displaystyle O}{\diagup\diagdown}}$$

The epoxy ring may then be readily attacked not only by active hydrogen and available ions but even by tertiary amines. For example with the latter it is believed that the reaction mechanism is as follows:

$$R_3N + \underset{\displaystyle CH_2-CH\text{\tiny\char`\~}}{\overset{\displaystyle O}{\diagup\diagdown}} \longrightarrow R_3N^{\oplus}\!-\!CH_2-\underset{\displaystyle \overset{|}{O^{\ominus}}}{CH\text{\tiny\char`\~}}$$

This ion may then open up a new epoxy group generating another ion

$$\text{\tiny\char`\~}CH_2-\underset{\displaystyle \overset{|}{O^{\ominus}}}{CH\text{\tiny\char`\~}} + \underset{\displaystyle CH_2-CH\text{\tiny\char`\~}}{\overset{\displaystyle O}{\diagup\diagdown}} \longrightarrow \begin{array}{c} \text{\tiny\char`\~}CH_2-CH\text{\tiny\char`\~} \\ | \\ O\text{---}CH\text{\tiny\char`\~} \\ | \\ O^{\ominus} \end{array}$$

which can in turn react with a further epoxy group.

Since this reaction may occur at both ends of the molecule (in case of glycidyl ether resins) a cross-linked structure will be built up.

The overall reaction is complicated by the fact that the epoxy group, particularly when catalysed, will react with hydroxyl groups. Such groups may be present due to the following circumstances:

1. They will be present in the higher molecular weight homologues of the diglycidyl ether of bis-phenol A.
2. They may be introduced by the curing agent or modifier.

3. They will be formed as epoxy resins are opened during cure.
4. In unreacted phenol type materials they are present as impurities.

The epoxy-hydroxyl reaction may be expressed as

$$R\text{—OH} + \overset{\displaystyle O}{\overset{\displaystyle \diagup\,\diagdown}{CH_2\text{—}CH}}\text{\textasciitilde} \longrightarrow RO\text{—}CH_2\text{—}\underset{\underset{OH}{|}}{CH}\text{\textasciitilde}$$

or

$$HO\text{—}CH_2\text{—}\underset{\underset{OR}{|}}{CH}\text{\textasciitilde}$$

This product will contain new hydroxyl groups that can react with other epoxy rings generating further active hydroxy groups

e.g. $R\text{—}CH_2\text{—}\underset{\underset{OH}{|}}{CH}\text{\textasciitilde}$ $+$ $\overset{\displaystyle O}{\overset{\displaystyle \diagup\,\diagdown}{CH_2\text{—}CH}}\text{\textasciitilde}$

$$\longrightarrow RO\text{—}CH_2\text{—}\underset{\underset{O\text{—}CH_2\text{—}\underset{\underset{OH}{|}}{CH}\text{\textasciitilde \ etc.}}{|}}{CH}\text{\textasciitilde}$$

The predominance of one reaction over the other is greatly influenced by the catalyst system employed. Tertiary amine systems are often used in practice.

In addition to the catalytic reactions the resins may be cross linked by agents which link across the epoxy molecules. These reactions may be via the epoxy ring or through the hydroxyl groups. Two examples of the former are:

1. With amines

$$\text{\textasciitilde}\overset{\displaystyle O}{\overset{\displaystyle \diagup\,\diagdown}{CH\text{—}CH_2}} + H\overset{\displaystyle \overset{R}{\underset{\diagup\,\diagdown}{N}}}{}H + \overset{\displaystyle O}{\overset{\displaystyle \diagup\,\diagdown}{CH_2\text{—}CH}}\text{\textasciitilde}$$

$$\longrightarrow \text{\textasciitilde}\underset{\underset{OH}{|}}{CH}\text{—}CH_2\text{—}\underset{\underset{R}{|}}{N}\text{—}CH_2\text{—}\underset{\underset{OH}{|}}{CH}\text{\textasciitilde}$$

2. With acids

$$\text{\textasciitilde}\overset{\displaystyle O}{\overset{\displaystyle \diagup\,\diagdown}{CH\text{—}CH_2}} + HOOC\cdot R\cdot COOH + \overset{\displaystyle O}{\overset{\displaystyle \diagup\,\diagdown}{CH_2\text{—}CH}}$$

$$\longrightarrow \text{\textasciitilde}\underset{\underset{OH}{|}}{CH}\text{—}CH_2\text{—}OOCRCOO\text{—}CH_2\text{—}\underset{\underset{OH}{|}}{CH}\text{\textasciitilde}$$

The reactions indicated above in fact only lead to chain extension. In practice however polyamines are used so that the number of active hydrogen atoms exceeds two and so cross linkage occurs.

In the case of acids and acid anhydrides reaction can also occur via the hydroxyl groups that are present, including those formed on opening of the epoxide ring

$$\text{\textasciitilde R} \cdot \text{COOH} + \text{HO---}| \longrightarrow \text{\textasciitilde R} \cdot \text{COO---}| + \text{H}_2\text{O}$$

Both amines and acid anhydrides are extensively used cross-linking agents. The resins may also be modified by reacting with other polymers containing hydroxyl or mercaptan groupings, e.g.

$$\text{\textasciitilde SH} + \overset{\text{O}}{\overset{/\backslash}{\text{CH}_2\text{---CH}}}\text{\textasciitilde} \longrightarrow \text{\textasciitilde S---CH}_2\text{---}\overset{\text{OH}}{\underset{|}{\text{CH}}}\text{\textasciitilde}$$

These various systems will be dealt with individually in the following sections.

22.3.1 Amine hardening systems

As indicated in the preceding section amine hardeners will cross link epoxide resins either by a catalytic mechanism or by bridging across epoxy molecules. In general the primary and secondary amines act as reactive hardeners whilst the tertiary amines are catalytic.

Diethylene triamine and *triethylene tetramine* are highly reactive primary aliphatic amines with 5 and 6 active hydrogen atoms available for cross linking respectively. Both materials will cure glycidyl ether at room temperature. In the case of diethylene triamine the exothermic temperature may reach as high as 250°C in 200g batches. With this amine 9–10 p.h.r., the stoichiometric quantity, is required and this will give a room temperature pot life of less than an hour. The actual time depends on the ambient temperature and the size of the batch. With triethylene tetramine 12–13 p.h.r. are required. Although both materials are widely used in small castings and in laminates because of their high reactivity they have the disadvantage of high volatility, pungency and of being skin sensitisers. Properties such as heat distortion temperature (H.D.T.) and volume resistivity are critically dependent on the amount of hardener used.

Similar properties are exhibited by *dimethylaminopropylamine* and *diethylaminopropylamine* which are sometimes preferred because they are slightly less reactive and allow a pot life (for a 1 lb batch) of about 140 minutes.

A number of modified amines have been marketed commercially. For example reaction of the amine with a mono- or poly-functional glycidyl material will give a larger molecule so that larger quantities are required for curing and thus help to reduce errors in metering the hardener.

$$R—CH_2—\overset{\displaystyle O}{\overset{\diagup\diagdown}{CH—CH_2}} + H_2NR_1NH_2$$

$$\longrightarrow R—CH_2—\overset{\displaystyle OH}{\overset{|}{CH}}—CH_2—\overset{\displaystyle H}{\overset{|}{N}}—R_1—NH_2$$

These hardeners are extremely active. The pot life for a 1 lb batch may be as little as 10 minutes.

The *glycidyl adducts* are skin irritants similar in behaviour in this respect to the parent amines. The skin sensitisation effects in the primary aliphatic amine may be reduced by addition of certain groups at the nitrogen atom. The hydroxyethyl group and its alkyl and aryl derivatives are the most effective found so far.

$$H_2N—R—NH_2 + CH_2—CH_2 \longrightarrow H_2N—R—NH—CH_2—CH_2—OH$$
$$\diagdown \;\; \diagup$$
$$O$$
$$\downarrow$$
$$HO—CH_2—CH_2—NH—R—NH—CH_2—CH_2—OH$$

Both ethylene and propylene oxide have been used in the preparation of adducts from a variety of amines including ethylene diamine and diethylene triamine. The latter amine provides adducts which appear free of skin sensitising effects.

A hardener consisting of a blend of the two reaction products shown in the above equation is a low viscosity liquid giving a 16–18 minute pot life for a 1 lb batch at room temperature.

Modification of the amine with acrylonitrile results in hardeners with reduced reactivity.

$$H_2NR\ NH_2 + CH_2{=}\overset{\displaystyle }{\underset{\displaystyle \underset{|}{CN}}{CH}} \longrightarrow H_2N \cdot R \cdot NH \cdot CH_2 \cdot CH_2 \cdot CN$$
$$\diagdown$$
$$CN \cdot CH_2 \cdot CH_2 \cdot NHR \cdot NH \cdot CH_2 \cdot CH_2 \cdot CN$$

The greater the degree of cyanoethylation the higher the viscosity of the adduct, the larger the pot life and the lower the peak exotherm. The products are skin sensitive.

It is thus seen that as a class the primarily aliphatic amines provide fast curing hardeners for use at room temperatures. With certain exceptions they are skin sensitisers. The chemical resistance of the hardened resins varies according to the hardener used but in the case of the unmodified amines is quite good. The hardened resins have quite low heat-distortion temperatures and except with diethylene triamine seldom exceed 100°C. The number of variations in the properties obtainable may be increased by using blends of hardeners.

A number of aromatic amines also function as cross-linking agents. By incorporating the rigid benzene ring structure into the cross-linked network products are obtained with significantly higher heat distortion temperatures than are obtainable with the aliphatic amines.

Metaphenylene diamine, a crystalline solid with a melting point of about 60°C, gives cured resins with a heat distortion temperature of 150°C and very good chemical resistance.   It has a pot life of 6 hours for a $\frac{1}{2}$ lb batch at room temperature whilst complete cures require cure times of four to six hours at 150°C.   About 14 p.h.r. are used with the liquid epoxies. The main disadvantages are the need to heat the components in order to mix them, the irritating nature of the amine and persistent yellow staining that can occur on skin and clothing.   The hardener finds use in the manufacture of chemical resistant laminates.

Higher heat distortion temperatures are achieved using *4,4'methylene dianiline* (diamino diphenyl methane) and *diamino diphenyl sulphone*, in conjunction with an accelerator, but this is at some expense to chemical resistance.

Many other amines are catalytic in their action.   One of these, *piperidine*, has been in use since the early patents of Castan.   5–7 p.h.r. of piperidine are used to give a system with a pot life of about 8 hours.   A typical cure schedule is 3 hours at 100°C.   Although it is a skin irritant it is still used for casting of larger masses than are possible with diethylene triamine and diethyl amino propylamine.

Tertiary amines form a further important class of catalytic hardeners. For example triethylamine has found use in adhesive formulations.   Also of value are the aromatic substituted tertiary amines such as benzyldimethylamine and dimethyldiaminophenol.   They have found uses in adhesive and coating applications.   A long pot life may be achieved by the use of salts of the aromatic substituted amines.

Typical amine hardeners are shown in Table 22.3 and their characteristics and behaviour are summarised in Table 22.4.

22.3.2.   Acid hardening systems

The use of acid hardening systems for epoxy resins was first described in Castan's early patent but use was restricted in many countries until the consummation of cross-licensing arrangements between resin suppliers in 1956.   Compared with amine-cured systems, they are less skin-sensitive and generally give lower exotherms on cure.   Some systems provide cured resins with very high heat distortion temperatures and with generally good physical, electrical and chemical properties.   The cured resins do however show less resistance to alkalis than amine cured systems.   In practice acid anhydrides are preferred to acids since the latter release more water on cure, leading to foaming of the product, and are also generally less soluble in the resin.   Care must however be taken over storage since the anhydrides generally are somewhat hygroscopic.

The mechanism of anhydride hardening is complex but as the first stage of the reaction is believed to be the opening of the anhydride ring by an alcoholic hydroxyl group (or salt or a trace of water), e.g. Fig. 22.6.

Table 22.3 TYPICAL AMINE HARDENERS FOR EPOXY RESINS

PRIMARY ALIPHATIC AMINES	
1. Diethylene triamine (d.e.t.)	$NH_2—CH_2—CH_2—NH—CH_2—CH_2—NH_2$
2. Triethylene tetramine (t.e.t.)	$NH_2—(CH_2)_2—NH—(CH_2)_2—NH—(CH_2)_2—NH_2$
3. Dimethylaminopropylamine (d.m.a.p.)	CH_3 $\quad\diagdown$ $\quad N—CH_2—CH_2—CH_2—NH_2$ $\quad\diagup$ CH_3
4. Diethylaminopropylamine (d.e.a.p.)	C_2H_5 $\quad\diagdown$ $\quad N—CH_2—CH_2—CH_2—NH_2$ $\quad\diagup$ C_2H_5
ALIPHATIC AMINE ADDUCTS	
5. Amine–glycidyl adducts e.g.	$R—CH_2—CH—CH_2—NH\ (CH_2)_2NH—(CH_2)_2—NH_2$ from diethylene triamine
6. Amine–ethylene oxide adducts e.g.	$HO—CH_2—CH_2—NH—(CH_2)_2—NH—(CH_2)_2—NH_2$
7. Cyanoethylation products e.g.	$CN—CH_2—CH_2—NH—(CH_2)_2—NH—(CH_2)_2—NH_2$
AROMATIC AMINES	
8. Metaphenylene diamine (m.p.d.)	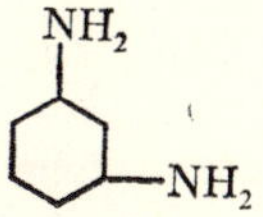
9. Diamino diphenyl methane (d.d.p.m.)	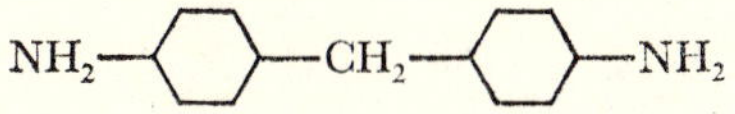
10. Diaminodiphenyl sulphone (d.d.p.s.)	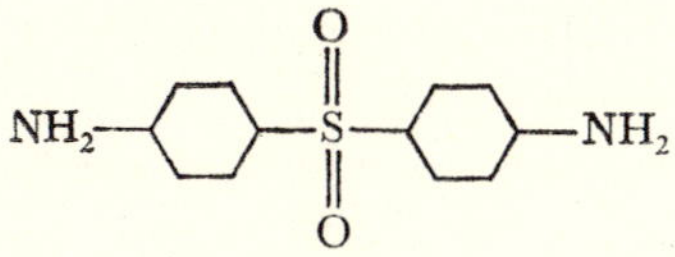
CYCLIC ALIPHATIC AMINES	
11. Piperidine	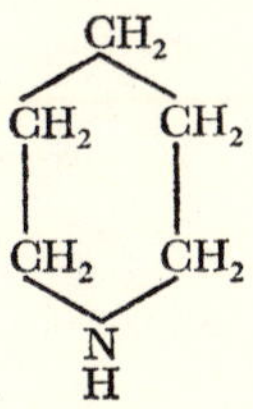
TERTIARY AMINES	
12. Triethylamine	$\qquad\qquad\quad CH_2—CH_3$ $\qquad\qquad\diagup$ $CH_3—CH_2—N$ $\qquad\qquad\diagdown$ $\qquad\qquad\quad CH_2—CH_3$

Table 22.3 (cont.)

13. Benzyldimethylamine
    (b.d.a.)

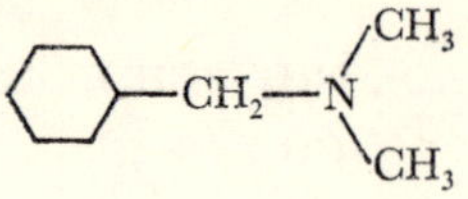

14. Dimethylaminomethyl
    phenol (d.m.a.m.p.)

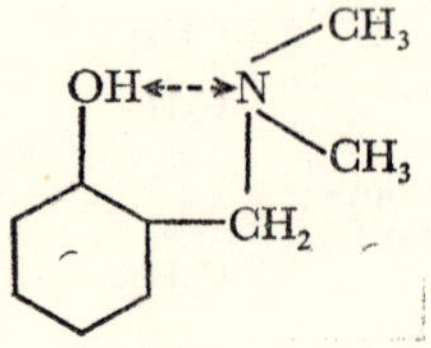

15. Tridimethylamino
    methyl phenol
    (t.d.m.a.m.p.)

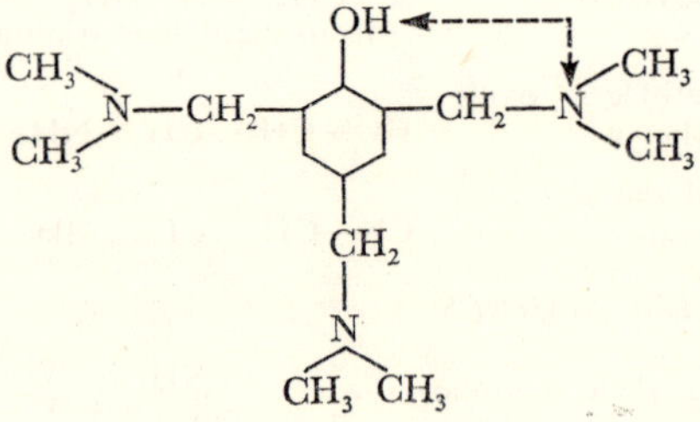

16. Tri-2-ethyl hexoate salt of tridimethylaminomethyl phenol

$$\overset{C_2H_5}{X[HOOC\!-\!CH_2\!-\!CH\!-\!CH_2\!-\!CH_2\!-\!CH_3]_3}$$

where X = tridimethyl amino methyl phenol

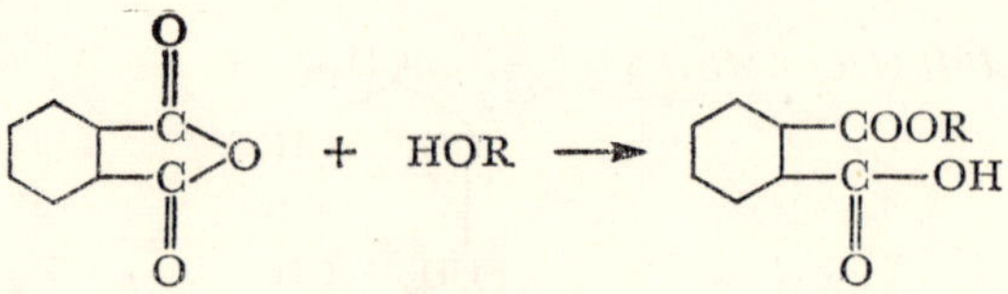

Fig. 22.6

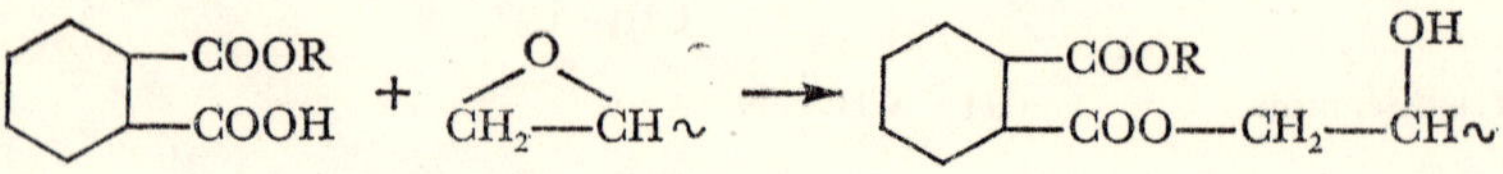

Fig. 22.7

Hydroxyl groups attached to the epoxy resin would suffice for this purpose. Five further reactions may then occur.

1. Reaction of the carboxylic group with the epoxy group (Fig. 22.7).
2. Etherification of the epoxy group by hydroxyl groups (Fig. 22.8).

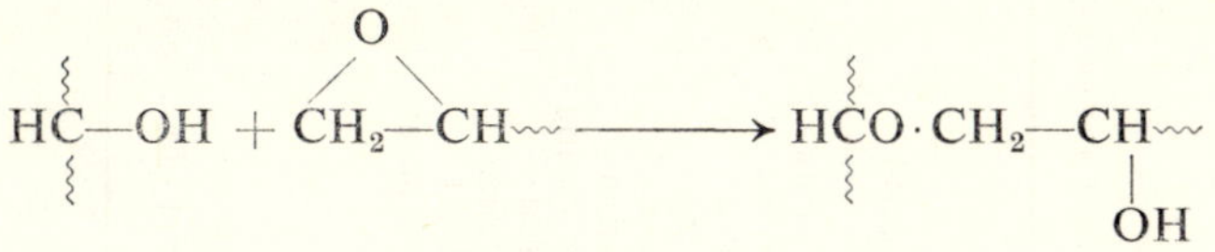

$$HC-OH + CH_2-CH\sim \longrightarrow HCO \cdot CH_2-CH\sim$$

Fig. 22.8

3. Reaction of the mono-ester with a hydroxyl group (Fig. 22.9).

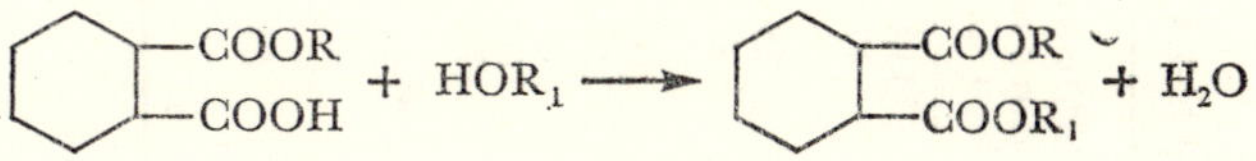

$$\text{—COOR} + HOR_1 \longrightarrow \text{—COOR} + H_2O$$

Fig. 22.9

4. Hydrolysis of the anhydride to acid by the water released in 3.
5. Hydrolysis of the monoester with water to give acid and alcohol.

In practice it is found that reactions 1 and 2 are of greatest importance and ester and ether linkages occur in roughly equal amounts. The reaction is modified in commercial practice by the use of organic bases, tertiary amines to catalyse the reaction.

The anhydrides are usually used at ratios of 0·85–1·1 moles anhydride carboxyl group per epoxy equivalent. Lower ratios down to 0·5/1 may however be used with some systems. The organic bases are used in amounts of 0·5–3%. These are usually tertiary amines such as α-methyl benzyl dimethylamine and *n*-butylamine.

Three classes of anhydride may be recognised, room temperature solids, room temperature liquids and chlorinated anhydrides.

Phthalic anhydride (Fig. 22.10 (I)) is an important example of the first class of hardener. It has a molecular weight of 148 and about 0·6–0·9 equivalent is used per epoxy group. For the lower molecular weight bis-phenol resins this works out at about 35–45 p.h.r. The hardener is usually added at elevated temperatures of about 120–140°C. It will precipitate out below 60°C but will again dissolve on reheating.

The resin is slow curing with phthalic anhydride and a typical cure schedule would be 4–8 hours at 150°C. Longer cures at lower temperatures tend to improve the heat distortion temperatures and reduce the curing shrinkage. As with the amine hardeners the heat distortion temperature is very dependent on the amount of anhydride added and reaches a maximum at about 0·75 equivalent. Maximum heat distortion temperatures quoted in the literature are of the order of 110°C, a not particularly exceptional figure, and the hardener is used primarily for large castings where the low exotherm is particularly advantageous.

Hexahydrophthalic anhydride (Fig. 22.10 (II)) (Mol. Wt. 154) has a melting

Table 22.4 SOME CHARACTERISTICS OF AMINE HARDENERS FOR USE IN LOW MOLECULAR WEIGHT GLYCIDYL ETHER RESINS

Hardener	Parts used per 100 pts resin	Pot life (1 lb batch)	Typical cure schedule	Skin irritant	Max h.d.t. cured resin (°C)	Features	Applications
D.E.T.	10–11	20 min	room temp.	yes	110	cold curing	general purpose
D.E.A.P.	7	140 min	room temp.	yes	97	slightly slower than d.e.t.	general purpose
D.E.T.-glycidyl adduct	25	10 min	room temp.	yes	75	fast cure	adhesives laminating
D.E.T.-ethylene oxide adduct	20	16 min	room temp.	reduced	92	minimum irritation	—
D.E.T.-cyanoethylation adduct†	~37·5	60–80 min	*	yes	100	slower curing	—
M.P.D.	14–15	>6 hr	4–6 hr at 150°C	yes	150	chemical resistance	laminates
D.D.P.M.	28·5	—	4–6 hr at 165°C	yes	160	high h.d.t.	laminates
D.D.P.S.	30	—	8 hr at 160°C	yes	175	use with accelerator	laminates
Piperidine	5–7	8 hr	3 hr at 100°C	yes	75	—	general purpose
Triethylamine	10	7 hr	room temp.	yes	—	—	adhesives
B.D.A.	15	75 min	room temp.	yes	—	—	adhesives
T.D.M.A.M.P.	6	30 min	room temp.	yes	64	—	adhesives coatings
2-ethyl hexoate salt of above	10–14	3–6	—	yes	—	long pot life	encapsulation

* 2 hr at 70°C, 3 hr at 100°C, 1 hr at 110°C.
† Results are for highly substituted amines.

point of 35–36°C and is soluble in the epoxy resin at room temperature. When 0·5% of a catalyst such as benzyldimethylamine is used the curing times are of the same order as with phthalic anhydride. About 80 p.h.r. are required. In addition to the somewhat improved ease of working the hardener gives slightly higher heat distortion temperatures (~ 120°C) than with phthalic anhydride. It is however more expensive.

Maleic anhydride (Fig. 22.10 (III)) is not usually used on its own because the cured resins are brittle, but it may be used in conjunction with pyromellitic dianhydridic.

In order to obtain cured products with higher heat distortion temperatures from bis-phenol epoxy resins, hardeners with higher functionality

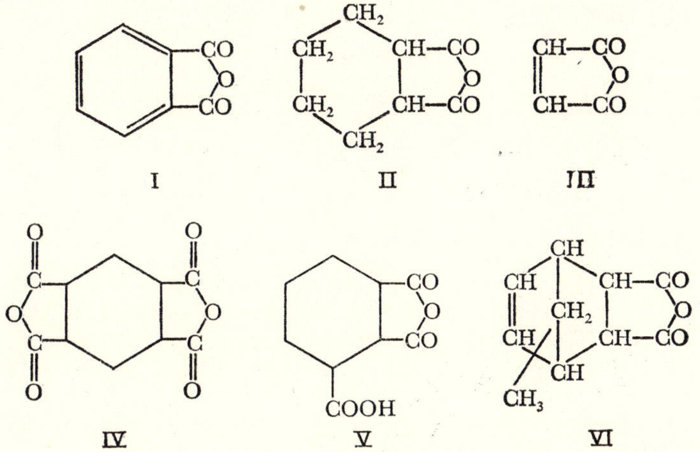

Fig. 22.10

have been used thus giving a higher degree of cross linking. These include *pyromellitic dianhydride* (IV), and *trimellitic anhydride* (V).

Heat distortion temperatures of resins cured with pyromellitic dianhydride are often quoted at above 200°C. The high heat distortion is no doubt also associated with the rigid linkages formed between epoxy molecules because of the nature of the anhydride. The use of these two anhydrides has however been restricted because of difficulties in incorporating them into the resin.

The methylated maleic acid adduct of phthalic anhydride, known as *methyl nadic anhydride* (VI), is somewhat more useful. Heat distortion temperatures as high as 202°C have been quoted whilst cured systems, with bis-phenol epoxides, have very good heat stability as measured by weight loss over a period of time at elevated temperatures. The other advantage of this hardener is that it is a liquid easily incorporated into the resin. About 80 p.h.r. are used but curing cycles are rather long. A typical schedule is 16 hours at 120°C and 1 hour at 180°C.

Other anhydrides that have been used include *dodecenyl succinic anhydride*, which imparts flexibility into the casting, and *chlorendic anhydride* where flame resistant formulations are called for.

Table 22.5 PROPERTIES OF SOME ANHYDRIDE HARDENERS USED IN LOW MOLECULAR WEIGHT DIGLYCIDYL ETHER RESINS

Anhydride hardener	*Parts used p.h.r.*	*Typical cure schedule*	*Physical form*	*Max. h.d.t. of cured resin °C*	*Use*
Phthalic	35–45	24 hr at 120°C	powder	110°C	casting
Hexahydrophthalic (+ accelerator)	80	24 hr at 120°C	glassy solid	130°C	casting
Maleic	—	—	solid	—	secondary hardener
Pyromellitic (dianhydride)	26	20 hr at 220°C	powder	290°C	high h.d.t.
Methyl nadic	80	16 hr at 120°C	liquid	202°C	high h.d.t.
Dodecenyl succinic (+ accel.)		2 hr at 100°C + 2 hr at 150°C	viscous oil	38°C	flexibilising
Chlorendic	100	24 hr at 180°C	white	180°C	flame retarding

Table 22.5 summarises the characteristics of some of the anhydride hardeners.

22.3.3   Miscellaneous hardening systems

In addition to the amine and anhydride hardeners other curing systems are available.   For example a number of amides which contain amine groups may be used.   These include the low molecular weight polyamide resins, dealt with in the section on modifiers and such materials as dicyandiamide

$$NH_2-\underset{\underset{NH}{\|}}{C}-NH-CN$$

This material is used in work involving solutions of the higher molecular weight resins.   Some mixtures have a very good pot life.

Better pot lives are obtained with complexes of boron trifluoride with various amines such as monoethylamine.   Some of these complexes have an almost indefinite shelf life and are marketed under proprietary names. The disadvantages of the materials are their hygroscopic behaviour and the corrosive effects of BF_3, liberated during cure, on electrical and other equipment.

22.3.4   Comparison of hardening systems

The number of hardening agents used commercially is very large and the final choice will depend on the relative importance of economics, ease of handling, pot life, cure rates, dermatitic effects and the mechanical, chemical, thermal and electrical properties of the cured products.   Since these will differ from application to application it is understandable that such a wide range of materials is employed.

As a very general rule it may be said that the amines are fast curing and give good chemical resistance but most are skin sensitive.   The organic anhydrides are less toxic and in some cases give cured resins with very high heat distortion temperatures. They do not cross link the resins at room temperature.

In addition to the considerable difference of the properties of the cured resins with different hardeners it must also be stressed that the time and temperatures of cure will also have an important effect on properties. As a very general rule, with the aliphatic amines and their adducts increasing the time of cure and temperature of cure (up to 120°C at least) will improve most properties.[10]

22.4 MISCELLANEOUS EPOXIDE RESINS

In addition to the resins based on bis-phenol A dealt with in preceding sections there are now available a number of other resins containing epoxide groups. These can be treated in two main groups:

1. Other glycidyl ether resins.
2. Non-glycidyl ether resins.

22.4.1 Miscellaneous glycidyl ether resins

Glycidyl ether resins will be formed by reacting epichlorhydrin with polyhydroxy compounds. Three examples of materials stated to be used in commercial resins are bis-phenol F (Fig. 22.11 (I)), glycerol (II) and the long chain bis-phenol derived from cashew nut oil (III)

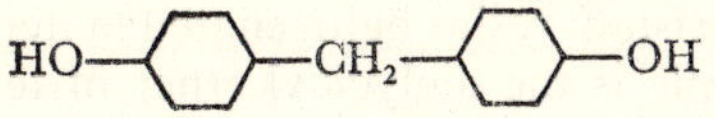

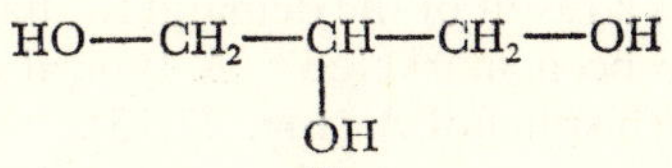

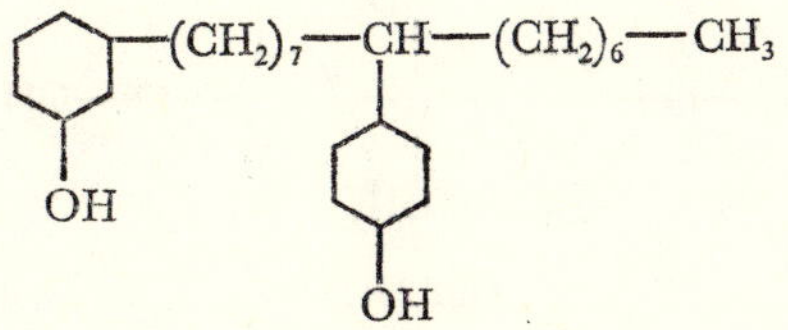

Fig. 22.11

In each case resins are prepared which when cured have lower softening temperatures than the conventional resins.

Novolak resins (Chapter 19) have also been epoxidised through their phenolic hydroxy groups. The resins have good heat resistance because of their higher functionality. The high viscosity and the difficulty in controlling the composition have limited application of these materials.

Low viscosity diglycidyl ether resins of undisclosed composition[11] have been marketed in the United States and in Britain. The materials are stated to be totally difunctional, i.e. free from monofunctional reactive diluents.   The cured resins have properties very similar to those of the standard diglycidyl ether resins.

To produce resins of high heat distortion temperature it is important to have a high density of cross linking and to have inflexible segments between the cross links.   This approach has been used with reasonable

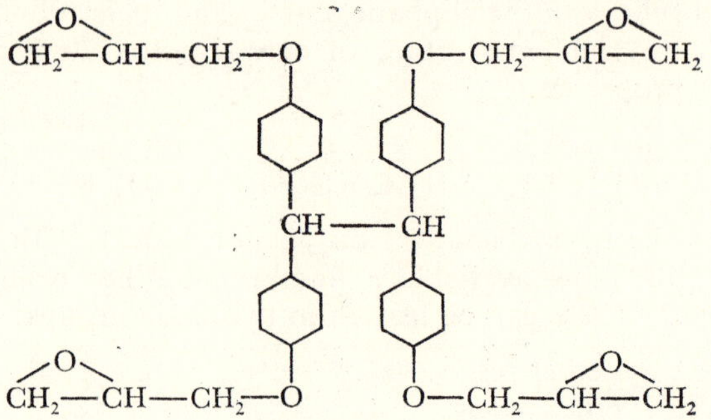

Fig. 22.12

success using certain anhydride hardeners such as pyromellitic dianhydride and with the cyclic aliphatic resins (Section 22.4.2).   Attempts have also been made to use glycidyl ether resins of higher functionality such as the tetrafunctional structure (Fig. 22.12).

Because of the higher viscosity of such resins their use has been restricted to applications where they may be used in solution.

As a result of the demand for flame resistant resins halogenated materials have been marketed.   A typical example is the diglycidyl ether of tetra-chlorbisphenol A (Fig. 22.13).

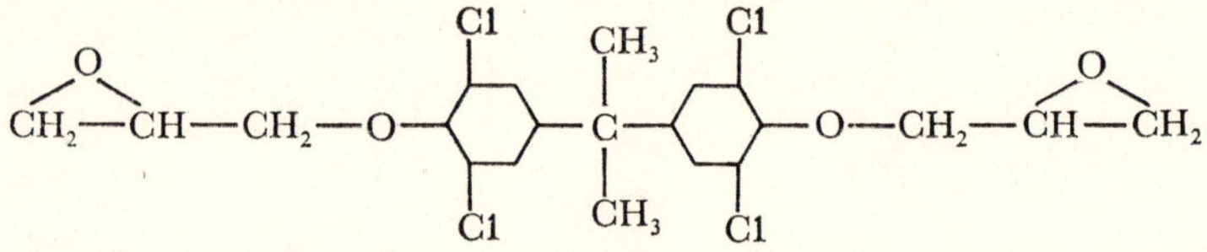

Fig. 22.13

The resin is a semisolid and must be used either in solution form or as blends.

Mention may also be made of mixed diethers some of which are unsaturated.   These materials may be cured by a variety of mechanisms. An example is the allyl glycidyl mixed ether of bis-phenol A (Fig. 22.14).

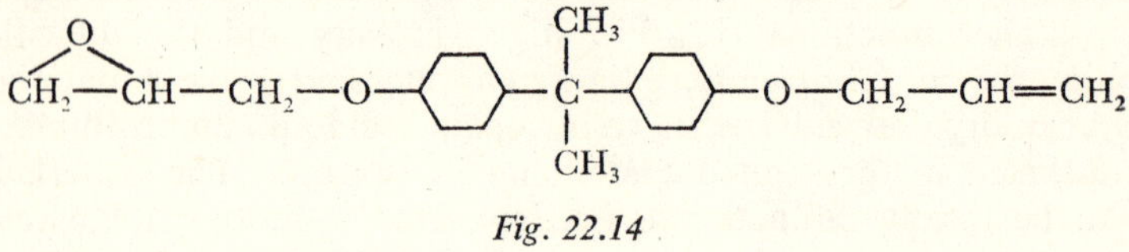

Fig. 22.14

22.4.2 Non-glycidyl ether epoxides

Although the first and still most important epoxide resins are of the glycidyl ether type other epoxide resins have been commercially marketed in recent years. These materials are generally prepared by epoxidising unsaturated compounds using hydrogen peroxide or peracetic acid.

Such materials may be considered in two classes:

1. Those which contain a ring structure as well as an epoxide group in the molecule—the cyclic aliphatic resins.
2. Those which have an essentially linear structure on to which are attached epoxide groups—the acyclic aliphatic epoxide resins.

Cyclic aliphatic resins

Cyclic aliphatic epoxide resins[11] were first introduced in the United States. Some typical examples of commercial materials are shown in Table 22.6.

Compared with standard diglycidyl ether resins, the liquid cyclic aliphatic resins are paler in colour and have a much lower viscosity. Whereas in general the cyclic aliphatic resins react more slowly with

Table 22.6 SOME COMMERCIALLY AVAILABLE CYCLIC ALIPHATIC EPOXIDE RESINS

Chemical name	Commercial reference	Approximate structure	Physical state
1. 3,4 epoxy-6-methyl cyclohexyl methyl 3,4 epoxy-6-methyl cyclohexane carboxylate	Unox epoxide 201		liquid
2. Vinyl cyclohexene dioxide	Unox epoxide 206		liquid
3. Dicyclopentadiene dioxide	Unox epoxide 207		solid

amines, there is less difference with acid anhydrides. Table 22.7 provides data illustrating this point.

Because of the compact structure of the cycloaliphatic resins the intensity of cross linking occuring after cure is greater than with the standard diglycidyl ethers. The lack of flexibility of the molecules also leads to more rigid segments between the cross links.

Table 22.7 SOME PROPERTIES OF CYCLIC ALIPHATIC RESINS

	Unox Epoxide 201	*Unox Epoxide 206*	*Unox Epoxide 207*	*Standard diglycidyl ether*
Appearance	pale straw liquid	water white liquid	white powder	straw liquid
Viscosity at 25°C (cP)	1,200	7·7	—	10,500
Specific gravity	1·121	1·099	1·330	1·16
Epoxide equivalent	145	76	82	185
Hardening time (100°C)				
using aliphatic polyamine (hr)	24	0·25	—	0·12
HHPA (hr)	15	6·75	—	7·00
HHPA + 0·5% BDA (hr)	1·25	0·75	—	0·75

HHPA, hexahydro phthalic anhydride, BDA benzyl dimethyl amine.

As a consequence the resins are rather brittle and generally unsuitable for potting and encapsulation. The high degree of cross linking does however lead to higher heat distortion temperatures than obtained with the normal diglycidyl ether resins.

Because of their low viscosity the liquid cyclic aliphatic resins find use in injection moulding and extrusion techniques as used for glass reinforced laminates. They are also very useful diluents for the standard glycidyl ether resins.

Acyclic aliphatic resins

These materials differ from the previous class of resin in that the basic structure of these molecules are long chains whereas as the cyclic aliphatics contain ring structures. Two subgroups may be distinguished, epoxidised diene polymers and epoxidised oils.

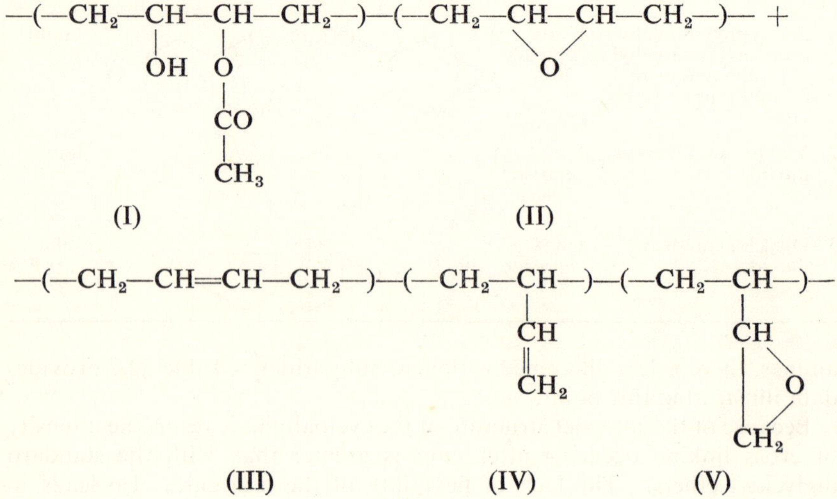

Fig. 22.15

Table 22.8 SOME PROPERTIES OF EPOXIDISED POLYBUTADIENE RESINS

	A	B	C*	Standard diglycidyl ether
Appearance	amber liquid	light yellow liquid	light yellow liquid	straw coloured liquid
Viscosity at 250°C (cP)	180,000	16,000	1,500	10,500
Specific gravity	1·010	1·014	0·985	1·16
Epoxide equivalent	177	145	232	185
Hardening time (100°C) (hr) using (a) aliphatic amine (hr)	1	1·3	1·7	0·12
(b) maleic anhydride (hr)	1	1·25	1·25	1·5

* Contains about 23% volatile matter.

Typical of the epoxidised diene polymers are products produced by treatment of polybutadiene with peracetic acid. The structure of a molecular segment (Fig. 22.15) indicates the chemical groupings that may be present.

Residue (I) is a hydroxy–acetate segment produced as a side reaction during the epoxidising process, (II) is an epoxide group in the main chain, (III) is an unreacted segment, (IV) is an unreacted pendant vinyl group present through a 1:2 addition mechanism whilst (V) is an epoxidised derivative of the vinyl group.

The epoxidised polybutadiene resins available to date are more viscous than the diglycidyl ethers except where volatile diluents are employed. They are less reactive with amines but have a similar reactivity with acid anhydride hardeners. Cured resins have heat distortion temperatures substantially higher than the conventional amine cured diglycidyl ether resins. A casting made from an epoxidised polybutadiene hardened with

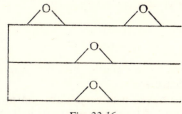

Fig. 22.16

maleic anhydride and cured for 2 hours at 50°C plus 3 hours at 155°C plus 24 hours at 200°C gave a heat distortion temperature of 250°C. Some typical characteristics of the resins are given in Table 22.8.[11]

Epoxidised drying oils have been available for several years as stabilisers for poly(vinyl chloride). They may be considered to have the skeletal structure shown in Fig. 22.16.

The number of epoxy groups per molecule will vary but for modified soya bean oils there are an average of about 4 where these are about 6 for epoxidised linseed oils.

As with the other non-glycidyl ether resins the absence of the ether oxygen near to the epoxide group results in low reactivity with amine hardeners whereas activity with acid anhydride proceeds at reasonable rates.

The epoxidised oils are seldom used in a cross-linked form as the products are rather soft and leathery. Exceptions to this are their occasional use as diluents for more viscous resins and some application in adhesive formulations.

22.5 DILUENTS, FLEXIBILISERS AND OTHER ADDITIVES

For a number of purposes the unmodified epoxide resins may be considered to have certain disadvantages. These disadvantages include high viscosity, high cost and too great a rigidity for specific applications. The resins are therefore often modified by incorporation of diluents, fillers, flexibilisers and sometimes, particularly for surface coating applications, blended with other resins.

Diluents are free flowing liquids incorporated to reduce the resin viscosity and simplify handling. At one time hydrocarbons such as xylene were used for this purpose but being non-reactive were lacking in permanence. Today, reactive diluents such as phenyl glycidyl ether (Fig. 22.17 (I)), butyl glycidyl ether (II) and octylene oxide (III) are employed.

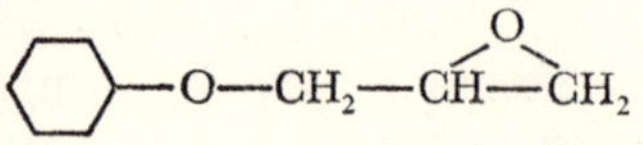

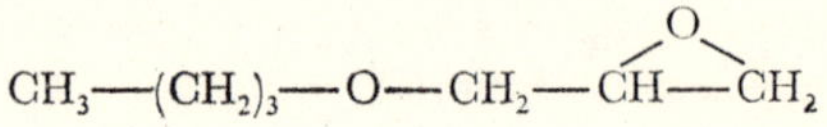

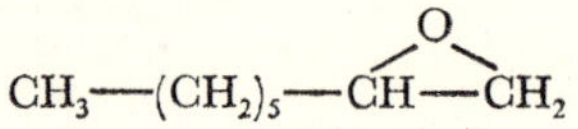

Fig. 22.17

Since however they are more volatile than the resin care must be used in vacuum potting applications.

The diluents tend to have an adverse effect on physical properties and also tend to retard cure. Many are also skin irritants and must be used with care. For this reason they are seldom used in amounts exceeding 10 p.h.r.

Fillers are used in tooling and casting application. Not only do they reduce cost but in diluting the resin content they also reduce curing shrinkage, lower the coefficient of expansion, reduce exotherms and may increase thermal conductivity. Sand is frequently used in inner cores whereas metal powders and metal oxide fillers are used in surface layers. Wire wool and asbestos are sometimes used to improve impact strength.

In order to increase the flexibility, and usually, in consequence the toughness of the resins, plasticisers and flexibilisers may be added. Non-reactive plasticisers such as the conventional phthalates and phosphates have proved unsuccessful. Monofunctional materials which in some cases also act as reactive diluents have been used but are not of great importance.

More interest has been shown in polymeric flexibilisers particularly the low molecular weight polyamides from dimer acid (see Chapter 15) and low molecular weight polysulphides (Chapter 16).

The low molecular weight polyamides are interesting in that they are not only flexibilisers but that they also act as non-irritating amine hardeners, reaction occurring across amine groups present. A certain amount of latitude is allowable in the ratio of polyamide to epoxy resin but the optimum amount depends on the epoxy equivalent of the epoxide resin and the amine value of the polyamide. (The amine value is the number of milligrams of potassium hydroxide equivalent to the base content of 1 gram polyamide as determined by titration with hydrochloric acid.) The polyamides are highly viscous and must be used in resin solutions or at elevated temperatures.

Elevated temperatures are necessary for cure and the chemical resistance of the laminates is inferior to those from unmodified resins. Because of problems in handling, the polyamides have found only limited use with epoxy resins, mainly for coating and adhesive applications.

The low molecular weight polysulphides have found somewhat greater use. Of general structure HS—R—SH and with molecular weights of approximately 1,000 they will react with the epoxy group to cause chain extension but not cross linking. The normal hardeners must therefore be employed in the usual amounts (Fig. 22.18).

The polysulphides used are relatively mobile liquids with viscosities of about 10 poise and are thus useful as reactive diluents. They may be em-

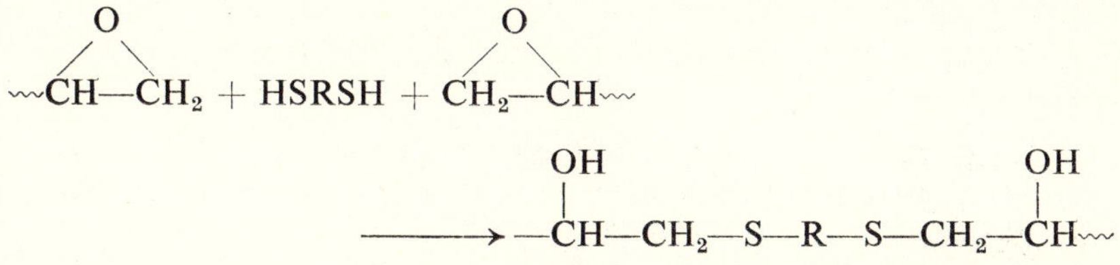

$$
\text{CH—CH}_2 + \text{HSRSH} + \text{CH}_2\text{—CH}
$$

$$
\longrightarrow \text{—CH—CH}_2\text{—S—R—S—CH}_2\text{—CH}
$$

Fig. 22.18

ployed in any ratio with the epoxide and products will range from soft rubbers, where only polysulphides are employed, to hard resins using only epoxide.

The more the polysulphide the higher will be the dielectric constant and the lower the volume resistivity. There will be a reduction in tensile strength and heat distortion temperature but an increase in flexibility and impact strength.

The polysulphides are frequently used in casting mixes and to a less extent in coating, laminating and adhesive applications. Their value in casting and encapsulation lies mainly with their low curing shrinkage and flexibility in the cured state. Their tendency to corrode copper and the somewhat inferior electric insulation properties of the blends does lead to certain limitations.

Interesting amine flexibilisers have also been described.[13] These materials are made by cyanoethylation of amine hardeners such as diethylene triamine, to such an extent that only two reactive hydrogens remain and the material is only difunctional, e.g. Fig. 22.19.

$$H_2N \cdot (CH_2)_2 \cdot NH \cdot (CH_2)_2 NH_2 + 3CH_2{=}CH$$
$$|$$
$$CN$$

$$\longrightarrow CN(CH_2)_2 NH \cdot (CH_2)_2 \cdot N \cdot (CH_2)_2 \cdot NH \cdot (CH_2)_2 CN$$
$$|$$
$$(CH_2)_2$$
$$|$$
$$CN$$

Fig. 22.19

Many amines have been examined and data for three of them[13] are given in Table 22.9.

The amine flexibilisers may be used in two ways:

1. Where allowance is made for the reactivity of the hardener.
2. Where the reactivity of the hardener is ignored.

Progressive replacement of amine hardener by a low viscosity flexibiliser will reduce mix viscosity, increase pot life and reduce the heat distortion temperature of the cured system. Higher impact strengths are achieved using approximately equivalent amounts of hardener and flexibiliser.

Using flexibilisers in addition to the usual amount of hardener very flexible products may be obtained.

Although in many respects similar to the liquid polysulphides, the amine flexibilisers differ in three important respects:

1. They reduce the reactivity of the system rather than increase it.
2. They are compatible with a different range of room temperature hardeners.
3. They have a low level of odour.

Table 22.10[13] compares the effect of the three classes of flexibiliser.

22.6   STRUCTURE AND PROPERTIES OF CURED RESIN

Since the characteristic grouping of the resins discussed in this chapter largely disappears on cross linking it is difficult to make simple generalisations relating structure to properties.

Being cross linked the resin will not dissolve without decomposition but will be swollen by liquids of similar solubility parameter to the cured resin. The chemical resistance is as much dependent on the hardener as on the resin since these two will determine the nature of the linkages formed. The acidic hardeners form ester groups which will be less resistant to alkalis.

The main skeleton of the resins themselves has generally good chemical resistance.

The thermal properties of the resin are dependent on the degree of cross linking, the flexibility of the resin molecule and the flexibility of the hardener molecule. Consequently the rigid structures obtained by using

Table 22.9

	A	B	C
Equivalency to epoxide resin*	56	69	120
Specific gravity (25°C)	0·98	1·08	1·05
Refractive index	1·479	1·495	1·507
Viscosity (cP at 25°C)	60	450	7,500

* The weight in grams to provide one reactive hydrogen atom for every epoxide group in 100 g of liquid epoxide resin of epoxide equivalent 190.

cycloaliphatic resins or hardeners such as pyromellitic dianhydride will raise the heat distortion temperatures.

The resins are somewhat polar and this is reflected in the comparatively high dielectric constant and power factor for an insulating material.

22.7 APPLICATIONS

The epoxide resins are used in a large number of fields including surface coatings, adhesives, in potting and encapsulation of electronic components, in tooling, for laminates and to a small extent in moulding powders and in road surfacing.

The encapsulation of electrical components provides an interesting extension to the use of plastics materials as insulators. Components of

Table 22.10 INFLUENCE OF FLEXIBILISERS ON EPOXY RESINS

	Difunctional amine				Polysulphide		Polyamide	
Flexibiliser	—	25	25	50	25	50	43	100
Epoxy resin	100	100	100	100	100	100	100	100
Amine hardeners	20	13·2	20	20	20	20	—	—
Pot life (1 lb) (min)	20	69	44	76	13	6	150	140
Viscosity (25°C) (cP)	3,700	1,070	870	490	—	—	210,000	210,000
Flexural strength (lb in^{-2})	16,000	14,400	17,710	—	15,300	—	10,700	11,670
Compressive yield stress (lb in^{-2})	15,000	13,900	14,330	—	12,300	—	12,800	10,700
Impact strength (ft lb/$\frac{1}{2}$ in. notch)	0·7	0·82	1·03	8·0	0·5	1·7	0·3	0·32
Heat distortion temperature (°C)	95	44	40	<25	53	32	81	49

electronic systems may be embedded in a single cast block of resin (the process of encapsulation). Such integrated systems are less sensitive to handling and humidity and in the event of failure the whole assembly may be replaced using seldom more than a simple plugging-in operation.

Encapsulation of miniaturised components has proved invaluable particularly in spacecraft.

The formulation of encapsulating systems involves a great deal of attention. In order to achieve adequate wetting and impregnation the resin viscosity must be low. Since the encapsulation operation is often carried out under vacuum it is also necessary that the mix be free of volatile components. Exothermic heat and shrinkage on cure may damage or affect the characteristics of the components to be potted. The coefficient of thermal expansion should be brought as near to that of the components as possible by judicious use of fillers. Alternatively the system used can be made more ductile by the use of a flexibiliser such as polysulphide resin. It is also important that components of the mix do not react with any of the materials forming the electronic components. Finally the cost should be at the minimum possible to give a satisfactory formulation.

Systems based on the epoxide resins may be provided which are closer to these requirements than can be obtained in other ways. The polyester resins are very restricted because of their high shrinkage, the corroding influence of polyester formulations on copper and on the volatility of components. There is however some application of flexible polyurethanes where good damping qualities are of importance. The low shrinkage and simplicity of fabrication make epoxy resins admirably suited for a number of tooling applications. Patterns, jigs, metal shaping moulds and vacuum forming moulds are frequently made from these materials. Since many of these products are quite large in bulk it is important that low exotherm curing systems are used. A reduction in exotherm is also achieved by using large quantities of fillers which in addition may substantially lower the cost. Large mouldings are often made by a two stage process. The inner surfaces of the casting mould are covered with about a $\frac{1}{2}$ in. deep layer of plasticine or some similar material. The residual space is then filled with a sand–resin–hardener mixture. When this has hardened it is removed, and the plasticine stripped from the mould and the resin sand core. The core is then replaced in its original position leaving a gap where there was previously the layer of plasticine. This gap is then filled with a resin mixture containing a fine filler and allowed to harden.

The choice of filler depends on the end use. Metal fillers will improve machineability, hardness and thermal conductivity but may in some cases inhibit cure.

The epoxide resins have not achieved the importance as laminating resins which has been reached by the phenolic, polyester and melamine resins. However they are favoured for certain special purpose applications and may be prepared by both wet lay-up and solvent processes.

Compared with the polyesters the epoxide resins generally have better mechanical properties and, using appropriate hardeners, better heat resistance and chemical resistance, in particular, resistance to alkalis.

The laminates are employed mainly where an intermediate degree of heat stability is required which does not justify the use of the more expensive silicone laminates. They have additional advantages over the silicones in their ease of forming by wet lay-up techniques and the greater strength of the laminates.

The main end-uses are in the aircraft industry and are found in ducting, tail sections, printed circuit bases and in instrument panels. The laminates also find some use in chemical engineering plant and in tooling.

The properties of the laminates will depend on a number of factors of which the following are the most important:

1. Resin used.
2. Hardener used.
3. Fillers and modifiers used.
4. Type of reinforcement.
5. Resin content of laminate.
6. Curing conditions.

Since these factors can have a considerable influence on properties it is difficult to give typical figures. Table 22.11 shows some quoted figures for

Table 22.11 MECHANICAL PROPERTIES OF EPOXY-GLASS CLOTH LAMINATES

	Resin A $Mol.\ Wt.$ 1,000	*Resin B* $Mol.\ Wt.$ 1,500
Tensile strength (10^3 lb in^{-2})	52–59	61–66
Tensile modulus (10^6 lb in^{-2})	2·9–3·4	3·5–4·0
Flexural strength (10^3 lb in^{-2}) 25°C	80–85	95–100
127°C	69–74	70–75
Flexural modulus (10^6 lb in^{-2}) 25°C	3·6–3·9	4·4–4·6
Hardness (Rockwell M)	0·04–0·06	0·04–0·08

glycidyl ether resin cured with diamino diphenyl methane. The laminates were pressed at 400 lb in^{-2} for one hour at 160°C and post cured for 8 hours at 60°C.

The electrical properties will also depend on the above factors as well as on the test conditions in particular temperature, test frequency and

Table 22.12 ELECTRICAL PROPERTIES OF LAMINATES

Property	*Unit*	*Range of values*
Power factor	—	0·008–0·04
Dielectric constant	—	3·4–5·7
Electric strength	V/0·001 in.	250–550
Volume resistivity	ohm cm	10^{14}–10^{16}

humidity. Table 22.12 quotes ranges for figures quoted in the literature for various electrical properties.

Moulding powders based on epoxy resins have been available on a small-commercial scale for several years. Their particular advantages are the very low shrinkage on cure and the high fluidity developed during the moulding operation. This makes them particularly suitable for moulding thin sections round relatively large metal inserts and for moulding around delicate pins and inserts.   Although some commercial grades are glass fibre-filled their low viscosity in the molten state allows them to be transfer moulded without difficulty.

The finished mouldings have high dimensional stability, low water absorption and good resistance to tracking.   They also exhibit good heat resistance and mouldings are said to have withstood temperatures of 200°C without undue deterioration.

The application of the moulding powders is limited by their cost which is several times that of general purpose phenolics.   Main end uses have been for electronic application where good electrical properties and heat resistance are required, particularly in mouldings containing inserts.

The moulding powders may be compression moulded under comparatively low pressures (700–2,100 lb in^{-2}) at temperatures of 150–165°C. A typical cure schedule is about 3 minutes for $\frac{1}{8}$ in. thick sections.   Transfer moulding pressures are of the order of $2\frac{1}{2}$–5 ton in^{-2}.   Little information is available about the composition of the moulding powders.   Solid resins are used either alone or in conjunction with liquid resins.   A number of fillers are suitable including glass fibre.   Hardeners quoted in the literature[14] include methylene dianiline and boron trifluoride–mono-ethylamine complexes.   Resins available to date have limited shelf-life and it is advisable to use refrigerated storage facilities if possible.

Some properties of two typical grades of epoxy moulding powder are given in Table 22.13.

The largest single end use of the resin is for surface coatings.   They may be blended with other resins such as alkyds, amino-resins and phenolics or

Table 22.13   PROPERTIES OF A TYPICAL EPOXIDE MOULDING COMPOSITION (B.S.771 Test Methods where applicable)

Property	Units	Value
Specific gravity	—	1·8–2·1
Flexural strength	10^3 lb in^{-2}	13–19
Mould shrinkage	in/in.	<0·002
After-shrinkage 48 hr at 105°C	%	neglible
Water absorption	mg	5–10
Dielectric constant 800 c/s	—	4·5–5·5
10^6 c/s	—	4·5–5·0
Power factor 800 c/s	—	0·01–0·02
10^6 c/s	—	0·01–0·02
Volume resistivity	ohm cm	10^{14}–10^{15}

they may be esterified by heating with resin acids or fatty acids.   They may be used in solution form or as solventless coatings, either liquid resins or powders, the latter being applied by fluid bed or electrostatic spray technique.   As a class the epoxides therefore have great versatility and this combined with excellent adhesion, good chemical resistance and flexibility has led to many industrial applications.

The excellent adhesion, high cohesion, low shrinkage on cure, absence of volatile solvents and low creep of the resins have led to important applications as adhesives particularly for metal-to-metal and metal-to-plastics bonding. As with the surface coatings there is a diversity of possible formulations available, selection being dependent on the requirements of the end product.

The resins have also found use in a number of other directions. The use of the resins in floorings and road surfacings is somewhat spectacular. In spite of the high initial cost such floorings have excellent chemical resistance and resistance to wear. The resins are claimed to be of particular value at road junctions and roundabouts where severe wear is experienced but where repairs and maintenance operations need to be kept to a minimum because of the resultant disruption in the flow of traffic.

Epoxide resins are available in a powder form that contains a suitable hardening system. The powder may be used for coating metals by fluidised bed or by electrostatic spraying techniques. Unlike nylon and polyolefins powder coatings it is necessary to bake the coating in order to cure the resin. The powder coatings are particularly useful for application of thick film to parts of a complicated or irregular shape and have good chemical and electrical resistance. The coatings are much harder and adhere more strongly to the substrate than the older more well established thermoplastic powders. The electrostatic spraying of epoxide powders to form surface coatings presents an important challenge to the usual methods using solutions.

22.8 PHENOXY RESINS[15]

In 1962 the Union Carbide Company announced that they were manufacturing thermoplastic epoxide resins which could be injection moulded, extruded and blow moulded, and which they designated phenoxy resins. These materials differ from the conventional glycidyl ethers largely in their molecular weight ($\sim$ 25,000) since their basic structure is virtually identical, viz Fig. 22.20.

The resin structure also has a strong resemblance to that of the bisphenol A polycarbonates so that it is not surprising to find a number of

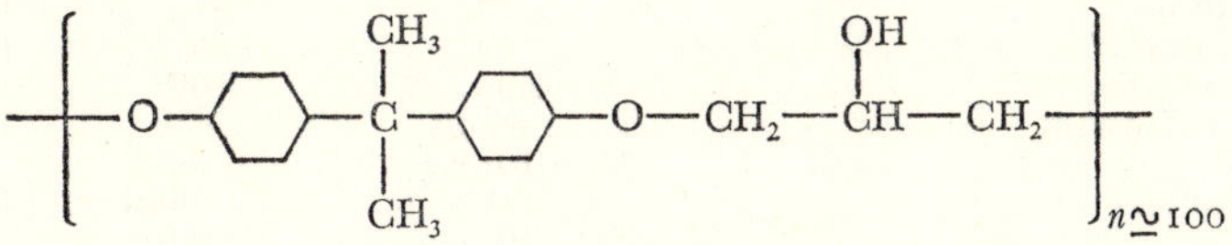

Fig. 22.20

similarities in the properties of the two materials. Because the phenoxy resins are moderately branched they are amorphous and are thus soluble in many solvents of similar solubility parameter ($\sim$ 9·4) such as tetrahydrofuran, mesityl oxide, diacetone alcohol and dioxane. Since the main chain is composed of stable C—C and C—O—C linkages the polymer is relatively stable to chemical attack particularly from acids and alkalis. The pendant hydroxyl groups are however reactive and provide a site for cross linking by diisocyanates and other agents. Being amorphous the material is transparent.

The outstanding properties of the resins are the high rigidity, tensile strength, ductility, impact strength and creep resistance, properties for which the polycarbonates are also outstanding. The phenoxies also show

Table 22.14 IMPACT STRENGTH OF $\frac{1}{8}$ IN. SHEET AS MEASURED BY FALLING DART TEST (Source: Union Carbide trade literature)

Polymer	Height of fall (in)	Nature of fracture
Phenoxy resin	54	dimple only
Phenoxy resin	60	puncture
ABS (type unspecified)	30	puncture
Impact polypropylene	12–18	shatters
G.P. polypropylene	6–12	shatters
High impact polystyrene	5–18	depends on type
Styrene–acrylonitrile ⎫ G.P. polystyrene ⎭	1	shatters

very good resistance to alkalies and very low gas permeability. Since the phenoxy resins are notch sensitive Izod-type impact strengths do not yield more than moderate values. On the other hand dart impact tests on flat unnotched sheets indicate a very high order of toughness. Table 22.14 quotes some impact strength figures from a test in which a 10 lb steel dart

Table 22.15 SOME TYPICAL PROPERTIES OF PHENOXY RESINS[15]

Property	Units	Value	ASTM test method
Specific gravity	—	1·18–1·3	—
Tensile strength	10^3 lb in^{-2}	9·0–9·5	D.638
Elongation at break	%	50–100	D.638
Flexural strength	10^3 lb in^{-2}	14	D.790
Flexural modulus	10^3 lb in^{-2}	410	D.790
Izod impact	ft lb/in notch	2·5	D.256
Heat distortion temperature:			
at 66 lb in^{-2}	°C	91	D.648
at 264 lb in^{-2}	°C	86	D.648
Volume resistivity	ohm cm	$>10^{18}$	D.257
Dielectric constant	60 c/s	4·1	D.150
	10^6 c/s	3·8	
Power factor	60 c/s	0·001	D.150
	10^6 c/s	0·03	
Water absorption (24 hr)	%	0·13	D.570
(equilibrium) (73°F)	%	1·5	

with a hemispherical nose was dropped on to an $\frac{1}{8}$ in. thick specimen. The height of the fall to break is a measure of the impact strength. Typical properties of phenoxy resins are given in Table 22.15.

Dried granules may be injection moulded at material temperatures of 220–260°C and pressures of 10–15,000 lb in^{-2} in ram machines. Because of the low moulding shrinkage (0·003–0·004 in/in) a somewhat greater than normal draft may be required in deep draw moulds. It is advisable to purge the cylinders after use.

Providing the granules are dry the resins may also be extruded and blow moulded without undue difficulty. Preliminary investigations indicate the resins will be of use as coatings and structural adhesives.

REFERENCES

1. *German Patent* 676,117
2. *U.S. Patent* 2,324,483
3. *U.S. Patent* 2,444,333
4. *U.S. Patent* 2,582,985
5. *U.S. Patent* 2,456,408
6. *U.S. Patent* 2,521,911; *U.S. Patent* 2,521,912
7. *U.S. Patent* 2,528,399; *U.S. Patent* 2,528,360
8. *U.S. Patent* 2,467,171
9. *U.S. Patent* 2,575,558
10. LEE, H., and NEVILLE, K., *Epoxy Resins in their Application and Technology*, McGraw-Hill, New York/Toronto/London (1957)
11. LEWIS, R. N., *Brit. Plastics*, **35,** 580 (1962)
12. SCHECHTER, J, WYNSTRA L., and KURKJY, R. P., *Ind. Eng. Chem.*, **48,** 94 (1956)
13. LEWIS, R. N., from *Plastics Progress* 1959, Ed. P. Morgan, p. 37, Iliffe, London (1960)
14. SKEIST, I., *Epoxy Resins*, Reinhold, New York (1958)
15. *Mod. Plastics*, **40** (3), 169 (1962)

BIBLIOGRAPHY

LEE, H., and NEVILLE, K., *Epoxy Resins in their application and Technology*, McGraw-Hill, New York/Toronto/London (1957)
SKEIST, I., *Epoxy Resins*, Reinhold, New York (1958)

Polyurethanes

23.1 INTRODUCTION

Reaction of an isocyanate and an alcohol results in the formation of a urethane.

$$R \cdot NCO + HOR_1 \longrightarrow R \cdot NH \cdot COOR_1$$

By the same reaction polyhydroxy materials will react with polyiso-cyanates to yield polyurethanes. For example, the reaction between 1,4-butane diol and hexamethylene diisocyanate is shown below

$$HO \cdot (CH_2)_4 \cdot OH \quad OCN \ (CH_2)_6 \ NCO \quad HO(CH_2)_4OH$$

$$\longrightarrow \sim O \cdot (CH_2)_4OOCNH \cdot (CH_2)_6NHCOO \cdot (CH_2)_4O \sim$$

This particular polymer is a fibre-forming material (Perlon U). Although in many respects this reaction resembles the formation of polyesters and polyamides it is not a condensation reaction but involves a transfer of hydrogen atoms and thus may be considered as an example of rearrange-ment polymerisation.

Although the first polyurethanes were similar to that shown above, several polymers currently used contain many linkages in addition to the urethane group. Because of this the term polyurethane is now generally extended to cover all the complex reaction products of isocyanates and polyhydroxy compounds (the latter frequently known in this context as polyols).

Commercial development of the polyurethanes arose from the work of German chemists attempting to circumvent the Du Pont patents on nylon 66. O. Bayer and his team of chemists were able to produce fibre-forming polymers by reacting aliphatic di-isocyanates and aliphatic diols (glycols). Subsequent work resulted in the production of useful products by reacting polymeric hydroxyl-containing compounds such as polyesters to give rubbers, foams, coatings and adhesives.

Although there have been developments in rubbers, fibres, adhesives and surface coatings, the most spectacular developments have occurred in flexible and rigid foams and these will be given prime consideration in this chapter.

23.2 ISOCYANATES

The first isocyanates were produced by Wurtz[1] in 1849 by reacting organic sulphates with cyanic acid salts.

$$R_2SO_4 + 2KCNO \longrightarrow 2RNCO + K_2SO_4$$

Current methods are based on the phosgenation of amines and their salts, a route first described by Hentschel[2] in 1884. Only a few isocyanates are used commercially and the most important are:

1. 80:20 mixtures of tolylene 2,4-di-isocyanate with tolylene 2,6-di-isocyanate (80:20 t.d.i.).
2. A 65:35 mixture of the above (65:35 t.d.i.).
3. 'Diphenylmethane' di-isocyanates.
4. Naphthylene di-isocyanate.
5. Hexamethylene di-isocyanate.
6. Triphenylmethane-*pp'p"*-triyl tri-isocyanate.

These materials are all liquids.

The t.d.i. mixtures are produced by a series of reactions starting from toluene. The first stage is the nitration of the toluene to yield 2-nitrotoluene and 4-nitrotoluene in roughly equal proportions. These can be further nitrated, the 4-isomer yielding only 2,4-dinitroluene and 2-nitrotoluene both the 2,4- and the 2,6-di-nitro compound (Fig. 23.1).

If toluene is dinitrated without separation a mixture of about 80% 2,4-dinitrotoluene and 20% 2,6-dinitrotoluene is obtained. Nitration of separated

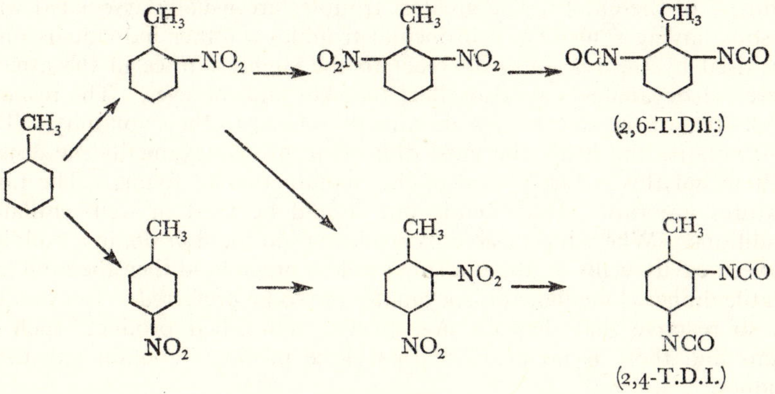

Fig. 23.1

2-nitrotoluene will yield a mixture of approximately 65% of the 2,4- and 35% of the 2,6-isomer. Both mixtures are used for the manufacture of commercial isocyanates.

The next stage is the reduction of the nitro compounds to amines using such reagents as iron dust and water. The resultant amines are then reacted with phosgene. A number of variations in the phosgenation process have been described in the literature.[3,4,5] The general method is to react a solvent slurry of the appropriate amine hydrochloride with phosgene at about 140°C, typical solvents being toluene, *o*-dichlorobenzene and nitrobenzene. The initial reactions are primarily

$$NH_2 \cdot R \cdot NH_2 + HCl \longrightarrow NH_2 \cdot R \cdot NH_2 \cdot HCl$$

$$NH_2 \cdot R \cdot NH_2 \cdot HCl + COCl_2 \longrightarrow Cl \cdot OC \cdot NH \cdot R \cdot NH_2 \cdot HCl + HCl$$

$$\longrightarrow Cl \cdot OC \cdot NH \cdot R \cdot NH \cdot CO \cdot Cl$$

The carbamoyl chloride formed may then be decomposed more or less simultaneously with the initial phosgene–amine reaction to produce di-isocyanate. A urea may be formed as the result of side reactions

$$Cl \cdot OC \cdot NH \cdot R \cdot NH \cdot CO \cdot Cl \longrightarrow OCNRNCO + 2HCl$$

1,5-naphthylene di-isocyanate, important in the production of certain rubbers, can be prepared by a similar route starting from naphthalene. Hexamethylene di-isocyanate is prepared from hexamethylene diamine, an intermediate readily available because of its large scale use in polyamide manufacture. Amine derivatives of diphenylmethane are made by reacting aniline or toluidines with formaldehyde. These can lead to a mixture of di-isocyanates, the 'diphenyl methane di-isocyanates' of commerce. Triphenylmethane *pp'p"*-triyl triisocyanate is produced from leucorosaniline.

Isocyanates are toxic materials and care should be exercied in their use. Their main effect is on the respiratory system and as a result persons exposed to them may suffer from sore throats, bronchial spasms and a tightness of the chest. The greatest troubles are usually associated with persons having a history of bronchial troubles. Other individuals once sensitised by the reagents will react to the slightest trace in the atmosphere. Isocyanates may also affect the skin and the eyes. The respiratory effects of the isocyanates are directly related to their volatility. The most volatile and hence the most difficult to use is hexamethylene di-isocyanate but this is rarely used in the manufacture of foams. The t.d.i. mixtures are rather less volatile but should be used in well-ventilated conditions. Where it is desired to produce rigid foam *in situ* in a building between cavity walls or about a ship's cold storage hold then the even less volatile diphenyl methane di-isocyanates are to be preferred. Isocyanates are so reactive that they do not survive in finished products such as foams and there is no evidence that these products contain any toxic residues.

Isocyanates are highly reactive materials and enter into a number of reactions with groups containing active hydrogen. The reactions of most importance in the formation of polyurethanes are:

1. $\sim$NCO $+$ H$_2$O $\longrightarrow$ $\sim$NHC$\cdot$OH $\longrightarrow$ $\sim$NH$_2$ $+$ CO$_2$

                      ||

                      O

             Unstable   Amine
        Carbamic Acid

2. $\sim$NCO $+$ NH$_2$$\sim$ $\longrightarrow$ $\sim$NH$\cdot$C$\cdot$NH$\sim$

                              ||

                              O

                A Urea Linkage

3. $\sim$NCO $+$ $\sim$NH$\cdot$C$\cdot$NH$\sim$ $\longrightarrow$ $\sim$NHC$-$N$-$C$\cdot$NH$\sim$

                  || || ||

                  O   O   O

             A Urea   A Biuret

4. $\sim$NCO $+$ HO$\sim$ $\longrightarrow$ $\sim$NHCO$\sim$

                                 ||

                                 O

    Hydroxy Compound   A Urethane
       or Polyol

5.

                                        |

   $\sim$NCO $+$ $\sim$NHCO$\sim$ $\longrightarrow$ $\sim$NHC$-$N$-$CO$\sim$

                  || || ||

                  O   O   O

        A Urethane   An Allophanate

The significance of these reactions will be discussed with the various type of product.

23.3 FIBRES AND MOULDING COMPOUNDS

As previously mentioned the initial research on polyurethanes was directed towards the preparation of fibre-forming polymers. Many polyhydroxy compounds and many di-isocyanates were used and the melting points of some of the more linear aliphatic polyurethanes produced are given in Table 23.1.

These figures in many ways resemble those of the aliphatic polyamides. Both types of polymer are capable of hydrogen bonding and in both types the greater the distance between amide or urethane links the lower the melting point, provided there is an even number of carbon atoms between the characteristic groupings. Both polyamides and polyurethanes with an odd number of carbon atoms have lower melting points than the polymer with one more, intermediate carbon atom (i.e. an even number of carbon atoms).

Although 4,4-polyurethane has the highest melting point this material was not produced commercially because of the difficulty of obtaining tetramethylene di-isocyanate with the desired degree of purity. The polymer with the next highest melting point, 6,4-polyurethane, the reaction product of hexamethylene diisocyanate and 1,4-butane diol, was thus chosen for commercial production by German chemists during the Second World War because of the availability of the isocyanate.

The polymer may be prepared by running the isocyanate into the glycol while the temperature is raised slowly to near 200°C. The reaction is

Table 23.1

No. of carbon atoms in		Melting
Di-isocyanate	*Glycol*	*point* (°C)
4	4	190
4	6	180
4	10	170
5	4	159
6	3	167
6	4	183
6	5	159
6	9	147
8	4	160
8	6	153
12	12	128

exothermic and carried out under a blanket of nitrogen. The polymers produced have a molecular weight of 10,000–15,000 and after filtration may be melt spun into fibres.

Compared with nylon 66 fibres, the polyurethane fibres (known as Perlon U) have a tensile strength at the higher end of the range quoted for nylon 66, they are less prone to discoloration in air, are more resistant to acid conditions and have a lower moisture absorption. On the debit side they are less easy to dye, are hard, wiry and harsh to handle and have too low a softening point for many applications. They are currently of little importance but have found some use in bristles, filter cloths, sieves and a few other miscellaneous applications.

The linear polyurethanes used to make fibres can also be used as thermoplastics and may be processed by injection moulding and extrusion techniques. A number of grades are available varying in hardness, softening point, water absorption and other properties. That with the highest melting point is based on 6,4-polyurethane but those with lower melting points are copolymers in which about 10–15% of the 1,4-butane diol is replaced by another diol such as hexamethylene glycol or methylhexamethylene glycol. The processing characteristics are very similar to the nylons, in particular the low melt viscosity requires the use of nylon-type injection nozzles. The polymers start to decompose at about 220°C and care should be taken to prevent overheating.

The properties of the polyurethane moulding compositions are also very similar to nylon 66. The greatest difference in properties is in water absorption, the 6,4-polyurethane absorbing only about $\frac{1}{6}$ of that nylon 66 under comparable conditions. This results in better dimensional stability and a good retention of electrical insulation properties in conditions of high humidity. Resistance to sulphuric acid is somewhat better than with nylon 66 but both types of polymer are dissolved by phenols and formic acid.

There is little call for these thermoplastics materials (marketed as Durethan U, Farbenfabriken Bayer) since they are about twice the price of nylons 66 and 6. Where a thermoplastic for light engineering purposes is required with a low water absorption, nylon 11, acetal resins and, in certain instances, polycarbonates are cheaper and, usually, at least as satisfactory.

23.4 RUBBERS

It was pointed out in Chapters 3 and 4 that rubbers are substantially amorphous polymers with glass transition temperatures below their service temperature. The greatest degree of elasticity is obtained with highly flexible segments, generally low intermolecular forces and little or no crystallinity. In order to reduce creep and high compression set it is usual to lightly cross link the polymers. For high tensile strength, tear resistance and abrasion resistance the above requirements for high elasticity and resilience may require some modification, in particular some ability to crystallise is often desirable.

By careful formulation it is possible to produce polyurethane rubbers with a number of desirable properties. The first rubbers were prepared by Pinten[6] in Germany about 1940. Known as I-Gummi they were produced by reacting a polyester with a di-isocyanate. These products had a high tensile strength and abrasion resistance but low tear strength and poor low temperature properties. Subsequently the variables in the formulation were systematically investigated by Bayer, Müller[7] and co-workers and this led to the advent of the Vulkollan rubbers.

The starting point in the preparation of these rubbers is a polyester prepared by reacting a glycol such as ethylene or propylene glycol with adipic acid. This is then reacted with an excess of a bulky di-isocyanate such as 1,5-naphthylene di-isocyanate (Fig. 23.2).

The molar excess of di-isocyanate is about 30% so that the number of polyesters joined together is only about 2–3 and the resulting unit has isocyanate end groups. A typical structure, with P for polyester groups, U for urethane and I for isocyanates would be

IPUPUPI

The resulting 'prepolymer' can then be chain extended with water, glycols or amines by linking across terminal isocyanate groups (Fig. 23.3).

The water reaction evolves carbon dioxide and is to be avoided with solid elastomers but is important in the manufacture of foams. The above reactions cause chain extension and by the formation of urea and urethane linkages they provide sites for cross linking since these groups

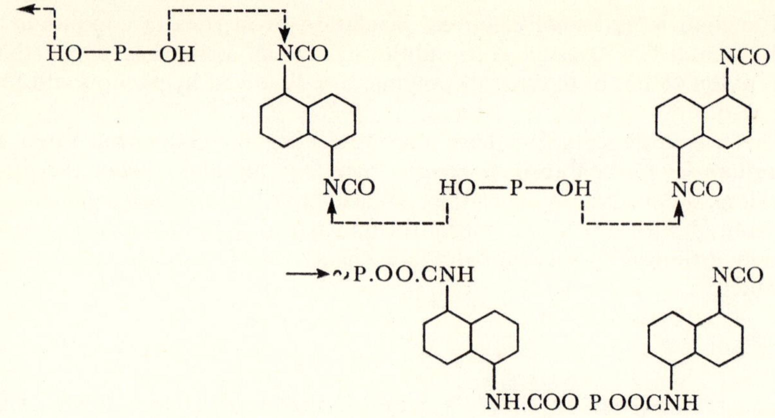

Fig. 23.2

can react with free isocyanate or terminal isocyanate groups to form biuret or allophanate linkages respectively (Fig 23.4).

Where urea and urethane groups are present in the polymer chain in approximately equal amounts most branch points are biuret since the urea group reacts faster than urethane links. For branching and cross linking to occur it is essential to have a slight excess of isocyanate over the glycol or amine chain extender so that isocyanate groups are available for formation of biuret and allophanate linkates. The degree of cross linking can to some extent be controlled by adjusting the amount of excess isocyanate, whilst more highly cross-linked structures may be produced by the use of a triol in the initial polyester.

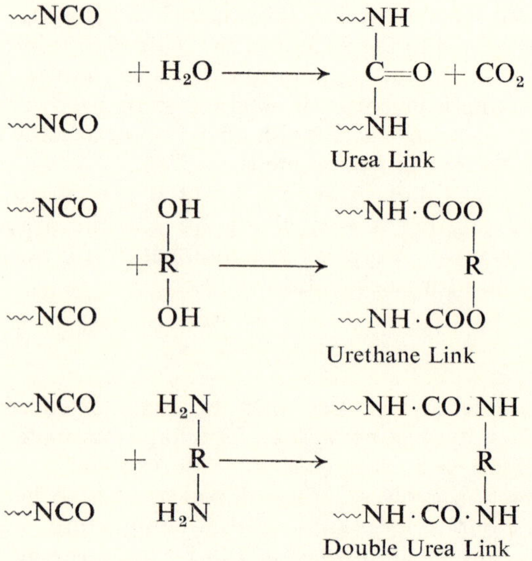

Fig. 23.3

In their classical researches Bayer and Müller[7] investigated the effect of the isocyanate type on the properties of the finished product. They found that 'bulky' aromatic isocyanates such as 1,5-naphthalene di-iso-cyanate and diphenyl methane di-isocyanates gave products of much higher tear strength and tensile strength than were obtained with either hexamethylene diisocyanate or t.d.i. Various polyesters were also pre-pared and, as a result, it was found that poly(ethylene adipate) with a molecular weight of about 2,000 and which allowed a moderate amount of crystallisation, gave the best balance of desirable properties in the product. A variety of chain extenders were also investigated. It was found that extenders which were 'bulky' and tended to 'stiffen' the rubber molecules gave the highest modulus and tear strength whereas extenders with flexible linkages gave the greatest elasticity. Aromatic diamines are examples of the first class and thiodiethylene glycol the latter.

Somewhat unexpected results are obtained when the degree of cross linking is increased by incorporating triols in the original polyester. In contrast to results with hydrocarbon rubbers the greater the degree of

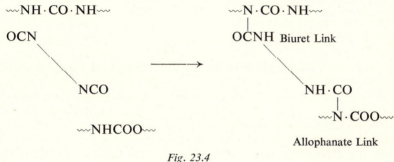

Fig. 23.4

cross linking the lower is the tensile strength and tear resistance but the higher the elasticity. This effect is believed to be due to the fact that in rubbers of the Vulkollan type much of the strength of the material is derived from secondary forces, hydrogen bonding in particular. Cross linking not only interferes with the effectiveness of such secondary forces but it also restricts crystallisation. Corroborating evidence for this is that at elevated temperatures, where secondary forces are greatly reduced, the more highly cross-linked polymers are rather stronger.

Vulkollan-type rubbers (produced in Britain as Duthane and Prescollan) suffer from the disadvantage that the prepolymer is unstable and must be used within a day or two of manufacture. In addition they cannot be processed on conventional rubber machinery and products are made by casting processes. This has resulted in the production of other rubbers of the polyurethane type which resemble other synthetic rubbers in their technology. One typical class is exemplified by Chemigum SL (Good-year), Daltoflex 1 (I.C.I.) and Desmophen A (Bayer). With these materials the isocyanate is reacted with a slight excess of polyester (or polyester–amide in the case of Daltoflex 1) so that terminal hydroxyl groups are produced in the prepolymer. The prepolymers are rubber-like gums which may be compounded on 2-roll mills with other ingredients. They may be cured by addition of a di-isocyanate or preferably a 'latent di-isocyanate', that is a substance which changes into an active isocyanate during moulding operations. This technique is reminiscent of

Table 23.2 COMPARISON OF PROPERTIES OF POLYURETHANE, NITRILE AND NATURAL RUBBERS

	Polyurethane raw rubber + black	Polyurethane casting rubber	Natural rubber + black	Nitrile rubber + black
Tensile strength (kg cm^{-2})	385	376	250	212
Elongation (%)	640	700	460	363
Resilience (%)	56	69	64	46
Hardness (BS°)	86	86	66	74
Tear strength (kg cm^{-2})	73	110	102	57
Abrasion resistance (Du Pont test) cm^3 loss)	15	16	128	70
Swell in				
aliphatic hydrocarbon (%)	3	6	150	3
aromatic hydrocarbon (%)	77	79	200	141
trichloroethylene (%)	95	120	330	152
lubricating oil (%)	0	0	330	0

the use of 'hexa' in phenolic resins and latent acid catalysts with the aminoplastics.

Another approach has been adopted by the Du Pont Company with Adiprene C. This is a urethane type polymer with unsaturated groups in the polymer. Because of the unsaturation the polymer may be vulcanised with sulphur, the standard vulcanising agent of the rubber industry. This is a clear-cut example of a product being modified to suit the processor rather than that of a processor adapting himself to meet new products. Whereas Adiprene C has poor tensile strength when unfilled the use of carbon black leads to appreciable reinforcement (as is the case with s.b.r. and to some extent natural rubber).

Polyurethane rubbers in general, and the Vulkollan types in particular, possess certain outstanding properties. They can have higher tensile strengths than any other rubber and have excellent tear and abrasion resistance. They tend to have a high hardness and a low resilience and in fact may be regarded as somewhat intermediate between conventional rubbers and flexible thermoplastics. The urethane rubbers also show outstanding resistance to ozone and oxygen (features lacking with the diene rubbers) and to aliphatic hydrocarbons. Reversible swelling occurs with aromatic hydrocarbons. One disadvantage of the materials is that hydrolytic decomposition occurs with acids, alkalis and the prolonged action of water and steam.

Some typical properties of a Vulkollan-type polyurethane cast rubber and a black-reinforced polyurethane rubber processed by conventional techniques are compared with black-reinforced natural and nitrile rubbers in Table 23.2.[8]

Urethane rubbers have found steadily increasing use for oil seals, shoe soles and heels, fork-lift truck tyres, diaphragms, chute linings and a variety of mechanical applications. Fabric coatings resistant to dry cleaning are a recent development. In many of these applications high elasticity is not an important prerequisite so that the polyurethane rubbers must be compared not only with other rubbers but also with a variety of thermoplastics.

If a branched polyol, usually either castor oil or a simple polyester is heated with an isocyanate but without chain extenders soft and weak rubbery products are obtained with very low resilience. These materials are useful for encapsulation of electronic components and for printers rollers.

Polyurethane rubber thread was introduced by two American Companies in 1959, Spandex fibre (Du Pont) and Vyrene (U.S. Rubber). These materials have a high modulus and are resistant to oxidation, machine washing and drying. They have found uses in foundation garments, swimwear and surgical hose.

Yet another class of polyurethanes which may be considered in this section are the elastomeric thermoplastics exemplified by Texin (Mobay) and Estane (Goodrich and British Geon). These may be extruded, injection moulded, calendered or compression moulded as a typical thermoplastic. Whereas it is usual to post-cure Texin in hot air this is

Table 23.3   SUMMARY OF TYPES OF SOLID POLYURETHANE ELASTOMERS

Processing route	Polyol	Method of cross linking	Examples
1. Tough casting rubbers	polyester	isocyanate + chain extenders	Vulkollan Prescollan Duthane
2. Soft casting rubbers	polyesters castor oil	isocyanate isocyanate	Daltorol PR1
3. Rubbers processed on conventional machinery	polyester–amide polyester	isocyanate isocyanate	Daltoflex 1 Chemigum SL Desmophen A
	polyether	isocyanate sulphur amines	Adiprene B Adiprene C Adiprene L
4. Rubber thread			Spandex Vyrene
5. Elastomeric (a) Thermoplastics (b)		unnecessary hot air	Estane Texin

not necessary with Estane. The latter material has excellent abrasion resistance, aliphatic hydrocarbon resistance and a low temperature brittleness temperature below −75°C. These materials cost about twice as much as the polyamides and initial interest has been mainly in the field of surface coatings.

The various types of solid polyurethane rubbers are summarised in Table 23.3.

23.5 FLEXIBLE FOAMS

Whereas the solid polyurethane rubbers are speciality products, polyurethane foams are widely-used and well-known materials. About half of the total weight of plastics in a small modern car is composed of such foams.

In many respects the chemistry of these foams is similar to that of the Vulkollan-type rubbers except that gas evolution reactions are allowed to proceed concurrently with chain lengthening and cross linking. Although volatile liquids are also used with rigid foams and for low density flexible foams the gas for flexible foam is usually carbon dioxide produced during reaction of the polyol, isocyanate and other additives. The earliest foams were produced by using polyesters containing carboxyl groups. These reacted with isocyanates thus

$$\sim\!\!COOH + OCN\!\!\sim \longrightarrow \sim\!\!\overset{\displaystyle O}{\overset{\displaystyle \|}{C}}\cdot NH\!\!\sim + CO_2$$

However subsequent polyesters were produced with low carboxyl values and gas evolution occurred by the reaction already mentioned when discussing the Vulkollans, that between isocyanate and water

$$
\begin{array}{c}
\sim\!\!NCO \\[4pt]
\\
+\ H_2O \longrightarrow \\[4pt]
\\
\sim\!\!NCO
\end{array}
\qquad
\begin{array}{c}
\sim\!\!NH \\
| \\
C\!=\!O + CO_2 \\
| \\
\sim\!\!NH
\end{array}
$$

The isocyanate group may be terminal on a polyester chain or may be part of the unreacted di-isocyanate. The density of the product, which depends on the amount of gas evolved, can be reduced by increasing the isocyanate content of the reaction mixture and by correspondingly increasing the amount of water to react with the excess isocyanate (that is excess over that required for chain extension and cross linking).

Polyurethane foams may be rigid, semirigid or flexible. They may be made from polyesters, polyethers or natural polyols such as castor oil (which contains approximately three hydroxyl groups in each molecule). Three general processes are available known as one-shot, prepolymer or quasi-prepolymer processes. These variations lead to twenty-seven basic types of product or process all of which have been used commercially. This section only deals with flexible foams (which are only made from polyesters and polyethers). Since prepolymers and quasi-prepolymer processes are no longer important with polyesters only the four following types will be considered here:

1. One-shot polyesters.
2. Polyether prepolymers.
3. Polyether quasi-prepolymers.
4. One-shot polyethers.

23.5.1 One-shot polyester foams

Until the late 1950's most flexible foams were based on polyester resins. These foams were developed in Germany during the Second World War and became well-known as 'Moltopren'. The polyesters commonly have a molecular weight of about 2,000 and are commonly produced from adipic acid and a glycol such as diethylene glycol together with a small proportion of a trifunctional ingredient such as trimethylol propane. They are viscous liquids rather similar to polyester laminating resins.

Foams may be produced from these resins by addition of 65:35 t.d.i., water, a catalyst, an emulsifier, a structure modifier and paraffin oil which helps to control pore size and prevents splitting of the foams.

Amongst the catalysts described in the literature may be mentioned dimethylbenzylamine, dimethylcyclohexylamine, diethylaminoethanol, *N*-alkyl morpholines and the adipic acid ester of *N*-diethylaminoethanol. A number of proprietary products of undisclosed composition have also been successfully employed. Emulsifiers include sulphonated castor oil and structure modifiers such substances as ammonium oleate and silicone oils.

The bulk of flexible polyester foam is produced in block form using machines of the Henecke type or some simple modification of it. In this machine polyester and isocyanate are fed to a mixing head which oscillates in a horizontal plane. The other ingredients, known as the 'activator mixture' are then injected or bled into the isocyanate–polyester blend and the whole mixture is vigorously stirred and forced out of the base of the mixing head. The emergent reacting mixture runs into a trough which is moving backwards at right angles to the direction of traverse of the reciprocating head. In this way the whole of the trough is evenly covered with the reacting mass which has frequently foamed within a minute or so of issuing from the mixing head. The principle of the Henecke machine is illustrated in Fig. 23.5.

23.5.2 Polyether prepolymers

As will be discussed later flexible polyester foams are not altogether satisfactory for upholstery applications and in the 1950's the attention of American chemists turned to the use of polyethers. These materials could be obtained more cheaply than the polyesters but the products were less reactive and with catalyst systems then available could not be directly converted into foams by a one-shot process. As a result a prepolymer technique, reminiscent of that used with Vulkollan and which had already been used with certain polyesters, was developed.

In this process the polyether is reacted with an excess of isocyanate to give an isocyanate-terminated prepolymer which is reasonably stable if kept in sealed tins in dry conditions. If water, catalysts and other ingredients are added to the product a foam will result. Where linear polyethers are used it is found that this foam has rather poor load bearing and cushioning properties and where this is important a low molecular weight triol, such as glycerol or trimethylol propane, is added to the polyether before reaction with isocyanate. This will then provide a site for chain branching. Alternatively a small amount of water could be added to the system. This would react with terminal isocyanate groups which link up to produce a urea link as mentioned previously. This urea

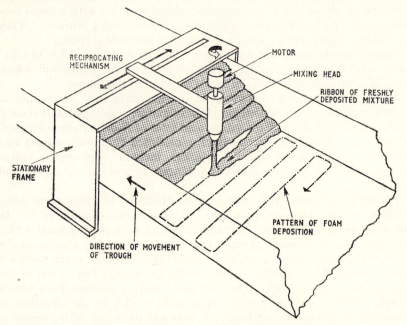

Fig. 23.5. Principle of the Henecke machine (Farbenfabrik Bayer). (After Phillips and Parker[4])

group is more reactive than a urethane link and reacts with isocyanates to give a biuret link as a site for chain branching. It is important that carbon dioxide evolved in the isocyanate–water reaction be allowed to escape and also that the reaction is kept down so that premature foaming does not occur.

Although prepolymer processes have become less important with the advent of the one-shot process they have certain advantages. Because there is less exotherm large blocks of foam can often be produced, there is often a greater flexibility in design of compounds, the reduced amount of free isocyanate reduces handling hazards and there is some evidence[9] that two-stage foams have slightly better cushioning properties. On the other hand prepolymers have limited stability, are often rather viscous to handle, and do involve an extra process.

23.5.3   Quasi-prepolymer polyether foams

This process, which is intermediate between the prepolymer and one-shot process, is useful where prepolymers are too viscous, where the resin does not easily adapt itself to one-shot processes and where the equipment available is more suitable for two-part systems. In principle a polyol is reacted with a large excess of isocyanate so that the prepolymer formed is of low molecular weight and there are a large number of free isocyanate groups. This product is then reacted at the time of foaming with additional hydroxy compound, water and catalyst to produce the foam. The additional hydroxy compound may be a polyol or a simple molecule such as ethylene glycol or glycerol which has the additional function of a viscosity depressant. The system has the advantage of flexibility and of having low viscosity components, but as with one-shot foams there are problems with high exotherms and with a high free isocyanate content. Quasi-prepolymer systems (also known as semi-prepolymer systems) are rarely used with flexible foam but are more important with the rigid and semirigid materials.

23.5.4 Polyether one-shot foams

The one-shot polyethers now form the bulk of the flexible polyurethane foam now being manufactured. This is a result of the favourable economics of polyethers, particularly when reacted in a one-shot process, and because the polyethers generally produce foams of better cushioning characteristics. A typical formulation for producing a one-shot polyether foam will comprise polyol, isocyanate, catalyst, surfactant and blowing agent and these will be considered in turn.

A variety of polyethers have been used and may be enumerated in their order of development as follows—

1. Polymers of tetrahydrofuran introduced by Du Pont as Teracol in 1955:

$$\longrightarrow HO \cdot [(CH_2)_4 O]_n H$$

These polyethers produced good foams but were rather expensive.
2. Polymers of ethylene oxide, cheaper than the tetrahydrofuran polymers, were found to be too hydrophilic for successful use.
3. Propylene oxide polymers are also low in cost and may be prepared by polymerising the oxide in the presence of propylene glycol as an initiator and a caustic catalyst at about 160°C. They have the general structure

$$CH_3 \cdot CH \cdot CH_2 \cdot O \cdot (CH_2 \cdot CH \cdot O)_n \cdot CH_2 \cdot CH \cdot CH_3$$
$$\quad\ | \qquad\qquad\qquad\quad | \qquad\qquad\quad\ |$$
$$\quad OH \qquad\qquad\qquad CH_3 \qquad\qquad OH$$

The secondary hydroxyl groups of these poly(oxypropylene) glycol diols are less reactive than the primary hydroxyl groups of the earlier polyesters.   At the time of the introduction of these polyethers, the catalysts then available were insufficiently powerful for one-shot processes to be practical and so these polymers have been used primarily in prepolymer processes.

4. Block copolymers of ethylene oxide and propylene oxide, less hydrophilic than poly(oxyethylene) glycol and more reactive than the propylene oxide polymers were introduced by Wyandotte Chemical (U.S.A.) under the trade name Pluronic.

5. Today the bulk of the polyethers used for flexible foams are produced by polymerising propylene oxide using trimethylol propane, 1,2,6-hexane triol or glycerol as initiator.

This leads to triols of the following general type:

$$HO \cdot (C_3H_6O)_n \cdot CH_2$$
$$|$$
$$CH \cdot OH$$
$$|$$
$$HO \cdot (C_3H_6O)_n \cdot CH_2$$

The higher functionality of these polymers leads to foams of better load bearing characteristics.   Polymers of molecular weights in the range of 3,000–3,500 are found to give the best balance of properties.

The second largest component of a foam formulation is the isocyanate. 80:20 t.d.i. is found to be the most suitable of the various isocyanates available and is used almost exclusively.

One-shot polyether processes became feasible with the advent of sufficiently powerful catalysts.   For many years tertiary amines had been used with both polyesters and the newer polyethers.   Examples included alkyl morpholines and triethylamine.   More recent catalysts such as triethylene diamine ('Dabco') and 4-dimethyl amine pyridine were rather more powerful but not satisfactory on their own.   In the late 1950's organo-tin catalysts such as dibutyl tin dilaurate and stannous octoate were found to be powerful catalysts for the chain extension reactions.   It was found that by use of varying combinations of a tin catalyst with a tertiary amine (which catalyse both the gas evolution and chain extension reaction) it was possible to produce highly active systems in which foaming and cross-linking reactions could be properly balanced.   Although stannous octoate is more susceptible to hydrolysis and oxidation than dibutyl tin dilaurate it does not cause such rapid aging of the foam, a problem with organo-metallic catalysts, and thus it is somewhat more popular.

The presence of silicones in the formulation is essential in obtaining foam stability during reaction and also in controlling pore size.   For best results a polyether miscible product is required and special polymers which are essentially block copolymers of silicones with a polyether have been developed.

The water present reacts with isocyanate to produce carbon dioxide and urea bridges. The more the water present (together with a corresponding additional amount of isocyanate) the more the gas evolved and the more the number of active urea points for cross linking. Thus the foams of lower-density do not necessarily have inferior load-bearing characteristics. When soft foams are required a volatile liquid such as fluorotrichloromethane may be incorporated. This will volatilise during the exothermic reaction and will increase the total gas present but not increase the degree of cross linking.

Formulations should be based on stoichiometric considerations. Based on a knowledge of the hydroxyl value of the polyol the amount of isocyanate necessary to cause chain growth should be calculated. The gas evolved will depend on the water content and additional isocyanate must be incorporated corresponding to the water present. When the isocyanate used equals the theoretical amount the system is said to have a t.d.i. index of 100. In practice a slight excess of isocyanate is used (t.d.i. index 105–110) to ensure complete reaction and to make available some free isocyanate for the biuret and allophanate reactions. A typical formulation would be

Polyether triol	100
80:20 t.d.i.	40
Water	3
Triethylene diamine	0·5
Stannous octoate	0·3
Silicone block copolymer	1·0

Most foam is produced on machines based on the Henecke process but in many cases it is necessary to have at least four streams to the mixing head; e.g. polyol and fluorocarbon (if any); isocyanate; water, amine, silicone; and tin catalyst. Reaction is carried out with slightly warmed components and foaming is generally complete within a minute of the mixture emerging from the head. Recent developments include direct moulding where the mixture is fed into warmed moulds which are subsequently heated to enable rapid curing of the outer layers of the foam. Foam-in-place techniques have also been developed for use in the manufacture of car seats, arm rests and crash pads.

23.5.5 Properties and applications of flexible foams

Flexible polyurethane foams are resilient open-cell structures. Compared with foams from natural rubber and s.b.r. latex they are less inflammable and have better resistance to oxidation and aging. The major interest of flexible polyurethane foams is for cushioning and other upholstery materials and for this reason the load–compression characteristics are of importance. People differ considerably in their opinions as to what constitutes an ideal cushioning material and, as a result, manufacturers have tended to try to reproduce the characteristics of natural rubber latex foam which has become widely accepted as a cushioning material. The early polyester foams unfortunately did not correspond well in their load–deflection characteristics for, although they had an initially high

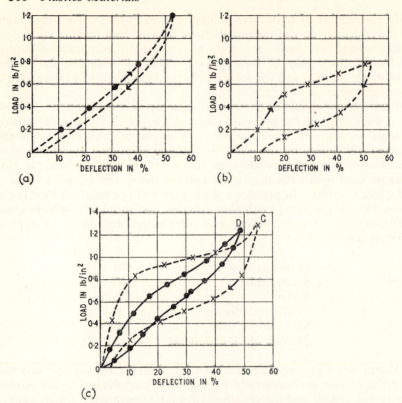

Fig. 23.6. *Typical load–deflection curves for (a) latex, (b) flexible p.v.c.,*
(c) polyester polyurethane, (d) polyether polyurethane foams. (Shell
Chemical Co.)

modulus, they tended to collapse or 'bottom out' above a certain loading.
Thus in many applications the foam became essentially a solid piece of
rubber. In addition the foam showed a slow recovery from compression
and a pronounced hysteresis loop in the load-compression curve. More
recent polyether foams tend to be much more in line with latex foam but
with a slightly greater damping capacity which in many instances may be
considered a desirable feature. Fig. 23.6 shows typical load–compression
curves for latex, p.v.c. and polyurethane foams.

Today polyether foam with a density of less than half that of rubber
latex foam is widely used as a cushioning material. Polyester foams,
although tending to be more expensive continue to have a number of
outlets particularly where a high initial modulus is desirable. In addition
to miscellaneous upholstery applications polyester foams are useful as
'foam backs', that is a foam backing in order to stiffen or shape some
softer fabric. Examples include car door and roof trim, quilting, shoulder
pads and coat interlinings. Amongst the many miscellaneous uses for
both types of foam are paint rollers, sponges, draught excluders and
packaging for delicate equipment.

23.6 RIGID FOAMS[10]

The flexible foams discussed in the previous section have polymer structures with a low degree of cross linking. If polyols of higher functionality, i.e. more hydroxyl groups per molecule, are used, tougher products may be obtained and in the case of material with a sufficiently high functionality rigid foams will result.

As with the flexible foams the early products were invariably based on polyesters, but more trifunctional alcohol such as glycerol or trimethylol propane was added to the initial polyester reaction mixture. These materials could then be reacted with isocyanate, catalyst, water and emulsifying agent in the presence of a flame-retarder such as tri-β-chloroethyl phosphate. Although t.d.i. was used initially the increasing use of rigid foams for *in situ* applications led to the development of less volatile and subsequently less unpleasant isocyanates such as the diphenylmethane di-isocyanates. These foams can be produced without difficulty using one-shot techniques either on large factory-installed machines of the Henecke type or alternatively on small portable equipment. In most systems the reaction is rather slower than with the flexible foam and conditions of manufacture rather less critical. In the United States prepolymer and quasi-prepolymer systems corresponding to those discussed under flexible foam were developed, largely to reduce the hazards involved in handling t.d.i. on portable equipment in places where there were severe ventilation problems.

As with the flexible foams there has been a trend in recent years to the use of polyethers. These are largely adducts based either on trifunctional hydroxy compounds, on tetrafunctional materials such as pentaerythritol or a hexafunctional material such as sorbitol. Where trifunctional materials are used these are of lower molecular weight (~ 500) than with the polyethers for flexible foams in order to reduce the distance between hydroxyl groups and hence increase the degree of cross linking.

Formulations for one-shot polyether systems are similar to those used for flexible foams and contain polyether, isocyanate, catalyst, surfactant and water. Trichloroethyl phosphate is also often used as a flame retardant. As with polyesters, diphenylmethane di-isocyanate is usually preferred to t.d.i. because of its lower volatility. Tertiary amines and organo-tin catalysts are used as with the flexible foams but not necessarily in combination. Silicone oil surfactants are again found to be good foam stabilisers. Volatile liquids such as trichlorofluoromethane are widely used as supplementary blowing agents and give products of low density and of very low thermal conductivity.

Halocarbons have the further advantages of reducing the viscosity of the reaction mixture and, where used as the main blowing agent instead of the carbon dioxide produced by the isocyanate-water reaction, cheaper foams are obtained since less isocyanate is used.

In addition to one-shot processes, quasi-prepolymer systems are used commercially with rigid polyether foams. The quasi-prepolymer is commonly produced using excess t.d.i. rather than diphenylmethane di-isocyanate.   Since the former isocyanate is light in colour and the latter dark, quasi-prepolymer foams are usually lighter in colour.   The quasi-prepolymer systems are also more tolerant to variations in processing conditions and often less careful control of the process can be tolerated.

Products intermediate to the flexible and rigid foams may be obtained from castor oil (a trihydroxy molecule), synthetic triols of moderate molecular weight and polyesters with a moderate amount of trifunctional hydroxy compound in the structure.   Such semirigid or high molecular foams are useful in the manufacture of car crash-pads and in packaging equipment.

Although some rigid foams are used in sandwich constructions for aircraft and building structures the major interest of rigid foams has been in the field of thermal insulation.   In such application the foams encounter competition from polystyrene and U–F foams.   With both the polystyrene and the polyurethane foams there has been intensive development in recent years leading to improved products of lower cost.   The polystyrene foams have the economic advantage of being made from cheaper starting materials, can be produced successfully at lower densities (1 lb ft^{-3} instead of 1·3 lb ft^{-3} for polyurethane foam) and are generally less friable.   One particular advantage of polyurethanes is that they may be formed *in situ* and themselves act as an adhesive to most cavity surrounds or skins.   At the present time where it is only necessary to lay a piece of foam in position expanded polystyrene is cheaper.   Where, however, it is necessary to bond the foam on to the skin material, such as in a sandwich construction, the cost of the adhesives necessary with polystyrene makes a substantial addition to the overall cost.   The relative economics of the two materials will therefore depend very much on the end-use in question.

For materials of equivalent density water-blown polyurethanes and the hydrocarbon-blown polystyrene foams have similar thermal conductivities. This is because the controlling factor determining the conductivity is the

Table 23.4

Gas	Thermal conductivity (Btu in. ft^{-2} hr^{-1} degF^{-1})
Air	0·168
CO$_2$	0·102
CCl$_3$F	0·058

nature of the gas present in the cavities.   In both of the above cases air, to all intents and purposes, normally replaces any residual blowing-gas either during manufacture or soon after.   Polyurethane foams produced using fluorocarbons have a lower thermal conductivity (0·12–0·15 Btu in. ft^{-2} hr^{-1} degF^{-1}) because of the lower conductivity of the gas.   The comparitive thermal conductivities for air, carbon dioxide and monofluorotrichloromethane are given in Table 23.4.

Except where the foam is surrounded by a skin of relatively impermeable material, it would be expected that the blowing gas would diffuse out and be replaced by air and that the thermal conductivities of the foams would increase until they approached that of expanded polystyrene of similar density. Whilst this is true of foams which generate carbon dioxide it is found that this does not happen when fluorocarbons are used. In this case diffusion of the fluorocarbon proceeds very slowly and it appears that an equilibrium is eventually reached when the ratio of air to fluorocarbon in the cell is about 1:1. For this reason fluorocarbon-blown foams have ultimate thermal conductivities significantly lower than those of CO_2-blown foams or expanded polystyrene of similar densities.

Whilst most interest in rigid foams has been centred round polyurethane and polystyrene products a number of other rigid foams have been produced on a small scale. The only one of these which is at all competitive with the two major materials is U–F foam. This material has a low cost and conductivity and can be found *in situ* in building cavity walls. The process is widely used in Britain for domestic house insulation.

23.7 COATINGS AND ADHESIVES

A wide range of polyurethane-type products have become available in recent years for coating applications. These include simple solutions of linear polyurethanes, two-pot alkyd-isocyanate and polyether-isocyanate systems and a variety of prepolymer and adduct systems. The coatings can vary considerably in hardness and flexibility and find use mainly because of their toughness, abrasion resistance and flexibility. Uses include metal finishes in chemical plant, wood finishes for boats and sports equipment, finishes for rubber goods and rain-erosion resistant coatings for aircraft. One type of coating is potentially competitive with p.v.c. leathercloth. Both alkyd/di-isocyanate and adduct/di-isocyanate compositions may be coated on to fabrics from solutions of controlled viscosity and solids content. Such coated fabrics are soft, flexible and unlike p.v.c. leathercloth, free from plasticisers.

Many isocyanates have good adhesive properties and one of them, triphenylmethane-*pp'p"*-triyl triisocyanate, has been successfully used for bonding of rubber. Isocyanates are however rather brittle and somewhat limited in application. Somewhat tougher products are obtained from adhesives involving both polyols and isocyanates, i.e. polyurethane-type materials. The major application of these materials to date is in the boot and shoe industry.

REFERENCES

1. WURTZ, A., *Ann.*, **71**, 326 (1849)
2. HENTSCHEL, W., *Ber.*, **17**, 1284 (1884)
3. SAUNDERS, J. H., and FRISCH, K. G., *Polyurethanes—Chemistry and Technology*; Pt 1—*Chemistry*, Interscience, New York (1962)
4. PHILLIPS, L. N., and PARKER, D. B. V., *Polyurethanes—Chemistry, technology and Properties*, Iliffe, London (1964)
5. ARNOLD, R. G., NELSON, J. A., and VERBANC, J. J., *Chemistry of Organic Isocyantes*, Du Pont Bulletin HR-2 (1–20–56)

6. PINTEN, H., *German Patent, Appl.* D-90, 260 (March 1942)
7. BAYER, O., MÜLLER, F., PETERSEN, S.,P IEPENBRINK, H. F., and WINDEMUTH, E., *Angew. Chem.* **62**, 57 (1950); *Rubber Chem. Technol.*, **23**, 812 (1950)
8. HAMPTON, H. A., and HURD, R., *Trans. Plastics Inst.*, **29**, 204 (1961)
9. BUIST, J. M., *Trans. Plastics Inst.*, **29**, 100 (1962)
10. FERRIGNO, T. H., *Rigid Plastics Foams*, Reinhold, New York (1963)

BIBLIOGRAPHY

DOMBROW, B. A., *Polyurethanes*, Reinhold, New York (1957)
FERRIGNO, T. H., *Rigid Plastics Foams*, Reinhold, New York (1963)
HEALY, T. T. (Ed.), *Polyurethane Foams*, Iliffe, London (1964)
PHILLIPS, L. N., and PARKER, D. B. V., *Polyurethanes—Chemistry, Technology and Properties*, Iliffe, London (1964)
SAUNDERS, J. H., and FRISCH, K. C., *Polyurethanes—Chemistry and Technology*; Pt 1—*Chemistry*; Pt 2—*Technology*, Interscience, New York (1962)

24

Furan Resins

24.1 INTRODUCTION

The furan resins have never achieved importance for mouldings or laminates but because of their good chemical resistance have found value in anti-corrosion applications.

24.2 PREPARATION OF INTERMEDIATES

The two intermediates of commercial furan resins are furfural and furfuryl alcohol. Furfural occurs in the free state in many plants but is obtained commercially by degradation of hemi-cellulose constituents present in these plants. There are a number of cheap sources of furfural and theoretical yields of over 20% (on a dry basis) may be obtained from both corn cobs and oat husks. In practice yields of slightly more than half these theoretical figures may be obtained. In the U.S.A. furfural is produced in large quantities by digestion of corn cobs with steam and sulphuric acid. The furfural is removed by steam distillation.

Furfural is a colourless liquid which darkens in air and has a boiling point of 161·7°C at atmospheric pressure. Its principal uses are as a selective solvent used in such operations as the purification of wood rosin and in the extraction of butadiene from other refinery gases. It is also used in the manufacture of phenol–furfural resins and as a raw material for the nylons. The material will resinify in the presence of acids but the product has little commercial value.

Catalytic hydrogenation of furfural in the presence of copper chromite leads to furfuryl alcohol, the major intermediate of the furan resins (Fig. 24.1).

The alcohol is a mobile liquid light in colour with a boiling point of 170°C. It is very reactive and will resinify if exposed to high temperatures, acidity, air or oxygen. Organic bases such as piperidine and n-butylamine are useful inhibitors.

505

24.3 RESINIFICATION

Comparatively little is known of the chemistry of resinification of either furfuryl alcohol or furfural.

It is suggested that the reaction shown in Fig. 24.2 occurs initially with furfuryl alcohol

The liberation of small amounts of formaldehyde has been detected in the initial stage but it has been observed that this is used up during later reaction. This does not necessarily indicate that formaldehyde is essential

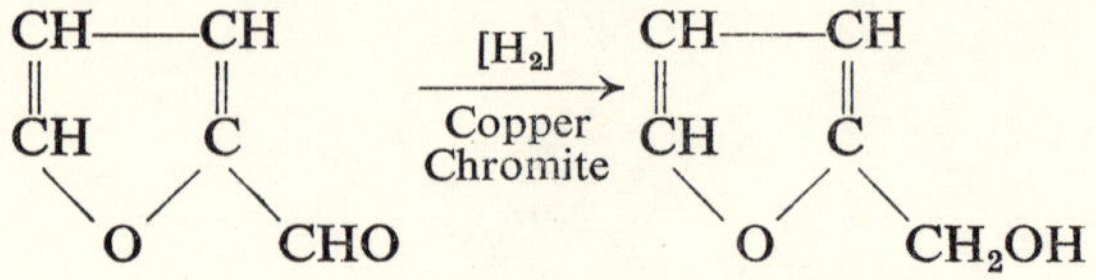

Fig. 24.1

to cross linking, and it would appear that its absorption is due to some minor side reaction.

Loss of unsaturation during cross linking indicates that this reaction is essentially a form of double bond polymerisation, viz Fig. 24.3.

This reaction, like the initial condensation, is favoured by acidic conditions and peroxides are ineffective.

The polymerisation of furfural is apparently more complex and less understood.

For commercial use a partially condensed furan resin is normally prepared which is in the form of a dark free-flowing liquid. Final cure is carried out *in situ*.

The liquid resins are prepared either by batch or continuous process by treating furfuryl alcohol with acid. Initially the reaction mixture is

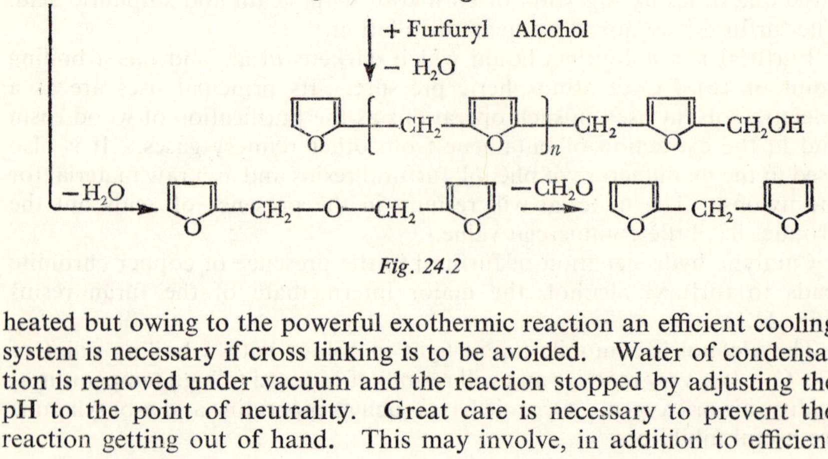

Fig. 24.2

heated but owing to the powerful exothermic reaction an efficient cooling system is necessary if cross linking is to be avoided. Water of condensation is removed under vacuum and the reaction stopped by adjusting the pH to the point of neutrality. Great care is necessary to prevent the reaction getting out of hand. This may involve, in addition to efficient cooling, a judicious choice of catalyst concentration, the use of a mixture of furfuryl alcohol and furfural which produces a slower reaction but gives a more brittle product and, possibly, reaction in dilute aqueous solution.

The resins are hardened *in situ* by mixing with an acidic substance just before application. A typical curing system would be 4 parts of *p*-toluene

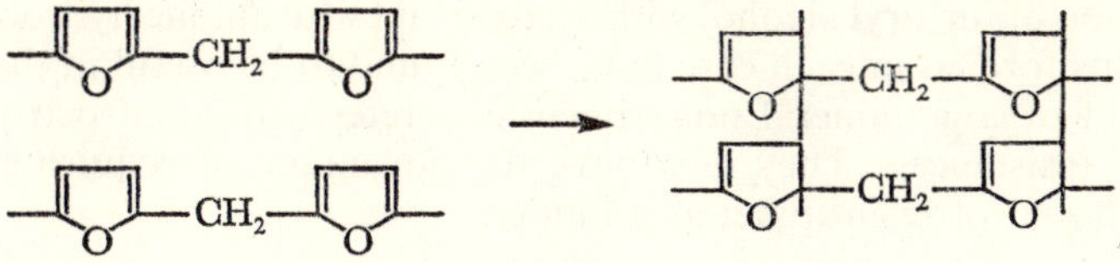

Fig. 24.3

sulphonic acid per 100 parts resin. The curing may take place at room temperature if the resin is in a bulk form but elevated temperatures cures will be often necessary when the material is being used in thin films or coatings.

24.4 PROPERTIES OF THE CURED RESINS

The resins are cross linked and the molecular segments between the cross links are rigid and inflexible. As a consequence the resins have an excellent heat resistance, as measured in terms of maintenance of rigidity on heating but are rather brittle.

Cured resins have excellent chemical resistance. This is probably because, although the resins have some reactive groupings, most of the reactions occurring do not result in the disintegration of the polymer molecules. Therefore, whilst surface layers of molecules may have undergone modification they effectively shield the molecules forming the mass of the resin. The resins have very good resistance to water penetration.

Compared with the phenolics and polyesters the resins have better heat resistance, better chemical resistance, particularly to alkalis, greater hardness and better water resistance. In these respects they are similar to, and often slightly superior to, the epoxide resins. Unlike the epoxies they have a poor adhesion to wood and metal, this being somewhat improved by incorporating plasticisers such as poly(vinyl acetate) and poly(vinyl formal) but with a consequent reduction in chemical resistance. The cured resins are black in colour.

24.5 APPLICATIONS

The principal applications for furan resins are in chemical plant. Specific uses include the lining of tanks and vats and piping and for alkali resistant tile cements. The property of moisture resistance is used when paper honeycomb structures are treated with furan resins and subsequently retain a good compression strength even after exposure to damp conditions.

Laminates have been prepared for the manufacture of chemical plant. They have better heat and chemical resistance than the polyester, epoxy, phenolic or aminoplastic-based laminates but because of the low viscosity of the resins are not easy to handle. Because they are also somewhat brittle furan-based laminates have been limited in their applications.

Recent work by Russian workers has led to interesting products formed by reaction of furfuryl alcohol with acetone and with aniline hydrochloride. The resins formed in each case have been found to be useful in the manufacture of organic-mineral non-cement concretes with good petrol, water and gas resistance. They also have the advantage of requiring only a small amount of resin to act as a binder.

BIBLIOGRAPHY

MCDOWALL, R., and LEWIS, P., *Trans. Plastics Inst.*, **22,** 189 (1954)
MORGAN, P., *Glass-reinforced Plastics*, 3rd Edn, Iliffe, London (1961)
  See also various articles by Itinskiï, Kamenskiï, Ungureau and others in Plasticheskie Massy from 1960 onwards. (Translations published as Soviet Plastics published by Rubber and Technical Press Ltd., London.)

25

Silicones and Inorganic Polymers

25.1 INTRODUCTION

To many polymer chemists one of the most fascinating developments of the last thirty years has been the discovery of a range of semi-inorganic and wholly inorganic polymers, and the attendant commercial development of the silicone polymers. Because of their general thermal stability, good electrical insulation characteristics, constancy of properties over a wide temperature range, water repellency and anti-adhesive properties, the silicone polymers find use in a very wide diversity of applications. Uses range from high temperature insulation materials and gaskets for jet engines to polish additives and water repellent treatments for leather. The polymers are available in a number of forms such as fluids, greases, rubbers and resins.

The possibility of the existence of organo-silicone compounds was first predicted by Dumas in 1840 and in 1857 Buff and Wöhler[1] found the substance now known to be trichlorsilane by passing hydrochloric acid gas over a heated mixture of silicone and carbon. In 1863 Friedel and Crafts[2] prepared tetraethyl silane by reacting zinc diethyl with silicon tetrachloride.

$$2Zn(C_2H_5)_2 + SiCl_4 \longrightarrow Si(C_2H_5)_4 + 2ZnCl_2$$

In 1872 Ladenburg[3] produced the first silicone polymer, a very viscous oil by reacting diethyl diethoxy silane with water in the presence of traces of acid.

$$C_2H_5-O-\underset{\underset{C_2H_5}{|}}{\overset{\overset{C_2H_5}{|}}{Si}}-O-C_2H_5 \xrightarrow[\text{Acid}]{H_2O} -(-O-\underset{\underset{C_2H_5}{|}}{\overset{\overset{C_2H_5}{|}}{Si}}-)- + C_2H_5OH$$

The basis of modern silicone chemistry was however laid by Professor F. S. Kipping at the University College, Nottingham, between the years

1899 and 1944.   During this period Kipping published a series of 51 main papers and some supplementary studies, mainly in the *Journal of the Chemical Society*.   The work was initiated with the object of preparing asymmetric tetrasubstituted silicon compounds for the study of optical rotation.   Kipping and his students were concerned primarily with the preparation and study of new non-polymeric compounds and they were troubled by oily and glue-like fractions that they were unable to crystallise. It does not appear that Kipping even foresaw the commercial value of his researches for in concluding the Bakerian Lecture delivered in 1937 he said

'We have considered all the known types of organic derivatives of silicon and we see how few is their number in comparison with the purely organic compounds.   Since the few which are known are very limited in their reactions, the prospect of any immediate and important advance in this section of chemistry does not seem very hopeful.'

Nevertheless Kipping made a number of contributions of value to the modern silicone industry.   In 1904 he introduced the use of Grignard reagents for the preparation of chlorsilanes and later discovered the principle of the intermolecular condensation of the silane diols, the basis of current polymerisation practice.   The term silicone was also designated by Kipping to the hydrolysis products of the disubstituted silicon chlorides because he at one time considered them as being analogous to the ketones. In 1931 J. F. Hyde of the Corning Glass Works was given the task of preparing polymers with properties intermediate between organic polymers and organic glasses.   The initial objective was a heat resistant resin to be used for impregnating glass fabric to give a flexible electrical insulating medium.   As a result silicone resins were produced.   In 1943 the Corning Glass Works and the Dow Chemical Company co-operated to form the Dow Corning Corporation which was to manufacture and develop the organo-silicon compounds.   In 1946 the General Electric Company of Schenectady, N.Y. also started production of silicone polymers using the then new 'Direct Process' of Rochow.   The Union Carbide Corporation started production of silicones in 1956.

The first full scale production of silicones in Great Britain commenced in 1954 at the Barry works of Albright and Wilson Limited on the basis of the Dow Corning patents.   In 1955 Imperial Chemical Industries began production of a process licensed by the General Electric Company.   In the early 1960's it has been estimated that world annual production (excluding the Soviet bloc) is about 20,000 tons with production in Great Britain of the order of 1,500 tons.

25.1.1   Nomenclature

Before discussing the chemistry and technology of silicone polymers it is necessary to consider the methods of nomenclature of the silicon compounds relevant to this chapter.   The terminology used will be that adopted by the International Union of Pure and Applied Chemistry.

The structure used as the basis of the nomenclature is *silane* SiH_4 corresponding to methane CH_4. Silicon hydrides of the type SiH_3 $(SiH_2)_n$ SiH_3 are referred to as disilane, trisilane, tetrasilane etc., according to the number of silicon atoms present.

Alkyl, aryl, alkoxy and halogen substituted silanes are referred to by prefixing 'silane' by the specific group present. The following are typical examples:

$(CH_3)_2SiH_2$	dimethylsilane
$CH_3 \cdot Si \cdot Cl_3$	methyltrichlorsilane
$(C_6H_5)_3 \cdot Si \cdot C_2H_5$	triphenyl ethyl silane

Compounds having the formula $SiH_3 \cdot (OSiH_2)_n O \cdot SiH_3$ are referred to as disiloxane, trisiloxane etc., according to the number of silicon atoms. Polymers in which the main chain consists of repeating —Si—O— groups together with predominately organic side groups are referred to as *polyorganosiloxanes* or more loosely as *silicones*.

Hydroxy derivatives of silicones in which the hydroxyl groups are attached to a silicon atom are named by adding the suffices -ol, -diol, -triol etc., to the name of the parent compound. Examples are:

H_3SiOH	silanol
$H_2Si(OH)_2$	silanediol
$(CH_3)_3SiOH$	trimethyl silanol
$(C_6H_5)_2(C_2H_5O)SiOH$	diphenylethoxysilanol

25.1.2 Nature of chemical bonds containing silicon

Silicon has an atomic number of 14 and an atomic weight of 28·06. It is a hard, brittle substance crystallising in a diamond lattice and has a specific gravity of 2·42. The elemental material is prepared commercially by the electrothermal reduction of silica.

Silicon is to be found in the fourth group and the second short period of the periodic table. It thus has a maximum covalency of six although it normally behaves as a tetravalent material. The silicon atom is more electropositive than the atoms of carbon or hydrogen. The electronegativity of silicon is 1·8, hydrogen 2·1, carbon 2·5 and oxygen 3·5. It has a marked tendency to oxidise, the scarcity of naturally occurring silicon providing an excellent demonstration of this fact.

At one time it was felt that it would be possible to produce silicon analogues of the multiplicity of carbon compounds which form the basis of organic chemistry. Because of the valency difference and the electropositive nature of the element this has long been known not to be the case. It is not even possible to prepare silanes higher than hexasilane because of the inherent instability of the silicon–silicon bond in the higher silanes.

The view has also existed in the past that the carbon–silicon bond should be similar in behaviour to the carbon–carbon bond and would have a similar average bond energy. There is some measure of truth in the assumption about average bond energy but because silicon is more electropositive than carbon the C—Si bond will be polar and its properties will be very dependent on the nature of groups attached to the carbon and silicon groups. For example the CH_3—Si group is particularly resistant to oxidation but C_6H_{13}—Si is not.

The polarity of the silicon–carbon bond will affect the manner in which the reaction with ions and molecules takes place. For example on reaction with alkali, or in some conditions with water, it is to be expected that the negative hydroxyl ion will attack the positive silicon atom rather than the negative carbon atom to form, initially, Si—OH bonds. Reaction with hydrogen chloride would lead similarly to silicon-chlorine and carbon–hydrogen bonds.

It is important to realise that the character of substituents on either the carbon or silicon atoms will greatly affect the reactivity of the carbon–silicon bond according to its effect on the polarity. Thus strongly negative substituents, e.g. trichloromethyl groups, attached to the carbon atom, will enhance the polarity of the bond and facilitate alkaline hydrolysis. A benzene ring attached to the carbon atom will also cause an electron shift towards the carbon atom and enhance polarity. Hydrogen chloride may then affect acid cleavage of the ring structure from the silicon by the electronegative chlorine attacking the silicon and the proton attacking the carbon.

The foregoing facts of relevance to the preparation and properties of silicone polymers may be summarised as follows:

1. Silicon is usually tetravalent but can assume hexavalent characteristics.
2. Silicon is more electropositive than carbon and hence silicon–carbon bonds are polar (12% ionic).
3. The reactivity of the Si—C bond depends on the substituent group attracted to the Si and C atoms.
4. The reactivity also depends on the nature of the attacking molecule.

Two further statements may also be made at this stage.

5. Inclusion of silicon into a polymer does not ensure by any means a good thermal stability.
6. The siloxane Si—O link has a number of interesting properties which are relevant to the properties of the polyorganosiloxanes. These will be dealt with later.

25.2 PREPARATION OF INTERMEDIATES

The polyorganosiloxanes are generally prepared by reacting chlorsilanes with water to give hydroxyl compounds which then condense to give the polymer structure, e.g.

$$
\underset{\underset{R_1}{|}}{\overset{\overset{R}{|}}{Cl-Si-Cl}} + H_2O \longrightarrow \underset{\underset{R_1}{|}}{\overset{\overset{R}{|}}{HO-Si-OH}} \longrightarrow \underset{\underset{R_1}{|}}{\overset{\overset{R}{|}}{-(-Si-O-)-}}
$$

Similar reactions can also be written for the alkoxysilanes but in commercial practice the chlorsilanes are favoured. These materials may be prepared by many routes of which four appear to be of commercial value, the Grignard process, the direct process, the olefin addition method and the sodium condensation method.

25.2.1 The Grignard method

The use of the Grignard reagents of the type RMgX for the production of alkyl and aryl-chlorsilanes was pioneered by Kipping in 1904 and has been for long the favoured laboratory method for producing these materials.

The reaction is carried out by first reacting the alkyl or aryl halide with magnesium shavings in an ether suspension and then treating with silicon tetrachloride (prepared by passing chlorine over heated silicon). With methyl chloride the following sequence of reactions occur:

$$CH_3Cl + Mg \xrightarrow{\text{Ether}} CH_3MgCl$$

$$CH_3MgCl + SiCl_4 \longrightarrow CH_3SiCl_3 + MgCl_2$$

$$CH_3MgCl + CH_3SiCl_3 \longrightarrow (CH_3)_2SiCl_2 + MgCl_2$$

$$CH_3MgCl + (CH_3)_2SiCl_2 \longrightarrow (CH_3)_3SiCl + MgCl_2$$

The reaction proceeds in a stepwise manner but because of the differences in the reactivities of the intermediates a high yield of dimethyldichlorsilane is produced.

The products are recovered from the reaction mixture by filtration to remove the magnesium chloride, followed by distillation. It is then necessary to distil fractionally the chlorsilanes produced. The fractional distillation is a difficult stage in the process because of the closeness of the boiling points of the chlorsilanes and some by-products (Table 25.1) and 80–100 theoretical plates are necessary to effect satisfactory separation.

Table 25.1 BOILING POINT OF SOME CHLORSILANES AND RELATED COMPOUNDS

Compound	Boil Pt (°C)
$(CH_3)_2SiCl_2$	70
CH_3SiCl_3	65·7
$(CH_3)_3SiCl$	57
CH_3SiHCl_2	41
$SiCl_4$	57·6
$(CH_3)_4Si$	26

The Grignard method was the first route used commercially in the production of silicone intermediates. Its great advantage is its extreme flexibility since a wide range of organic groups may be attached to the silicon in this method. Because of the need to use ether or other inflammable solvents considerable production hazards arise. On economic grounds the main drawbacks of the process are the multiplicity of steps and the dependence on silicon tetrachloride, which contains only 16% Si and is thus a rather inefficient source of this element.

25.2.2   The direct process

The bulk of the methyl silicones are today manufactured via the direct process. In 1945 Rochow[4] found that a variety of alkyl and aryl halides may be made to react with elementary silicon to produce the corresponding organosilicon halides.

$$Si + RX \longrightarrow R_nSiX_{n-4} \ (n\ 0{=}{-}4)$$

The hydrocarbon can be in either the liquid or vapour phase and the silicon is finely divided. The inclusion of certain solid catalysts in the reactive mass may in some instances greatly facilitate the reaction. A mixture of powdered silicon and copper in the ratio 90:10 is used in the manufacture of alkyl chlorsilanes.

In practice vapours of the hydrocarbon halide, e.g. methyl chloride, are passed through a heated mixture of the silicon and copper in a reaction tube at a temperature favourable to obtaining the optimum yield of the dichlorsilane, usually 250–280°C. The catalyst not only improves the reactivity and yield but also makes the reaction more reproducible. Presintering of the copper and silicon or alternatively deposition of copper on to the silicon grains by reduction of copper (I) chloride is more effective than using simple mixtures of the two elements. The copper appears to function by forming unstable copper methyl, $CuCH_3$, on reaction with the methyl chloride. The copper methyl then decomposes into free methyl radicals which react with the silicon.

Under the most favourable reaction conditions when methyl chloride is used the crude product from the reaction tube will be composed of about 73·5% dimethyldichlorsilane, 9% of methyltrichlorsilane and 6% of trimethylchlorsilane together with small amounts of other silanes, silicon tetrachloride and high boiling residues.

The reaction products must then be fractionated as in the Grignard process.

The direct process is less flexible than the Grignard process and is restricted primarily to the production of the, nevertheless all important, methyl- and phenyl-chlorsilanes. The main reason for this is that higher alkyl halides than methyl chloride decompose at the reaction temperature and give poor yields of the desired products and also the fact that the copper catalyst is only really effective with methyl chloride.

In the case of phenyl chlorsilanes some modifications are made to the process. Chlorobenzene is passed through the reaction tube which contains a mixture of powdered silicon and silver (10% Ag), the latter as catalyst. Reaction temperatures of 375–425°C are significantly higher than for the methyl chlorsilanes. An excess of chlorobenzene is used which sweeps out the high-boiling phenylchlorsilanes of which the dichlorsilanes are predominant. The unused chlorobenzene is fractionated and recycled.

The direct process involves significantly fewer steps than the Grignard process and is more economical in the use of raw materials. This may be seen by considering the production of chlorsilanes by both processes starting from the basic raw materials. For the Grignard process the following basic materials will normally be sand, coke, chlorine and methane and the following steps will be necessary before the actual Grignard reaction:

$$SiO_2 + 2C \longrightarrow Si + 2CO \tag{1}$$

$$Si + 2Cl_2 \longrightarrow SiCl_4 \tag{2}$$

$$CH_3OH + HCl \longrightarrow CH_3Cl + H_2 \tag{3}$$

$$MgCl_2 \longrightarrow Mg + Cl_2 \tag{4}$$

Rochow[5] has summed the entire Grignard process from basic raw material to polymer as:

Formula $SiO_2 + 2C + 2CH_3OH + 2Cl_2 + 2Mg$
Mol. Wt.   60   24   64   142   48·6

$$\longrightarrow (CH_3)_2SiO + 2MgCl_2 + H_2O + 2CO$$
$$\qquad\qquad\quad 74 \qquad\quad 190·6 \qquad 18 \qquad 56$$

On the other hand only the additional steps (1) and (3) will be required in the direct process which gives the summarised equation

Formula $SiO_2 + 2C + 2CH_3OH \longrightarrow (CH_3)_2SiO + H_2O + 2CO$
Mol. Wt.   60   24   64   74   18   56

25.2.3 The Olefin addition method

The basis of this method is to react a compound containing Si—H groups with unsaturated organic compounds. For example ethylene may be reacted with trichlorsilane

$$CH_2{=}CH_2 + SiHCl_3 \longrightarrow CH_3 \cdot CH_2 \cdot SiCl_3$$

The method may also be used for the introduction of vinyl groups

$$CH{\equiv}CH + SiHCl_3 \longrightarrow CH_2{=}CH \cdot SiCl_3$$

The trichlorsilane may be obtained by reacting hydrogen chloride with silicon in yields of 70% and thus is obtainable at moderate cost. As the olefins are also low cost materials this method provides a relatively cheap route to the intermediates. It is, of course, not possible to produce methylchlorsilanes by this method.

25.2.4   Sodium condensation method

This method depends on the reaction of an organic chloride with silicon tetrachloride in the presence of sodium, lithium or potassium.

$$4RCl + SiCl_4 + 8Na \longrightarrow SiR_4 + 8NaCl$$

This reaction, based on the Wurtz reaction tends to go to completion and the yield of technically useful chlorsilane is low.

The commercial value of this method is also limited by the hazards associated with the handling of sodium.

25.2.5   Rearrangement of organochlorosilanes

A number of techniques have been devised which provide useful methods of converting by-product chlorosilanes into more useful intermediates. A typical example useful in technical scale work is the redistribution of trimethylchlorsilane and methyltrichlorsilane to the dichlorsilane by reacting at 200–400°C in the presence of aluminium chloride.

$$(CH_3)_3SiCl + CH_3SiCl_3 \rightleftharpoons 2(CH_3)_2SiCl_2$$

25.3   GENERAL METHOD OF PREPARATION AND PROPERTIES OF SILICONES

A variety of silicone polymers has been prepared varying from low viscosity fluids to rigid cross-linked resins. The bulk of such materials are based on methylchlorsilanes and the gross differences in physical states depend largely on the functionality of the intermediate.

Reaction of trimethylchlorsilane with water will produce a mono-hydroxy compound which condenses spontaneously to form hexamethyldisiloxane.

$$2(CH_3)_3SiCl + 2H_2O \longrightarrow 2(CH_3)_3SiOH$$

$$\longrightarrow (CH_3)_3Si \cdot O \cdot Si(CH_3)_3 + H_2O$$

Hydrolysis of dimethyldichlorsilane will yield a linear polymer.

$$\underset{\underset{CH_3}{|}}{\overset{\overset{CH_3}{|}}{Cl-Si-Cl}} \xrightarrow{H_2O} \sim\!\!\sim(\!\sim\!\!\sim\underset{\underset{CH_3}{|}}{\overset{\overset{CH_3}{|}}{Si}}-O\!\sim\!\!\sim)\!\sim\!\!\sim$$

Hydrolysis of monomethyltrichlorsilane yields a network structure

$$
\underset{\underset{\displaystyle Cl}{|}}{\overset{\overset{\displaystyle CH_3}{|}}{Cl-Si-Cl}} \longrightarrow \underset{\underset{\displaystyle OH}{|}}{\overset{\overset{\displaystyle CH_3}{|}}{HO-Si-OH}} \longrightarrow \underset{\underset{\displaystyle O}{|}}{\overset{\overset{\displaystyle CH_3}{|}}{-Si-O-}}
$$

For convenience a shorthand nomenclature is frequently used in silicone literature where

$(CH_3)_3$—Si—O— is designated M (for monofunctional)

$$
\underset{\underset{\displaystyle CH_3}{|}}{\overset{\overset{\displaystyle CH_3}{|}}{-Si-O-}} \text{ is designated D (for difunctional)}
$$

and

$$
\underset{\underset{\displaystyle O}{|}}{\overset{\overset{\displaystyle CH_3}{|}}{-Si-O-}} \text{ is designated T (for trifunctional)}
$$

The tetrafunctional
$$
\underset{\underset{\displaystyle O}{|}}{\overset{\overset{\displaystyle O}{|}}{-O-Si-O-}}
$$

of silica, which may be considered the derivative of silicon tetrachloride is designated Q.

Thus hexamethyl silane may be referred to as M-M or M_2. A linear silicone polymer with a degree of polymerisation of n would be referred to as $MD_{n-2}M$. The compound

$$
\begin{array}{c}
CH_3 \\
| \\
CH_3-Si-CH_3 \\
| \\
CH_3 \quad O \\
| \quad\quad | \\
CH_3-Si-O-Si-CH_3 \\
| \quad\quad | \\
CH_3 \quad O \\
| \\
CH_3-Si-CH_3 \\
| \\
CH_3
\end{array}
$$

would be referred to as TM_3.

Difficulties arise in characterising commercial branched and network structures in this way because of their heterogeneity. In these cases the R/Si ratio (or specifically the CH_3/Si ratio in methyl silicones) is a useful parameter. On this basis the R/Si ratio of four types is given in Fig. 25.1.

Since both Si—O and Si—CH_3 bonds are thermally stable it is predictable that the polydimethyl siloxanes (dimethyl silicones) will have good

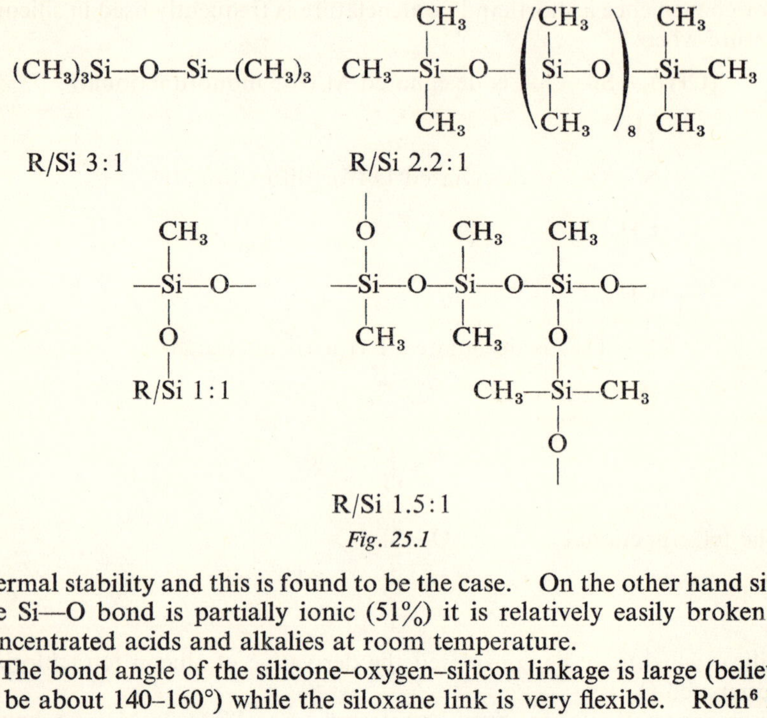

$$(CH_3)_3Si—O—Si—(CH_3)_3$$

R/Si 3:1

$$CH_3—\underset{\underset{CH_3}{|}}{\overset{\overset{CH_3}{|}}{Si}}—O—\left(\underset{\underset{CH_3}{|}}{\overset{\overset{CH_3}{|}}{Si}}—O\right)_8—\underset{\underset{CH_3}{|}}{\overset{\overset{CH_3}{|}}{Si}}—CH_3$$

R/Si 2.2:1

R/Si 1:1

R/Si 1.5:1

Fig. 25.1

thermal stability and this is found to be the case. On the other hand since the Si—O bond is partially ionic (51%) it is relatively easily broken by concentrated acids and alkalies at room temperature.

The bond angle of the silicone–oxygen–silicon linkage is large (believed to be about 140–160°) while the siloxane link is very flexible. Roth[6] has stated that

'The softness of the bond angle plus the favourable geometry reducing steric attractions of attached groups, should result in a negligible barrier to (very) free rotation about the Si—O bonds in the linear polymers. Consequently the low boiling points and low temperature coefficients of viscosity may be attributed to the rotation preventing chains from packing sufficiently closely for the short range intermolecular forces to be strongly operative.'

There is evidence to indicate that intermolecular forces between silicone chains are very low. They include the low boiling points of organosilicon polymers, the low tensile strength of high molecular weight polymers even when lightly cross linked to produce elastomers, the solubility data which indicates a low cohesive energy density and low temperature coefficient of viscosity. The position of the polymers in the triboelectric series and the non-stick properties give similar indications. On the other hand Scott and co-workers[7] have measured the height of the rotational barriers about the Si—O bond and believe that the peculiar properties are due to the very free rotation about the Si—O bond and not due to low intermolecular forces. By studying gas imperfection data of hexamethyldisiloxane they consider that in fact normal intermolecular forces exist.

25.4 SILICONE FLUIDS

The silicone fluids form a range of colourless liquids with viscosities from 1 to 1,000,000 centistokes. High molecular weight materials also exist but these may be more conveniently considered as gums and rubbers (see Section 25.6). It is convenient to consider the fluids in two classes:

1. Dimethyl silicone fluids.
2. Other fluids. These other fluids are used only for specialised purposes and will be considered only in the section on applications.

25.4.1 Preparation

As indicated in Section 25.3, the conversion of the chlorsilane intermediates into polymers is accomplished by hydrolysis with water followed by spontaneous condensation. In practice there are three important stages:

1. Hydrolysis, condensation and neutralisation by either a batchwise or continuous process.
2. Catalytic equilibration.
3. Devolatilisation.

When batch hydrolysis is being employed a weighed excess amount of water is placed in a glass lined jacketed reactor.[8] Dimethyldichlorsilane is run in through a subsurface dispersion nozzle and the contents are vigorously agitated. The reaction is carried out under reflux to prevent loss of volatile components. Although the hydrolysis reaction itself is endothermic the absorption of the HCl evolved on hydrolysis generates enough heat to render the overall reaction exothermic and it is necessary to control the reaction temperature by circulating a coolant through the jacket of the reactor. When hydrolysis is complete the agitation is stopped and the oily polymer layer is allowed to separate from the dilute acid phase which is then drawn off. The oil is then neutralised in a separate operation by washing with sodium carbonate solution, decantation and filtration. The condensate at this stage consists of a mixture of cyclic and linear polymers and careful control of reactant ratios, acid concentration, reaction temperature and oil–acid contact time should be maintained since these will affect the composition of the product which should be as constant as possible for further processing. The batch process has the advantage that these variables are controlled without undue difficulty.

In the continuous process the chlorsilane and the water are run into the suction side of a centrifugal pump. The reacting mixture is then passed through a loop of borosilicate glass pipe where the hydrolysis is completed and from there back to the pump.   The mixture then passes to a decanter to allow separation of the two ingredients.   The decanting stage is critical and care must be taken in order to avoid low yields and difficulties in the neutralisation stage which is carried out as in the batch process.

The products of the hydrolysis reaction under normal conditions will consist of an approximately equal mixture of cyclic compounds, mainly the tetramer, and linear polymer.   In order to achieve a more linear polymer, but with a random molecular weight distribution, and also to stabilise the viscosity it is common practice to equilibrate the fluid by heating with a catalyst such as dilute sulphuric acid.   This starts a series of reactions which would lead to the formation of higher molecular weight polymer except that controlled amounts of the monofunctional trimethylchlorsilane or more usually the dimer, hexamethyl disiloxane ($Me_3Si \cdot O \cdot SiMe_3$), are added as a 'chain stopper' to control molecular weight, the latter functioning by a trans-etherification mechanism.   The more the chain stopper added the lower is the average molecular weight of the equilibrated product.   When assessing the amount of chain stopper to add it is necessary to calculate the amount of trifunctional material present as an impurity in the fluid before equilibration.

In practice, for fluids of viscosities below 1,000 centistokes, the equilibration reaction will take a number of hours at 100–150°C.   Residual esters and silicones which may occur during the reaction are hydrolysed by addition of water and the oil is separated from the aqueous acid layer and neutralised as before.

For some applications it is desirable that the fluids be free from the volatile low molecular products that result from the randomising equilibration reaction.   This operation may be carried out either batchwise or continuously using a vacuum still.   Commercial 'non-volatile' fluids have a weight loss of less than 0·5% after 24 hours at 150°C.

25.4.2   General properties

As a class dimethyl silicone fluids are colourless, odourless, of low volatility and non-toxic.   They have a high order of thermal stability and a fair constancy of physical properties over a wide range of temperature ($-70°C$ to 200°C).   Although fluids have prolonged stability at 150°C they will oxidise at 250°C with an increase in viscosity and eventual gelling.   The oxidation rate may however be retarded by conventional antioxidants such as phenyl-β-naphthylamine and diphenyl-p-phenylene diamine.

Table 25.2   SOME PHYSICAL PROPERTIES OF DIMETHYLSILOXANE POLYMERS OF THE TYPE $(CH_3)_3SiO[Si(CH_3)_2O]_nSi(CH_3)_3$[9]

Value of n	1	3	6	14	90	210	350
Viscosity (centistokes)	1·04	2·06	3·88	10	100	350	1,000
Specific gravity d_{25}^{25}	0·818	0·871	0·908	0·937	0·965	0·969	0·970
Refractive index n_D^{25}	1·382	1·390	1·395	1·399	1·403	1·403	1·404

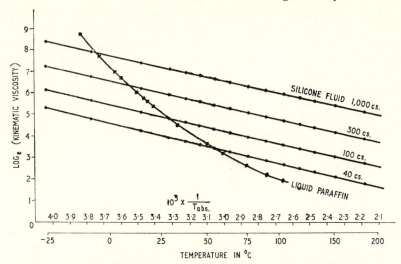

Fig. 25.2. Viscosity–temperature curves for four commercial dimethyl polysiloxane fluids and for liquid paraffin. The numbers 1,000, 300, 100 and 40 indicate the viscosities in centistokes at 38°C. (After Freeman[10])

The fluids have reasonably good chemical resistance but are attacked by concentrated mineral acids and alkalis. They are soluble in aliphatic aromatic and chlorinated hydrocarbons which is to be expected from the low solubility parameter of 7·3. They are insoluble in solvents of higher solubility parameter such as acetone, ethylene glycol and water. They are themselves very poor solvents. Some physical properties of the dimethyl silicone fluids are summarised in Table 25.2.

Barry[11] has shown that for linear dimethyl silicones the viscosity (η) in centistokes at 25°C and the number (n) of dimethyl siloxy groups are connected by the surprisingly simple relationship

$$\log \eta = 0\cdot1 \sqrt{n} + 1\cdot1$$

It has been shown[12] that branched polymers have lower melting points and viscosities than linear polymers of the same molecular weight. The viscosity of the silicone fluids is much less affected by temperature than with the corresponding paraffins (see Fig. 25.2).

25.4.3 Applications

Silicone fluids find a very wide variety of applications mainly because of their water-repellency, anti-stick properties, low surface tension and thermal properties.

Polish additives

One of the most important applications of the dimethyl silicone fluids, and without doubt the most well known to the general public, is as a polish additive.   The polishes contain normally 2–4% of silicone together with a wax which has been formulated either into an aqueous emulsion or a solution in a volatile solvent.   The value of the silicone fluid is not due to such factors as water-repellency or anti-stick properties but due to its ability to lubricate, without softening, the micro-crystalline wax plates and enable them to slide past each other, this being the basis of the polishing process.   The effort in polishing a car with a polish containing silicone fluid is claimed to be less than half that required with a conventional wax polish.   The protective action is at least as good if not slightly superior.

Release agents

Dilute solutions or emulsions containing $\frac{1}{2}$–1% of a silicone fluid have been extensively used as a release agent for rubber moulding, having replaced the older traditional materials such as soap.   Similar fluids have also been found to be of value in the die-casting of metals.   Silicones have not found extensive application in the moulding of thermosetting materials since the common use of plated moulds and of internal lubricants in the moulding powder obviate the need.   Their use has also been restricted with thermoplastics because of the tendency of the fluids to cause stress cracking in many polymers.

Silicone greases do however have uses in extrusion for coating dies etc., to facilitate stripping-down.   Greases have also found uses in the laboratory for lubricating stop-cocks and for high vacuum work.

Water repellent applications

The silicones have established their value as water-repellent finishes for a range of natural and synthetic textiles.   A number of techniques have been devised which result in the pick up of 1–3% of silicone resin on the cloth. The polymer may be added as a solution, an emulsion or by spraying a fine mist, alternatively intermediates may be added which either polymerise *in situ* or attach themselves to the fibre molecules.

In one variation of the process the textile fabric is treated with either a solution or emulsion of a polymer containing active hydrogen groups such as the polymer of methyldichlorsilane.   If the impregnated fabric is heated in the presence of a catalyst such as the zinc salt of an organic acid or an organo-tin compound for about 5 minutes at 100–150°C the hydrogen atoms are replaced by hydroxyl groups which then condense so that individual molecules cross link to form a flexible water repellent shell round each of the fibres (Fig. 25.3).

Leather may similarly be made water repellent by treatment with solutions or emulsions of silicone fluids.   A variety of techniques are available, the method chosen depending to some extent on the type of leather to be treated.   The water repellency may be obtained without appreciably affecting the ability of a leather to transpire.

Silicone fluids containing Si—H groups are also used for paper treatment.   The paper is immersed in a solution or dilute emulsion of the polymer containing either a zinc salt or organo-tin compound. The paper is then air-dried and heated for 2 minutes at 80°C to cure the resin. The treated paper has a measure of water repellency and in addition some anti-adhesive properties.

Lubricants and greases

Silicone fluids and greases have proved of use as lubricants for high temperature operation for applications depending on rolling friction. Their use as boundary lubricants, particularly between steel surfaces is, however, somewhat limited although improvement may be obtained by incorporating halogenated phenyl groups in the polymer. Higher working temperatures are possible if phenyl–methyl silicones are used.

Greases may be made by blending the polymer with an inert filler such as a fine silica, carbon black or a metallic soap. The silicone–silica greases

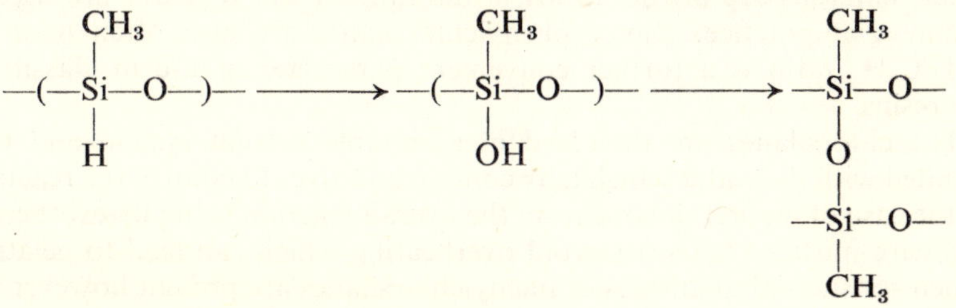

$$-(-\underset{\underset{H}{|}}{\overset{\overset{CH_3}{|}}{Si}}-O-)- \longrightarrow -(-\underset{\underset{OH}{|}}{\overset{\overset{CH_3}{|}}{Si}}-O-)- \longrightarrow -\underset{\underset{O}{|}}{\overset{\overset{CH_3}{|}}{Si}}-O-$$

$$-\underset{\underset{CH_3}{|}}{Si}-O-$$

Fig. 25.3

are used primarily as electrical greases for such applications as aircraft and car ignition systems.

The fluids are also used in shock absorbers, hydraulic fluids, dashpots and in other damping systems designed for high temperature operation.

Miscellaneous

Dimethyl silicone fluids are used extensively as antifoams although the concentration used in any one system is normally only a few parts per million. They are useful in many chemical and food production operations and in sewage disposal.

The use of small amounts of the material in paints and surface coatings is claimed to help in eliminating faults such as 'silking' in dipping applications and 'orange peel' in stoved finishes.

Interesting graft polymers based on silicone polymers are finding use in the manufacture of polyurethane foams particularly of the polyether type (see Chapter 23) because of their value as cell structure modifiers.

The columns in vapour phase chromotographic apparatus use incorporate high molecular weight dimethyl silicone fluids as the stationary phase.

The fluids have also found a number of uses in medicine. Barrier creams based on silicone fluids have been found to be particularly useful against the cutting oils in metal machinery processes which are common industrial irritants.   The serious and often fatal frothy bloat suffered by ruminants can now often be countered by the use of small quantities of silicone fluid acting as an antifoam.

25.5 SILICONE RESINS

25.5.1   Preparation

On the commercial scale silicone resins are prepared batchwise by hydrolysis of a blend of chlorsilanes.   In order that the final product shall be cross linked, a quantity of trichlorsilanes must be incorporated into the blend. A measure of the functionality of the blend is given by the R/Si ratio (see Section 25.3).   Whereas a linear polymer will have an R/Si ratio of just over 2:1 the ratio when using trichlorsilane alone will be 1:1.   Since these latter materials are brittle, ratios in the range 1·2:1 to 1·6:1 are used in commercial practice.   Since phenylchlorsilanes are also often used the CH_3/C_6H_5 ratio is a further convenient parameter of use in classifying the resins.

The chlorsilanes are dissolved in a suitable solvent system and then blended with the water which may contain additives to control the reaction. In the case of methyl silicone resin the overall reaction is highly exothermic and care must be taken to avoid overheating which can lead to gelation. When substantial quantities of phenylchlorsilanes are present however it is often necessary to raise the temperature to 70–75°C to effect a satisfactory degree of hydrolysis.

At the end of the reaction the polymer–solvent layer is separated from the aqueous acid layer and neutralised.   A portion of the solvent is then distilled off until the correct solids content is reached.

The resin at this stage consists of a mixture of cyclic, linear, branched and cross-linked polymers rich in hydroxyl end-groups, but of a low average molecular weight.   This is increased somewhat through 'bodying' the solution by heating with a catalyst such as zinc octoate at 100°C until the viscosity, a measure of molecular weight at constant solids content, reaches the desired value.

The resins are then cooled and stored in containers which do not catalyse further condensation of the resins.

The cross linking of the resin is, of course, not carried out until it is *in situ* in the finished product.   This will take place by heating the resin at elevated temperatures with a catalyst several of which are described in the literature, e.g. triethanolamine and metal octoates.   The selection of the type and amount of resin has a critical influence on the rate of cure and on the properties of the finished resin.

25.5.2   Properties

The general properties of the resins are much as to be expected.   They have very good heat resistance but are mechanically much weaker than the corresponding organic cross-linked materials. This weakness may be ascribed to the tendency of the polymers to form ring structures with consequent low cross-linking efficiency and also to the low intermolecular forces.

High phenyl content resins are compatible with organic resins of the P–F, U–F, M–F, epoxy–ester and oil-modified alkyd types but are not compatible with non-modified alkyds. Silicone resins are highly water-repellent.

The resins are good electrical insulators particularly at elevated temperatures and under damp conditions. This aspect is discussed more fully in the next section.

25.5.3 Applications

Laminates

Methyl–phenyl silicone resins are used in the manufacture of heat resistant glass-cloth laminates particularly for electrical applications. The glass cloth is first cleaned of size either by washing with hot trichloroethylene followed by a hot detergent solution or alternatively by heat cleaning. The cloth is then dipped into a solution of the resin in an aromatic solvent, the solvent is evaporated and the resin is partially cured by a short heating period so that the resin no longer remains tacky. Resin pick-up is usually in the order of 35–45% for high pressure laminates and 25–35% for low pressure laminates.

The pieces of cloth are then plied up and moulded at about 170°C for 30–60 minutes. Whilst flat sheets are moulded in a press at about 1,000 lb in^{-3} pressure, complex shapes may be moulded by rubber bag or similar techniques at much lower pressures ($\sim$ 15 lb in^{-2}) if the correct choice of resin is made. A number of curing catalysts have been used including triethanolamine, zinc octoate and dibutyl tin diacetate. The laminates are then given a further prolonged curing period in order to develop the most desirable properties.

The properties of the laminate are dependent on the resin and type of glass cloth used, the method of arranging the plies, the resin content and the curing schedule. Fig. 25.4 shows how the flexural strength may be affected by the nature of the resin and by the resin content.

A number of different resins are available and the ultimate choice will depend on the end-use and proposed method of fabrication. For example one resin will be recommended for maximum strength and fastest cures whilst another will have the best electrical properties. Some may be suitable for low pressure laminating whilst others will require moulding pressure of 1,000 lb in^{-2}.

Of particular importance are the electrical properties of the laminates. These are generally superior to P–F and M–F glass cloth laminates as may be seen from Table 25.3.[13]

The dielectric constant is normally in the range 3·6–4·4 and decreases with an increase in resin content.

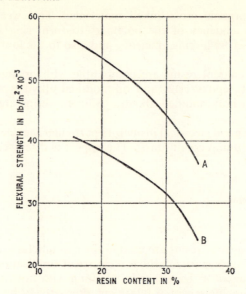

Fig. 25.4. *Influence of resin content on the flexural strength of glass cloth laminates made with two silicone resins A and B. (After Gale[14])*

The dielectric properties are reasonably constant over a fair range of temperature and frequency.

The power factor of typical glass cloth laminates decreases with aging at about 250°C which is the main reason for post curing (Fig. 25.5). A power factor drift is however observed[13] under wet conditions and the ratio of power factors between wet and dry conditions is about 3:1.

The mechanical properties of the laminates are somewhat poorer than observed with phenolic and melamine laminates. Tensile and flexural strength figures are typically about 20% less than for the corresponding P–F and M–F materials and about 60% of values for epoxy laminates.

Silicone–asbestos laminates are inferior mechanically to the glass reinforced laminates and have not found wide commercial use. Interesting laminates have however been introduced based on mica paper and it is expected that their use will increase during the next few years.

Silicone laminates are used principally in electrical applications such as slot wedges in electric motors, particularly class H motors, terminal

Table 25.3   TYPICAL PROPERTIES OF GLASS FIBRE LAMINATES[13]

Property	Unit	Test method	P–F	M–F	Silicone
Power factor (1 Mc/s)	—	B.S.1137	0·06	0·08	0·0002
Electric strength	V/0·001 in.	B.S.1137	150–200	150–200	250–300
Insulation resistance (dry)	ohm	B.S.1137	10,000	20,000	500,000
Insulation resistance (after water immersion)	ohm	B.S.1137	10	10	10,000

boards, printed circuit boards and transformer formers. There is also some application in aircraft including use in firewalls and ducts.

Moulding compositions

Compression moulding powders based on silicone resins have been available on a small scale from manufacturers for a number of years. They consist of mixtures of a heat resistant fibrous filler (e.g. glass fibre or asbestos) with a resin and catalyst. Non-fibrous inorganic fillers may also be included. They may be moulded, typically, at temperatures of about

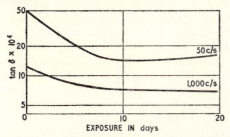

Fig. 25.5. *Effect of aging at 250°C on the power factor of silicone-bonded, glass-cloth laminates.* (*After Newland*[13])

160°C for 5–20 minutes using pressures of $\frac{1}{2}$–2 ton in^{-2}. Post-curing is necessary for several hours in order to develop the best properties. Materials currently available suffer from a short storage life of the order of 3–6 months but have been used in the moulding of brush rings holders, switch parts and other electrical applications that need to withstand high temperatures. They are extremely expensive and are of even greater volume cost than p.t.f.e.

Some typical properties of a cured silicone moulding composition are given in Table 25.4.

Table 25.4 SOME TYPICAL PROPERTIES OF A SILICONE MOULDING COMPOUND

	Units	*ASTM test*	*Value*
Specific gravity (250°C)	—	—	1·65
Flexural strength			
23°C	lb in^{-2}	D.790	14,000
200°C	lb in^{-2}		5,000
Flexural modulus			
23°C	lb in^{-2}	D.790	$1·8 \times 10^6$
200°C	lb in^{-2}		$0·9 \times 10^6$
Tensile strength			
23°C	lb in^{-2}	D.651	4,400
200°C	lb in^{-2}	D.652	1,300
Dielectric constant			
10^3–10^6 c/s		D.150	3·6
Power factor			
10^3–10^6 c/s		D.150	~0·005

Miscellaneous applications

Like the fluids the silicone resins form useful release agents and although more expensive initially are more durable.   The resin is applied in solution form and the coated surface is then dried and the resin cured by heating for about 2 hours at 200–230°C.   The bakery industry has found a particular use for these materials in aiding the release of bread from baking pans.

Resins, usually in a partially condensed form are used to provide a water-repellent treatment for brickwork and masonry.   Methyl–phenyl silicone resins are used as coatings for electrical equipment and in the impregnation of class H electrical equipment.   Dimethyl silicone fluids are also used as water repellent coatings for class A or class B insulation.

The heat resistance and water resistance of the resins are attractive properties for surface coatings but the poor scratch resistance of the materials has limited applications of straight silicone resins.

Blends with alkyd or other organic resins have however been prepared and these show heat resistance intermediate between those of the organic resins and the silicones.   Of particular interest is the use of silicone-organic resin blends filled with aluminium powder for the coating of metal chimneys and furnace doors.   At the operating temperatures the resins are destroyed leaving a layer of aluminium film.

25.6 SILICONE RUBBERS

In spite of their high cost silicone rubbers have over the last twenty years established themselves in a variety of applications where heat resistance and retention of properties over a wide range of temperatures are required.

25.6.1   Preparation of dimethyl silicone rubbers

The elastomers consist of a very high molecular weight ($\sim 0{\cdot}5 \times 10^6$) linear gums cross linked after fabrication.   In order to achieve such polymers it is necessary that very pure difunctional monomers be employed since the presence of monofunctional material will limit the molecular weight while trifunctional material will lead to cross linking.   Where dimethyl silicone rubbers are being prepared the cyclic tetramer, octamethyl cyclotetrasiloxane, which may be obtained free from mono- and tri-functional impurities, is used.   This tetramer occurs to the extent of about 25% during the hydrolysis of dichlorsilanes into polymers.

To obtain high molecular weight polymers the tetramer is equilibrated with a trace of alkaline catalyst for several hours at 150–200°C.   The product is a viscous gum with no elastic properties.   The molecular weight is controlled by careful addition of monofunctional material.

25.6.2   Special purpose rubbers

Dimethyl silicone rubbers show a high compression set which can be reduced to some extent by additives such as mercurous oxide and cadmium

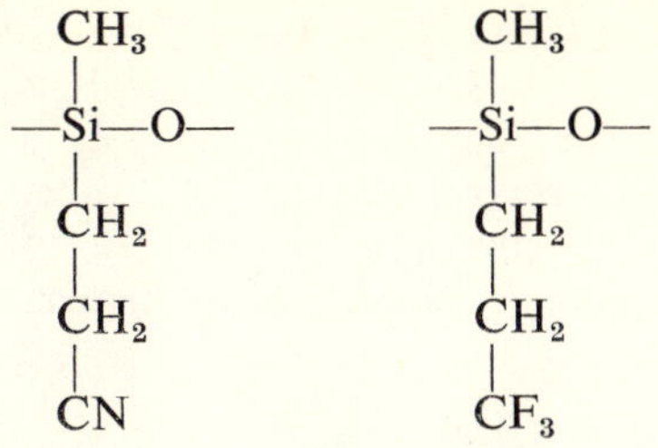

$$
\begin{array}{cc}
\text{CH}_3 & \text{CH}_3 \\
| & | \\
-\text{Si}-\text{O}- & -\text{Si}-\text{O}- \\
| & | \\
\text{CH}_2 & \text{CH}_2 \\
| & | \\
\text{CH}_2 & \text{CH}_2 \\
| & | \\
\text{CN} & \text{CF}_3 \\
\text{(I)} & \text{(II)}
\end{array}
$$

Fig. 25.6

oxide. These materials are undesirable however because of their toxicity. Substantially reduced compression set values may be obtained by using a polymer containing small amounts of methyl vinyl siloxane ($\sim 0 \cdot 1\%$). These materials may be vulcanised with less reactive peroxides than usual and may also be reinforced with carbon black if desired.

Oil resistant rubbers containing nitrile (Fig. 25.6 (I)) or fluorinated alkyl (II) groupings have become available.

Possessing both oil and temperature resistance these materials are competitive with the fluoro rubbers for aircraft fuel seals and gaskets.

Improved low temperature resistance is obtained by replacing 5–15% of the methyl groups in a dimethyl silicone rubber with phenyl groups. In a typical example the brittle point of cured rubbers will be reduced from $-50°C$ to $-100°C$. Similar improvements may also be achieved by the incorporation of cyanoalkyl groups such as in Fig. 25.6 (I).

Linear polymers containing silanol end groups may be cured at room temperature with a silicone ester in the presence of a catalyst such as dibutyl tin dilaurate (Fig. 25.7). These materials have little mechanical strength but retain the good electrical and thermal characteristics of the conventional rubbers. They are used for encapsulating and caulking.

Non-tacky self-adhesive rubbers (fusible rubbers) are obtained if small traces of boron are incorporated into the polymer chain (~ 1 B atom per 300 Si atoms). They may be obtained by condensing dialkylpolysiloxanes end blocked with silanol groups with boric acid or alternatively reacting ethoxyl end-blocked polymers with boron triacetate.

'Bouncing putty' is somewhat similar in that the Si—O—B bond occurs occasionally along the chain. This material flows on storage and in extension appears quite plastic in its behaviour. Small pieces dropped on a hard surface however show a high elastic rebound whilst on sudden striking they shatter. Some use is made of this material in electrical equipment.

25.6.3 Compounding

Before fabrication it is necessary to compound the gum with fillers, vulcanising agent and other special additives on a 2-roll mill or in an internal mixer.

Incorporation of fine fillers is necessary if the vulcanisates are to have any strength. Unfilled polymers have negligible strength whereas reinforced silicone rubbers may have strengths up to 2,000 lb in^{-2}. Since carbon blacks do not give outstanding reinforcement, adversely affect electrical insulation properties and may interfere with the curing action, fine silica fillers are generally used. These materials have particle sizes

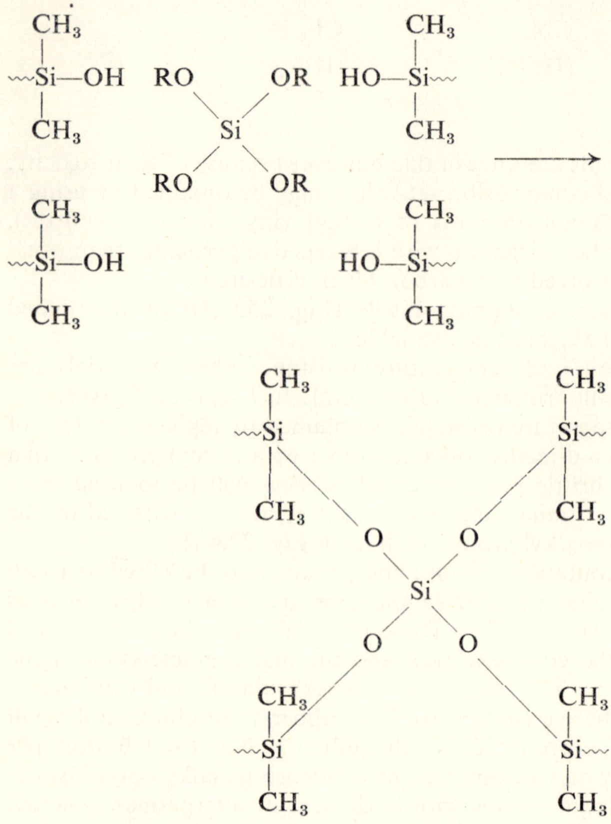

Fig. 25.7

in the range 30–300 Å and are prepared by combustion of silicon tetra-chloride (fume silicas), by precipitation or as an aerogel. Dilution of the rubber by filler will also reduce cost, shrinkage on cure, thermal expansion and may aid processing. It is to be noted that different silica fillers may lead to large differences in processing behaviour, cure rates and in the properties of the finished product.

Silicone rubbers are normally cured with peroxide. Benzoyl peroxide, 2,4-dichlorbenzoyl peroxide and t-butyl perbenzoate being used for the dimethyl silicones in quantities of 0·5–3%. These materials are stable in the compounds for several months at room temperature but will start to cure at about 70°C.

Polymers containing vinyl side groups may be cured with less reactive peroxides such as dicumyl peroxide and t-butyl peracetate.

Heat aging characteristics may be improved by the addition of a few per cent of ferric oxide and barium zirconate to name but two materials mentioned in the literature.

25.6.4 Fabrication and cross linking

Compounded rubbers may be fabricated by the normal techniques employed in rubber technology, e.g. extrusion, calendering and compression moulding. In order to develop rubbery properties it is necessary to cross link (vulcanise) the compound after shaping. With moulded articles this may be accomplished by heating in a press for 5–25 minutes at a range of temperatures from 115–175°C according to the composition of the compound. Calendered and extruded materials are generally cured in a hot air or steam pan. Prolonged post curing at temperatures up to 250°C may be necessary in order to achieve the best mechanical and electrical properties.

Indications are that cross linking of dimethyl silicone rubbers occurs by the sequence of reaction shown in Fig. 25.8.

The peroxide decomposes at elevated temperatures to give free radicals which then abstract a hydrogen atom from the methyl group. The

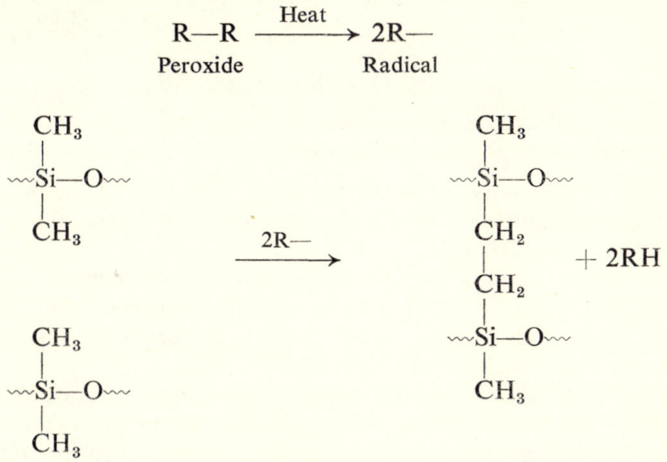

$$R—R \xrightarrow{\text{Heat}} 2R—$$

Peroxide   Radical

Fig. 25.8

radicals formed then combine to form a hydrocarbon linkage. Results obtained by reacting model systems with benzoyl peroxide and analysing the reaction products are consistent with this type of mechanism.[15]

Rubbers containing traces of vinyl groups can be cross linked by weaker peroxide catalysts, the reaction involving a vinyl group. It is however unlikely that vinyl-to-vinyl linking occurs. Where there is a high vinyl content (4–5% molar) it is possible to vulcanise with sulphur.

25.6.5   Properties and applications

The important properties of the rubbers are their temperature stability, retention of elasticity at low temperatures and good electrical properties. They are much more expensive than the conventional rubbers (e.g. natural rubber and s.b.r.) and have inferior mechanical properties at room temperature.

The temperature range of general purpose material is approximately -50 to $+250°C$ but both ends of the range may be extended by the use of special purpose materials.   Whereas the general purpose silicone compounds have a tensile strength of 500–1,000 lb in^{-2} it is possible using fumed silicas to achieve values of up to 2,000 lb in^{-2}.   Similarly, whereas the normal cured compounds have a compression set of 20–50% after 24 hours at 150°C, values of as low as 6% may be obtained with the special rubbers.

Compared with organic rubbers the silicones have a very high air permeability, being 10–20 times as permeable as the organic rubbers.   The

Table 25.5   PHYSICAL PROPERTIES OF GENERAL PURPOSE SILICONE RUBBERS
(Values determined at 20°C after curing for 24 hours at 250°C)[10]

Tensile strength (lb in^{-2})	500–1,000
Elongation at break (%)	100–400
Hardness (B.S.°)	40–50
Compression set (% after 24 hr at 150°C)	20–50
Minimum useful temperature (°C)	-55
Maximum useful temperature (°C)	250
Linear shrinkage (%)	2–6
Thermal conductivity (c.g.s. units)	7×10^{-4}
Volume resistivity (ohm cm)	10^{16}
Dielectric strength (V/0·001 in. at 50% RH)	500
Power factor (60 c/s)	0·002
Dielectric constant (60 c/s)	3–6

thermal conductivity is also high, about twice that of the natural rubber. Some typical figures for the physical properties of a general purpose rubber are given in Table 25.5.

The applications of the rubbers stem from the thermal stability, good electrical insulation properties, non-stick properties and physiological inertness but they remain special purpose materials because of their high cost and poor mechanical properties.

It is stated[10] that modern passenger and military aircraft each use about 1,000 lb of silicone rubber.   This is to be found in gaskets and sealing rings for jet engines, ducting, sealing strips, vibration dampers and insulation equipment.

Silicone cable insulation is also used extensively in naval craft since the insulation is not destroyed in the event of a fire but forms a protective and insulating layer of silica.

The rubbers are also used for such diverse applications as blood transfusion tubing capable of sterilisation, antibiotic container closures, electric iron gaskets, domestic refrigerators and for non-adhesive rubber covered rollers for handling such materials as confectionery and adhesive tape. The cold curing rubbers are of value in potting and encapsulation.

25.7 INORGANIC AND OTHER TEMPERATURE RESISTING POLYMERS

In recent years there has been an intensive effort to find polymers which would be of use at temperatures above 250°C particularly in order to meet certain requirements of the aircraft and heavy electrical industries. Such polymers would normally need to possess as a minimum the properties of good oxidation resistance, hydrolytic stability and tractability (capability of being processed) in addition to the requirement of high heat resistance. Whilst a large number of temperature-resistant polymers have now been produced few meet all of the three other common requirements.

At the present time organic polymers offer the most promise. The carbon–carbon bond has a good thermal stability, particularly when it is part of an aromatic or other ring structure and its principal limitation is a lack of oxidation stability at elevated temperatures. The organic heat resistant polymers fall roughly into three groups:

1. Fluorine-containing polymers.
2. Polymers in which the backbone is largely composed of aromatic rings.
3. Polymers with nitrogen-containing ring structures.

Of the fluorine-containing polymers those containing only C—C and C—F bonds, such as p.t.f.e., are the most thermally stable. Both types of bond have a high bond energy whilst the fluorine atoms also appear to shield the C—C backbone. These polymers were discussed in Chapter 10.

The presence of benzene rings in the backbone or in the cross link of a polymer has a marked effect in improving the softening point and the heat resistance. This effect has already been noted with the polyesters, the polycarbonates, certain epoxides and the polypyromellitimides. In

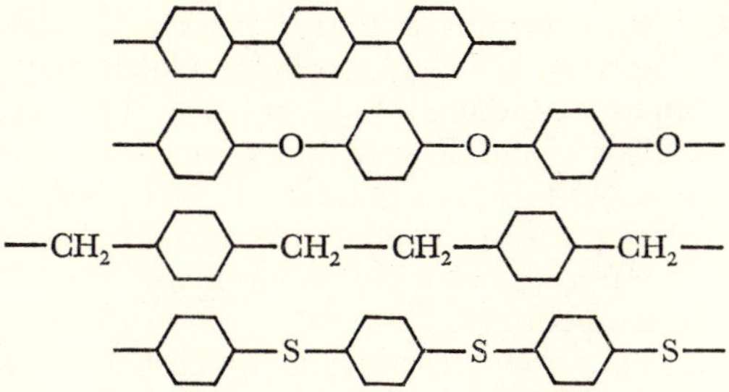

Polyphenylene

Poly(Phenylene Oxide)

Poly-*p*-Xylylene

Poly(Phenylene Sulphide)

Fig. 25.9

general the higher the aromatic content the higher the softening point and the greater the heat resistance. An extreme example is graphite, a layer–lattice carbon structure which has a melting point of 3,600°C. Of the linear polymers examined polyphenylene has a high thermal stability but is very brittle, insoluble and infusible (Fig. 25.9). A compromise may be effected by spacing the rings by flexible bonds and this had led to the investigation of poly(phenylene oxide), poly-*p*-xylylene, poly(phenylene sulphide) and a number of other similar materials.

The first of these materials to be introduced commercially was poly-(phenylene oxide) by General Electric in 1964.   Production in Europe by General Electric in conjunction with the Dutch A.K.U. company was also being planned at the same time.   Some properties of poly(phenylene oxide) are given in Table 25.5.

Table 25.5   SOME PROPERTIES OF POLY(PHENYLENE OXIDE)

Specific gravity	1·06
Tensile strength	10–11 × 10^3 lb in^{-2}
Tensile modulus	3·5–3·8 × 10^5 lb in^{-2}
Dielectric constant	2·58 (60–10^6 c/s
Dissipation factor	0·00035 (60 c/s)
	0·0009 (10^6 c/s)
Deflection temperature under load (heat distortion temp.)	190° (264 lb in^{-2})
Colour	opaque light amber

The polymer is resistant to hydrolysis, acids and alkalis.   It may be injection moulded and extruded on conventional equipment.   The main applications envisaged are for medical and surgical instruments, in domestic appliances, electronic equipment and in piping.

Poly-*p*-xylene was introduced under the name Parylene by Union Carbide in 1965, initially as a coating and film-forming material for use in electronics applications.   Using special techniques devised for the material, coatings varying in thickness from 100 Å to 0·012 in. may be produced.

Poly(*p*-phenylene sulphide), produced by solid state or solution polymerisation of *p*-halothiophenoxides, would appear most promising.[16] The polymer has a heat distortion temperature of 250°C and the crystalline structure melts at 290°C.   The polymer is stable in air up to 400°C.   At somewhat higher temperatures a friable glass is obtained which is stable up to 900°C.   Poly(*p*-phenylene sulphide), which may be moulded and processed by other common techniques, is a stiff, easily machined polymer.   By thermal or chemical cross linking an even better degree of heat stability can be achieved.   The polymer has a high degree of adhesion to glass, chrome plating and steel and it is expected to become of value as an adhesive, laminating resin, for wire insulation and for other electric parts subjected to high temperature.

A number of polymers containing ring structures in which C—N links are present have been found to have good heat resistance.   Of these the most interesting are the polypyromellitimides which have been dealt with earlier.   Also of interest are certain polymers reminiscent of the melamine-formaldehyde resins in that they are based on triazine rings.   A perfluoro derivative, Fig. 25.10, is said to be rubbery with a high thermal stability.

Controlled heating of polyacrylonitrile at 200°C can also lead to products of good heat resistance and which glow in a Bunsen flame (Fig. 25.11).

The well-known thermal stability of most minerals and glasses, many of which are themselves polymeric, has led to intensive research into synthetic

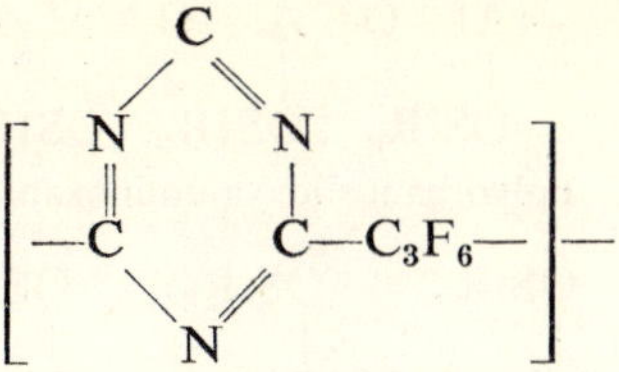

$$\left[\begin{array}{c} \overset{\displaystyle C}{\diagup \ \diagdown} \\ N \qquad N \\ \| \qquad \ \ \| \\ -C \qquad C-C_3F_6- \\ \diagdown \ \diagup \\ N \end{array} \right]-$$

Fig. 25.10

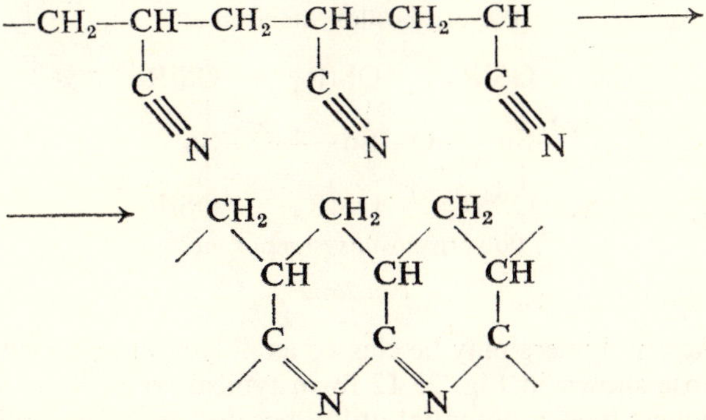

$$-CH_2-CH-CH_2-CH-CH_2-CH \longrightarrow$$
$$\begin{array}{ccc} C & C & C \\ \| & \| & \| \\ N & N & N \end{array}$$

$$\longrightarrow \quad \begin{array}{ccc} CH_2 & CH_2 & CH_2 \\ \diagup \ \diagdown & \diagup \ \diagdown & \diagup \ \diagdown \\ CH & CH & CH \\ | & | & | \\ C & C & C \\ \diagup \ \diagdown & \diagup \ \diagdown & \diagup \ \diagdown \\ & N & N \end{array}$$

Fig. 25.11

inorganic and semi-inorganic polymers. These materials can be classified into the following groups:

1. Polymers containing main-chain silicon atoms.
2. Polymetallosiloxanes.
3. Polymetalloxanes.
4. Phosphorus-containing polymers.
5. Boron-containing polymers.
6. Sulphur-containing polymers.
7. Miscellaneous polymers.

In addition to the silicates and silicones, a number of polymers which contain silicon atoms in the main chain have been studied in recent years. These include the silicon-nitrogen structure (a) and silicon sulphides (b).

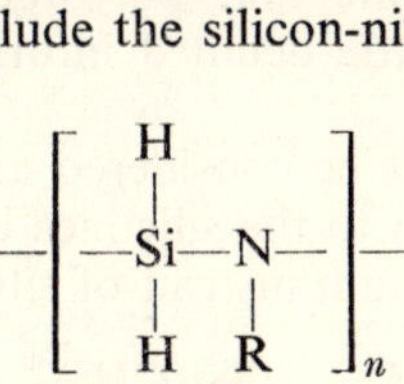

$$\left[\begin{array}{c} H \\ | \\ -Si-N- \\ | \quad | \\ H \quad R \end{array} \right]_n \qquad\qquad \begin{array}{ccccc} S & & S & & \\ \diagup \ \diagdown & \diagup \ \diagdown & \diagup \ \diagdown & \diagup \\ Si & Si & Si & \\ \diagup \ \diagdown & \diagup \ \diagdown & \diagdown \\ & S & & S & \end{array}$$

(a) 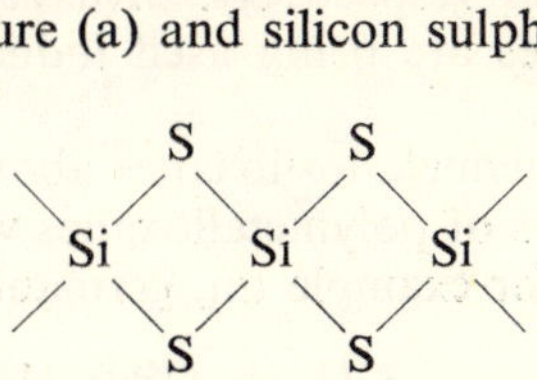   (b)

Unfortunately neither of these materials show hydrolytic stability.

Of greater interest and potential value are the polymetallosiloxanes which have been investigated by Andrianov and co-workers in the Soviet

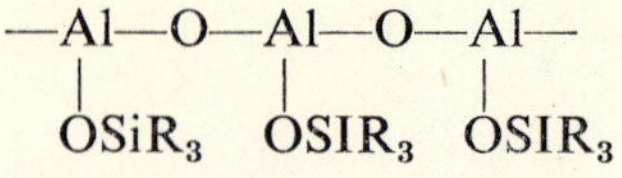

$$\begin{array}{ccc}
-\mathrm{Al}-\mathrm{O}-\mathrm{Al}-\mathrm{O}-\mathrm{Al}- \\
| & | & | \\
\mathrm{OSiR_3} & \mathrm{OSIR_3} & \mathrm{OSIR_3}
\end{array}$$

Polyorganosiloxyaluminoxanes

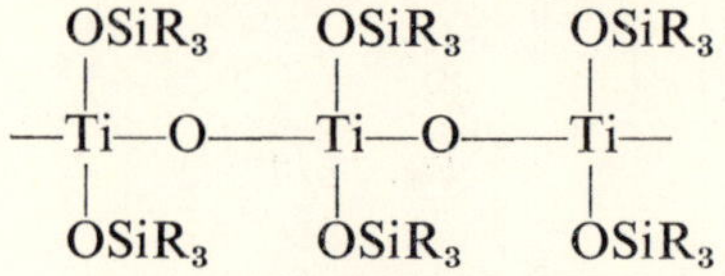

$$\begin{array}{ccc}
\mathrm{OSiR_3} & \mathrm{OSiR_3} & \mathrm{OSiR_3} \\
| & | & | \\
-\mathrm{Ti}-\mathrm{O}-\!\!\!-\mathrm{Ti}-\mathrm{O}-\!\!\!-\mathrm{Ti}- \\
| & | & | \\
\mathrm{OSiR_3} & \mathrm{OSiR_3} & \mathrm{OSiR_3}
\end{array}$$

Polyorganosiloxytitanoxanes

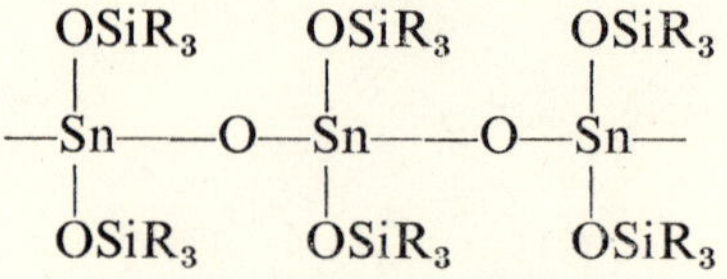

$$\begin{array}{ccc}
\mathrm{OSiR_3} & \mathrm{OSiR_3} & \mathrm{OSiR_3} \\
| & | & | \\
-\mathrm{Sn}-\!\!\!-\mathrm{O}-\mathrm{Sn}-\!\!\!-\mathrm{O}-\mathrm{Sn}- \\
| & | & | \\
\mathrm{OSiR_3} & \mathrm{OSiR_3} & \mathrm{OSiR_3}
\end{array}$$

Polyorganosiloxystannoxanes

Fig. 25.12

Union. These polymers may be classed as polyorganosiloxymetalloxanes of which those shown in Fig. 25.12 form typical types.

Such polymers have a potential utility because experience indicates that thermal stability is often increased by increasing the polarity of the bond. Therefore where the metal is more electropositive than silicon the metal–oxygen bond will be more polar than the silicon–oxygen bond and greater thermal stability may be expected. By surrounding the inorganic skeleton of the molecules Andrianov has produced thermally stable materials which are soluble in organic solvents and may therefore be cast on films, lacquers or used in laminating resins.

American workers have produced a different class of polymetal-losiloxanes in which the metal atom and the silicon atom are both incorporated into the main chain (e.g. Fig. 25.13).

The metallosiloxanes are liable to hydrolysis but the rate of hydrolysis is very dependent on the metal used in the polymer. Examination of three low molecular weight metallosiloxanes indicate the relative rates of hydrolysis to be $2220:27\cdot2:1$ for the tin, aluminium and titanium derivatives respectively. Russian workers state that polyorganometal-losiloxanes are being used industrially but little detailed information is available.

The polymetallosiloxanes above may in fact be considered as variants of a series of polymetalloxanes which are akin to the silicones but which contain for example tin, germanium and titanium instead of silicon. Of

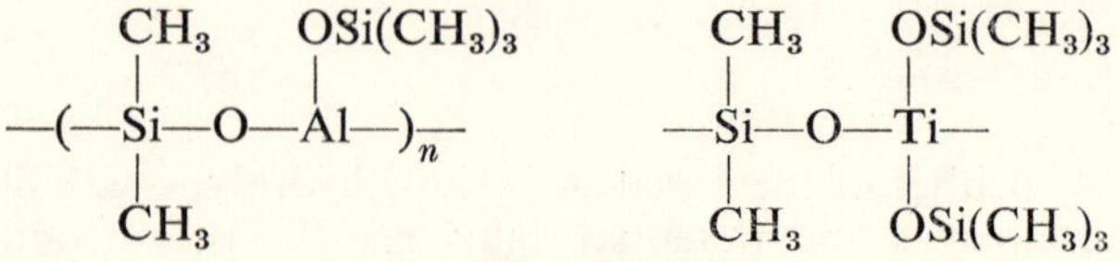

$$\begin{array}{cc}
\mathrm{CH_3} & \mathrm{OSi(CH_3)_3} \\
| & | \\
-(\!-\mathrm{Si}-\mathrm{O}-\mathrm{Al}-)_{\overline{n}} \\
| \\
\mathrm{CH_3}
\end{array}
\qquad
\begin{array}{cc}
\mathrm{CH_3} & \mathrm{OSi(CH_3)_3} \\
| & | \\
-\mathrm{Si}-\mathrm{O}-\mathrm{Ti}- \\
| & | \\
\mathrm{CH_3} & \mathrm{OSi(CH_3)_3}
\end{array}$$

Fig. 25.13

the polyorganostannoxanes, dibutyl tin oxide finds use as a stabiliser for p.v.c. and as a silicone cross-linking agent. Polyorganogermanoxanes have also been prepared (Fig. 25.14).

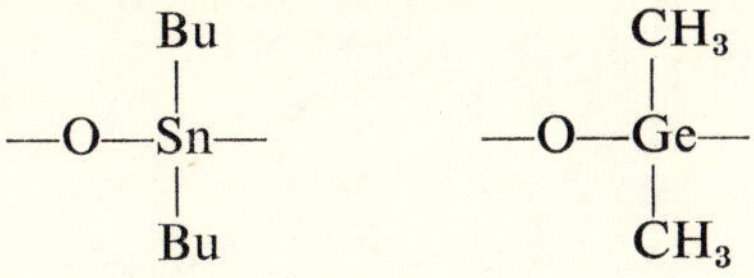

$$
\begin{array}{ccc}
& \mathrm{Bu} & & \mathrm{CH_3} \\
& | & & | \\
\mathrm{-O-Sn-} & & \mathrm{-O-Ge-} \\
& | & & | \\
& \mathrm{Bu} & & \mathrm{CH_3}
\end{array}
$$

Fig. 25.14

Although titanium does not form a stable bond with carbon, organo-titanium polymers have been formed such as polyalkoxy titanoxanes and polymeric titanate esters (Fig. 25.15).

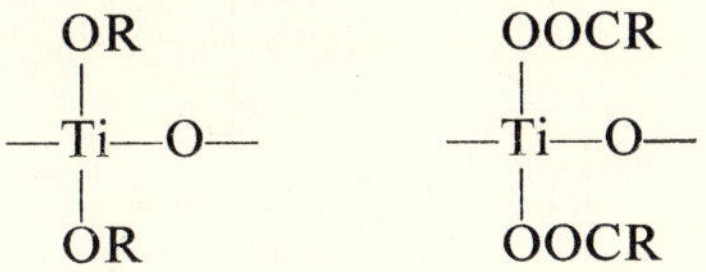

$$
\begin{array}{ccc}
& \mathrm{OR} & & \mathrm{OOCR} \\
& | & & | \\
\mathrm{-Ti-O-} & & \mathrm{-Ti-O-} \\
& | & & | \\
& \mathrm{OR} & & \mathrm{OOCR}
\end{array}
$$

Fig. 25.15

Butyl titanate polymers find use in surface coatings[17] but as a class titanium polymers lack hydrolytic stability.

Of the phosphorus-containing polymers the polyphosphates have been known for many years. Aluminium phosphate has been used in the manufacture of heat-resistant silica-fibre reinforced laminates.

If phosphorus pentachloride is reacted with ammonium chloride in an inert solvent such as sym-tetrachloroethane, polymers will be formed by the following equation

$$n\ \mathrm{PCl_5} + n\ \mathrm{NH_4Cl} \longrightarrow (\mathrm{PNCl_2})_n + 4n\ \mathrm{HCl}$$

If an excess of pentachloride is employed linear polymers of the type (I) will be produced but if the pentachloride is added slowly to a suspension of ammonium chloride then cyclic polymers such as the trimer (II) and tetramer (III) will be produced.

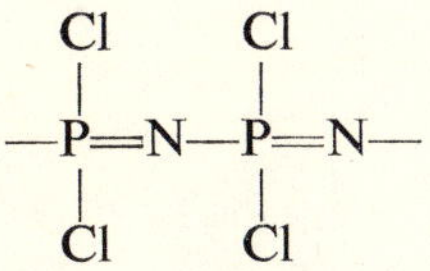

$$
\begin{array}{ccc}
\begin{array}{cc}
\mathrm{Cl} & \mathrm{Cl} \\
| & | \\
\mathrm{-P{=}N-P{=}N-} \\
| & | \\
\mathrm{Cl} & \mathrm{Cl}
\end{array}
&
\begin{array}{c}
\mathrm{PCl_2} \\
\diagup\ \diagdown \\
\mathrm{N}\qquad\mathrm{N} \\
\mathrm{Cl_2{-}P}\qquad\mathrm{P{-}Cl_2} \\
\diagdown\ \diagup \\
\mathrm{N}
\end{array}
&
\begin{array}{c}
\mathrm{PCl_2{=}N{-}PCl_2} \\
|\qquad\quad || \\
\mathrm{N}\qquad\quad\mathrm{N} \\
||\qquad\quad | \\
\mathrm{PCl_2{-}N{=}PCl_2}
\end{array} \\
\text{(I)} & \text{(II)} & \text{(III)}
\end{array}
$$

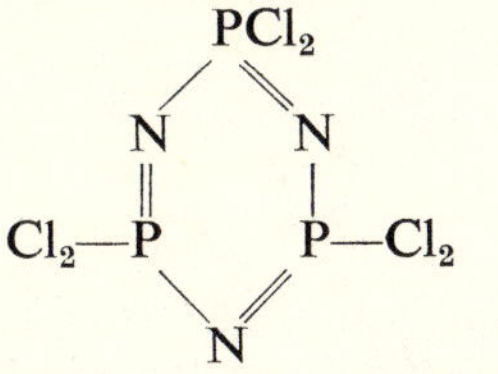

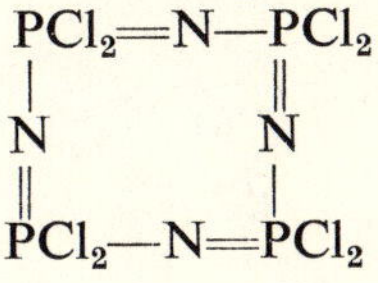

Heating of the cyclic polymer at 250°C will also lead to the production of the linear polymer which is rubbery and stable to 350°C. On standing, however, the material hydrolyses and after a few days loses its elastic properties and becomes hard and covered with drops of hydrochloric acid solution.

In attempts to improve the hydrolytic stability of these materials the active chlorine atom has been replaced by other groups such as F, CF_3, NH_2, CH_3 and phenyl to give a range of polymers of general form PNX_2.

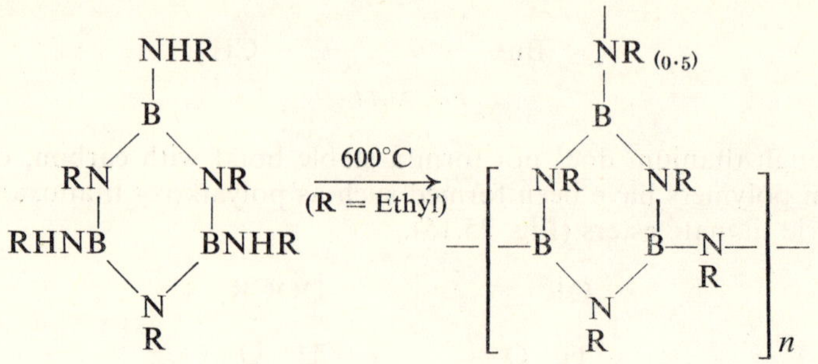

Fig. 25.16

One heat resistant resin based on phosphonitrilic polymers has been marketed (Inorganic Resin 251—Albright and Wilson). The uncured resin is somewhat soluble in water and aqueous solutions may be used for impregnating asbestos fabric for subsequent lamination. Glass fibre, silica fibre and mineral wool are unsuitable as they are degraded during cure. Lamination is carried out at 300°C for 30 minutes under pressure of

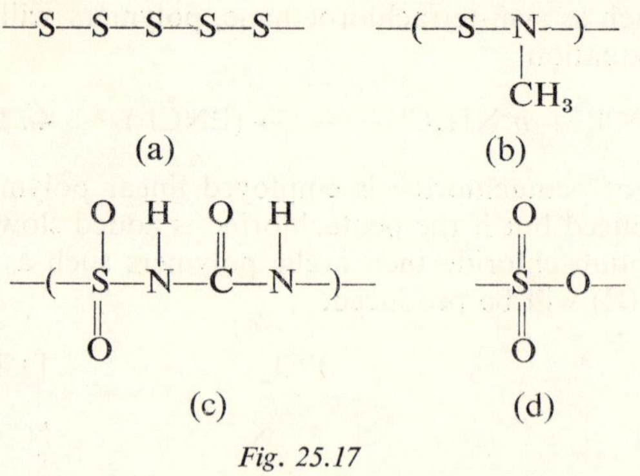

Fig. 25.17

about 2 ton in⁻². The cured laminates have flexural strengths of 13,000–15,000 lb in⁻² and retain 45% of their strength after 500 hours at 300°C.

Other phosphorus-based polymers have been investigated and have proved to be of some interest. An example is phosphorus oxynitride which forms a glass above 1000°C.

A large number of polymers containing boron in the main chain have been prepared but most of them have either been of low molecular weight or intractable cross-linked structures. Some interest has been shown in the tri-β-amino borazoles which polymerise on heating (Fig. 26.16).[18]

These materials are intractable and easily hydrolysed in hot water but are stable to heating to 600°C.

Hydrolytic instability or, alternatively, a tendency to revert to simple forms is shown by a number of sulphur-containing polymers. Some examples are shown in Fig. 25.17: (a) 'plastic sulphur', (b) a polymer unstable at room temperature, (c) and (d) the β form of sulphur trioxide.

In some respects the results of research to date on inorganic polymers have been disappointing. A high thermal stability has often been accompanied by hydrolytic instability whilst, in other cases, the polymers have been brittle or intractable. On the credit side some products have found limited use as laminating resins, surface coating resins and wire enamels, and a background of knowledge is being built up of the factors which influence thermal, oxidative and hydrolytic stability. One object of current research is to produce rubbers and tough flexible polymers for use at temperatures well above 250°C. At the present time it seems that a great deal more research work will need to be carried out if inorganic polymers are to be of value in these fields.

REFERENCES

1. BUFF, H., and WOHLER, F., *Liebigs Ann.*, **104,** 94 (1857)
2. FRIEDEL, C., and CRAFTS, J. M., *Compt Rend.*, **56,** 592 (1863)
3. LADENBURG, A., *Liebigs Ann.*, **164,** 300 (1872)
4. ROCHOW, E. G., *J. Am. Chem. Soc.*, **67,** 963 (1945)
5. ROCHOW, E. G., *An Introduction to the Chemistry of the Silicones*, 2nd Edn, Chapman and Hall Ltd., London (1951)
6. ROTH, J., *J. Am. Chem. Soc.*, **69,** 474 (1947)
7. SCOTT, D. W., et al. *J. Phys. Chem.*, **65,** 1320 (1961)
8. GUTOFF, R., *Ind. Eng. Chem.*, **49,** 1807 (1957)
9. HARDY, D. V. N., and MEGSON, N. J. L., *Quart. Rev.* (London), **2,** 25 (1948)
10. FREEMAN, G. G., *Silicones*, Iliffe (1962) (also published as Plastics Institute Monograph No. C16)
11. BARRY, A. J., *J. Appl. Phys.*, **17,** 1020 (1946)
12. WILCOCK, D. J., *J. Am. Chem. Soc.*, **68,** 691 (1946)
13. NEWLAND, J. J., *Trans. Plastics Inst.*, **25,** 311 (1957)
14. GALE, P. A. J., *Trans. Plastics Inst.*, **28,** 194 (1960)
15. NITZSCHE, S., and WICK, M., *Kunstoffe Plastics*, **47,** 431 (1957)
16. SMITH, H. A., *Rubber Plastics Age*, **44,** 1048 (1963)
17. BRADLEY, D. C., *Metal–Organic Compounds, Advances in Chemistry Series* Vol. 23, p. 10, American Chemical Society, Washington (1959)
18. AUBREY, D. W., and LAPPERT, M. F., *J. Chem. Soc.*, 2927 (1959)

BIBLIOGRAPHY

FORDHAM, G. (Ed), *Silicones*, Newnes, London (1961)
FREEMAN, G. G., *Silicones*, Iliffe (1962) (also published as Plastics Institute Monograph No. C16)
GIMBLETT, F. G. R., *Inorganic Polymer Chemistry*, Butterworths, London (1963)
HUNTER, D. N., *Inorganic Polymers*, Blackwell, Oxford (1963)
Inorganic Polymers, Special Publication No. 15 of the Chemical Society, London (1961)
LAPPERT, M. F., and LEIGH, G. J. (Eds), *Developments in Inorganic Polymer Chemistry*, Elsevier, Amsterdam (1962)
MCGREGOR, R. R., *Silicones and their Uses*, McGraw Hill, London (1954)
ROCHOW, F. G., *An Introduction to the Chemistry of the Silicones*, 2nd Edn, Chapman and Hall Ltd., London (1951)
STONE, F. G. A., and GRAHAM, W. A. G. (Eds), *Inorganic Polymers*, Academic Press, London, New York (1962)

26

Miscellaneous Plastics Materials

26.1 INTRODUCTION

The materials dealt with in this chapter may be classed chemically into two groups

1. Plastics derived from natural polymers.
2. Non-polymeric plastics whose resinous behaviour stems from the colloidal complexity of a mixture of heterogeneous molecules.

With the exception of some of the natural rubber derivatives these materials were available during the first decade of this century and together with celluloid actually completed the range of plastics materials then in commercial use. In spite of being ousted from important markets they have continued to find use in specialised applications, details of which will be given in subsequent sections of this chapter. The historical significance of these materials was dealt with in the first chapter of this book.

26.2 CASEIN

Casein is a protein found in a number of animal and vegetable materials but only one source is of commercial interest, cow's skimmed milk. The amount of casein in milk will vary but a typical analysis of cow's milk is:

Water	87%
Fat	3·5–4%
Lactose	5%
Casein	3%
Globulin + albumin	0·5%
Other ingredients	0·5–1%

The butter fat is a coarse dispersion readily removable on standing or by a centrifuging operation. The casein will be present in the skimmed milk

540

as colloidally dispersed micelles of diameter of the order of 10^{-5} cm, and is associated with calcium and phosphate ions.

Plastics materials may be produced from casein by plasticising with water, extrusion and then cross linking with formaldehyde (formolisation). The resultant products have a pleasant horn-like texture and are useful for decorative purposes. The amount of casein produced has decreased since the Second World War but is still one of the preferred materials for use in the button industry.

26.2.1 Chemical nature

Casein is one member of the important group of natural polymers, the proteins. These materials bear a formal resemblance to the polyamides in that they contain repeating —CONH— groups and could be formally considered as polymers of amino-acids. However unlike polymers such as nylon 6 and nylon 11 a number of different α-amino acids are found in

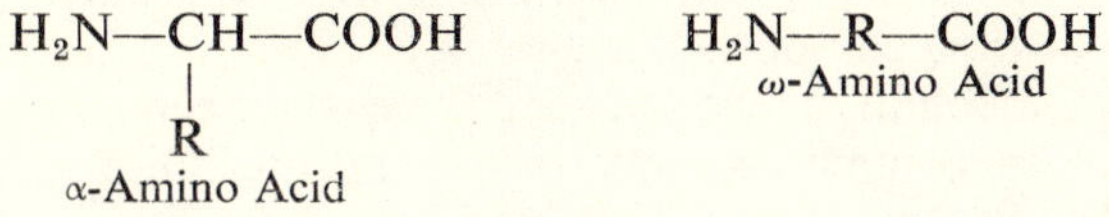

$$\begin{array}{cc} \text{H}_2\text{N—CH—COOH} & \text{H}_2\text{N—R—COOH} \\ | & \omega\text{-Amino Acid} \\ \text{R} & \\ \alpha\text{-Amino Acid} & \end{array}$$

Fig. 26.1

each molecule whereas the nylons 6 and 11 have only one ω-amino acid molecule (Fig. 26.1).

Over 30 amino-acids have been identified in the hydrolysis product of casein of which glutamic acid, hydroxy glutamic acid, proline, valine, leucine and lysine comprise about 60%. The residues of the amino-acid arginine also appear to be of importance in the cross linking of casein with formaldehyde.

The protein polymers are highly stereospecific, the amino-acid residues always adopting the L-configuration, i.e. the same configuration as the reference substance L-malic acid.

It is interesting to note that the amino-acid side chains may be either neutral as in valine, acidic as in glutamic acid or basic as in lysine. The

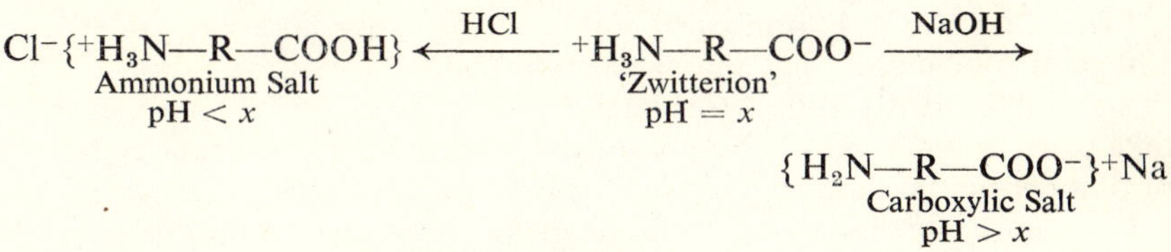

$$\text{Cl}^-\{^+\text{H}_3\text{N—R—COOH}\} \xleftarrow{\text{HCl}} {}^+\text{H}_3\text{N—R—COO}^- \xrightarrow{\text{NaOH}}$$

$$\begin{array}{ccc} \text{Ammonium Salt} & \text{'Zwitterion'} & \\ \text{pH} < x & \text{pH} = x & \end{array}$$

$$\{\text{H}_2\text{N—R—COO}^-\}^+\text{Na}$$
$$\text{Carboxylic Salt}$$
$$\text{pH} > x$$

Fig. 26.2

presence of both acidic and basic side chains leads to proteins such as casein acting as amphoteric electrolytes and their physical behaviour will depend on the pH of the environment in which the molecules exist. This is indicated by Fig. 26.2 showing a simplified protein molecule with just one acidic and one basic side group.

At the pH $= x$ there is a balance of charge and there is no migration in an electric field.   This is referred to as the isoelectric point and is determined by the relative dissociation constants of the acidic and basic side groups and does not necessarily correspond to neutrality on the pH scale. The isoelectric point for casein is about pH $= 4\cdot6$ and at this point colloidal stability is at a minimum.   This fact is utilised in the acid coagulation techniques for separating casein from skimmed milk.

Casein may be considered to be a conjugated protein, that is the protein is associated in nature with certain non-protein matter known as prosthetic groups.   In the case of casein the prosthetic group is phosphoric acid. The protein molecule is also associated in some way with calcium.   The presence of these inorganic materials has an important bearing on the processability and subsequent use of casein polymers.

26.2.2   Isolation of casein from milk

Destruction of the casein micelles in the milk with subsequent precipitation of the casein can be accomplished in a number of ways.   The action of heat or the action of alcohols, acids, salts and the enzyme rennet all bring about precipitation.   In commercial practice the two techniques used employ either acid coagulation or rennet coagulation mechanisms.

Addition of acetic or mineral acid to skimmed milk to reduce the pH value to 4.6, the isoelectric point, will cause the casein to precipitate.   As calcium salts have a buffer action on the pH, somewhat more than the theoretical amount of acid must be used.   Lactic acid produced in the process of milk 'souring' by fermentation of the lactoses present by the bacterium *Streptococcus lactis* will lead to a similar precipitation.

Although acid caseins are employed for a number of purposes, rennet caseins in which the protein remains associated with calcium and phosphate are preferred for plastics applications.   Rennet is the dried extract of rennin, obtained from the inner lining of the fourth stomach of calves and is a very powerful coagulant.   As little as $0\cdot2$ parts per million are said to be sufficient to coagulate slightly acidic milk.   Its coagulating power is destroyed at 100°C.

In the rennet coagulation process fresh skimmed milk is adjusted to a pH of 6 and about 40 ounces of a 10% solution of rennet are added per 100 gallons of milk.   The initial reaction temperature is about 35°C and this is subsequently raised to about 60°C.   The coagulation appears to take place in two stages.   Firstly the calcium caseinate is converted to the insoluble calcium paracaseinate and this then coagulates.

Great care is essential in controlling the temperature and the coagulation process otherwise impurities, particularly other proteins will be brought down with the casein.   Such impurities will adversely affect the transparency of the product.

The condition of the curd on precipitation is important.   As the milk starts to gel, agitators in the coagulation tanks are started as the temperature is raised to about 65°C.   Under these conditions the protein is thrown out in fine particles.   Too slow an agitation will produce large

Table 26.1 PROPERTIES OF RENNET CASEIN
Determined according B.S. 1416 : 1961

Grade	Premium	First	Standard
Moisture content %	11–13	11–13	11–13
Ash content %	7–8	7–8	7–8
Fat content %	0·5–0·7	0·7–1·0	1·0–1·5
pH	7·0–7·2	6·9–7·2	6·8–7·2
Gross contamination	0–50	50–100	>100
Fine contamination	0–50	50–100	>100
Colour	A	B	C
Wet heat resistance	A	B	C

clots difficult to wash whilst too fine a curd also presents washing problems. In order to obtain the requisite consistency of the precipitate it may be necessary to add inorganic material to the skimmed milk. For example the addition of phosphate ions will prevent undesirable flaky polymer. Similarly calcium-deficient casein will not coagulate satisfactorily and the addition of calcium ions may be necessary.

The coagulated particles are given four to five preliminary washings after precipitation and the washed curd is then passed through a curd press to extract the maximum amount of water.

The casein must be dried with considerable care. If dried too rapidly an impervious layer is formed on the outside of the particles which prevents the inside of the particles from drying out. Too slow a drying operation will lead to souring.

The drying of the curd is completed by passing through a rotary drier such as the Pillet–Bordeaux. In this machine the curd passes down a chute with hot air running counter to the direction of flow of material. Strict temperature control is essential, the maximum discharge temperature being below 65°. If such control is not made then the resin will be dark in colour.

Dried casein is normally imported into Great Britain largely from New Zealand and to some extent France and Scandinavia. Some typical properties are the proteins as imported and shown in Table 26.1.

26.2.3 Production of casein plastics

Casein plastics are today produced by the 'dry process'. Although a wet process was used originally in Great Britain it has been obsolete for 50 years and need not be discussed here.

In the dry process the casein is ground so that it will pass through a 30 mesh sieve but be retained by one of a 100 mesh. The powder is then loaded into a dough mixer, usually of the Artofex type. Water is fed slowly into the mixer until the moisture present forms about 20% of the total. The water has a plasticising effect on the casein and heat is generated during the mixing operation. Any large lumps formed must be broken down individually and returned to the mixer. Mixing times for a 60 kg batch are usually about 30 minutes.

In addition to the casein and water, other ingredients are added at this stage.   They may include dyes or pigments, titanium dioxide as a white pigment or colour base and 'clearing agents' to enhance the transparency. Typical clearing (or 'clarifying') agents are ethyl benzyl aniline, tritolyl phosphate, trixylyl phosphate and chlorinated diphenyls.   The incorporation of 1% trixylyl phosphate may increase the percentage light transmission from 40 to 80 through a 2 mm thick sample.

The resultant compound, in spite of the high water content, is a free flowing powder.   It should be processed soon after mixing since it will tend to putrefy.

In commercial operations the next stage of the process involves passing the compound through a small extruder so that under heat and pressure the granular powder is converted to a rubbery material.   Extruders, somewhat simpler in construction than those generally employed in the plastics industry but very similar to rubber extruders are used.   A typical machine would have a 2:1 compression ratio and 8:1 length–diameter ratio screw.   The barrel is heated by steam or water, temperatures at the feed end being at about 30°C with temperatures at the die end at about 75°C.   The die itself may be heated with a simple gas burner.   Breaker plates and screens have the function not only of building up back-pressure but of helping to develop the pattern in a coloured extrudate.

The extrudate is cut up into appropriate lengths and cooled by plunging into cold water.   Subsequent operations depend on the end product required.

When rods are required they are placed in wooden trays in a formolising bath.   If the requirement is for a disc or 'blank' such as used by the button trade the extrudate is cut up by an automatic guillotine and the blanks are immersed in the formalin solution.   For manufacture of sheets the rods are placed in moulds and pressed into sheets before formolising.   Many attractive patterns may be made by pressing different coloured rods into grooves set on the bias to the rods thus forming new multi-coloured rods. This operation may be repeated a number of times in order to produce complex patterns.

The rod, blank or sheet must then be cured by a formolising process. They are immersed into a 4–5% solution of formaldehyde in water (formalin) for anything from 2 days to several months according to the thickness of the section.   The formolising temperature is kept at about 60°F and the pH between 4 and 7.

Too high a concentration of formaldehyde will cause rapid curing of the surface of the section thus reducing its permeability to formaldehyde and as a result it is extremely difficult to cure adequately the centre of the casein. Too low a concentration will unneccessarily prolong the time of cure.

By the time that the desired degree of formolisation has been reached the casein will contain a large quantity of water and free formaldehyde which has to be removed.   This must be done slowly to prevent cracking and warping of the material.   Warm drying cabinets or rotating perforated cylinders are used with slightly humid circulating air.   The drying operation is also a very lengthy process and may take anything from three days to as long as three weeks.

The finished articles may be obtained by machining and polishing the dried formolised sections.

The precise mode of interaction between the casein and the formaldehyde has not been fully elucidated but the following reactions are believed to occur:

1. Reaction across the peptide (—CONH—) groups

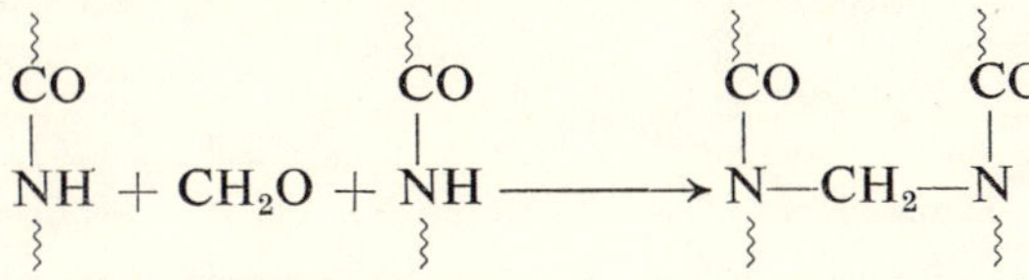

$$CO \qquad\qquad CO \qquad\qquad CO \qquad CO$$
$$|\qquad\qquad\qquad |\qquad\qquad\qquad |\qquad\quad |$$
$$NH + CH_2O + NH \longrightarrow N—CH_2—N$$

2. Reaction between a peptide group and a lysine side chain

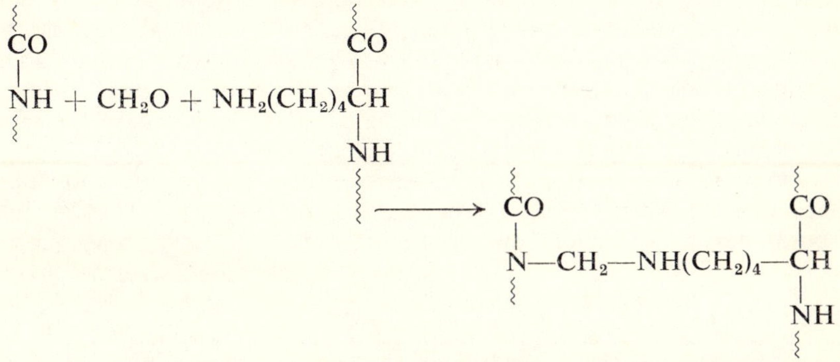

$$CO \qquad\qquad\qquad\qquad CO$$
$$|\qquad\qquad\qquad\qquad\quad |$$
$$NH + CH_2O + NH_2(CH_2)_4CH$$
$$\qquad\qquad\qquad\qquad\qquad |$$
$$\qquad\qquad\qquad\qquad\qquad NH$$

$$\longrightarrow CO \qquad\qquad\qquad\qquad CO$$
$$\qquad\quad |\qquad\qquad\qquad\qquad\qquad |$$
$$\qquad\quad N—CH_2—NH(CH_2)_4—CH$$
$$\qquad\qquad\qquad\qquad\qquad\qquad\qquad |$$
$$\qquad\qquad\qquad\qquad\qquad\qquad\qquad NH$$

3. Similar reactions between amino groups on both side chains.

These three reactions fail to account for all the observed phenomena that occur in curing casein and therefore a number of alternative mechanisms no doubt also occur.

The very long times required for the formolising stages have promoted research to discover methods of speeding the process and also of finding alternative cross-linking techniques. The most effective modification to date is to incorporate small quantities ($\sim 1\%$) of a water soluble cellulose material such as sodium carboxymethyl cellulose into the casein mix. During the formolisation process a fibrous cellulose compound appears to be formed acting as a multiple wick along which the formalin may pass much more rapidly. By this technique formolisation times may be cut by as much as 25%. The use of the cellulose compound will slightly impair the clarity of the finished product.

Attempts have also been made to avoid the formolising process by incorporating formaldehyde donors, i.e. materials that will evolve formaldehyde at elevated temperatures after shaping has been completed. Amongst materials for which patents have been taken out are paraformaldehyde and dimethylol urea. Since premature evolution of only small quantities of formaldehyde will seriously affect the flow properties

Table 26.2 SOME PHYSICAL PROPERTIES OF CASEIN PLASTICS

Specific gravity	1·35
Tensile strength	8,000–10,000 lb in^{-2}
Elongation at break	2·5–4%
Modulus of elasticity (in tension)	6 × 10^5 lb in^{-2}
Izod impact strength	1·0–1·5 ft lb in^{-1}
Brinell hardness	18–24
Rockwell hardness	M75
Breakdown voltage (50 c/s)	200 V/0·001 in.
Power factor (10^6 c/s)	0·06
Water absorption (4 mm thickness)	
24 hr	5–7%
28 days	30%
Heat distortion temperature (ASTM) (66 lb in^{-2})	80–85°C

of the mix this technique causes further problems. The use of formaldehyde evolution inhibitors such as carbamide compounds and amides has been suggested in the patent literature. Although processable moulding powders may be made in this way the products are both dark and brittle and offer no attractive advantage over either normal casein products or conventional thermosetting moulding powders. They have therefore failed to achieve any commercial significance.

A number of alternative curing agents to formaldehyde are known but once again have achieved no commercial significance. They include isocyanates, benzoquinone and chromium salts.

26.2.4   Properties of casein

Formolised casein is a rigid horn-like material which may be made available in a wide variety of colours and patterns. By common consent it is recognised as having a pleasant feel and appearance and it is these properties which are relevant to current commercial application.

Mechanical properties are typical of a rigid plastics material and numerical values (Table 26.2) are similar to those for poly(methyl methacrylate). Although thermosetting it has a low heat distortion temperature ($\sim$80°C) and is not particularly useful at elevated temperatures.

When dry, casein is a good electrical insulator but is seriously affected by humid conditions. For this reason it can no longer compete with the many alternative plastics materials now available for electrical applications.

Both acids and alkalis will adversely affect the material. Strong alkalis and acids will cause decomposition. The water absorption is high and consequently casein is easily stained. As a corollary to this it may be dyed without difficulty. Acidic and basic water soluble dyes are normally used. Typical properties of casein plastics are given in Table 26.2.

26.2.5   Applications

The use of casein has been curtailed by the extensive development of synthetic polymers over the past twenty years. Today its outlets are largely restricted to decorative applications where its pleasant appearance is of value. Specific applications include buttons, buckles, slides and hair pins, the button industry being by far the largest user. In this field it continues to be valuable in spite of the competition of acrylic and polyester materials because of its great colour possibilities, its ability to be machined and dyed easily and its 'natural' appearance.

The once considerable application in knitting pins, pens and pencils is no longer of importance although a few propelling pencils were still being produced with casein barrels in 1964.

Casein is also used for a number of miscellaneous purposes in which formolisation is not required. These include adhesives, stabilisers for rubber latex, paper finishing agents and miscellaneous uses in the textile industry. Mention may also be made of casein fibres, available in Italy between the two world wars under the name of Lanital.

26.3 MISCELLANEOUS PROTEIN PLASTICS

Casein is the only protein that has achieved commercial significance as a plastics raw material. Many other proteins are readily available in many vegetable material residues which arise from such processes as the extraction of oils and starches from seeds. It would be advantageous to countries possessing such residues if plastics could be successfully exploited commercially. Although plastics materials have been produced they have failed to be of value since they are invariably dark in colour and still have the water susceptibility and long curing times both of which are severe limitations of casein.

Of these materials zein, the maize protein has been used for plastics on a small scale. It can be cross linked by formaldehyde but curing times are very long. Complicated bleaching processes have led to the production of almost colourless samples in the laboratory but the process cannot readily be extended to large scale operation. The cured product has a greater water resistance than casein. Proteins from soya bean, castor bean and blood have also been converted into plastic masses but each have the attendant dark colour.

In addition to the natural protein fibres wool and silk, fibres have been produced commercially from other proteins. These materials were introduced as wool substitutes but today have little or no significance. Mention may however be made of 'Ardil' products from the groundnut protein and marketed for some years after the Second World War by I.C.I. Compared with wool it had inferior wet and dry strength and abrasion resistance. The inclusion of up to 20% 'Ardil' into wool however yielded a product with negligible loss in wearing properties.

26.4 DERIVATIVES OF NATURAL RUBBER

The chemistry and technology of natural rubber are outside the scope of this work and have been dealt with extensively elsewhere.[1,2,3] As, however, it is possible to produce non-elastic derivatives, some of which are of interest as plastics materials, a brief summary will be given here.

Natural rubber may be obtained from the latex of many hundreds of plants but virtually all commercial natural rubber is obtained from the *Hevea Brasiliensis*.   This is now cultivated mainly in Malaysia, Indonesia, Ceylon and in parts of Central Africa.   The composition of the latex varies but contains about 35% rubber hydrocarbon and 55% water.   The residue is made up largely of proteins (4·5%) and resinous material (4%). The rubber is normally obtained from the latex by acid coagulation but a certain amount is concentrated and shipped to producing countries for subsequent conversion into 'latex foam', 'latex thread', etc.

The natural rubber hydrocarbon is a linear cis-1,4-polyisoprene with a regular head-to-tail structure.

$$-CH_2-\underset{\underset{CH_3}{|}}{C}=CH-CH_2-CH_2-\underset{\underset{CH_3}{|}}{C}=CH-CH_2-$$

The dried coagulated material is highly elastic at room temperature but becomes 'boardy' in cold weather and sticky in hot weather.   It is soluble in aromatic and chlorinated hydrocarbons.   Because of its highly elastic behaviour it cannot be shaped by moulding or extrusion.   The glass transition temperature is about $-70°C$.

In order to achieve a processable polymer it is necessary to reduce the molecular weight from about 10^6 to about 70,000 by a process known as 'mastication'.   The reduction in molecular weight is brought about either by mechanical rupture of the chains (predominant at low temperatures) or by oxidative scission (predominant at elevated temperatures).   The resultant masticated rubber is essentially plastic and may then be compounded and processed.

Regeneration of elastic properties will occur if the rubber is suitably treated with a vulcanising agent which will bring about light cross linking. In practice about 3% of sulphur is compounded into the rubber together with vulcanising aids such as accelerators and activators.   The shaped rubber may then be vulcanised on heating.   Using mercaptobenzothiazole as an accelerator and zinc oxide and stearic acid as the activator system the curing time will be of the order of 15 minutes at 140°C.   The curing schedule will be considerably affected by the accelerator employed.

The sulphur vulcanisation mechanism is complex and Fig. 26.3 indicates some of the structural features believed to exist in vulcanised natural rubber.

The main feature is the cross link between the α-methylene carbon on one chain to the double bond of the other.   Certain, generally undesirable, side reactions may however take place leading for example to the formation of cyclic sulphide structures and to main chain scission.

Alternative vulcanising systems are occasionally employed.   These include tetramethyl thiuram disulphide (also an accelerator), sulphur chloride, selenium, *p*-quinone dioxime and dicumyl peroxide.   Many other reagents have also been found to be effective.

By far the bulk of natural rubber produced is used in a conventionally vulcanised form, the largest outlet being for tyres and other automotive applications. A small quantity is however converted to other chemical forms some of which are of interest as plastics.

The most important of these is *ebonite* which may be considered as the world's earlier thermosetting plastics material. It is obtained by vulcanising (natural) rubber with large quantities of sulphur. Whereas ordinary vulcanised rubber as used in tyres contains normally only 2–3% of sulphur

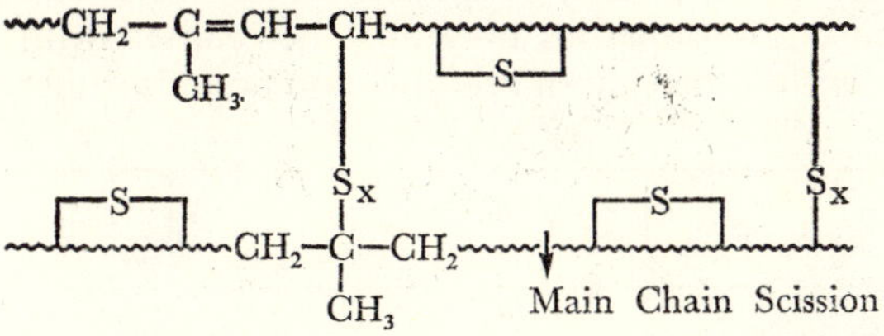

Fig. 26.3

a typical rubber–sulphur ratio for ebonite is 68:32. Compared with ordinary vulcanisates, ebonite is more rigid, shows less swelling in hydrocarbon solvents and has a higher density. These factors indicate a fairly high degree of cross linking. As the vulcanisation reaction proceeds it is observed that the non-extractable sulphur content steadily increases to reach a maximum of 32% and at this point the unsaturation of the composition falls to zero. The sulphur content is in accord with the empirical formula C_5H_8S so that in effect for each atom of sulphur combined there is a loss of one double bond.

The detailed structure of ebonite is not known but it is believed that the same structures occur in the rigid material as have been suggested for vulcanised rubber. There will, however, be far more S-containing structures per unit volume and the ratios of the various structures may differ. The curing reaction is highly exothermic the heat of reaction being about 300 cal g^{-1} and since the specific heat of the mix is about 0·33 the compound on cure could in theory rise in temperature by 1,000°C!

Ebonite compositions may be prepared without difficulty either in an internal mixer or on a two-roll mill. In addition to the rubber and sulphur, fillers are invariably present in commercial mixes. These materials have the important function of diluting the rubber phase. Because of this the exotherm will be diluted and there will be less shrinkage or cure. Ebonite dust is very useful for such diluent purposes as its use has the minimum adverse effect on the properties of the finished product. For many purposes mineral fillers may be incorporated to reduce cost. The vulcanisation time may be reduced by addition of 2–3 parts p.h.r. of an accelerator such as diphenyl guanidine or butyraldehyde-aniline and this is common practice. Softeners and processing aids may also be added.

The ebonite compound before cure is a rather soft plastic mass which may be extruded, calendered and moulded on the simple equipment of the type that has been in use in the rubber industry for the last century. In the case of extruded and calendered products vulcanisation is carried out in an air or steam pan. There has been a progressive reduction in the cure times for ebonite mixes over the years from 4–5 hours down to 7–8 minutes. This has been brought out by considerable dilution of the reactive rubber and sulphur by inert fillers, by use of accelerators and an increase in cure temperatures up to 170–180°C. The valuable effect of ebonite dust in reducing the exotherm is shown graphically in Fig. 26.4.

Ebonite, or hard rubber as it is often known, is black in colour and has a specific gravity, in the absence of mineral fillers of about 1·18.

The best physical properties may only be realised with unfilled compounds. As with other thermosetting compounds different properties

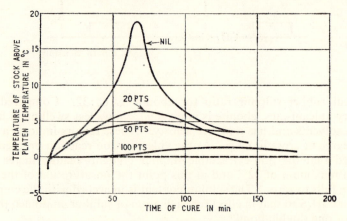

Fig. 26.4.   Variation of internal temperature during cure of ebonite stocks containing 0, 20, 50 and 100 parts of ebonite dust per 100 parts (rubber and sulphur).   (After Scott[4])

will be at their best after different amounts of cure. Care should also be taken in the selection of the rubber–sulphur ratio. High sulphur contents, up to 40% may be used for greatest resistance to swelling and for minimum dielectric loss. The best mechanical properties and the greatest heat resistance are generally obtained with about 35% sulphur whereas the best impact strength is obtained with somewhat less combined sulphur. Hot air and steam cures usually result in poorer ebonites than with press cures since volatilisation of sulphur from the surface layers of the product leads to surface undercure.

Typical properties of a high quality ebonite are given in Table 26.3.

On exposure of ebonite to light there is a rapid deterioration in surface resistivity. It is believed that this is due to the formation of sulphuric acid through oxidation of the rubber–sulphur complex. The sulphuric acid sweats out into droplets on the surface of the polymer and eventually a stage is reached where the droplets link up into a continuous film forming a conductive path along the surface. The influence of this phenomenon on the surface resistivity is shown in Fig. 26.5.

Ebonite has a good resistance to a range of inorganic liquids including most non-oxidising acids. It is severely swollen by aromatic and chlorinated hydrocarbons.

Table 26.3 PROPERTIES OF UNFILLED EBONITE VULCANISATES
(as B.S.903 where applicable)

Specific gravity	1·18
Tensile strength	9,000 lb in^{-2}
Elongation at break	3%
Cross-breaking strength	12,000–16,000 lb in^{-2}
Impact strength	0·5 ft lb/in. notch
Yield temperature	85°C
Dielectric constant	2·7–3·0 (10^3–10^6 c/s, 25–75°C)
Power factor (10^6 c/s)	∼0·01
Electric strength	1·0–1·5 × 10^5 kV/mm^{-1}
Volume resistivity	∼10^{17} ohm cm
Surface resistivity (of unexposed samples)	10^{18} ohms
Equilibrium water absorption	∼0·25%

The continuing use of ebonite is due to the good insulating characteristics, good chemical resistance and ease of machining. Its use is limited by the long curing times, the limited temperature range, its colour limitations and its poor resistance to air, light and oxidising chemicals. Its principal applications are in chemical plant and car batteries which are largely based on very low cost reclaimed rubber. Miscellaneous uses include the manufacture of water meters and pipe stems. It is, however, under continual challenge in all of these applications from synthetic plastics materials.

The terms ebonite and hard rubber are now extended to cover hard products made from synthetic rubbers. S.B.R. is now replacing the

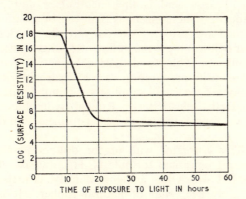

Fig. 26.5. Influence of exposure to light on the surface resistivity of ebonite. (After Scott[4]*)*

natural materials in many ebonite applications whilst nitrile rubber ebonites are of interest where oil resistance is required.

Hard products may also be made by vulcanising rubber (natural or synthetic) using only about 2 parts of sulphur per 100 parts of rubber. In these cases either the so-called 'high styrene resins' or phenolic rubber compounding resins are incorporated into the formulation. These compounds are processed using the methods of rubber technology but, like those of ebonite, the products are more akin to plastics than to rubbers. Examples of the usage of these materials are to be found in battery boxes, shoe heels and car washer brushes.

A number of other natural rubber derivatives may be prepared by addition reactions at the double bond.

Hydrogenation may be accomplished by heating a dispersion of rubber and Ni–Kieselguhr catalyst in cyclohexane with hydrogen under pressure at 170–250°C in the complete absence of oxygen. The hydrogenation is accompanied by degradation and cyclisation and the product has not been commercially exploited. It is to be noted however that hydrogenated polybutadiene, of the same fundamental structure as polyethylene, is used for some specialised rubber-to-metal bonding operations.

Reaction of the natural rubber hydrocarbon with hydrochloric acid yields *rubber hydrochloride*. The hydrogen chloride adds on according to Markownikoff's rule (that the halogen atom attaches itself to the carbon atom with the least number of hydrogen atoms).

$$\text{---CH}_2\text{---}\underset{\underset{}{\overset{\overset{\text{CH}_3}{|}}{}}}{\text{C}}\text{=CH---CH}_2\text{---} \longrightarrow \text{---CH}_2\text{---}\underset{\underset{\text{Cl}}{|}}{\overset{\overset{\text{CH}_3}{|}}{\text{C}}}\text{---CH}_2\text{---CH}_2\text{---}$$

The hydrohalide is usually prepared by passing hydrogen chloride into a solution of masticated high grade raw rubber in benzene at 10°C for about 6 hours. Excess acid is then neutralised and plasticisers and stabilisers are added. The benzene is removed by steam distillation and the product washed and dried. Alternatively the solution is cast on to a Neoprene rubber belt leaving a tough film after evaporation of the solvent.

The hydrohalide is liable to dehydrochlorination particularly when moist acid is used in its preparation so that hydrochloric acid acceptors such as lead carbonate are useful stabilisers. Dibutyl phthlate and tritolyl phosphate are effective plasticisers. Rubber hydrochloride is used as a packaging film (Pliofilm) and as a rubber-to-metal bonding agent (e.g. Typly).

If natural rubber is treated with proton donors a product is formed which has the same empirical formula (C_5H_8), and is soluble in hydrocarbon solvents but which has a higher density, is inelastic and whose unsaturation is only 57% that of natural rubber. It is believed that intramolecular ring formation occurs to give products containing the following segments:

$$\begin{array}{ccc}
 & \text{CH}_2\text{---CH}_2 & \\
 & \diagup \qquad \diagdown & \\
\text{CH}_2 & & \text{C---CH}_3 \\
 \diagdown & & \diagup \\
 & \text{C-----C} & \\
 \diagup \quad & | & \quad \diagdown \\
\text{---CH}_2 \quad & \text{CH}_3 & \quad \text{CH}_2\text{---}
\end{array}$$

Known as *cyclised rubber* it may be prepared by treating rubber, on a mill, in solvent or in a latex with materials such as sulphuric acid or stannic chloride.

Attempts have been made to popularise cyclised rubber as an additive for use in shoe soling compounds but they have not been able to compete with the 'high styrene resins' used extensively for this purpose. They have found a small use for rubber-to-metal bonding and as an ingredient in surface coatings.

Treatment of natural rubber with chlorine gives a product, *chlorinated rubber*, with a maximum chlorine content of 65% corresponding to the empirical formula $C_{10}H_{11}Cl_7$. Such a compound corresponds neither to a hypothetical simple addition to the double bond (Fig. 26.6 (I)) nor to a product with α-methylenic substitution in addition (II).

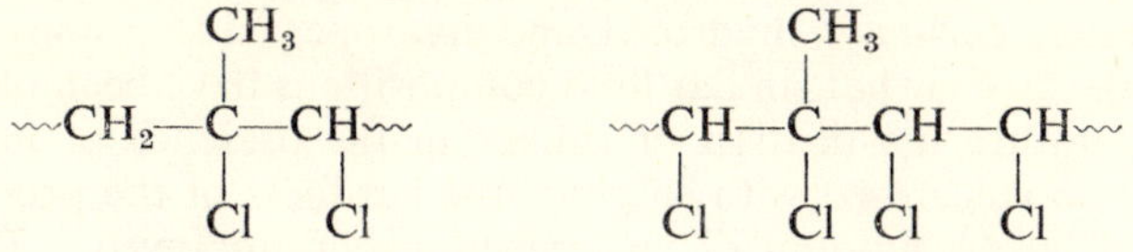

$$\begin{array}{cc}
\text{CH}_3 & \text{CH}_3 \\
| & | \\
\sim\!\text{CH}_2\!-\!\overset{|}{\text{C}}\!-\!\text{CH}\!\sim & \sim\!\text{CH}\!-\!\overset{|}{\text{C}}\!-\!\text{CH}\!-\!\text{CH}\!\sim \\
\quad\ \ | \quad | & \ \ | \quad | \quad | \quad\ | \\
\quad\ \ \text{Cl} \quad \text{Cl} & \ \ \text{Cl} \quad \text{Cl} \ \text{Cl} \quad \text{Cl} \\
51\%\ \text{Cl} & 68\%\ \text{Cl} \\
C_{10}H_{16}Cl_4 & C_{10}H_{16}Cl_8 \\
\text{(I)} & \text{(II)}
\end{array}$$

Fig. 26.6

It has been shown that the reaction occurs in three stages, illustrated by the following empirical equation:

$$\begin{array}{lr}
 & \%\text{Cl} \\
C_{10}H_{16} + 2Cl_2 \longrightarrow C_{10}H_{14}Cl_2 + 2HCl & 35 \\
C_{10}H_{14}Cl_2 + 2Cl_2 \longrightarrow C_{10}H_{13}Cl_5 + HCl & 57 \\
C_{10}H_{13}Cl_5 + 2Cl_2 \longrightarrow C_{10}H_{11}Cl_7 + 2HCl & 65
\end{array}$$

The first and third stages involve substitution only whilst the second stage involves simultaneous addition and substitution.

The structure of the completely chlorinated product has not been fully elucidated but one suggested structure for the repeating unit is shown in Fig. 26.7.

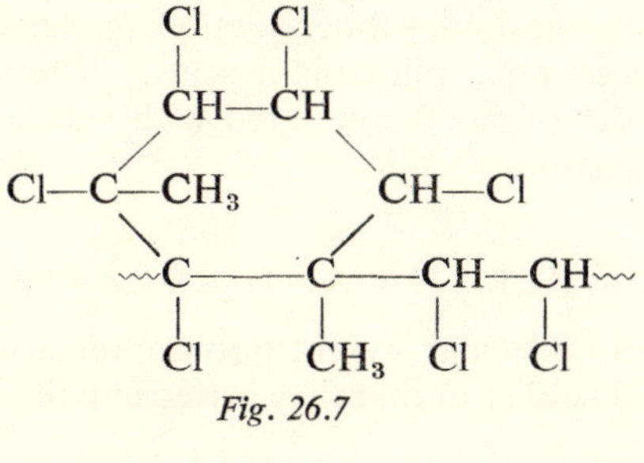

$$\begin{array}{c}
\text{Cl} \quad\ \text{Cl} \\
| \qquad | \\
\text{CH}\!-\!\text{CH} \\
\diagup \qquad\quad \diagdown \\
\text{Cl}\!-\!\text{C}\!-\!\text{CH}_3 \qquad \text{CH}\!-\!\text{Cl} \\
\diagdown \qquad\qquad \diagup \\
\sim\!\text{C}\!-\!-\!-\!\text{C}\!-\!-\!\text{CH}\!-\!\text{CH}\!\sim \\
| \qquad | \qquad | \quad\ | \\
\text{Cl} \qquad \text{CH}_3 \ \text{Cl} \quad \text{Cl}
\end{array}$$

Fig. 26.7

Chlorinated rubber is usually prepared by bubbling chlorine into a solution of masticated rubber in a chlorinated hydrocarbon solvent such as carbon tetrachloride. Hydrochloric acid is removed during the reaction. The solvent may be removed by vacuum or steam distillation or by precipitation of the derivative by a non-solvent such as petroleum.

Chlorinated rubber is extensively employed in industrial corrosion resistant surface coatings for which purpose it is marketed by I.C.I. under the trade name Alloprene. Although thermoplastic moulding compositions have been made by plasticising with the common ester plasticisers such as tritolyl phosphate they are of no commercial importance.

A number of oxidation products of natural rubber have been prepared. Of some interest at one time were the *rubbones*, produced by the degradation of rubber using cobalt linoleate in conjunction with a cellulosic material. These materials are very complex in structure and the presence of acid, ester, carbonyl, hydroxyl and hydroperoxide groups has been established. Somewhat similar fluid compositions have been obtained by high temperature mastication of rubber in the presence of mineral oil. They may be vulcanised with sulphur, the hardness of the product being determined by the amount of vulcanising agent employed. These compositions have been used for embedding bristles and to a minor extent for casting.

Mention may finally be made of graft polymers derived from natural rubber which have been the subject of intensive investigation for the past decade but which have not achieved commercial significance. It has been

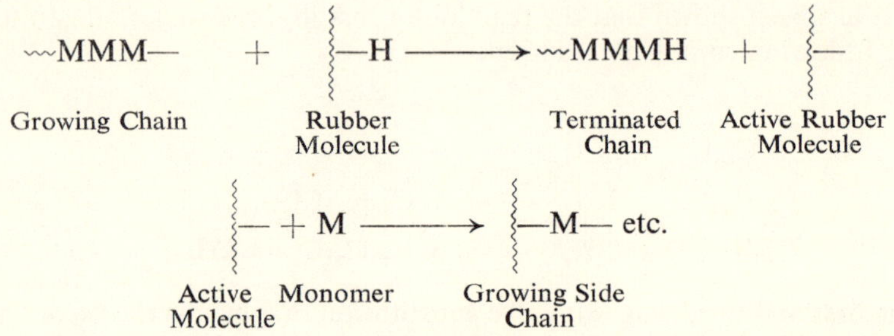

Fig. 26.8

found that natural rubber is an efficient chain transfer agent for free radical polymerisation and that grafting appears to occur by the mechanism shown in Fig. 26.8.

Both rubber–styrene and rubber–methyl methacrylate graft polymers have been produced on a pilot plant scale. The side chains have unit weights of the order of 5,000 compared with values of 70,000–270,000 for the main rubber chain.

26.5 GUTTA PERCHA AND RELATED MATERIALS[2]

A number of high molecular weight polyisoprenes occur in nature which differ from natural rubber in that they are essentially non-elastic. As with natural rubber they are obtained from the latex of certain plants but differ in that they are either trans-1:4,polyisoprenes and/or are associated with large quantities of resinous matter.

Of these the most interesting, from the point of view of the plastics technologist, is gutta percha. This material may be obtained from trees of the genera ·*Palaquium* and *Payena* of which *Pal. oblongifolium* is the most important. This plant is found in Malaysia and Indonesia and is cultivated mainly in Sumatra. The latex is more viscous than natural rubber latex and is now most commonly extracted from the leaves rather than by making incisions in the bark. The coagulated latex would give a typical analysis of hydrocarbon 70%, resin 11%, dirt 3% and moisture 16%. The trans-1,4-polyisoprene is reported to have molecular weight of 30,000 which is somewhat lower than obtained with the natural rubber hydrocarbon. X-ray evidence indicates that gutta percha may exist in one of two crystalline states. As produced from the tree the gutta percha crystallites are in the α-form which melts at 65°C. On slow cooling the α-form is reproduced but rapid cooling will produce the β-form which melts at 56°C. The latter form slowly reverts to the α-form. About 60% of gutta percha is crystalline at room temperature when it has the character of a hard inelastic solid.

As with cis-polyisoprene, the gutta molecule may be hydrogenated, hydrochlorinated and vulcanised with sulphur. Ozone will cause rapid degradation. It is also seriously affected by both air (oxygen) and light and is therefore stored under water. Anti-oxidants such as those used in natural rubber retard oxidative deterioration. If the material is subjected to heat and mechanical working when dry there is additional deterioration so that it is important to maintain a minimum moisture content of 1%.

Gutta percha has a lower water absorption than natural rubber and is a good dielectric. It is dissolved by carbon disulphide, chloroform and benzene but alkaline solutions and dilute acids do not affect it. It is destroyed by nitric acid and charred by warm concentrated sulphuric acid but resists hydrofluoric acid.

At one time gutta percha had an important outlet in undersea cable insulation and a lesser use in chemical plant, in particular for storage of hydrofluoric acid. The most important application at the present time is for golf ball covers and although gutta has withstood the challenge of synthetic resins for some years it is unlikely that it will be able to withstand competition from synthetic trans-1,4-polyisoprenes which have been shown to be quite suitable as replacements. It is not usual to vulcanise the polymers.

An alternative source of the trans-1,4,polyisoprene is *balata*, obtained from the *Mimosups balata* occurring in Venezuela, Barbados and British Guiana. The latex is thin and may be tapped in the same way as natural rubber.

The coagulated material consists of about 50% polyisoprene and the remainder is primarily composed of resins. Deresinated balata has been used as an alternative to gutta percha which it resembles in properties and also in belting applications. It is no longer of any real importance.

The latex of the *Sapota achras* yields a thermoplastic material, *chicle*, consisting of about 17·4% hydrocarbon, 40% acetone soluble resin and 35% occluded water. The hydrocarbon appears to contain both trans- and cis-polyisoprene. Although originally introduced as gutta percha and natural rubber substitutes deresinated chicle has become important as the base for chewing gum. As with other polyisoprenes it is meeting competition from synthetic polymers.

An alternative chewing gum base is obtained from *jelutong*, a mixture of polyisoprene and resin obtained from latex of the *Dyera costulata*. This tree is found in many countries but Borneo is the principal commercial source. At one time jelutong was an important rubber substitute and 40,000 tons were produced in 1910. Production in recent years has been of the order of 5,000 tons per annum mainly for chewing gum.

26.6 SHELLAC

The importance of shellac to the plastics industry has declined rapidly since 1950. Before that time it was the principal resin employed in 78 rev/min gramophone records. The advent of the long playing micro-groove record meant that mineral fillers could no longer be tolerated because any imperfections in the microgroove led to a high background noise on the record. The record industry therefore turned towards alternative materials which required no mineral filler and vinyl chloride–vinyl acetate copolymers eventually became pre-eminent. Today the use of shellac in the plastics industry is restricted to the manufacture of mica laminates. It is however still used for a number of purposes outside the normal realm of plastics.

26.6.1 Occurrence and preparation

Shellac is the refined form of lac, the secretion of the lac insect parasitic on certain trees in India, Burma, Thailand and to a minor extent in other Asian countries.

The larvae of the lac insect, *Laccifer lacca (Kerr)*, swarm around the branches and twigs of the host trees for 2–3 days before inserting their probosces into the phloem tissues to reach the sap juices. There may be as many as 100–150 larvae on each inch of twig. This is followed by secretion of the lac surrounding the cells. Whereas the male insects subsequently move out of their cells the female insects become entombed for life. After about eight weeks of life the male insects fertilise the females and die within a few days. The fertilised females subsequently exude large quantities of lac and shed eyes and limbs. The female gives birth to 200–500 further insects and finally dies.

In commercial practice the crop is taken from the tree shortly before emergence of the new brood. Some of these twigs are then tied to new trees to provide future sources of lac but the rest, *sticklac*, is subjected to further processing. The average yield per tree is about 20 lb per annum, usually one crop being allowed per tree per year.

Subsequent treatment of the sticklac carried out by hand or by mechanical methods first involves removal of woody matter and washing to remove the associated lac dye to produce seed lac, containing 3–8% of impurities. This may be further refined by various methods to produce the shellac flakes of commerce.

The hand process for producing shellac has been used since ancient times and is carried on largely as a cottage industry. It has been estimated that 3–4 million people are dependent for their livelihood on this process. The lac encrustation is first separated from woody matter by pounding with a smooth stone, the latter being removed by a winnowing process. The lac dye is then removed by placing the lac in a pot together with a quantity of water. A man, known as a ghasander, then stands in the pot and with bare feet treads out the dye from the resin. At one time lac dye was of commercial value but is today a worthless by-product. The product, *seedlac*, is then dried in the sun.

The next stage may best be described as a primitive hot-filtration process. Two members of the village sit across the front of a simple fire resembling a Dutch oven holding between them a bag about 30 feet long and about two inches in diameter. The lac inside the bag melts and, through one of the operators twisting the end of the bag, the lac is squeezed out. The lac is then removed from the outside of the bag and collected into a molten lump which is then stretched out by another operator using both hands and feet until a brittle sheet is produced. This is then broken up to produce the shellac of commerce. (A model of an Indian village producing shellac has been an exhibit for many years at Kew Gardens, Surrey.)

In the factory processes the sticklac is first passed through crushing rollers and sieved. The lac passes through the sieve but retains the bulk of the woody matter. The sieved lac is then washed by a stream of water and dried by a current of hot air. A second mechanical cleaning process removes small sticks which have not been removed in the earlier roller process. The product, seedlac, now contains 3–8% of impurities.

The seedlac may then be converted to shellac by either a heat process or by solvent processes. In the heat process the resin is heated to a melt which is then forced through a filter cloth which retains woody and insoluble matter. In the solvent process the lac is dissolved in a solvent, usually ethyl alcohol. The solution is filtered through a fine cloth and the solvent recovered by distillation.

Variation in the details of the solvent processes will produce different grades of shellac. For example when cold alcohol is used lac wax which is associated with the resin remains insoluble and a shellac is obtained free from wax. Thermally processed shellacs were greatly favoured for gramophone records as they were free from residual solvent and also contained a small quantity of lac wax which proved a useful plasticiser.

26.6.2 Chemical composition

The lac resin is associated with two lac dyes, lac wax and an odiferous substance, and these materials may be present to a variable extent in shellac. The resin itself appears to be a polycondensate of aldehydic and hydroxy acids either as lactides or inter-esters. The resin constituents can be placed into two groups, an ether soluble fraction (25% of the total) with an acid value of 100 and molecular weight of about 550, and an insoluble fraction with an acid value of 55 and a molecular weight of about 2,000.

Hydrolysis of the resins will produce aldehydic acids at mild concentration of alkali ($\sim \frac{1}{2}N$); using more concentrated alkalies ($5N$) hydroxy acids are produced, probably via the aldehydic acids. Unfortunately most of the work done in order to analyse the lac resin was carried out before the significance of the hydrolysis conditions was fully appreciated. It does however appear to be agreed that one of the major constituents is aleuritic acid (Fig. 26.9)

$$HO \cdot (CH_2)_6 \cdot CH - CH - (CH_2)_7 COOH$$
$$\underset{\displaystyle OH}{|} \quad \underset{\displaystyle OH}{|}$$

Fig. 26.9

which is present to the extent of about 30–40% and is found in both the ether soluble and insoluble fractions. Both free hydroxyl and free carboxyl groups are to be found in the resin.

26.6.3 Properties

The presence of free hydroxy and carboxyl groups in lac resin makes it very reactive, in particular to esterification involving either type of group. Of particular interest is the inter-esterification that occurs at elevated temperatures ($> 70°C$) and leads to an insoluble 'polymerised' product. Whereas ordinary shellac melts at about 75°C prolonged heating at 125–150°C will cause the material to change from a viscous liquid, via a rubbery state to a hard horny solid. One of the indications that the reaction involved is esterification is that water is evolved. The reaction is reversible and if heated in the presence of water the polymerised resin will revert to the soluble form. Thus shellac cannot be polymerised under pressure in a mould since it is not possible for the water to escape. 'Polymerisation' may be retarded by basic materials some of which are useful when the shellac is subjected to repeated heating operations. These include sodium hydroxide, sodium acetate and diphenyl urea. 'Polymerisation' may be completely inhibited by esterifying the resin with monobasic saturated acids. A number of accelerators are also known, such as oxalic acid and urea nitrate. Unmodified lac polymerises in about 45 minutes at 150°C and 15 minutes at 175°C.

Shellac is soluble in a very wide range of solvents of which ethyl alcohol is most commonly employed. Aqueous solutions may be prepared by warming shellac in a dilute caustic solution.

The resin is too brittle to give a true meaning to mechanical properties. The thermal properties are interesting in that there appears to be a transition point at 46°C. Above this temperature specific heat and

Table 26.4 SOME PROPERTIES OF SHELLAC

Property	Condition	Units	Value
Specific gravity	15·5°C	—	1·20
Refractive index	—	—	1·51–1·53
Colour	—	—	pale yellow–dark red
Specific heat	10–40°C	c.g.s.	0·36–0·38
	45–50°C	c.g.s.	0·56
	(heat hardened)		
Volume resistivity	20°C	ohm cm	$1·8 \times 10^{16}$
Surface resistivity	20% RH	ohm	$2·2 \times 10^{14}$
	40% RH	ohm	$1·1 \times 10^{14}$
Dielectric constant	30°C	—	3·91
(50 c/s)	80°C	—	7·85
Dielectric loss	30°C	—	0·02
(50 c/s)	80°C	—	0·435

temperature coefficient of expansion are much greater than below it. The specific heat of hardened shellac at 50°C is lower than that of unhardened material, this no doubt reflecting the disappearance, or at least the elevation of the transition temperature.

From the point of view of the plastics technologist the most important properties of shellac are the electrical ones. The material is an excellent room-temperature, low-frequency insulator and particular mention should be made of the resistance to tracking.

Some typical physical properties of shellac are given in Table 26.4.

26.6.4 Applications

Until 1950 the principal application of shellac was in gramophone records. The resin acted as a binder for about three times its weight of mineral filler, e.g. slate dust. The compound had a very low moulding shrinkage and was hard-wearing but not suitable for the microgroove records because of the effect of the filler on the background noise.

Today the most important applications are in surface coatings, including some use as French polish, as adhesives and cements including valve capping and optical cements, for playing card finishes and floor polishes. The material also continues to be used for hat stiffening and in the manufacture of sealing wax.

Although development work on shellac in blends with other synthetic resins has been carried out over a period of time the only current use in the plastics industry is in the manufacture of electrical insulators. At one time electrical insulators and like equipment were fabricated from mica but with increase in both the size and quantity of such equipment shellac was introduced as a binder for mica flake. For commutator work the amount of shellac used is only 3–5% of the mica but in hot moulding Micanite for V-rings, transformer rings etc., more than 10% may be used. The structures after assembly are pressed and cured, typically for 2 hours at 150–160°C under pressure.

In recent years the dominance of shellac in mica-based laminates has met an increasing challenge from the silicone resins but, because of its low cost, continues to be used for the many purposes for which it is quite adequate.

26.7 AMBER

In addition to shellac a number of other natural resins find use in modern industry.   They include rosins, copals, kauri gum and pontianak.   Such materials are either gums or very brittle solids and, although suitable as ingredients in surface coating formulations and a miscellany of other uses, are of no value in the massive form, i.e. as plastics in the most common sense of the word.

One resin, however, can be considered as an exception to this.   Although rarely recognised as a plastics material it can be fabricated into pipe mouthpieces, cigarette holders and various forms of jewellery.   It may also be compression moulded and extruded.   It is the fossil resin amber.

Amber is of both historical and etymological interest as its property of attracting dust was known over 2,000 years ago.   From the Greek word for amber, *elektron*, has come the word electricity.   Pliny[5] in his works makes an interesting and informative dissertation on the occurrence and properties of amber.

Amber is a fossil resin produced in the Oligocene age by exudation from a now extinct species of pine.   It occurs principally in the region known before the Second World War as East Prussia but which is today part of Russia.   It may be obtained by mining and also by collecting along the seashore.   Small amounts of amber may also be found off the coasts of England, Sweden, Holland and Denmark.   Similar resins are found in Burma, Rumania and Sicily but only the Baltic variety, known also as succinite is considered a true amber.   At one time a Royal Amber Works existed in Königsberg (now Kaliningrad) and in 1900 annual production was approximately 500 tons.

26.7.1   Composition and properties

The chemical nature of amber is complex and not fully elucidated.   It is believed not to be a high polymer, the resinous state being accounted for by the complexity of materials present.   The empirical formula is $C_{10}H_{16}O$ and true amber yields on distillation 3–8% of succinic acid.

The resin is fairly soluble in alcohol, ether and chloroform and is decomposed by nitric acid.   It becomes thermoplastic at temperatures above 150°C and decomposes at a temperature rather below 300°C yielding an oil of amber and leaving a residue known as amber colophony or amber pitch.

X-ray evidence shows the material to be completely amorphous as might be expected from such a complex mixture.   The specific gravity ranges from 1·05 to 1·10.   It is slightly harder than gypsum and therefore just not possible to scratch with a fingernail.   Yellow in colour it is less brittle than other hard natural resins and may therefore be carved or machined with little difficulty.   The refractive index is 1·54.

Amber has been a much prized gem material for many millenia and has been found at Stonehenge, in Mycenaen tombs and in ancient European lake dwellings. In modern times it is used for beads and other ornaments, cigarette holders and pipe mouthpieces.

At one time the small fragments of amber produced during the fabrication and machining operations were used to produce varnishes. In 1880 they were first used in the production of Ambroid. This is made by pressing the fragments in a hydraulic press at temperatures somewhat above 160°C. The moulded product has a close resemblance to amber. A form of extrusion has also been used to produce amber rods for subsequent conversion into pipe and cigarette mouthpieces.

26.8 BITUMINOUS PLASTICS

Although generally ignored in plastics literature the bituminous plastics are still important for specific applications. The moulding compositions consist of fibrous and mineral fillers held together by a bituminous binder together with a number of minor ingredients.

A number of types of bituminous material exist and terminology is still somewhat confusing. The term *bitumens* in its widest sense includes liquid and solid hydrocarbons but its popular meaning is restricted to the solid and semisolid materials. The bitumens occur widely in nature and may be considered to be derived from petroleum either by evaporation of the lighter fraction under atmospheric conditions or by a deeper seated metamorphism. The purer native bitumens are generally known as *asphaltites* and include Gilsonite, extensively used for moulding, which occurs in Utah.

Where the bitumens are associated with mineral matter the mixture is referred to as *native asphalt*. These are widely distributed in nature of which the best known deposit is the asphalt lake in Trinidad which covers an area of about 100 acres. The terms *asphalt* or *asphaltic bitumen* are applied to petroleum distillation residues and these today form the bulk of commercial bituminous matter. Related chemically and in application but not in origin are the *pitches*. These are the industrial distillation residues. They include wood tar, stearin pitch, palm oil pitch and coal tar pitch. The last varies from soft semi-solid to hard brittle products. Of these materials those most useful in moulding compositions are coal tar pitches with a softening range of 115–130°C and natural bitumens such as Gilsonite and Rafaelite with softening points in the range 130–160°C.

The bitumens are complex mixtures of paraffinic, aromatic and naphthenic hydrocarbons. A small amount of unsaturation is usually present which accounts for the slow oxidation which occurs on exposure to ultraviolet light and the ability to bring about a form of vulcanisation on heating with sulphur.

The bulk of bituminous materials are used for road making and building applications which are outside the scope of this book. Only a very small percentage is used in moulding compositions and little data have been made publicly available concerning the properties of these compositions.

The bitumens have a good order of chemical corrosion resistance, have reasonably good electrical insulation properties and are very cheap. Their main disadvantage are their black colour and their somewhat brittle nature.

Moulding compositions contain a number of ingredients. These may include:

1. Bituminous binder.
2. Fibrous filler.
3. Inert filler.
4. Softener.
5. Drying oil and drier.

Of the fibrous fillers, which greatly reduce the brittleness blue asbestos fibre is normally used for battery boxes, the principal outlet. Other materials that may be used include cotton fibres, ground wood, slag wool and ground cork.

Mineral fibres are incorporated to reduce cost and to raise the softening point. China clay, natural silicas, talc and slate dust are frequently used.

To facilitate moulding a softener is incorporated. These may include soft industrial pitches or heavy tars, coumarone–indene resins or waxes.

In the United States softer stocks have been employed using a drying oil which is incorporated with a drier such as cobalt naphthenate to harden the oil.

The compositions are mixed in heated trough mixers the mixing temperature being in the range of 150–200°C. Skill is required in order to achieve good dispersion of the fibrous filler without charring the bituminous matter. Moulding is carried in compression moulds using prewarmed doughs. For battery boxes the mould temperature on changing the composition is about 100°C which is reduced to at least 50°C before extraction of the moulding. Some simple mouldings can be carried out using pre-warmed mixes but cold moulds.

The largest outlet for the bituminous plastics is in automobile battery boxes. Bituminous battery boxes do however have a susceptibility to electrical breakdown between the cells and in Great Britain their use is mainly confined to the cheaper batteries installed initially in new cars. Ebonite battery boxes are preferred for replacement purposes. Bituminous compositions are also used for toilet cisterns and to some extent for cheap containers.

REFERENCES

1. DAVIS, C. C., and BLAKE, J. (Eds), *The Chemistry and Technology of Rubber*, American Chemical Society, Monograph, Reinhold, New York (1937)
2. NAUNTON, W. J. S. (Ed), *The Applied Science of Rubber*, Arnold, London (1961)
3. MORTON, M. (Ed.), *Introduction to Rubber Technology*, Reinhold, New York (1959)
4. SCOTT, J. R., *Ebonite*, Maclaren, London (1958)
5. PLINY, Book 37, Chapter 3

BIBLIOGRAPHY
Casein

COLLINS, J. H., *Casein Plastics*, Plastics Institute Monograph No. C5, 2nd Edn, London (1952)
PINNER, S. H., *Brit. Plastics*, **18**, 313 (1946)

SUTERMEISTER, F., and BROWNE, F. L., *Casein and its Industrial Applications*, American Chemical Society, Monograph No 30, New York (1939)

Rubber, Derivatives of Rubber and similar polymers

DAVIES, B. L., and GLAZER, J., *Plastics derived from Natural Rubber*, Plastics Institute Monograph No. C8, London (1955)
DAVIS, C. C., and BLAKE, J. (Eds), *The Chemistry and Technology of Rubber*, American Chemical Society Monograph, Reinhold, New York (1937)
NAUNTON, W. J. S. (Ed), *The Applied Science of Rubber*, Arnold, London (1961)
SCOTT, J. R., *Ebonite*, MacLaren, London (1958)

Shellac

GIDVANI, B. S., *Shellac and Other Natural Resins*, Plastics Institute Monograph No. S1 2nd Edn, London (1954)
Shellac, Angelo Brothers Ltd., Calcutta (1956)

Amber

HERBERT SMITH, G. F., *Gemstones*, Methuen, London (1952)
LEY, WILLY, *Dragons in Amber*, Sidgwick and Jackson, London (1951)
Pliny, Book 37, Chapter 3

Index

565